INTERNATIONAL ASTRONOMICAL UNION

COMMISSIONS 27 AND 42

IVTH COLLOQUIUM ON VARIABLE STARS
HELD IN BUDAPEST, HUNGARY, 5—9 SEPTEMBER 1968

CHAIRMAN OF SCIENTIFIC ORGANIZING COMMITTEE:
G. H. HERBIG
(Lick Observatory)

NON-PERIODIC PHENOMENA IN
VARIABLE STARS

Edited by

L. DETRE
Konkoly Observatory, Budapest

1969

ACADEMIC PRESS — BUDAPEST

ISBN-13: 978-94-010-3059-5   e-ISBN-13: 978-94-010-3057-1
DOI: 10.1007/978-94-010-3057-1

PREFACE

From September 5 until September 9, 1968, the IVth Colloquium on Variable Stars was held in Budapest, Hungary. The Colloquium was organized by a committee consisting of G. H. Herbig (President), A. Boyarchuk, M. W. Feast, D. McNamara, J. E. Merrill, D. J. K. O'Connell, V. Tsessevich, W. Wenzel.

The local organization was placed in the hands of a Committee consisting of members of the Konkoly Observatory, Budapest: L. Detre (Chairman), I. Almár, Julia Balázs-Detre, K. Barlai, M. Ill, S. Kanyó, M. Lovas and of J. Kovács (Hungarian Academy of Sciences).

The Colloquium was attended by about 90 scientists representing Argentina, Austria, Bulgaria, Canada, France, GDR, GFR, Hungary, Italy, The Netherlands, Poland, Roumania, South Africa, Sweden, United Kingdom, U.S.A., U.S.S.R.

As chairmen acted at the sessions: M. W. Feast, G. H. Herbig, J. Sahade, A. Boyarchuk, W. Wenzel, F.B. Wood and L. Rosino.

The contents of the present volume parallel closely the programme of the individual sessions of the Colloquium.

L. DETRE

PARTICIPANTS

ARGENTINA

Dr. J. Sahade, Observatorio Astronomico, La Plata

AUSTRIA

Dr. P. L. Fischer, Universitäts-Sternwarte Wien
Dr. F. V. Prochazka, Universitäts-Sternwarte Wien
Dr. A. Schnell, Universitäts-Sternwarte Wien

BULGARIA

Dr. B. Kowatchew, Section d'Astronomie, Sofia
Dr. N. S. Nikolov, Dept. of Physics, Sofia

CANADA

Prof. G. A. Bakos, University of Waterloo
C. Coutts, David Dunlap Observatory
Dr. J. D. Fernie, David Dunlap Observatory
Dr. J. B. Hutchings, Dominion Astrophysical Observatory, Victoria
Prof. Dr. H. B. Sawyer-Hogg, David Dunlap Observatory

CZECHOSLOVAKIA

Dr. J. Tremko, Skalnate Pleso Observatory

ENGLAND

Dr. M. J. Penston, Royal Greenwich Observatory
Dr. M. Penston, Royal Greenwich Observatory
Dr. S. H. Plagemann, Institute of Theoretical Astronomy, Cambridge

FRANCE

Dr. A. Baglin, Paris
Dr. C. Chevalier, Institut d'Astrophysique, Paris
Dr. A. M. Fringant, Institut d'Astrophysique, Paris
Dr. R. Herman, Observatoire de Paris, Meudon
Dr. M. Lacoarret, Observatoire de Nice
Dr. J. M. le Contel, Paris

DR. M. C. LORTET-ZUCKERMANN, Institut d'Astrophysique, Paris
DR. C. OUNNAS, Observatoire de Nice
M. PETIT, Drancy

GERMAN
DEMOCRATIC REPUBLIC

DR. P. AHNERT, Sternwarte Sonneberg
DR. W. GÖTZ, Sternwarte Sonneberg
H. HUTH, Sternwarte Sonneberg
I. MEINUNGER, Sternwarte Sonneberg
DR. L. MEINUNGER, Sternwarte Sonneberg
DR. G. RICHTER, Sternwarte Sonneberg
DR. W. SCHÖNEICH, Astrophysikalisches Observatorium Potsdam
W. THÄNERT, Sternwarte Sonneberg
DR. W. WENZEL, Sternwarte Sonneberg

GERMAN
FEDERAL REPUBLIC

DR. E. H. GEYER, Observatorium Hoher List
DR. H. MAUDER, Remeis-Sternwarte Bamberg
DR. E. POHL, Sternwarte Nürnberg
E. SCHÖFFEL, Remeis-Sternwarte Bamberg
DR. W. SEGGEWISS, Observatorium Hoher List
DR. W. C. SEITTER, Universitäts-Sternwarte Bonn
Prof. DR. W. STROHMEIER, Remeis Sternwarte Bamberg

HUNGARY

DR. I. ALMÁR, Konkoly Observatory Budapest
DR. B. BALÁZS, Konkoly Observatory Budapest
DR. J. BALÁZS-DETRE, Konkoly Observatory Budapest
K. BARLAI, Konkoly Observatory Budapest
Prof. DR. L. DETRE, Konkoly Observatory Budapest
E. ILLÉS, Konkoly Observatory Budapest
S. KANYÓ, Konkoly Observatory Budapest
M. LOVAS, Konkoly Observatory Budapest
DR. B. SZEIDL, Konkoly Observatory Budapest

ITALY

DR. P. BROGLIA, Osservatorio Astronomico Merate
DR. C. BLANCO, Osservatorio Astrrofisico Catania
DR. F. CATALANO, Osservatorio Astrofisico Catania
DR. S. CRISTALDI, Osservatorio Astrofisico Catania
Prof. DR. G. GODOLI, Osservatorio Astrofisico Catania
DR. A. MAMMANO, Osservatorio Astrofisico Asiago
DR. R. MARGONI, Osservatorio Astrofisico Asiago
Prof. DR. A. MASANI, Osservatorio Astrofisico di Brera
DR. M. RODONÒ, Osservatorio Astrofisico Catania
Prof. DR. L. ROSINO, Astrophysical Observatory Asiago

NETHERLANDS

DR. M. DE GROOT, University of Utrecht

POLAND

DR. P. FLIN, Astronomical Observatory Wroclaw
DR. T. JARZĘBOWSKI, Astronomical Observatory Wroclaw
DR. J. M. KREINER, Astronomical Observatory Cracow
DR. W. KRZEMIŃSKI, Astronomical Observatory Warsaw
DR. J. SMAK, Astr. Obs. of Warsaw University
DR. K. STĘPIEŃ, Astronomical Observatory Warsaw
DR. R. SZAFRANIEC, Astronomical Observatory Cracow

ROUMANIA

DR. Á. PÁL, Observatoire astronomique Cluj
DR. I. TODORAN, Observatoire Astronomique Cluj

SOUTH AFRICA

M. W. FEAST, Radcliffe Observatory

SWEDEN

DR. G. LARSSON-LEANDER, Astr. Institute, Lund

U. S. A.

DR. L. ANDERSON, University of California, Berkeley
R. W. AVERY, Flower and Cook Observatory
DR. W. S. FITCH, Steward Observatory
DR. G. H. HERBIG, Lick Observatory
DR. M. W. MAYALL, Cambridge
DR. E. F. MILONE, Gettysburg College
J. F. SIEVERS, Flower and Cook Observatory
Prof. DR. A. SLETTEBAK, Perkins Observatory
Prof. DR. F. B. WOOD, University of Florida, Gainesville

U. S. S. R.

DR. M. A. ARAKELIAN, Byurakan Astrophysical Observatory
DR. A. A. BOYARCHUK, Crimean Astrophysical Observatory
DR. E. A. DIBAJ, Sternberg State Astronomical Institute
DR. YU. S. EFIMOV, Crimean Astrophysical Observatory
DR. G. E. ERLEKSOVA, Astrophysical Institute Dushanbe
DR. R. E. GERSHBERG, Crimean Astrophysical Observatory
DR. I. D. KUPO, Alma Ata Observatory
DR. L. LUUD, Estonian Academy of Sciences, Tartu
DR. V. S. OSKANIAN, Byurakan Astrophysical Observatory
DR. E. S. PARSAMIAN, Byurakan Astrophysical Observatory
DR. U. V. PROKOFJEVA, Crimean Astrophysical Observatory
DR. I. B. PUSTYLNIK, Byurakan Astrophysical Observatory
DR. N. SHAKOVSKY, Crimean Astrophpical Observatory
DR. R. A. VARDANIAN, Byurakan Astrophysical Observatory

TABLE OF CONTENTS

PART V
MISCELLANEOUS PROBLEMS

PART I

STATISTICAL AND PHYSICAL INTERPRETATION OF IRREGULARITIES
MODERN TECHNIQUES OF OBSERVATION OR ANALYSIS

STATISTICAL AND PHYSICAL INTERPRETATION OF NON-PERIODIC PHENOMENA IN VARIABLE STARS

Introductory Report by

L. DETRE

Konkoly Observatory, Budapest

The subject of this Colloquium is similar to that of the third IAU Symposium on Non-Stable Stars, held 13 years ago in Dublin. At that time, the subject was limited to certain areas of particular interest. Now, we are trying to pay attention to the complex of non-periodic phenomena in variable stars. Dr. Herbig (1968), in his announcement of this Colloquium, has given an excellent summary of the topics in which we are concerned in these days.

Since then, the pulsating radio sources are added to our field, as they are most likely stars and they show in the amplitude of the pulses random fluctuations, that according to recent spaced receiver observations by Australian radio astronomers (Slee et al. 1968) take their origin predominantly at or near the sources themselves, and not in the intervening interplanetary or interstellar media.

Not long before, it was generally believed that stellar variability is only significant, if the variation exhibits a sizeable amplitude and (or) a certain amount of regularity. We shall adopt a different approach: random phenomena occurring both in variable and non-variable stars are of the same importance as regular large-scale phenomena, not only in the case when they result in such spectacular events as eruptions of novae or novoids, stellar flares, or the phenomena connected with the R Coronae Borealis stars, but also when they appear as small changes in the shape or position of spectral lines or in the periods of periodic variables, because these minute effects might be the manifestations of fundamental hydrodynamic or magnetic circulations in the star or signs of a star's rotational instability. In this way the frequency, intensity, or extent of random stellar phenomena may show cycles or pseudo-periods, as for example the solar magnetic activity with all its random manifestations like spots, plages, prominences and flares, has a 22 year cycle, and the irregular velocity and brightness oscillations in the photosphere of the sun have a pseudo-period of about 5 minutes. Other periodicities may sometimes be imposed on the observable effects of stellar random phenomena by stellar rotation, binary motion, or by some interaction with pulsation.

Any observed data representing a physical phenomenon, e.g. a light curve or radial velocity curve of a variable star, can be classified as being either deterministic or nondeterministic. Deterministic data can be described by an explicit mathematical relationship. They are either periodic or nonperiodic. The simplest periodic phenomenon has a sinusoidal time history and a frequency spectrum (that is an amplitude-frequency plot) consisting of a single frequency. Generally the spectrum of periodic data contains besides a fundamental frequency, f, its multiples. Almost periodic data, when the effects of two or more unrelated periodic phenomena are mixed, can be similarly characterized by a discrete frequency spectrum. For the determination of these frequencies different methods of harmonic analysis can be applied. For transient nonperiodic data, as e.g. the light curve of a flare, a discrete spectral representation is not possible. However, a continuous spectral representation can be obtained from a Fourier integral given by

$$X(f) = \int_{-\infty}^{+\infty} m(t) \, e^{-2\pi f t i} \, dt \,, \tag{1}$$

where $m(t)$ is the light curve, $X(f)$ is the Fourier spectrum.

We can consider irregular stellar variability as the observable effect of random succession of transitory events. On the sun and some types of variable stars, e.g. novae, U Geminorum, R Coronae Borealis and flare stars, these events can be observed separately. Solar activity can be followed even spatially separated. But generally, only the intermingling of many local and global transitory events can be observed in the stars as a continuously varying irregular light curve.

Also such a random time series $m = m(t)$ can be represented by a complicated mathematical relationship over a time interval $(0, T)$. But the formula will not hold for $t > T$. We obtain for different time intervals different formulae, different sample records of the same random process. Therefore, it is more practical to characterize a random time series by some simple parameters:

1. Taking the *mean value* of m for a time interval $(0, T)$

$$\overline{m} = \frac{1}{T} \int_0^T m(t) \, dt \tag{2}$$

and putting $\overline{m} = 0$ we can define the *variance*, as the mean square value about the mean, by

$$\sigma^2 = \frac{1}{T} \int_0^T [m(t)]^2 \, dt \tag{3}$$

The positive square root of the variance is called the *standard deviation*.

2. *The probability density function* describes the probability that the data will assume a value within some defined range at any instant of time. The probability that $m(t)$ assumes a value within the range between m and $m + \Delta m$ may be obtained by taking the ratio $T_{(m, m+\Delta m)}/T$, where $T_{(m, m+\Delta m)}$ is the total amount of time that $m(t)$ falls inside the range $(m, m + \Delta m)$ during an observation time T.

3. *The autocorrelation function* describes the general dependence of the values of the data at one time on the values at another time. An estimate for the autocorrelation between the values of $m(t)$ at times t and $t + \tau$ may be

obtained by taking the product of the two values and averaging over the observation time T. In equation form:

$$R(\tau) = \frac{1}{T} \int\limits_0^T m(t)\, m(t + \tau)\, dt \tag{4}$$

$R(\tau)$ is an even function with a maximum at $\tau = 0$.

4. For stationary data, i.e. for data characterized by time-independent parameters, we can construct the Fourier transform of the autocorrelation function

$$\Pi(f) = \int\limits_{-\infty}^{+\infty} R(\tau)\, e^{-2\pi i f \tau}\, d\tau \,. \tag{5}$$

$\Pi(f)$ is called the *power spectral density function*. That is a breakdown of the light curve into sinusoidal components and gives the mean squared amplitude of each component.

In the first column of Fig. 1. we see four special light curves, a sine wave (a), a sine wave with superposed irregularity (b), a narrow-band random light curve having cycles of nearly equal length (c), and a wide-band random light curve with strongly different cycles(d).

In the next column of Fig. 1. we see the corresponding probability density function plots. For the sine wave we have the maxima for $p(m)$ at the extremities, because the curve varies slowly there.

Next to the right we see the *autocorrelograms*. The sharply peaked autocorrelogram diminishing rapidly to zero (d) is typical of wide-band random data with a zero mean value. The autocorrelogram for the sine wave with random noise is simply the sum of the autocorrelograms for the sine wave and random noise separately (b). On the other hand, the autocorrelogram for the narrow-band random light curve appears like a decaying version of a sine wave autocorrelogram.

Finally we see the corresponding power spectra. A discrete power spectrum for a sine wave and a relatively smooth and broad power spectrum for the wide-band random light curve. The power spectrum for the sine wave with irregularities is the sum of the power spectra for the sine wave and the random case separately. On the other hand, the power spectra for the narrow band random light curve is sharply peaked, but still smoothly continuous as for random light-curve. The period corresponding to the peak may be called as pseudo-period. The four examples illustrate a definite trend in all the three parameters going from the sine wave to the wide-band noise case.

The principal application for an autocorrelation and for a power spectral density function is the detection of periodicities which might be masked in a random background. Any periodicity in the light variation will manifest itself as a series of peaks corresponding to a fundamental and its harmonics.

Such method of analysis requires enormous amount of computation, hence it has not been popular in the past. With the aid of high speed computers this is no longer a problem. However, the requirements of accuracy, extent,

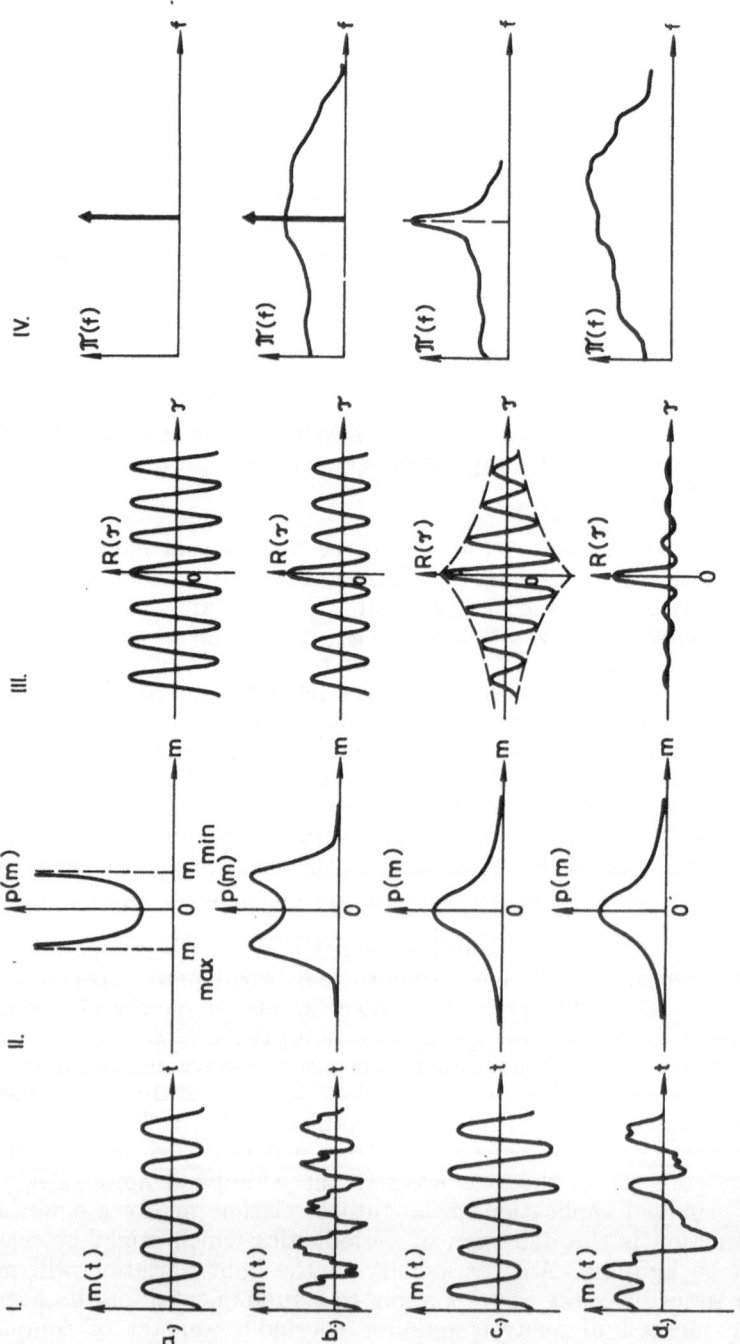

Fig. 1. Light curves, corresponding probability density functions, autocorrelograms and power spectra (s. text).

continuity, and reasonable homogeneity for the light curve to be analysed restrict considerably the applicability of the method to semiregular or irregular variable stars. Only one single semiregular variable was till now treated by this way, μ Cephei, on the one hand by Ashbrook, Duncombe and Woerkom (1954), who found the light curve to result from stochastic rather than harmonic processes, on the other hand by Sharpless, Riegel and Williams (1966) with the conclusion that the light variations are characterized by a much greater degree of regularity than is generally attributed to stars classed as semiregular variables.

Lukatskaya (1966) has investigated the autocorrelation and spectral functions of seven T Tauri-type variables and those of AE Aqr (1968), Kurochkin (1962) those of T Orionis, and on this colloquium we shall hear Dr. Plagemann on the same topic.

A very important question is, are the parameters of the light curves of irregular variables constant in time, or are they changing. Tsessevitch and Dragomineskaya (1967) investigated the light variation of 10 RW Aurigae stars on sky patrol photographs of the Odessa, Harvard, Dushanbe and Sonneberg observatories. They prepared probability density functions for several stars, finding longterm variations of this function, with cycles of 25 to 60 years. Hence, light curves of some irregular variables represent *nonstationary data*.

A particular random process is the so called *Markov process*. The property that distinguishes Markov processes from more general random ones can be described in non-mathematical form like this: If we know the present state of the process, and want to make predictions about its future, then information about the past has no predictive value, i.e. the process has no memory, its relationship to the past does not extend beyond the immediately preceding observation. We have a beautiful example for Markov processes in astronomy, the O—C diagrams for periodic variables, if the period or phase fluctuations are random, and independent from the preceding ones.

If we are able to determine the length of many individual cycles of a variable star, as for example in the case of continuously observed Mira-variables, we can test directly whether successive periods fluctuate accidentally or not. Moreover we can also determine the probability density function of the phase-fluctuations: $\psi(f)$. For cepheids, eclipsing binaries and every kind of short-period variables we must get every information for period-changes from the study of O—C diagrams. These are for most types of periodic variables determined by the cumulative effects of random phase fluctuations. At the Bamberg Colloquium we have shown how the probability structure of the O—C diagram could be determined using the central limit theorem of probability theory (Balázs-Detre and Detre, 1965). The structure depends on the mean square value of the phase fluctuations, σ, where

$$\sigma = + \sqrt{\int_{-\infty}^{+\infty} f^2 \, \psi(f) \, df}, \tag{6}$$

but it is highly independent from $\psi(f)$. O—C diagrams resulting from random phase-fluctuations consist of cycles of different lengths and amplitudes. (See Fig. 2, showing the O—C diagrams of three W Virginis-type variables: RU Cam, AP Herculis and $\varkappa$ Pavonis.)

2*

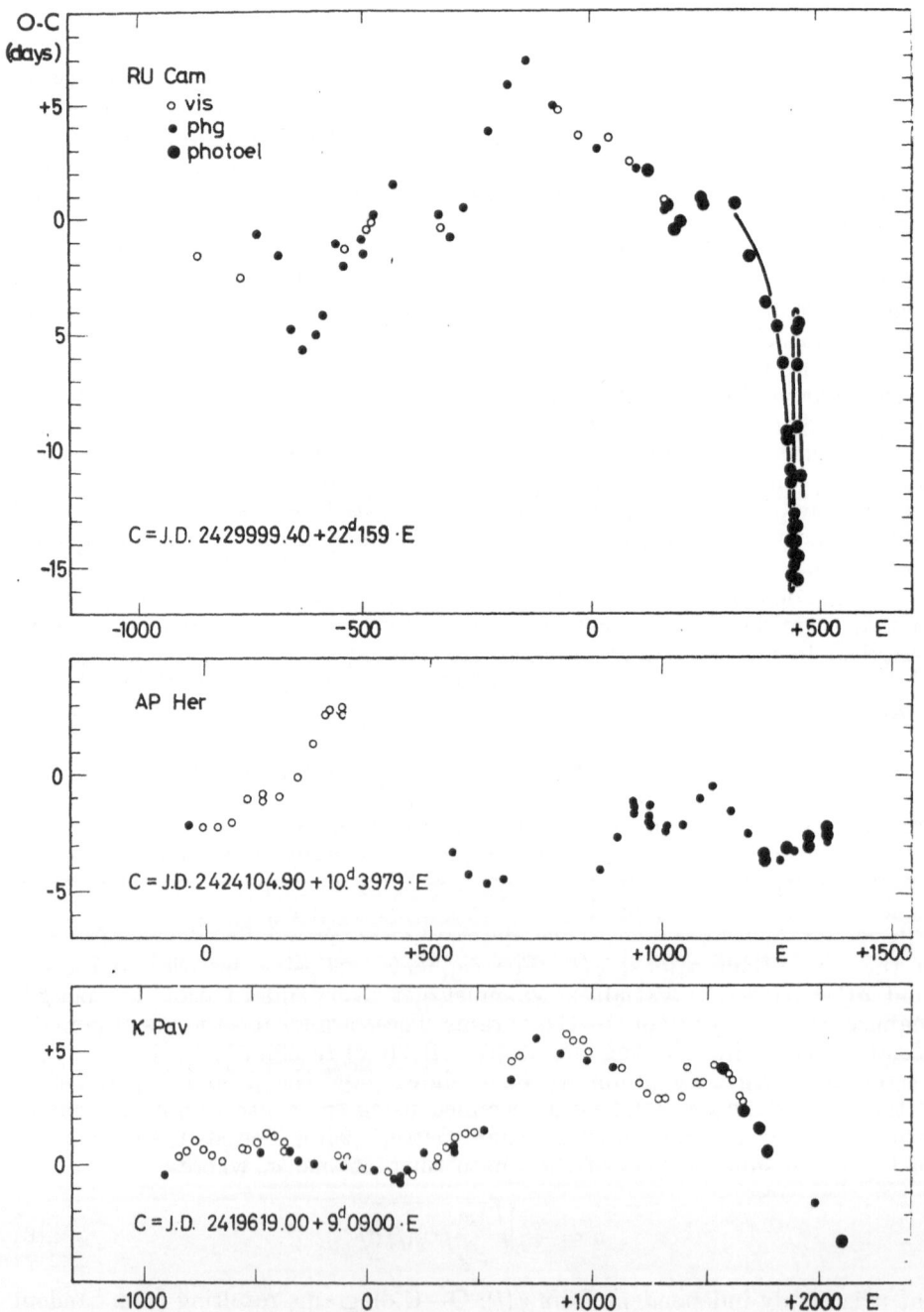

Fig. 2. O—C diagrams for three W Virginis-type variables, RU Cam, AP Her and κ Pav. That for AP Her is taken from Kwee (1967). For RU Cam, the strongly oscillating phase corresponds to the recent semiregular behaviour of the star.

Till now we have no evidence of evolutionary period-changes for most kinds of variables. Some cyclic terms in the O—C diagrams can be interpreted as due to binary or apsidal motion, but generally they must be treated as cummulative effects of random fluctuations. The value of σ can be determined from the O—C diagrams.

Several attempts have been made to represent the time sequence of explosions of eruptive variables by a Markov chain. Yet, as Mme Lortet-Zuckermann (1966) has stated, the Markov chain seemed poorly adapted for the representation of the sequence of the various sorts of explosions of the SS Cygni stars. Stellar variability refuses compliance with simple mathematical models.

The maxima or minima of an ideally irregular variable would be distributed at random, and the cycle lengths l would follow a Poisson-distribution:

$$f(l)\, dl = ne^{-nl}\, dl \tag{7}$$

Sterne (1934) has shown that the minima of R Coronae Borealis fulfil this condition. But this conclusion is only true if the minima are independent events, and this is certainly not the case, if the minima are not well separated.

We see that all analyses of this kind, with the exception of the search of hidden periodicities or changes of the statistical parameters in time, are rather formal and do not say much about the physical nature of the stars, since objects of the most various kinds may show similar light curves. Combined spectroscopic and photometric, sometimes radio observations are needed to reveal the real nature of some objects with irregular light variation. Such combined efforts often lead to surprising results. I mention the beautiful interpretation of V Sagittae by Herbig, Preston, Smak and Paczynski (1965), who resolved the complex light variations of this star into three apparently independent activities, showing that the star is a peculiar nova-like eclipsing binary. VV Puppis, a star formerly classified as an RR Lyrae variable, was interpreted by Herbig (1960) likewise as a nova-like double star. In June 1967 Deutsch (1967) has reported that the spectrum of CH Cygni, classified earlier as a semiregular a-type variable, changed from a normal M6 type into that of a symbiotic nova-like star. The star now shows rapid light variations in the ultraviolet (Wallerstein 1968b). Using new high quality spectrograms, Herbig (1966) succeeded in interpreting the 1936 flare-up of Wachmann's star, FU Orionis, as a phenomenon of early stellar evolution, a pre-main-sequence collapse in conformity with Hayashi and Cameron's ideas of early stellar evolution. Two irregular variables have recently been identified as radio sources: BW Tauri by Penston (1968) and BL Lacertae by Schmitt (1968)*.

These examples go to show that we need in many cases special interpretations for individual objects. Yet, we also have some general principles for trying the physical interpretation of broad classes of non-periodic phenomena in variable stars. These attempts can be classified into four categories:

1. *Solar analogies.* An increasing convergence is apparent between the fields of stellar physics and solar physics, stellar analogues of solar phenomena

* BW Tau = 3C120, BL Lac = VRO 42.22.01.

are becoming the subjects of specific researches. I mention a recent interesting paper by Godoli (1967) at the Padova 1967 conference.

2. Irregular phenomena connected with or caused by the *binary nature* of the star, as eruptions of different kinds or other irregularities associated with gaseous material streaming between the components in very short-period binaries and in symbiotic variables, further, period variations in all kinds of eclipsing and spectroscopic binaries.

3. Irregularities connected with *rotational instability* of the equatorial region of a rapidly rotating star, as in Be, Of and Wolf-Rayet stars.

4. *Veiling theories,* put forward by Merrill for long-period variables* and considered by Loreta (1934) and O'Keefe (1939) in connection with R Coronae Borealis in terms of solid carbon particles.

The closest similarity between solar activity and irregular light variation in stars is that between solar flares and extremely sudden increases in the integrated brightness of some stars, mainly of flare and T Tau stars. Since Sir Lovell (1964) discovered that optical stellar flares are accompanied by radio bursts of the I, II and III solar type, it became very probable that flare stars show in a gigantic form the same kind of activity as the sun. The quite irregular light curve of T Tauri stars could be due to the superposition of very many flares, with a variation of the activity of the star, intermingled with effects of the neighbouring circumstellar material. We shall have introductory papers on this theme by Wenzel and Gershberg. I refer to a recent excellent review on flare stars by Haro (1967). Since the Prague meeting of the IAU, Commission 27 has under Chugainov's leadership a very well organized working group on UV Ceti type stars for cooperative radio and optical observations.

One of the most interesting possibilities for solar analogies is the extension of our concept of the chromosphere and of its activity to stars. The discovery of the Wilson—Bappu (1957) effect, a correlation over a range of nearly 16 absolute magnitudes between the widths (and not the intensity) of the emission cores of the H and K-lines of Calcium II and the visual absolute magnitude for stars of types G, K and M, has engendered considerable effort to interpret this effect in terms of chromospheric macroturbulence including all irregular, non-periodic or pseudo-periodic motions of the atoms in a stellar atmosphere. Kraft, Preston and Wolff (1964) showed that a similar correlation exists between the width of the hydrogen (H_α) absorption line and the ultraviolet absolute magnitude. Recently Vaugham and Zirin (1968) studied the infrared He line at λ 10830 A which is the only line from 3000 to 11 000 A that originates solely in the chromosphere, free of changes in an underlying photospheric line. Since the line is excited only at high temperatures, its presence is an excellent test for hot chromospheres in late-type stars.

The sun fits the Wilson—Bappu relation, but the intensity of K_2 emission in the integrated light of the sun is very small and can be observed with high dispersion only. In the spectra of many stars K_2 emission is observable even

* Merrill, P. W., Stellar atmospheres. The University of Chicago Press 1960. p. 512.

with rather small dispersion, indicating that some stars possess much more active chromospheres than does the sun.

Leighton (1964) has shown that the K_2 emission on the sun occurs at the edge of supergranulation cells, where photospheric magnetic fields are sometimes found to be strengthened to the order of 100 gauss. There is a point-to-point correlation between chromospheric activity and the photospheric magnetic field strength. The Ca II network is not due to a circulation of matter in the chromosphere but due to a more general circulation which underlies the chromosphere. To the same effect points Bonsack and Culver's (1966) result that in K-type stars the widths of weak lines which do not have a chromospheric origin, are well correlated with the widths of K_2 emission or the strength of the infrared He-line.

Because this emission and the strength of the infrared He-line appear greatly enchanced in the region of solar plages and in this way it is well correlated with the 11-year solar cycle, a study of the nature of variability of the K_2 emission or of the He-line in other stars should add substantially to our understanding of both sun and stars.

That K_2 does indeed vary, has been established by Wilson and Bappu, by Griffin (1964), Deutsch, Vaugham, and most recently by Liller (1968), especially in the stars α Bootis, α Tauri and ε Geminorum. The type of variation noted has usually been a change in the relative intensities of the violet and red components of the K_2 emission, but there was little evidence of periodicity analogue to the solar cycle.

Transitory Ca II emission develops at the phase of minimum radius in cepheids and longperiod variables. The study of this phenomenon by Herbig (1952), Jacobsen (1956) and Kraft (1957) led Kraft (1967) at the IAU Symposium 28 to the interesting suggestion, that the behaviour of cepheids at this phase is an exaggeration of the disturbed sun. At the time of minimum radius the surface of the cepheids becomes covered with something like plages. As the cycle progresses, a shock wave moves through the atmosphere and all such solar-like disturbances disappear: the cepheid becomes an F-type star.

Some non-periodic secondary variations in eclipsing binaries were attributed to star spots (Kron, 1947, 1952). But these stars are not adapted for such investigations, because gas streams between and around the components may cause irregularities in the light curve.

Prominence activity was found in supergiant stars, for example in 31 Cygni, which are components of eclipsing systems. When the star goes behind the atmosphere of the supergiant K3 star, at times several absorption components due to Calcium II H and K are seen, providing unmistakable evidence that bodies of gas moving with discrete velocities exist in its atmosphere (S. Underhill, 1960).

Mass loss in stars might bear a relation to the solar wind, which is a plasma extension of the solar corona moving outward at the velocity of about 500 km/sec carrying away a mass of about 10^{-13} solar mass per year and the frozen-in magnetic fields from the sun. The solar wind has a steady continuous and an irregularly varying component. The evidence that considerable mass loss occours in stars apart from novae, supernovae and close binaries, came from Deutsch's (1956) remarkable discovery of a set of circumstellar lines in the visual companion of the M supergiant α Herculis. There is now ample spectroscopic evidence for the efflux of cool gas from the surfaces of all giant stars with spectral types later than M0 (Deutsch, 1966), at a rate of some 10^{-9} solar mass per year. Weyman (1962) pointed out the difficulties in the way of a solar wind explanation for these phenomena. More violent mass losses from stars are certainly not of the solar wind type. In some pulsating stars the pulsation shock can be so violent that the surface layer may be driven away from the star in a relatively small number of periods, as was shown by Christy (1965) for W Virginis stars. According to Paczynski and Ziótkowski (1968) Mira type variables may throw out their envelopes and in this way planetary nebulae might be formed. Mass loss may be the dominating factor in horizontal branch evolution rather than nuclear burning. Kuhi (1964, 1966) estimated the rate of mass loss from T Tauri stars at about 10^{-7} solar mass per year. Spectra secured from rocket flights provided first evidence for the extremely violent ejection processes in the atmospheres of O and B-type supergiants and bright giants (Jenkins and Morton, 1967). We shall hear more on this subject next week in Trieste, where a Colloquium will be held on mass loss from stars.

The weak point of solar analogies is that solar phenomena are not yet quite understood. Yet, we can be certain of the magnetic nature of all processes of solar activity and that all its accompanying phenomena like spots, faculae, flares, the irregular component of solar wind, etc. are connected with local concentration as well as annihilation of magnetic fields. Hence it is very probable that also the analogous stellar phenomena are of magnetic origin.

Moreover, it becomes increasingly evident that magnetic fields may have a share also in other aspects of stellar irregularities. E.g., Merrill's veiling theory is supported by Serkowski's (1966a, b) recent discovery of large amounts of plane polarization in some Mira stars at minimum light. This polarization can be explained by graphite flakes, condensed in the atmosphere of these stars, presuming that they are aligned by stellar magnetic fields (Donn et al. 1966; Wickramasinghe 1968). Magnetic forces may play an important role in the formation of the envelopes of Be stars. Of course, Struve's (1931) suggestion of rotationally forced ejection in a star rotating at the rotational limit in which its equatorial rotational velocity is first sufficient to balance by centrifugal effects the gravitational attraction of the star at its equator, is correct. But an additional force is required to move the matter outward from the region just above the star's equator. The complex kinematic behaviour of the shell, the occurrence of stars such as Pleione, which seem able to lose and reform their shell at intervals, is particularly suggestive of the presence of forces which trend to drive the gases away from the star. Even quite weak magnetic fields could produce significant dynamical effects in such a shell (Crampin and Hoyle 1960; Limber and Marlborough 1968). Hazlehurst (1967) studied in a recent paper the magnetic release of a circumstellar ring, and he found that the gases describe a decelerated motion, compatible with the observed spectral properties of circumstellar shells. From the ultimate velocity of the material an observational determination of the magnetic field in the stellar photosphere will be possible.*

We would have a better understanding of the observed period-variations in eclipsing binaries, if an adequate electromagnetic theory of the gaseous

---

* About problems of irregular variations in light and radial velocity of Be, Of and WR stars I refer to the excellent book The Early Type Stars by Anne B. Underhill (Reidel Publishing Company, 1966).

streams in the systems had been elaborated. It appears from the work by Plavec and Schneller that the most erratic O—C diagrams are obtained for contact and undetached systems. If an O—C diagram has random walk propperties, then the underlying physical processes that give rise to the random period fluctuations, are themselves random processes. Wood's hypothesis of mass ejection for the explanation of the period fluctuations, if the areas of ejection are distributed over the surface at random, fits the criterion of randomness, but the required masses are too high. However, we might have a very efficient agent for angular momentum changes in the interaction of the ionized gaseous streams moving around the components with the magnetic fields of the stars.

Magnetic fields might play an even greater role in hot short-period eruptive binaries and in symbiotic stars. Babcock measured a magnetic field of 1000 gauss in the symbiotic variable AG Pegasi. The configuration in eruptive binaries, a highly ionized disk, a strongly flickering hot component ejecting highly ionized material into the disk, might be extremely unstable, especially if the components have strong magnetic fields. It is just possible that the magnetic and gravitational instability of such a configuration might lead from time to time to major eruptions. According to my opinion the seat of the eruptions might be the plasma surrounding the stars, not a stellar component.

According to Ambarzumjan's (1954) hypothesis, the continuous emission observed in the spectra of the 'T Tauri type variables and UV Ceti type stars during their outbursts originates from relativistic electrons in the magnetic fields of these stars.

Random processes may influence the pulsation of the stars, giving rise to irregular fluctuations in the light and radial velocity curve and in the period. The triggering mechanism of the pulsation, which is sought in the convective layers of the stars, may especially be sensitive to magnetic activity.

Epstein (1950) has shown in an important paper that in highly centrally concentrated stellar models the period of the fundamental mode is determined primarily by conditions in the envelope and that the period is almost independent of conditions in the central regions where most of the mass is located. This result suggests that stellar pulsation, at least in giant and supergiant-like stars, is a fairly superficial phenomenon effecting only the outer stellar layers. The higher modes are even more sensitive to properties of the most external layers of the star, since these modes have higher relative amplitudes near the surface.

Indeed, red variables, where the outer layers play a great role, have very erratic light variation, whereas classical Cepheids and most RR Lyrae stars show very little if any irregularities.

Zhevakin introduced the peripheral zone of He II critical ionization as the excitation mechanism of the pulsation. He (1959) developed an interesting theory of semiregular and irregular variables. The period of oscillation of the inner region of the star is constant to a high degree of accuracy. The nonadiabatic oscillations of the atmosphere show relative to the adiabatic oscillations of the inner regions a phase shift, whose value depends primarily on how close is the ionization zone to the stellar surface. Random fluctuations in the position of the zone change the phase shift, and in this way the period of the outer zones will fluctuate about the period of oscillations of the inner region.

If the driving mechanism of the pulsation is affected by random perturbations, we may expect a suppression of the amplitude of the pulsation relative to stars free from such perturbations. As it is well known, semiregular red variables differ from the longperiod variables only in their smaller amplitudes (Fig. 3).

The RR Lyrae-variables with the Blashko-effect have very complicated O—C diagrams. Though the Blashko-effect is a periodic phenomenon, it causes great random fluctuations both in the fundamental and in the secondary period (Fig. 4). As a period-amplitude diagram for RRab-stars in M3, taken from a paper by Szeidl (1965), shows (Fig. 5), the mean amplitudes of the RRab stars with Blashko-effect are much smaller than the amplitudes of RRab stars with stable light curves.

Babcock discovered a strongly variable magnetic field in RR Lyrae which ranges from +1200 to —1600 gauss. Julia Balázs (1959) has shown that there was some correlation between Babcock's measures and the Blashko-effect of this star: the maximum positive and maximum negative fields were associated with the maximum and minimum light amplitudes, respectively.

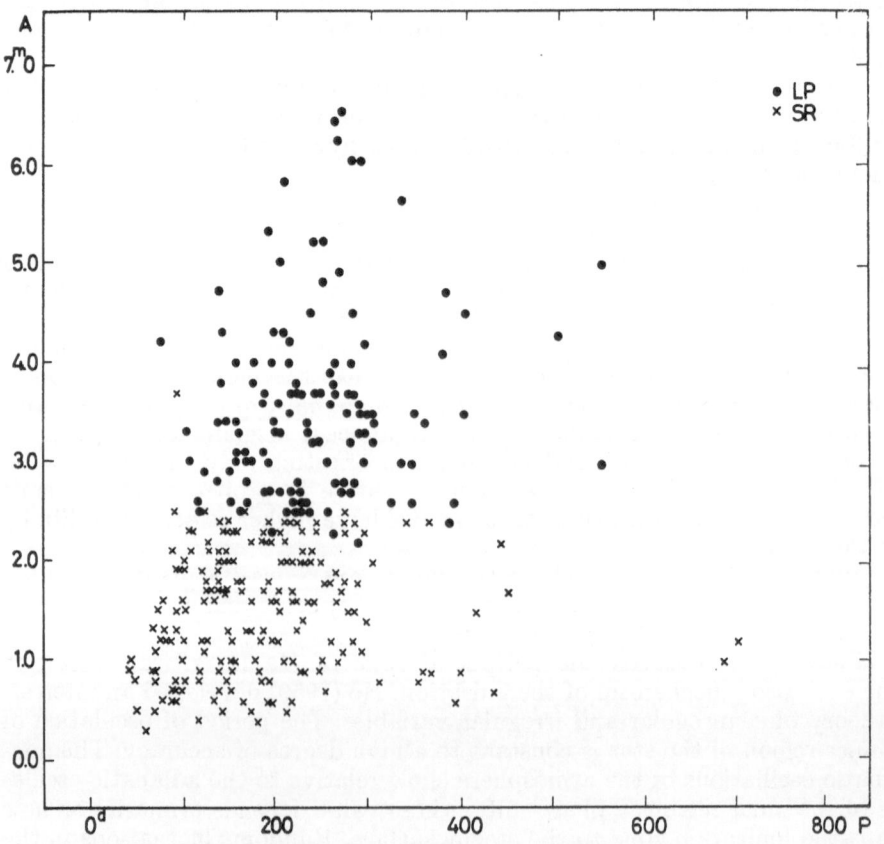

Fig. 3. Period-amplitude relation for longperiod (points) and semiregular (crosses) variables in Sagittarius.

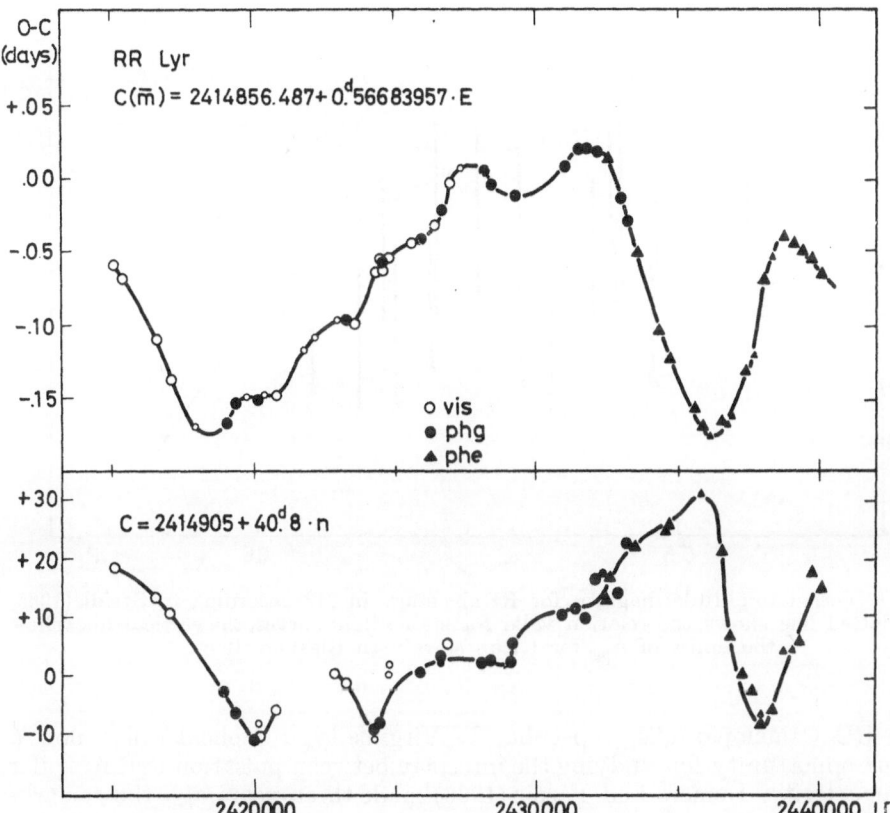

Fig. 4. O—C diagrams for the fundamental and secondary periods of RR Lyrae.

She made the proposal that RR Lyrae is an oblique rotator with a rotation period of 41 days, which is the period of the Blashko-effect. Preston (1967) has observed RR Lyrae for the Zeeman effect in 1963 and 1964 some 50 times and has not once found a measurable field. Yet, this negative result does not disprove, that the Blashko-effect and the irregularities connected with it are of magnetic origin. As Fig. 6 prepared by Szeidl shows, the Blashko-effect is a very erratic phenomenon, the amplitude of the phase variations and that of the maximum light variations are changing strongly from time to time, they are sometimes scarcely observable.* The same may happen with the magnetic field. In any case, the greatest part of magnetic activity might take place, as on the sun, below the photosphere.

For Delta Scuti variables and dwarf Cepheids we do not observe the suppression of the amplitude in stars with secondary periodicities, and such stars have the same simple O—C diagrams as variables with stable light curves. Here, the secondary periods may originate from non random influences, e.g. from tidal effects as proposed by Fitch (1962).

* In RR Gem a strong Blashko-effect was observed till about 1937 which disappeared later.

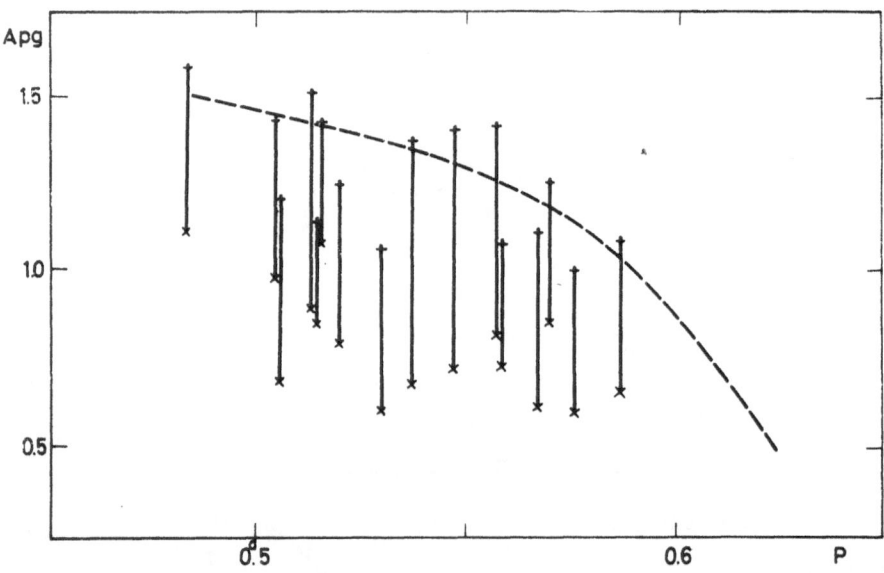

Fig. 5. Period-amplitude diagram for RRab stars in M3 according to Szeidl (1965)
The dotted line shows the relation valid for stable light curves, the vertical lines show
the limits of A_{pg} for RRab stars with Blashko-effect.

RU Camelopardalis, a peculiar W Virginis-type Cepheid offers now a
unique opportunity for studying the interplay between pulsation and irregular
stellar activity. Demers and Fernie (1966) made three years ago the remarkable discovery that the star nearly stopped its light variation. Considering
all available photoelectric observations I was able to show (Detre 1966) that
the star exhibited cyclic amplitude variations with a mean cycle of about
5 years. Therefore, I expected an increase in the amplitude for the year 1967.
Indeed the amplitude began to increase in the spring 1967, and in the summer
it reached 0.3 mg. in V and nearly half a magnitude in B. But immediately,
after Fernie and myself have reported about this amplitude increase at the
Prague IAU meeting, the amplitude came back very rapidly to the small value
it had in 1966 (Fig. 7). At first the light minimum, a little later the maximum
passed to its former value. The star needed in both cases only four cycles
of its 22 day-variation to restore the small amplitude.

The most important point we should know, how the spectrum changed.
Faraggiana and Hack (1967) studied 11 high dispersion spectrograms taken
by Prof. Deutsch between 1956 and 1961 when the light amplitude was normal.
They observed hydrogen emission-lines from minimum to maximum and
emission cores in the H and K lines on all the spectrograms sufficiently exposed
in this region. The radial velocity curve obtained from these lines was shifted
with respect to the curve for the absorption lines by about —70 km/sec, suggesting that the chromosphere of the star was in expansion. Demers and Crampton
(1966) taking spectra during the small amplitude stage, state that no emission
lines are visible. To the same conclusion comes Wallerstein (1967, 1968) using
Lick coudé spectra. But unfortunately, these spectra do not contain the H,

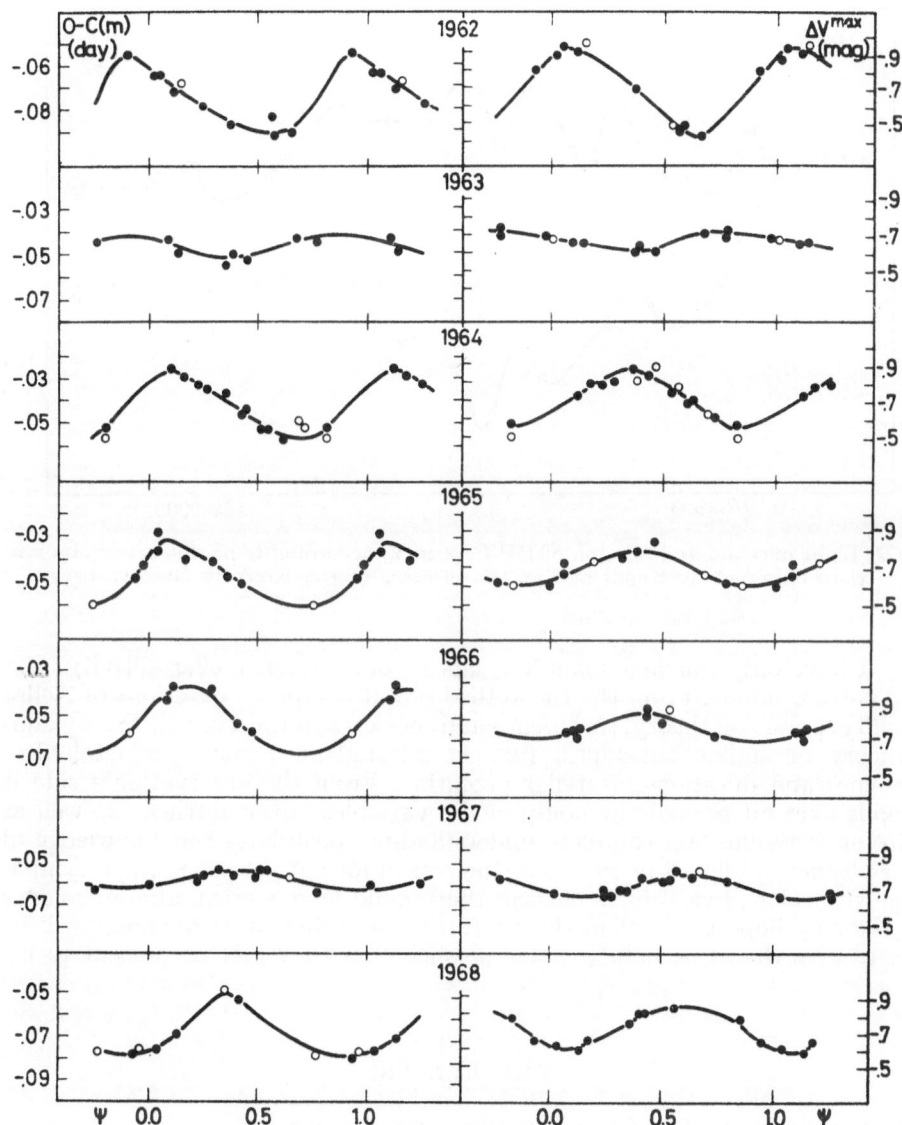

Fig. 6. Variations in the Blashko-effect of RR Lyrae.

K lines region. I wonder if there are any spectra taken during the quiescent or temporary recovery stage similar to those obtained formerly by Prof. Deutsch.

According to my opinion the pulsation mechanism of the star is very sensitive to changes of its magnetic field and we witness the effects of such changes in the last time. I hope, the star will yet give opportunity to study this question. At present it shows irregular light variations with a V-amplitude smaller than 0.1 magnitude.

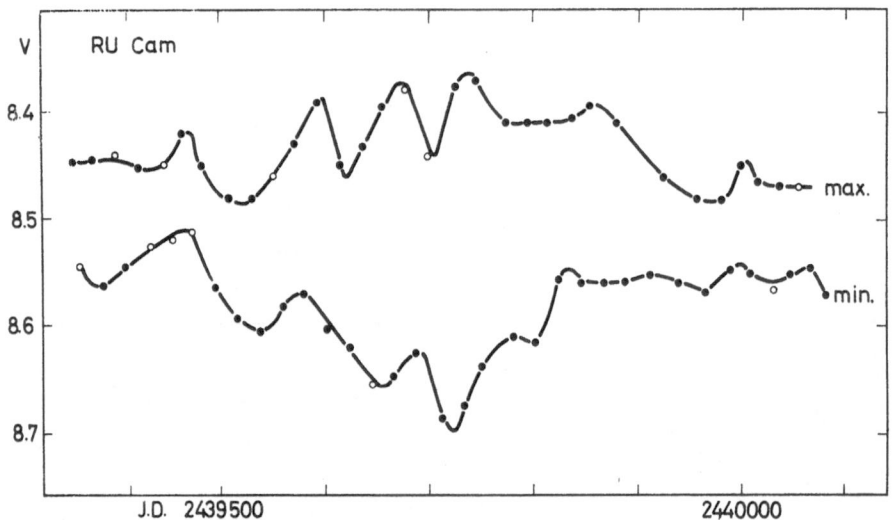

Fig. 7. Light maxima and minima of RU Cam in V, according to photoelectric observations obtained by Szeidl at the 24″ reflector of the Konkoly Observatory.

I have only touched some few aspects of irregular stellar activity. Our field is tremendous. From the theoretical side it comprises questions of stellar stability, pulsation theory, turbulence and shock wave theory, i.e., the dynamic theory of stellar envelopes, further, celestial mechanics, magnetohydrodynamics and questions of stellar evolution. From the observational side it extends over all periodic or non-periodic variables, over intrinsic as well as eclipsing variables. Its complete understanding postulates the knowledge of solar physics in highest degree. And the reason for not yet proposing a symposium with solar physicists, is because they would have a great advantage over us. We may hope that not in the far future variable star astronomers will be able to give the same help to solar physicists as they give at present to us.

REFERENCES

Ambarzumjan, V. A., 1954. Comm. Byurakan Olss. No. 13.
Ashbrook, J., Duncombe, R. L. and Van Woerkom, A. J. J., 1954, Astr. J. **59**, 12.
Balázs-Detre, Julia and Detre, L., 1965, Kl. Veröff. Remeis-Sternw. Bamberg IV, No. 40 p. 184.
Balázs-Detre, Julia, 1959, Kl. Veröff. Remeis-Sternw. Bamberg **27**, 26.
Bonsack W. K. and Culver, R. B., 1966, Astrophys. J. **145**, 767.
Christy, R. F., 1965, Kl. Veröff. Remeis-Sternw. Bamberg, **40**, 77.
Crampin, J. and Hoyle, F., 1960, Mon. Not. R. astr. Soc. **120**, 33.
Demers, S. and Crampton, D., 1966, Astr. J. **71**, 852.
Demers, S. and Fernie, J. D., 1966, Astrophys. J. **144**, 440.
Detre, L., 1966, Inf. Bul. Var. Stars No. 152.
Deutsch, A. J., 1956, Astrophys. J. **123**, 210.
Deutsch, A. J., 1966, Stellar Evolution. Plenum Press New York p. 377.
Deutsch, A. J., 1967, IAU Circ. No. 2020.

Donn, B., Stecher, T. P., Wickramasinghe, N. C. and Williams, D. A., 1966, Astrophys. J. **145**, 949.
Durney, B. R., Faulkner, J., Gribbin, J. R. and Roxburgh, I. W., 1968, Nature **218**, 20.
Epstein, J., 1950, Astrophys. J. **112**, 6.
Faraggiana, R. and Hack, M., 1967, Z. Astrophys. **66**, 343.
Godoli, G., 1967, Atti dell'XI. Convegno, Padova, p. 224.
Griffin, R. F., 1964, Observatory **83**, 255.
Haro, G., 1967, Stars and Stellar Systems. Vol. VIII. 141.
Hazlehurst, J., 1967, Z. Astrophys. **65**, 311.
Herbig, G. H., 1952, Astrophys. J. **116**, 369.
Herbig, G. H., 1960, Astrophys. J. **132**, 76.
Herbig, G. H., 1966, Vistas in Astr. **8**, 109.
Herbig, G. H., Preston, G. W., Smak, J. and Paczyński, B., 1967, Astrophys. J. **141**, 617.
Herbig, G. H., 1968. Inf. Bul. Var. Stars No. 243.
Jacobsen, T. S., 1960, Publ. Dom. Astrophys. Obs. Victoria **10**, 145.
Jenkins, E. B. and Morton, D. C., 1967, Nature **215**, 1257.
Kraft, R. P., 1957, Astrophys. J. **125**, 336.
Kraft, R. P., Preston, G. W. and Wolff, S. C., 1964, Astrophys. J. **140**, 235.
Kraft, R. P., 1967, IAU Symp. 28, p. 236.
Kron, G. E., 1947, Publ. astr. Soc. Pacific **59**, 261.
Kron, G. E., 1952, Astrophys. J. **115**, 301.
Kuhi, L. V., 1964, Astrophys. J. **140**, 1408.
Kuhi, L. V., 1966, Stellar Evolution. Plenum Press New York, p. 373.
Kurochkin, N. E., 1962, Perem. Zvezdy **14**, 284.
Kwee, K. K., 1967, Bull. astr. Inst. Netherl. Suppl. **2**, 97.
Leighton, R. B., 1964, Astrophys. J. **140**, 1120.
Liller, W., 1968, Astrophys. J. **151**, 589.
Limber, D. N. and Marlborough, J. M., 1968, Astrophys. J. **152**, 181.
Loreta, E., 1934, Astr. Nachr. **254**, 151.
Lortet—Zuckermann, M. C., 1966, Ann. Astrophys. **29**, 205.
Lovell, B., 1964, Observatory **84**, 191.
Lukatskaya, F. I., 1966, Per. Zvezdy **16**, 168.
Lukatskaya, F. I., 1968, Inf. Bul. Var. Stars No. 295.
Merrill, P. W., 1960, Stellar Atmospheres. The University of Chicago Press, p. 512.
O'Keefe, J. A., 1939, Astrophys. J. **90**, 294.
Paczynski, B. and Ziótkowski, J., 1968, Acta Astr. **18**, 255.
Penston, M. V., 1968, Inf. Bul. Var. Stars 255.
Preston, G. W., 1967, The Magnetic and Related Stars. Mono Book Corp. Baltimore, p. 3.
Schmitt, J. L., 1968, Nature **218**, 663.
Serkowski, K., 1966a, Astrophys. J. **144**, 857.
Serkowski, K., 1966b, Inf. Bul. Var. Stars No. 141.
Slee, O. B., Komesaroff, M. M. and McCullock, P. M., 1968, Nature **219**, 342.
Sharpless, S., Riegel, K. and Williams, J. O., 1966, R. astr. Soc. Can. **60**, 15.
Sterne, T. E., 1934, Harvard Bul. 896, 16 and 17.
Struve, O., 1931, Astrophys. J. **73**, 94.
Szeidl, B., 1965, Mitt. Sternwarte Budapest No. 58
Tsessevich, V. P. and Dragominezkaya, B. A., 1967, Sky Tel. **34**, 366.
Underhill, A. B., 1960, IAU Symp. No. 12, p. 88.
Vaugham, A. H. Jr., and Zirin, H., 1968, Astrophys. J. **152**, 122.
Wallerstein, G. and Crampton, D., 1967, Astrophys. J. **149**, 225.
Wallerstein, G., 1968(a), Astrophys. J. **151**, 1011.
Wallerstein, G., 1968(b), Observatory **88**, 111.
Weyman, R., 1962, Astrophys. J. **136**, 476 and 844.
Wickramasinghe, N. C., 1968, Mon. Not. R. astr. Soc. **140**, 273.
Wilson, O. C. and Bappu, M. K. Vainu, 1957, Astrophys. J. **125**, 661.
Zhevakin, S. A., 1959, Astr. Zu. **36**, 394.

SPECTRAL ANALYSIS OF THE LIGHT CURVES OF T TAURI STARS AND OTHER OBJECTS

STEPHEN PLAGEMANN

Institute of Theoretical Astronomy, Cambridge, England

LIST OF SYMBOLS

(1) f — frequency in units of cycles per day.

(2) f_N — Nyquist frequency, where $fn = \frac{1}{2}\Delta t$

(3) ω — angular frequency in radians, $f_N = \pi/\Delta t$

(4) Δt — sampling interval.

(5) n — number of observations.

(6) $\lambda(k)$ — lag of time domain window, also known as the covariance averaging kernel.

(7) k — number of lags in the time domain.

(8) m — the total number of lags.

(9) $X_i(t)$ — observed time series (light curve).

(10) σ^2 — variance.

(11) μ_i — mean of time series.

(12) $\gamma_{11}(k)$ — sample estimate of the autocovariance of lag k uncorrected for the mean.

(13) $\gamma_{12}(k)$ — crosscovariance of $X_1(t)$ and $X_2(t)$.

(14) $\varrho_{11}(k)$ — autocorrelation coefficient.

(15) $c_{11}(k)$ — sample estimate of the autocovariance.

(16) $r_{12}(k)$ — ordinary product-moment correlation between $X_1(t)$ and $X_2(t)$.

(17) U_k — series used as filter in time domain with convolution.

(18) $T(\omega)$ — spectral window or kernel.

(19) $b(\omega)$ — bias.

(20) ν — degrese of freedom.

(21) $B\omega$ — bandwidth.

(22) $A_{11}(k)$ — autocorrelation function of $X_1(t)$.

(23) $A_{21}(k)$ — crosscorrelation function of $X_2(t)$ and with $X_1(t)$.

(24) $I_{11}(\omega)$ — spectral estimate for harmonic analysis.

(25) $\eta(\omega\!-\!y)$ — spectral window in frequency domain.

(26) $f(\omega)$ — spectral density function of $X_1(t)$; the cosine transformation of $X_1(t)$.

(27) R_{12} — contribution to the sample variance from the jth harmonic.

(28) $E(k)$ — the energy per unit interval of $f \cdot (am\Delta t)$ centred upon $k = 0$, 1, ..., m.

(29) $Z_{(K)}$ — the co-spectrum; the cosine transformation of the crosscorrelation

(30) $W(k)$ — the quadratine spectrum; the sine transformation of the crosscorrelation.

(31) $R(k)$ — the coherence.

(32) $\Theta(k)$ — phase difference.

(33) $G(\omega)$ — the gain.

(34) $T(i\omega)$ — transfer function.

I. INTRODUCTION

The purpose of computing power spectra from the light curves of irregular stars is both theoretical and practical. A power spectrum of a star is an ideal method to study a complicated light or velocity curve whose inherent periodicities may be obscured by noise. If one lacks theoretical light or velocity curves, power spetcrum analysis of the empirical data can reveal the presence of peaks above the noise level, giving an exact value for the periodicities and their amplitude, and the distribution of energy at different vibrational frequencies. We can also study the changes in time of the period, amplitude and energy distribution. This information can in turn be compared with the T Tauri variables, such as their peculiar absorption and emission spectra, their spectral type and their age and mass when compared to a theoretical H—R diagram.

The preparation of power spectra of T Tauri stars involved a research scheme carried out in three stages. Stage I began with an extensive search of the literature over the past fifty years for published light curves of all the 130 T Tauri stars given in a table by Herbig (1962), plus some additions. This task has been eased by references in two biographical works, "Geschichte und Literatur des Lichtwechsels der Veränderlichen Sterne" and the "Astronomischer Jahresbericht". It was necessary to check through each reference to discover which contained the useful data. The light curves were then put on to punched cards, each card having a single Julian Date and magnitude. After the light curves were punched on IBM cards (some 27,000 readings in all) they were sorted out serially and counted on an IBM 1401 computer, the light curves were printed out and checked visually to reveal the presence of flares. The criteria used was an increase in brightness greater than $0^m \cdot 4$ in the course of one hour passage. Stage II involved the calculation of pilot power spectra of T Tauri stars for different sampling intervals Δt and lags K in order to select stable spectral estimates. Checks for stationarity and normality were made by breaking up long data sequences in halves and comparing the power spectra of each.

Stage III of estimating the power spectra of T Tauri stars was done using optimal sampling intervals and lags, i.e. those that all the essential details appeared in the spectrum. An important factor in revealing the data was a prewhitening or filtering of the data in order to reduce the effects of power leakage from higher frequencies that tend to distort the power spectrum. As a further check of the power spectrum's stability the light curves were converted to a new time series by the addition of a Gaussian distributed random numbers within a standard deviation of the mean, and the resulting power spectrum was then computed. Definitions and procedural details will be presented below.

All of the computations involving time series used a system of programmes (BOMM) developed by Bullard et al. (1966) for geophysical applications. The calculations were carried out by the IBM 7090 computer at Imperial College, London. Total computation times for each light curve analysed will be provided and other details will be presented in a later paper.

II. TIME DOMAIN AUTOCORRELATION

In the following pages, two forms of spectral analysis will be discussed:

a) auto-spectral analysis of a single time series, and

b) cross-spectral analysis of pairs of time series for the purpose of comparison of sets of empirical data with theoretical models.

We must first consider in the time domain some statistical limitations upon the data. The spectrum of the discreet series $X_1(t)$ is only defined for frequencies up to $f_N = \frac{1}{2}\varDelta t$ cycles per day. As we shall see, some arrangement must be made so that the process contains negligible power at frequencies higher than f_N. If there are no obvious linear trends in the data we can assume *stationarity*, which is a type of statistical equilibrium which implies that the random variables, X_1 and X_2 have the same probability density function, i.e. $f(X_1)$ and $f(X_2)$ are dependant of t. Practically we can estimate $f_1(X_1)$ and $f_2(X_2)$ by forming a histogram of the observed values of $X_1(t)$ and $X_2(t)$. A small visual difference is sufficient to confirm the stationarity assumption (Jenkins, 1965). Tukey (1961) observed that unless one is sure about the processes, it is wise not to average spectrum over too long a time, since stationarity may hold only for a short time. In fact frequently we are more interested in changes in the energy distribution at each frequency than the power spectrum itself, since this may give physical insight about the mechanism producing the irregular light curves. Another way of formulating stationarity is to insist that all multivariate distributions involving X_1 and X_2 depend only on time differences $t—t'$ and not on t and t' separately. The joint distribution of $X_1(t)$ and $X_2(t)$ is then the same for time points K lags apart, allowing us to construct a scatter diagram of $X_1(t)$ and $X_2(t—k)$. If this joint distribution can be approximated within a reasonably close measure by means of a multivariate normal distribution, then the joint distribution is characterized by its means and its covariance matrix. If the errors are Gaussian or normal, then μ_i and σ^2 characterize the distribution completely, and the condition of *normality* is fulfilled.

We must now more carefully define the covariance matrix, its components and relations to other statistical qualities. The variance σ^2 is:

$$\sigma^2 = \varepsilon(x_i - \mu_i)^2 \tag{2.1}$$

where

$$\varepsilon(x_i) = \frac{1}{n} \sum_{t=1}^{n} x_i(t) = \mu_i \tag{2.2}$$

and also

$$\sigma^2 = \int_{-\infty}^{\infty} (x_i - \mu_i)\, p(x)\, dx \tag{2.3}$$

where $p(x)$ is the probability density function of the errors and μ_i is the mean value of X and $p(x)$. We can also write the sample estimate $c_{11}(k)$ of the autocovariance of lag K uncorrected for the mean.

$$c_{11}(k) = \frac{1}{n} \sum_{t=1}^{n-k} (x_{1t} - \mu_i)(x_{1t+k} - \mu_i). \tag{2.4}$$

Since $c_{11}(k)$ is even $$c_{11}(k) = c_{11}(-k). \tag{2.5}$$

3*

If the number of observations n increases indefinitely, $c_{11}(k)$ becomes the ensemble autocovariance $\gamma_{11}(k)$. Priestley (1965) points out that $c_{11}(k)$ has a bias $b(\omega)$ the order m/n, but since most estimates of the spectral density use only the first $m(\ll n)$ values of $\gamma_{11}(k)$, so the bias in $\gamma_{11}(k)$ is unimportant. If we wish to compute only the autocorrelation, then we use an unbiased estimate of $c_{11}(k)$. However, to calculate the power spectrum of $X_1(t)$ requires a biased estimate of $c_{11}(k)$ (see designation 3.14 and the immediate discussion), thus a weighted autocorrelation becomes:

$$\varrho_{11}(k) = \frac{1}{n-k} \sum_{t=1}^{n-k} (x_{1t} - \mu_1)(x_{1t+K} - \mu_1). \tag{2.6}$$

The autocorrelation coefficients $\varrho_{11}(k)$ are $\varphi_{11}(k) = \gamma_{11}(k)/\gamma_{11}(0)$; here the same as the regression coefficients of $X_1(t)$ on $X_1(t-k)$. The cross correlation coefficients $\varrho_{12}(k)$ become

$$\varrho_{12}(k) = \gamma_{12}(k)/\sqrt{\gamma_{11}(0)\,\gamma_{22}(0)}. \tag{2.7}$$

Thus if the time series is stationary and normal, its statistical process, i.e. the joint distribution of $X_1(t)$ for time $t_1, t_2, \ldots, t_n$ is characterized sufficiently by μ_i, σ^2 and $\varrho_{12}(k)$ (Jenkins, 1961). More formally the estimators $\gamma_{12}(k)$ for $\varrho_{12}(k)$ are the ordinary product-moment correlations between sequences

$$r_{12}(k) = \frac{\sum_{1}^{n-k} [(x_1(t) - \mu_1)][x_2(t+k) - \mu_1]}{\left\{ \sum_{1}^{n-k} [x_1(t) - \mu_1]^2 \sum_{1}^{n-k} [x_2(t+k) - \mu_2]^2 \right\}^{1/2}} \tag{2.8}$$

where μ_1 and μ_2 are means calculated from the first $n-k$ and last $n-k$ terms, respectively. The plots of $r_{12}(k)$ versus the lag k are called the autocorrelation functions $A_{11}(k)$ and $r_{12}(k)$ against k the cross-correlation functions. It is worth noting that the means μ_1 and μ_2 must be subtracted out to produce proper autocorrelation coefficients. If we do not remove the sample mean before calculating the autocovariances it will dominate the contribution to its cosine transform not only at zero frequency, but also at frequencies close to zero. A Fourier transform of $\varrho_{11}(k)$ will dominate only the zero frequency, but if there is linear trend present in addition, there will be an additional contribution to the power at all frequencies, but predominantly at low frequencies. We note also that if the lag k becomes too large, $R_{12}(k)$ becomes erratic in appearance since fewer observations fall into each group, so that the variance of the estimate increases. Thus resolution cum number of lags k works in the opposite directions to the variance, and a compromise between the two must be effected to accurately estimate autocorrelation coefficients.

III. HARMONIC ANALYSIS OF STRICTLY PERIODIC TIME SERIES

The sample autocorrelation function is a difficult quantity to interpret physically since neighbouring correlations tend to be correlated. We shall try to look at the time series transformed into the frequency domain. We shall first decompose a time series that is valid only for strictly periodic phenomena.

After we show how certain types of noisy spectra can be represented, i.e. we shall estimate a power spectrum of a time series containing both periodic signals and noise.

Following Jenkins (1965), we shall unite $N = 2n$ sampling intervals of a strictly periodic time series (light curve) which will be fitted to a harmonic regression of the form

$$x_1(t) = A_{10} + \sum_{j=1}^{n-1} A_{1j} \cos(\omega_j t) + B_{1j} \sin(\omega_j t) + A_{1n} \cos(\pi t) \quad (3.1)$$

or

$$x_1(t) = A_{10} + \sum_{j=1}^{n-1} R_{1j} \cos(\omega_j t + \Phi_{1j}) + A_{1n} \cos(\pi t) \quad (3.2)$$

where

$$\tan \Phi_{1j} = -B_{1j}/A_{1j}, \quad \omega_j = 2\pi j/N \quad \text{and} \quad R_{1j}^2 = A_{1j}^2 + B_{1j}^2$$

The observations are represented by a mixture of sine and cosine waves whose frequencies are multiples of the fundamental frequency $2\pi/N$. Then

$$A_{10} = \frac{1}{N} \sum_{t=1}^{N} X_{1t} = \bar{X}_{1t} \quad (3.3)$$

$$A_{1j} = \frac{2}{N} \sum_{t=1}^{N} X_{1t} \cos(\omega_j t) \quad (3.4) \quad\quad (3.4)$$

and

$$B_{1j} = \frac{2}{N} \sum_{t=1}^{N} X_1(t) \sin(\omega_j t) . \quad (3.5)$$

If only a few of the harmonics are used, and the remainder are regarded as error, then (3.4) and (3.5) give the usual least square estimates of the constants. The total sum of the squares (variances) about the mean can be decomposed.

$$\sum_{t=1}^{N} (X_{1t} - \bar{X}_1)^2 = \frac{N}{2} \sum_{j=1}^{n-1} R_{1j}^2 + N A_{1n}^2 . \quad (3.6)$$

If we assume no harmonic terms are present, then $\frac{N}{2} R_{ij}^2$ is distributed as chi-square with two degress of freedom.

If we define the sample estimate of the autocovariance of lag k uncorrected for the mean $c_{11}(k)$ then the sample spectral density function is

$$f(\omega) = \frac{1}{2\pi T} \left| \sum_{t=1}^{T} X_1(t) e^{-i\omega t} \right|^2, \quad -\pi \leq \omega \leq \pi \quad (3.7)$$

We can now carry out a harmonic analysis for strictly periodic phenomena as

$$\frac{1}{2} R_{1j}^2 = \frac{2}{N} \sum_{k=-N+1}^{N-1} c_{11}(k) \cos(\omega_j k) . \quad (3.8)$$

If the series $X_1(t)$ is stationary, and if the number of observations N increases indefinitely the histogram R_{ij} becomes a continuous curve known as the spectral density function $f(w)$ and, $C_{11}(k)$ is replaced by $\gamma_{11}(k)$ so that

$$\sigma^2 f(\omega) = Re\left\{\frac{1}{\pi} \sum_{k=-\infty}^{\infty} \gamma_{11}(k) e^{-i\omega k}\right\} = \frac{1}{\pi} \sum_{k=-\infty}^{\infty} \gamma_{11}(k) \cos \omega k \qquad (3.9)$$

It is worth noting that $f(w)$ and $\gamma_{11}(k)$ are Fourier transforms of each other. We have

$$\gamma_{11}(k) = \int_{-\infty}^{\infty} e^{i\omega k} f(\omega) \, d\omega \qquad (3.10)$$

and we write $\varrho_{11}(k) = \gamma_{11}(k)/\sigma^2$

$$f(\omega) = \frac{1}{\pi}\left\{1 + 2 \sum_{k=1}^{\infty} \varrho_{11}(k) \cos (\omega k)\right\}. \qquad (3.11)$$

The Weiner-Khintchine theorem states this more formally: that a function $f(\omega)$ being the integrated spectrum of $X(t)$ has the physical interpretation that follows from the Fourier expansion of $X_1(t)_1$ i.e.

$$x_1(t) = \int_{-\infty}^{\infty} e^{i\omega t} \, dM(\omega) \qquad (3.12)$$

where $dM(\omega)$ is an orthogonal stochastic process in the interval $(-\infty, \infty)$ with the estimate

$$\varepsilon\{|dM(\omega)|^2\} = df(\omega). \qquad (3.13)$$

Thus $df(\omega)$ represents the average "power" associated with the components of $X_1(t)$ whose frequencies lie between ω and $\omega + d\omega$.

We shall now see why this easy progression in the proof does not yield valid results for physically meaningful spectra, since simply, the tenants of the basic assumption of strictly periodic $X_1(t)$ are violated for stars with irregularly varying light curves.

As we have stated harmonic analysis works only for strictly periodic phenomena. For mixed time series harmonic analysis does not give a natural estimate of the spectra. In fact engineers and statisticians know harmonic analysis of noise or random numbers produces a highly spiked spectrum. For large sample sizes the mean periodgram $I_{11}(\omega)$ does tend to the spectral density, but the variance of the fluctuations of $I_{11}(\omega)$ about $f(\omega)$ does not tend to zero as $n \to \infty$. For a large n, the distribution of $I_{11}(\omega)$ is a multiple of a chi-squared distribution with two degrees of freedom, independently of n. Another way to see that for mixed spectra harmonic estimates of the periodgram $I_{11}(\omega)$ are spurious is that

$$\varepsilon(I_{11}(\omega)) = \frac{\sigma^2}{\pi} \sum_{k=-n+1}^{n-1} \left(1 - \frac{|k|}{n}\right) \varrho_{11}(k) \cos (\omega_j(k)) \qquad (3.14)$$

where

$$\underset{n \to \infty}{\text{Limit}} \, \varepsilon(I_{11}(\omega_j)) \neq \sigma^2 f(\omega).$$

As we see, the weights tend to zero as k tends to N, giving erratic and incorrect estimates. We shall now see a true method of estimating time series as developed by Weiner (1967), Blackman and Tukey (1958), and others.

IV. ESTIMATION OF MIXED AUTOSPECTRA

We here define a mixed spectra by decomposing $F(\omega)$, integrated spectrum of $X_1(t)$ as

$$F(\omega) = a_1 F_1(\omega) + a_2 F_2(\omega) \tag{4.1}$$

where $F_1(\omega)$ is an absolutely continuous function having a derivative $f(\omega$ also called the spectral density function, and $F_2(\omega)$ is a step function at certain frequencies ω_n, $n = 1, 2, \ldots$ If $a = 1$, $a_2 = 0$ the spectrum is purely continuous; if $a_1 = 0$, $a_2 = 1$, it is purely discreet. If $a_1 \neq 0$, and $a_2 \neq 0$ are not constant, then the time series is said to possess a mixed spectrum. If a process contains periodic terms and residual processes which inthemselves have continuous spectra, then we have to estimate (1) the number of sine waves, (2) their amplitudes and frequencies, and (3) the spectral density function of the residual process.

To estimate mixed spectra one defines both a series $\lambda(k)$ as a time domain window and a spectral window $\tau(\omega)$ as a window in the frequency domain. We choose for mixed spectra to define spectral estimates in the form

$$f(\omega) = \frac{1}{\pi} \sum_{k=-n+1}^{n-1} \lambda(k)\, \gamma_{11}(k)\, \cos\,(\omega\, k) \tag{4.2}$$

where we assume that the width of the spectral window associated with $\lambda(k)$ is so small that the spectrum is small over the window. We can also write for $\gamma_{11}(k)$

$$\gamma_{11}(k) = \frac{\sigma^2}{\pi} \int_{-\pi}^{\pi} f(\omega)\, e^{-i\omega k}\, d\omega \tag{4.3}$$

Another expression of (4.2) is

$$f(\omega) = \int_0^\pi \tau(\omega - y)\, f(y)\, dy \tag{4.4}$$

where

$$\tau(\omega - y) = \frac{1}{2}\{\eta(\omega - y) + \eta(\omega + y)\} \tag{4.5}$$

with

$$\int_0^\pi \tau(\omega - y)\, d\omega = 1 \tag{4.6}$$

and

$$\eta(\omega - y) = \frac{1}{\pi} \sum_{k=-\infty}^{\infty} \lambda_k\, e^{-i(\omega - y)k} \tag{4.7}$$

The $\tau(\omega - y)$ is the spectral window. In changing ω we move the slit along through the entire frequency range, with a bandwidth $\pm \dfrac{2\pi}{N}$ about $y = \omega$. Jenkins (1961) gives several physical definitions of bandwidth which approximate a rectangular band pass window. In general we will replace the autocovariances

$\gamma_{11}(k)$ with the autocorrelations $\varrho_{11}(k)$ so that

$$f(\omega) = \frac{1}{\pi} \sum_{k=-N+1}^{N-1} \lambda_k \, \Gamma_{11}(k) \cos \omega \, k \, . \qquad (4.8)$$

For the choice of moving weights λ_k, we can choose from a number of cosine-like series, all nearly the same. We have used

$$\lambda_k = \frac{1}{2} \left(1 + \cos \frac{\pi k}{m} \right) . \qquad (4.9)$$

This weight gives variance $f(\omega) \sim \dfrac{3m}{4n}$, thus making m small increases the band-width, hence decreasing the variance for this choice of λ_{12}.

This operation can be regarded as a filtering process using a rectangular-like bandpass with small side lobes in order to increase the accuracy of the spectral density $f(\omega)$. Unless one is faced with peaky or a steeply slanted spectrum, the exact shape of the window is not too critical.

This window (4.9) and others like it tend to a normal distribution, hence its transform is the same as itself. It retains its shape in both time and frequency domains. For his particular window, the $\eta(\omega - y)$ in (4.1) is given by Jenkins (1961) as

$$\eta(\omega - y) = \frac{1}{\pi} \left\{ \frac{\sin(m+1/2)\omega}{\sin \omega/2} + \right.$$
$$\left. + \frac{1}{2} \left[\frac{\sin (m + 1/2) \, (\omega + \pi/m)}{\sin 1/2 \, (\omega + \pi/m)} + \frac{\sin (m + 1/2) \, (\omega - \pi/m)}{\sin 1/2 \, (\omega - \pi/m)} \right] \right\} . \quad (4.10)$$

Most windows are similar to those generated from the window $\lambda_{12} = 1$, giving a shape similar to the probability density function of a rectangular random variable. This particular window (4.9) has both small bandwidths and small side lobes, cutting the leakage of spectral power from frequencies distant from $y = \omega$, which would distort the true picture at $y = \omega_0$. It must be emphasized that there are a number of spectral windows, each with different bandwidths. We shall see later that filters similar to λ_{12} are used to suppress certain undesirable features in the spectrum such as power leakage. More important than the choice of window shape is the choice of the total number of lag k steps, which we call m. The nature of the analysis attempted here is to use a fixed run of observations and a fixed sampling interval Δt. Priestley (1962) has attempted to establish quantitative expressions for designing an optimum spectral estimate, involving the relation of variance σ^2 and bias $b(\omega)$. As we have mentioned earlier, one must first fix the form of the weight sequence and the number of lags m. A not always realizable suggestion by Priestley (1962) is to have the bandwidth $B\omega$ not greater than the width of the narrowest peak of $f(\omega)$, more specifically $B\omega$ is the distance between the 'half-power' points in the main lobe.

Blackman and Tukey (1958) discuss an optimal design in terms of sampling errors of $f(\omega)$ assuming a chi-squared distribution with ν degrees of freedom, including a measure of $f(\omega)$ in terms of variance, but not bias $b(\omega)$. We shall discuss these points later when considering the errors inherent in any estimate of $f(\omega)$. These considerations of σ^2 and $b(\omega)$ are not strictly applicable to *this* particular type of power spectrum of light curves analysis of light curves of irregular variable stars taken from the astronomy literature. Unless one studies only short period fluctuations of variable stars, or has several observatories in the world keeping constant watch on certain stars, one's sampling rate will always be disturbed by the day-night variation. The weather will also confound the most diligent observer, and tend to frustrate any decrease of Δt even for very long records of light curves.

Jenkins (1965) suggests an empirical method which has been followed in this work, and which will be illustrated by a few examples and diagrams. Those who try to determine an optimal m always find that the answers depend on the spectral density $f(\omega)$. We therefore choose an m small (or large bandwidth). We then increase m and decrease the bandwidth improving the resolution, stopping when we are satisfied about the detail of the spectrum (following the suggestion of Priestley). The matter may be judged in the light of the consideration that the variance tends to increase with m; we must in fact balance three considerations: accuracy of spectral estimate (the variance) against the number of lags m, i.e. the resolution of fine structure in the spectrum. The ideal method to increase resolution at low frequencies which would amount to an increase in the total length of data, hence a range of possible periods of up to 1,000 days. Along with this increase in n we could safely increase m and still retain a high level of confidence in our statistical estimates of the spectra. The diurnal variation of the Earth, infrequent bad weather and the new Moon puts a very real limitation on an accurate $f(\omega)$, without an enormous increase in time and effort of the observational astronomers.

The irregular nature of the observations that make up a light curve is dealt with by a subroutine of BOMM that interpolates the time series, using second differences, at regular intervals Δt over the range of observations that one chooses. In pilot studies, we usually choose several values of Δt.

We are led finally to conclude this section with the specific formulation of the spectral density as computed by a BOMM subroutine. The actual cosine transformation of the autocorrelation is written: —

$$f(\omega) = \frac{2}{2m+1} \sum_{i=0}^{n} \epsilon_i \gamma_{11}(k) \cos \pi i \, (\omega' + j/2\Delta t) \qquad (4.11)$$

where $\epsilon_1 = {}^1/_2$ if $i = 0$, $\epsilon_1 = 1$ otherwise; $j = 0, 1, \ldots, 1'$, $1'$ is the greatest integer such that $\omega' + \dfrac{l'}{2\Delta t} \leq \omega_N$ where ω' is the low frequency limit, usually 0 and ω_N is the Nyquist frequency.

Jenkins (1961) observes that the logarithmic transformation of the spectral density $f(\omega)$ produces a distribution whose variance is nearly independent of the frequency. We see that

$$\text{Variance } \{f(\omega)\} \approx f^2(\omega)\beta_i \; m/n \qquad (4.12)$$

where β_i depends on the window used for the spectral estimate and n is the total number of observations. If we then write an expression independent of the frequency

$$\text{Variance}\,\{\log_{10} f(\omega)\} \approx \beta_i \frac{m}{n}\,. \tag{4.13}$$

$\text{Log}_{10} f(\omega)$ produces estimates that do not rely so heavily upon normality, in fact it produces distributions closer to normality. It is assumed that the errors here are produced by the variance and not by leakage of power from other frequencies or by aliasing. It is worth noting that this leads us easily to the method engineers use for estimating spectral power since

$$\text{Decibals} = 10\,\log_{10}\,(\text{ratio of power}) \tag{4.14}$$

We see that an increase of power by a factor of 2 is equivalent of a change of 3 decibals.

V. ESTIMATION OF MIXED CROSS SPECTRA

Often we need either to compare two empirical time series $X_1(t)$ and $X_2(t)$ that arise in similar fashions. To do this, we shall show how one estimates the cross spectrum. We shall show now that if $X_1(t)$ and $X_2(t)$ do not have similar origins, then one can estimate the gain of a system, and perhaps the coherent energy. Let one of the time series $X_1(t)$ be a theoretical complex input and $X_2(t)$ the actual recording of the light curve. We can write

$$X_2(t) = f(X_1(t)) + n(t) \tag{5.1}$$

where $n(t)$ is the noise term which arises because other input variables are not well controlled. If the fluctuations in $X_1(t)$ are not too large, then we linearize the above equation to:

$$X_2(t) = \int_0^\infty \omega(u)\,x\,(t-u)\,du + n(t) \tag{5.2}$$

where $n(t)$ now contains quadratic and higher terms. This relationship is a linear dynamic equation, the regression or impulse response $\omega(u)$ can be estimated in several ways, by changing $X_1(t)$ by a well defined step pattern or sinusoidally. In the case of the latter we let $X_1(t) = \delta\cos\omega t$ and

$$X_2(t) = \delta G(\omega)\,\cos\,(\omega t + \Phi(\omega)) + n(t) \tag{5.3}$$

then the transfer function becomes

$$T(i\,\omega) = G(\omega)e^{-i\Phi(\omega)} = \int_0^\infty \omega(u)e^{-i\omega k}\,du\,. \tag{5.4}$$

$T(i\omega)$ is the transfer function, $G(\omega)$ the gain and $\Phi(\omega)$ the phase shift. If $n(t)$ is small, the gain $G(\omega)$ may be obtained from the ratio of the amplitudes of the output and input, and $\Phi(\omega)$ by matching up the two waves. In this case we used a sine wave generator to actuate $X_1(t)$ and sweep out a range of frequencies to generate all the information.

We shall now estimate $w(u)$ from the existing noisy fluctuations in $X_1(t)$ and $X_2(t)$. These two series are now given to be stationary time series, and

$X_1(t)$ and $n(t)$ are uncorrelated, so we can write the crosscovariance by multiply-ing (5.2) by $X_1(t-k)$ and averaging

$$\gamma_{12}(\tau) = \int_0^\infty \omega(u)\,\gamma_{11}(k-u)\,du \qquad (5.5)$$

where $\gamma_{12}(k)$ is the crosscovariance function of lag k between $X_1(t)$ and $X_2(t)$, and $\gamma_{11}(k-u)$ is the autocovariance function. If $\gamma_{12}(k)$ and $\gamma_{11}(k-u)$ are known, then $\omega(u)$ may be estimated by solving (5.5), usually called the Weiner—Hopf equation. If we calculate the Fourier transform of both sides, $T(\omega)$ is then

$$T(\omega) = \frac{\sigma_2}{\sigma_1}\frac{f_{12}(\omega)}{f_{11}(\omega)} \qquad (5.6)$$

where $f_{12}(\omega)$ is the cross spectral density function, and σ_1 and σ_2 are the standard deviations of $X_1(t)$ and $X_2(t)$. The spectral density $f_{12}(\omega)$ is given by

$$f_{12}(\omega) = \frac{1}{2\pi}\sum_{k=-(n-1)}^{n-1}\lambda_\tau\gamma_{12}(k)\,e^{-i\omega k}. \qquad (5.7)$$

We have shown that the frequency response $\tau(\omega)$ is given by the ratio of the cross spectral density and the input auto spectrum. We now write $\gamma_{12}(k)$ as:

$$\gamma_{12}(k) = \frac{1}{n}\sum_{t=1}^{n-k}\left(x_1(t)-\mu_1\right)\left(x_2(t+k)-\mu_2\right) \qquad (5.8)$$

We have already defined $\gamma_{11}(k)$ and $f_{11}(\omega)$ in an earlier section of the paper. Since $\gamma_{12}(k) \neq \gamma_{21}(k)$ in general, then (5.7) gives both a cosine and sine trans-form of the crosscorrelation

$$Z_{12}(\omega) = \frac{2}{\pi}\int_0^\infty \alpha_{12}(k)\cos\omega k\,dk \qquad (5.9)$$

$$W_{12}(\omega) = \frac{2}{\pi}\int_0^\infty \beta_{12}(k)\sin k\,dk \qquad (5.10)$$

where

$$f_{12}(\omega) = Z_{12}(\omega) - i\,W_{12}(\omega) \qquad (5.11)$$

$$\alpha_{12}(k) = \frac{1}{2}\{\varrho_{12}(k) + \varrho_{12}(k)\} \qquad (5.12)$$

and

$$\beta_{12}(k) = \frac{1}{2}\{\varrho_{12}(k) - \varrho_{12}(-k)\}. \qquad (5.13)$$

In this case $W_{12}(\omega)$ is the quadrature or out-of-phase spectrum, and $Z_{12}(\omega)$ is called the co- or in-phase spectrum. Using (5.6) and (5.8) we can write the gain $G(\omega)$ and phase $\Phi(\omega)$ as

$$G(\omega) = \frac{\sigma_2}{\sigma_1} \sqrt{Z_{12}^2(\omega) + W_{12}^2(\omega)}/f_{11}(\omega) \qquad (5.14)$$

$$\tan\{\Phi(\omega)\} = \frac{W_{12}(\omega)}{Z_{12}(\omega)}. \qquad (5.15)$$

Tan $\Phi_i(\omega)$ here is the phase difference. This formulation is due to Munk, et al. (1959). If we are examining the covariance between two stationary time series $X_1(t)$ and $X_2(t)$ and not as input and output of some linear system, one usually plots the coherence $R^2(\dot{w})$, which is normalized to run between 0 and 1 so that

$$R^2(\omega) = \frac{W_{12}^2(\omega) + Z_{12}^2(\omega)}{f_{11}(\omega) f_{22}(\omega)}. \qquad (5.16)$$

The coherency $R^2(\omega)$ is analogous to the correlation coefficient calculated at each frequency ω. The spectrum of the noise term $n(t)$ in (5.2) is now

$$\sigma_1^2 f_{nn}(\omega) = \sigma_2^2 f_{22}(\omega) \left(1 - R^2(\omega)\right) \qquad (5.17)$$

which is analogous to the residual sum of squares in linear regression.

$$\sigma_n^2 = \sigma_2^2(1 - \varrho^2). \qquad (5.18)$$

We see if $R^2(\omega)$ is large, the noise spectral density is small relative to the output spectral density and vice versa. We can also write $R^2(\omega)$ as

$$R^2(\omega) = 1 \bigg/ \left\{1 + \frac{\sigma_n^2 f_{nn}(\omega)}{\sigma_1^2 G(\omega) f_{11}(\omega)}\right\}. \qquad (5.19)$$

If $G(\omega)$ is specified, the coherency is large if the signal-to-noise ratio is large, that is $S(\omega)/N(\omega)$

$$\frac{S(\omega)}{N(\omega)} = \frac{\sigma_1^2 f_{11}(\omega)}{\sigma_n^2 f_{in}(\omega)}. \qquad (5.20)$$

Munk, Snodgrass and Tucker (1959) point out that, if both $X_1(t)$ and $X_2(t)$ are identical and simultaneous, $Z_{12}(\omega) = +1$, $W_{12}(\omega) = 0$. If they are identical but there is a time lag in one record corresponding to a phase difference $\Phi(\omega)$, then the coherence is $+1$.

Jenkins (1965) points out that positive cross correlations will result in a high frequency cross amplitude spectrum with most of its power at low frequencies; negative cross correlations will results in a high-frequency cross amplitude spectrum.

We follow Jenkins (1963, 1965) to choose $w(n)$ in (5.2) so that the integrated squared error is minimized, and that (5.21) is small compared to some ϵ

$$\int_0^T n^2(t)\, dt \ll \epsilon \qquad (5.21)$$

and

$$Z_{12}(k) = \int_0^T w(u) Z_{11}(k - u)\, du. \qquad (5.22)$$

Here the ensemble cross covariances $\gamma_{12}(k)$ are replaced by the sample cross covariances $Z_{12}(k)$. Since $X_1(t)$ and $X_2(t)$ are not phase shifted sine waves, but stationary time series, we must introduce a weight function $\lambda(k)$ as in the univariate spectral analysis. We then write the estimates for the co-spectra, quadrature and auto-spectra as

$$Z_{12}(\omega) = \frac{2}{\pi} \int_0^T \lambda(k) \, \alpha_{12}(k) \cos \omega k \, dk \qquad (5.23)$$

$$W_{12}(\omega) = \frac{2}{\pi} \int_0^T \lambda(k) \, \beta_{12}(k) \sin \omega k \, dk \qquad (5.24)$$

$$f_{11}(\omega) = \frac{2}{\pi} \int_0^T \lambda(k) \varrho_{11}(\omega) \cos \omega \, k \, dk \qquad (5.25)$$

where $\alpha_{12}(k)$ and $\beta_{12}(k)$ are estimates of α_{12} and β_{12} as defined earlier in (5.12) and (5.13).

It has been pointed out by Jenkins (1965) that coherencies may sometimes be low because of the influences of the individual autospectra. One can take two independant time series and produce large coherence between them arising from peaks in the individual spectra or if the bandwidth is too small. However, we can calculate the autospectra first, then pre-filter each series to make their spectra move nearly uniform or white. This leads to a large increase in accuracy in the estimation of coherency. If one likes to think of an input function into the atmosphere of a star as some type of periodic wave, then following Munk and Cartwright (1966) we can separate the estimate of the energy into two parts, namely

$$\text{"coherent energy"} = R^2(\omega)f_{11}(\omega) \qquad (5.26)$$

and
$$\text{"noncoherent energy} = \{1 - R^2(\omega)\} f_{11}(\omega) \qquad (5.27)$$

The latter will be effectively noise energy, but may contain energy coherent with other input functions.

VI. FILTERS

Once the basic spectrum has been created and stabilized for a fixed m and Δt it becomes important to examine further the processes that tend to distort the true nature of $f(\omega)$. We shall then show it is possible to construct a digital filter U_k so that convolution upon the data in the time domain will eliminate unwanted spectral energy at some specified frequency. If we generate a series which we shall term our filter, and given our light curve $X_1(t)$, we then create a new series $W_j(t)$ by the process described. We can write $W_j(t)$ as

$$W_j(t) = \frac{1}{l' + 1} \sum_{i=0}^{l'} U_{k+i} x_i \qquad (6.1)$$

where

$$j = 1, 2, \ldots, [1 + (m - l' - 1)/p']; \quad k = p'(j-1) + 1$$

p' being the number of terms by which the weight factors are advanced for unit increase in $1/(l' + 1)$ is a normalization factor. If we want weights that are symmetrical, then W_j assumes the following form

$$W_j(t) = \frac{1}{2\,l' + 1} \left[U_k x_0 + \sum_{i=1}^{l'} (U_{k+i} + U_{k-i}) \, x_i \right] \tag{6.2}$$

where

$$j = 1, 2, \ldots [1 + (m - 2l' - 1)/p']; \quad k = l' + (j-1) + 1$$

We can construct also a frequency response function $T(\omega)$ to accept or reject any range of frequencies so that

$$T(\omega) = \sum_{t=-k}^{k} U_k e^{i\omega t} . \tag{6.3}$$

The BOMM programme generates these series of moving weights either for use as lag windows or as filters. We have the choice of symmetrical filters or those fading to the left or right. If we want a high-pass or low-pass filter, then for some frequency f, the terms are multiplied by a factor which gives a filter whose transmission is 6 db ($1/2$ amplitude). Thus: —

$$V_j = U_j, \qquad\qquad f = 0 \tag{6.4}$$

$$V_j = U_j \frac{\sin \pi j f}{\pi j f}, \quad f \neq 0 \tag{6.5}$$

If we want a low pass filter to be convolved in equation (6.1), then we write

$$S_j = U_j \frac{2\,m + 1}{2\sum\limits_{j=0} \epsilon_j U_j} \tag{6.6}$$

and for a high pass filter

$$S_j = - U_j \frac{2m + 1}{2\sum\limits_{j=0} \epsilon_j U_j} + (2\,m + 1)\, S\,(j) \tag{6.7}$$

where $\epsilon_j = 1/2$ for $j = 0$ and $\epsilon_j = 1$ otherwise; $\delta(j) = 1$ for $j = 0$ and 0 otherwise.

As we stated earlier, the filter type is determined by the sharpness of cut-off and by the side band characteristics. The amplitude response of any filter can be obtained taking its cosine transform. It may be worthwhile explaining some of the conditions which necessitate the use of filters, such as the removal of trends, leakage of power from high frequencies due to aliasing. Often non-stationarity of time series may be due to the presence of a linear trend. In general, one can use a high-pass digital filter to suppress low frequency trends and then go on to perform a spectral analysis of the residual series. Another use of filters in time series analysis is where one suspects that a series consists of a mixture of stationary series, i.e. differing frequency bands may have different physical origins. In such a case, we can construct a series of digital band pass filters to separate the original spectrum into several distinct spectra.

A more complete discussion of the use of filters of all types will come in the next section with a discussion of the errors involved in calculating the spectral density.

VII. ERRORS IN SPECTRAL DENSITY ESTIMATES

Munk, Snodgrass and Tucker (1959) have given several quantitative estimates of errors in spectral density. We shall deal with those errors that are due to random errors, the properties of the spectral window $\lambda(k)$, leakage of power due to aliasing, and give some treatment of systematic errors. Within the range of this examination is the problem of detecting signals in the presence of noise, i.e. estimation of the range of error allowable for any peak that appears in the mixed spectra.

Parzen (1961) gives some design relations in terms of the signal to noise estimate, i.e.

$$SNR\left[f(\omega)\right] = \frac{\varepsilon\left[f(\omega)\right]}{\sigma\left[f(\omega)\right]} \tag{7.1}$$

We see the $SNR[f(\omega)]$ is the mean divided by the standard deviation (or the reciprocal) of the coefficient of variation of the random variable $f(\omega)$. If $f(\omega)$ is normally distributed and we seek to obtain a 95% confidence level for our spectral estimates, then we can estimate $SNR[f(\omega)]$ in terms of equivalent degrees of freedom. If $X_1(t)$ is a random variable (i.e. Gaussian noise) which has a chi-squared distribution with ν degrees of freedom, then

$$SNR\left[x_1(t)\right] = \sqrt{\nu/2} \tag{7.2}$$

and

$$\nu = 2\left\{SNR\left[x_1(t)\right]\right\}^2 = 2\left\{\frac{\varepsilon\left[f(\omega)\right]}{\sigma\left[f(\omega)\right]}\right\} \tag{7.3}$$

If $X_1(t)$ is a positive random variable, then ν is just the equivalent degree of freedom.

Several authors put forward their views on design relations to maximize $SNR[f(\omega)]$ under various circumstances (Parzen, 1961; Priestley, 1962; 1965), but we shall have to outline this in more detail later because of the peculiar nature of the light curves of T Tauri stars and the fixed amount and nature of the data available. The assumption of a chi-squared distribution is approximately correct where $f(\omega)$ can be expressed as a weighted sum of χ^2-variables. It seems a similar method of treatment is to sum the squares of amplitudes of the components of a Fourier series.

$$x_1(t) = \sum_0^n \left[p_n \cos\left(\frac{2\pi nt}{T}\right) + q_n \sin\left(\frac{2\pi nt}{T}\right)\right] \tag{7.4}$$

where $n = 0, 1, 2 \ldots$, and the energy density varies slowly with the frequency separation $1/T$ of the harmonics. Then p_n and q_n are normal and $p_n^2 = q_n^2 =$ = true energy density$/T$, i.e.

$$S(\omega) = \text{Estimate energy} (\omega) = \left(\sum_p \frac{1}{2} a_n^2 \right) \frac{T}{p} \qquad (7.5)$$

Thus for p harmonics, the number of degrees of freedom $\nu = 2p$, and from the chi-squared distribution "confidence limits" can be calculated. This is because the chi-squared distribution gives

$$\frac{[S(\omega) - E(\omega)]^2}{E^2(\omega)} == 2/\nu \qquad (7.6)$$

The effective number of degrees of a record is approximately ν. These ideas can be used with tables given in Blackman and Tukey (1957) to calculate the random errors based on the assumptions above. Thus

$$\nu \approx \frac{2n}{m} \qquad (7.7)$$

It seems likely that one must consider both the bias and variance when carrying out practical estimates of the spectral density $f(\omega)$, but it is difficult to formulate quantitatively any practical numerical formulae due to the presence in $f(\omega)$ of non-random statistical errors, i.e. side band leakage for which $f(\omega)$ differs with the type of spectral window chosen, aliasing. The problems of power leakage from peaks at frequencies $> \omega_N$ confuse the estimate at neighbouring frequencies $\omega_N \pm \Delta\omega$. Parzen (1961) and Priestley (1962) both use a formula

$$1.96 \, \eta(\omega) + b(\omega) = 0.1 \cdot f(\omega) \qquad (7.8)$$

to give an estimate at a particular frequency that does not have more than a 10% proportional error at the 95% confidence level. In this matter the variance is defined by

$$\eta(\omega) \equiv \varepsilon \left[f(\omega) - \varepsilon \left(f(\omega) \right) \right]^2 \approx \frac{4\pi f^2(\omega)}{n} \int\limits_{-\pi}^{\pi} \lambda^2(\theta) \, d\theta \qquad (7.9)$$

and the bias is given as

$$b(\omega) \equiv \varepsilon \left[f(\omega) \right] - f(\omega) \approx \int\limits_{-\pi}^{\pi} \{ f(\theta) - f(\omega) \} \, \lambda(\theta - \omega) \, d\theta \qquad (7.10)$$

where $\lambda(\theta - \omega)$ is the weight function.

Implimentation of these formulae for $\eta(\omega)$ and $b(\omega)$ rest critically on ones choice of the width of the weight function $\lambda(k)$. In fact since both n and b are fixed for the light curves, it is possible to calculate optimum lags m for a given spectra, if we accept the mean square error at frequency ω as

$$M(\omega) = \eta^2 (\omega) + b^2 (\omega) \qquad (7.11)$$

But it cannot be denied that other sources of error creep in as we have mentioned earlier, and the chi-squared error estimates only account for one source of error, those of a random nature.

The difficulty is that both $\eta(\omega)$ and $b(\omega)$ are asymptotically proportional to $f(\omega)$. The method we have used is to vary the resolution until increasing m any further does not reveal any more features in the spectra, while at the same time attempting to prewhiten the spectra as mush as possible, as suggested by Blackman and Tukey (1957). This is done during Stage II of the estitmation process, which is mostly exploratory.

Pre-whitening is useful in eliminating the systematic errors that occur due to the curvature of $f(\omega)$. If $f(\omega)$ is flat, or varies linearly with frequency, then $\varepsilon[f(\omega)] = f(\omega)$. If not, then spectral energy diffuses down toward lower spectral levels (Munk et al., 1959) and one must worry about leakage of power from higher frequencies to lower frequencies. We shall now discuss the question of leakage in the estimates of spectral density. The window we choose $\lambda(k)$ is usually some kind of polynomial which vanishes at a finite number of points. Its main lobe centering on a band of frequencies over which we wish to estimate $f(\omega)$ is inevitably accompanied by side bands or minor lobes, whish allow leakage from the part of the spectrum outside the desired band to affect our estimate inside the band. This can be cured by taking two estimates, one with the side lobes positive, the other negative.

One of the more critical problems recognised by Tukey is aliasing; or the problem of sampling at discreet intervals Δt. It is equivalent to multiplying the record with spikes with a frequency $1/\Delta t$, so that sum and difference frequencies are produced. An analysis for energy density at any frequency f- will also include the densities near frequencies:

$$1/\Delta t - f_0, \; 1/\Delta t + f_0, \; 2/\Delta t + f_0, \; \ldots \; \text{etc}$$

One of the major uses of filters comes in dealing with leakage of powers asso⁻ ciated with aliasing caused by the loss of information at frequencies above the Nyquist frequency, namely when $\omega_N = \pi/\Delta t$, or $f = 1/2 \, \Delta t$ cycles per day.

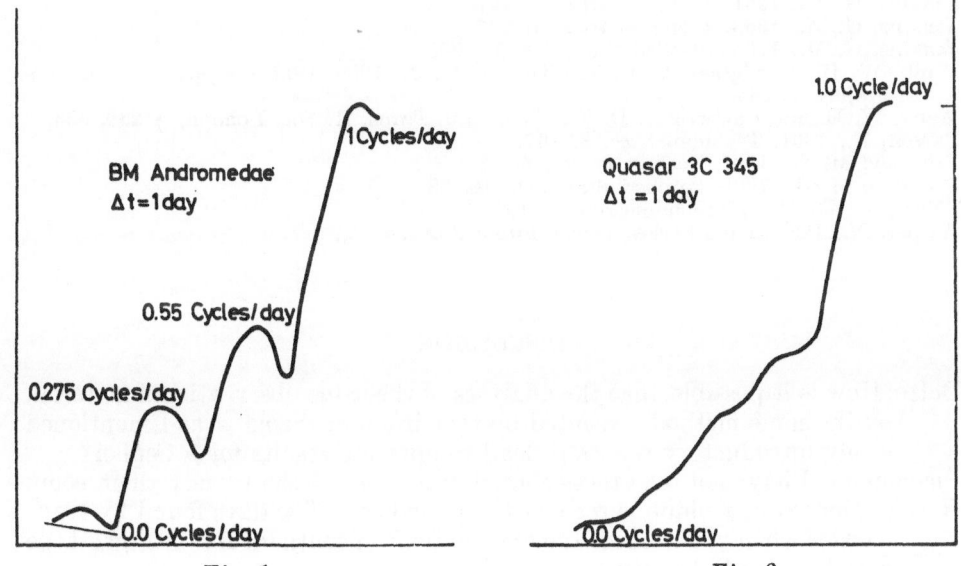

Fig. 1   Fig. 2

Since there is no information available about $X_1(t)$ between the data points, there is no means of estimating the amplitude of frequencies higher than the Nyquist frequency. In fact at frequency f_N we are unable to estimate, since at the Nyquist frequency, $f(\omega_N)$ is confused with all the frequencies that are indistinguishable from ω_N. If $f^*(\omega)$ is the spectral density corresponding to $X(t)$, then the spectral density of the sampled trace is

$$f(\omega) = \sum_{k=0}^{\infty} \left\{ f\left(\frac{2\pi k}{\Delta t} + \omega\right) + f\left(\frac{2\pi k}{\Delta t} - \omega\right) \right\}$$ (7.12)

We see the sampled spectrum is obtained by folding the unsampled spectrum about even multiples $\dfrac{2\pi k}{\Delta t}$ of f and adding this contribution in the range $(0, \omega_N)$ Thus to be able to measure $f^*(\omega)$ in $(0, \omega_N)$ we must hope that that $f(\omega) \approx 0$ for $\omega > \omega_N$ In our case it is certain that there is a need to use a low pass filter to cut any peaks that may occur near f_N. The details of this process and the practical results of it will be shown when we present the details of the power spectra for the particular T Tauri stars. The use of different Δt's in calculating pilot spectra will tell us if we are in danger of confusing $f(\omega)$ due to higher energies at higher frequencies.

REFERENCES

Blackman, R. B. and Tukey, J. W., 1958, The Measurement of Power Spectra. Dover Publications.
Bullard, E. C. et al, 1966, A User's Guide to BOMM. A Series of Programs for the Analysis of Time Series. Institute of Geophysics and Planetary Physics, La Jolla.
Herbig, G. H., 1962, Adv. Astr. Astrophys. 1, 47.
Jenkins, G. M., 1961, Technometrics, 3, 167.
Jenkins, G. M., 1963, Technometrics, 5, 227.
Jenkins, G. M., 1965, Applied Statistics, 14, 205.
Munk, W. H., Snodgrass, F. E. and Tucker, M. J., 1959, Bull. Scripps Inst. Oceanography, 7, 283.
Munk, W. H. and Cartwright, D. E., 1966, Phil. Trans. R. Soc. London, A 259, 533.
Parzen, E., 1961, Technometrics, 3, 167.
Priestley, M. B., 1962, Technometrics, 4, 551.
Priestley, M. B., 1965, Applied Statistics, 14, 33.
Tukey, J. W., 1961, Technometrics, 3, 191.
Weiner, N., 1967, Time Series, Dover Publications.

DISCUSSION

Detre: How is it possible, that the analyses of the same observational material, by the same method, executed by two different teams — as I mentioned in my introductory report — lead to opposite results for μ Cephei?
Plagemann: I have not read these contributions, but I shall study them soon.
Herbig: Does your technique recover the periods of a few days found by Hoffmeister when you analyze the observations of southern RW Aurigae-type variables made by him?

Plagemann: I have not yet analyzed his data, although the material is already on computer cards. It will be done when I return to Cambridge.

Penston: Do I understand that you find Kinman's 80 day period for 3C 345 is not statistically significant?

Plagemann: That is correct -- my preliminary results do not show his 80^d period.

4*

ON THE PRESUMED PRESUPERNOVA STAGE FOR TYPE II SUPER-NOVAE

G. BARBARO, N. DALLAPORTA, C. SUMMA

Istituto di Fisica dell'Università, Padova

(presented by Prof. L. Rosino)

ABSTRACT

The physical conditions of stars in presupernova type II stage when the outburst is expected to be due to the Fe—He transition occurring in its core are reviewed. The arguments showing that the star must preserve a large envelope in this stage and therefore appear as a red supergiant are stressed, and a lower mass limit of about $10 \sim 14$ $M_\odot$ for stars undergoing the outburst is confirmed on the basis of the more recent evaluations. Finally, the possibility that the presupernova type II stage could be represented by the small amplitude irregular and semiregular red variables with large masses belonging to young population I is briefly indicated.

This paper aims partly to summarize the present situation concerning the usually accepted interpretation of type II supernovae; and partly to focus the main phenomenological aspects which could allow to test some consequences of this theory. The opportunity for such a clarification is required by the fact that not unfrequently theoretical investigations on this subject neglect to connect the happenings in the core of the star to its more external characteristics, so that some supplementary considerations are necessary to bridge the gap between the two aspects of the problem.

According to present data (Minkowski, 1964), type II supernovae occur only in arms of spiral galaxies, and are therefore typical for early population I. The process giving rise to the outburst must affect only stars of relatively conspicuous mass, owing to the large values generally quoted for the amount of matter ejected (several solar masses); moreover, the abundance of hydrogen in the spectrum during the explosion seems to indicate that the ejected matter is largely formed by the envelope of the star.

Hoyle and Fowler (1960) have proposed the following mechanism as triggering the outburst: after having evolved through the whole series of thermonuclear reactions building heavier and heavier elements in its inner part, the star reaches the formation of an iron core for a central temperature of the order of a few 10^9 °K; as no heavier nuclei may be built with energy gain, for further contraction of the core, at temperatures of the order of $8 \cdot 10^9$ °K iron is transformed endothermically into helium plus neutrons; and in order to provide the energy necessary for such a transformation, the central part of the star collapses in practically free fall, giving thus rise to the outburst observed as a supernova explosion. Fowler and Hoyle have applied these ideas to a model of a 30 $M_\odot$ star with a core of 20 $M_\odot$; its evolution in the central density ϱ_c — central temperature T_c plane and its crossing the Fe—He transformation line are schematically indicated in Fig. 1.

It has been argued by Chiu (1961) and others that several neutrino production processes, according to the current-current interaction theory with universal constant for weak interactions could occur for temperatures of the order of 10^9 °K with such intensity as to compel the whole structure of the star to collapse, owing to the enormous amounts of energy subtracted by neutrinos in its center; so the question arose whether this neutrino collapse should prevent the Fe—He collapse. Although no definite word has been said on this subject, it is generally considered that the usual first order calculations done to evaluate the neutrino losses are inaccurate enough as not to allow to draw such a conclusion, and it is implicitly supposed, on the whole, that neutrinos contribute to accelerate the evolutionary process but do not prevent the star to reach the Fe—He conversion line; this point of view has been assumed in what follows.

In order to test the Fowler—Hoyle scheme on a more realistic model than the one used by them, Barbon and al. (1965) have tried to identify the presupernova stage with the red supergiant phase of large mass stars, and have considered the core to which the Fowler—Hoyle considerations apply as being only the 15% of the total mass of the star. Further, by studying the evolutionary sequence of the central core according to polytropic models and with different mass values allowing for the degeneracy of the gas, it was found that for cores with mass lower than a limiting value M_l, degeneracy would stop the increase of temperature in order to forbid for such stars the reaching of the Fe—He transition line, as may be seen in Fig. 1. It turned thus out that only for masses higher than M_l the type II supernova outburst was possible. The M_l value for the core is practically the Chandrasekhar limit for white dwarfs ~ 1.4 $M_\odot$; so that, keeping in mind the assumed proportion in mass between core and envelope, it resulted that only stars with total mass higher than about ~ 10 $M_\odot$ were expected to undergo an Fe—He supernova outburst.

In a more recent and detailed research, Rakavy and Shaviv (1966) have quite independently redetermined the evolutionary tracks of degenerate polytropes, obtaining exactly the same results as Barbon et al. However, they have considered in their work some other possible causes of collapse, among others, a dynamical instability interesting for the actual problem, occurring for very massive stars and due to the $e^+ - e^-$ annihilation process, whose domain is shown also in Fig. 1. It thus further appears that polytropic models with mass higher than 30 $M_\odot$ may be prevented to reach the Fe—He line because encountering the $e^+ - e^-$ instability domain in an earlier stage of their evolution.

In a second paper, Rakavy and Shaviv (1967) have reconsidered the problem according to a more accurate point of view; they integrate the equilibrium equations for the core and determine the evolution of its material by calculating in detail a number of reactions, allowing them to follow the transformation from carbon to iron. The results of their investigation show that, although the trajectories in the $\varrho_c - T_c$ plane for any of the model stars considered are much more complicated than those obtained with the simplified polytropic models, still they do not discard too much from them, the polytropic evolutionary curve acting as a kind of average behaviour in respect to the more exact one, and being thus confirmed as qualitatively reliable enough. Rakavy and Shaviv, however, do not consider at all the envelope of the star in their investigation; this may lead to wrong predictions when using their results for deducing the mass range of stars able to become type II supernovae.

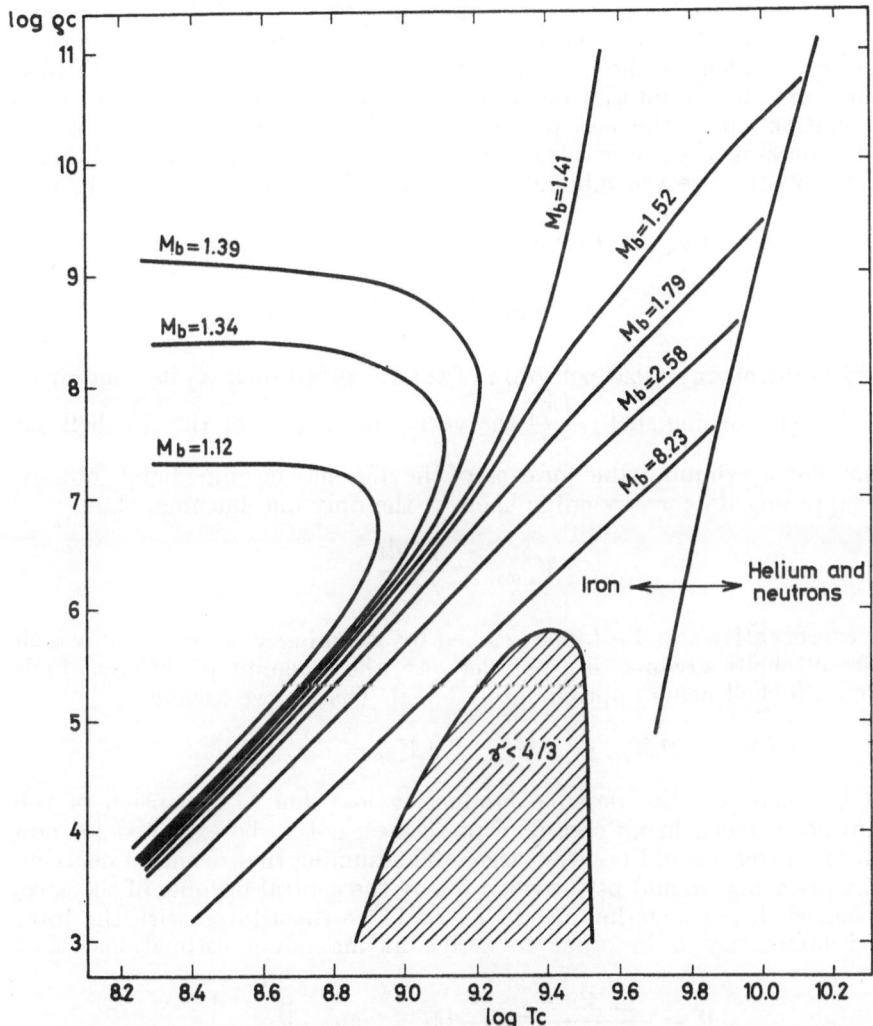

Fig. 1. Evolutionary tracks for the core. M_b represents the mass of the core; only for $M_b > 1.41$ the track crosses the Fe—He transition line. Dashed region corresponds to the region of dynamical instability due to the $e^+—e^-$ annihilation process.

An adequate investigation of the problem would require the detailed treatment of core models of Rakavy—Shaviv's type with an hydrogen envelope. In prevision of such a work, we try here to stress some points showing, on very general arguments, the likelihood that the hydrogen envelope persists throughout the whole presupernova stage and therefore cannot be ignored for comparison with data. Finally, we discuss some possible red supergiant types which could be suspected of being presupernova stages.

Concerning the first point, we first rely on the evolved models for large mass stars (i.e. the 15.6 $M_\odot$ star of Hayashi and Cameron (1962)) followed from the main sequence to the initial phase of carbon burning. At this stage, the star is left with a carbonoxygen core including the 18% of the total mass. Assuming, as usual, that no conspicuous mass loss should alter the evolution, one may try to calculate the maximum possible amount of nuclear fuel which should be burnt in the interior of the star in the remaining time of its supergiant evolution up to the last presupernova stage; this should give us the maximum amount of hydrogen transformed into heavier elements, and therefore allow to calculate the minimum envelope which the star preserves just before its outburst.

The luminosity L_n due to nuclear shell burnings, is expressed by:

$$L_n = \sum_i E_i^* X_i \frac{\Delta M_i}{\Delta t}$$

where E_i^* is the energy yield per gram of the ith given fuel, X_i its concentration in the ith burning shell, $\frac{\Delta M_i}{\Delta t}$ the variation of mass of the ith shell per unit time. The maximum value for each of the ΔM_i may be immediately obtained by supposing its corresponding shell as the only one burning, thus

$$\Delta M_{i\ max} = \frac{L_n \Delta t}{E_i^* X_i}$$

A more conservative assumption is reached by assuming an evolution in which the different shells advance in a parallel way, the amounts of different fuels burnt in each shell being approximately equal. That is, we assume:

$$\Delta M_H = \Delta M_{He} = \Delta M_{CO} = \Delta M_{Si} = \ldots = \Delta M_{eq}.$$

If we disregard the possible luminosity loss due to expansion of the outer envelope which in no case (except flashes, not to be expected for non degenerate matter) should be very large, and assuming that eventual neutrino losses, however big, should be provided for by the central burning of the core, the nuclear shell burnings luminosity L_n could be substituted with the total observed luminosity L in order to arrive at maximum estimations; thus we get:

$$\Delta M = \frac{L \Delta t}{\sum_i E_i^* X_i} \quad \text{with} \quad \Delta M < \text{any } M_{i\ max}.$$

Assuming the data of the Hayashi model

$$\log \frac{L}{L_\odot} = 5 \quad (\text{during } C \text{ burning})$$

$$\Delta t = 8 \cdot 10^5 \text{ years from } C \text{ burning to the explosion}$$

and the constants tabulated in Table I, we obtain for the ΔMs the results given also in Table I. These are probably rather insensitive to errors of L and Δt. Should in fact the luminosity increase due i.e. to neutrino emission, then the evolution time Δt should correspondingly decrease, so that the product $L\Delta t$ would not change much.

If we define the core of the star as the central portion of it for which $\mu = \text{const} \sim 2$, that is the whole portion inside the He burning shell, then according to data of Table I, the medium increase of it is $\dfrac{\Delta M}{M} = 0.06$, with extreme possibilities ranging from no increase at all (should the He burning shell stop burning) and maximum increase of about 0.23 (should the He burn alone). The core fraction therefore should increase from the 0.18 value in the last Hayashi model to $0.24 ^{+0,17}_{-0,06}$

Table I

$E^*_H = 6 \cdot 10^{18}$	$\overline{X}_H = 0.6$	$(\Delta M/M)_{H\ max} = 0.085$	$(\Delta M/M)_{eq} = 0.063$
$E^*_{He} = 6 \cdot 10^{17}$	$\overline{X}_{He} = 1.0$	$(\Delta M/M)_{He\ max} = 0.514$	
$E^*_C = 5.6 \cdot 10^{17}$	$\overline{X}_C = 0.5$	$(\Delta M/M)_{C-O\ max} = 0.58$	
$E^*_O = 5.0 \cdot 10^{17}$	$\overline{X}_O = 0.5$		
$E^*_{Si} = 1.9 \cdot 10^{17}$	$\overline{X}_{Si} = 1.0$	$(\Delta M/M)_{Si\ max} = 1.60$	

Should we instead consider the core as being only the innermost part inside the deeper burning shell (at the end of the iron core), then only a fraction of the previous increase is expected, which almost justifies the assumption of a constant core made by Barbon et al.

If now, according to Rakavy and Shaviv, we consider that only cores with mass greater than 2 $M_\odot$ can surely evolve towards the Fe—He transition line (the case of masses between 1.4 $M_\odot$ and 2 $M_\odot$ has not yet adequately studied by these authors), then we obtain for the lower limit of the total mass of the presupernova the value $M_l \simeq 8.3\ M_\odot {}^{+2,7}_{-3,3}$ with the first definition of the core, and the value $M_l \sim 11\ M_\odot$ with the second one. Moreover, if we accept Weymann's data (1961) on mass loss of red supergiants (20 per cent of the total mass for αOri), we obtain for the lower limits of the initial masses of future type II supernovae the values $M_l = 10.5 ^{+3,5}_{-4,3}\ M_\odot$ for the first type core definition, and $M \sim 14\ M_\odot$ for the other. Therefore, the results of Barbon et al., on the mass range of type II supernovae, are practically confirmed by the present analysis.

The absolute visual magnitude of a main sequence star with mass around 10—14 $M_\odot$ lies in the range —3 to —4. According to Limber's (1960) original luminosity function for the sun's neighbourhood, this gives us for the number of stars per cubic parsec with luminosity higher than this limit, the value $\sim 1.10^{-4}$. Assuming a mean lifetime for these stars on the main sequence of $T = 1.5 \cdot 10^7$ years, and equilibrium between birth rate and death rate functions, we obtain some $0.7 \cdot 10^{-11}$ type II supernovae per cubic parsec per year. Considering such events to occur possibly in all the outer disc portion of the Galaxy, we arrive at a frequency of a few events per year. Compared with data, this value is too high by a factor of about 100. Considering, however, the enormous uncertainty of the present evaluation, especially concerning the volume of the galaxy occupied by population I, and the extrapolation of the luminosity function for the sun's neighbourhood to the whole volume, one cannot conclude that the present disagreement is sufficient to disprove the theory. Moreover, it must be stressed that the present evaluation, although too large greatly improves the figure obtained by lowering the mass limit M_l for supernovae to about 2 $M_\odot$ as frequently done. Probably, should the Fe—He conversion mechanism be true for triggering supernovae, there should still be some other reason to further increase the lower mass limit for their occurrence.

As a further remark, according to Rakavy and Shaviv, cores with mass higher than ~ 30 $M_\odot$ fail to reach the Fe—He transition line, as they are stopped earlier in their evolutionary path by the $e^+ - e^-$ pair creation zone, in which the star grows unstable. The fate of such huge stars has been investigated by Fraley (1968) and found to lead them to a kind of softer collapse, which should perhaps show in a slower increase of the light output at the beginning of the explosion. Such a situation has been observed in some anomalous supernovae such as SN 96 in NGC 1058 discussed by Zwicky (1964) and Bertola (1963), which, moreover, appears also at minimum to be an exceptionally luminous star $(M_v \simeq -9)$; another example of the same type of event might have been η Car. The $e^+ - e^-$ collapse might perhaps be taken into consideration for interpreting such kind of events.

Concerning the second question of trying to identify the red supergiant types which could be considered as last presupernova stages, we have focussed our attention on the red irregular and semiregular variables. Although the difficulties of determining their low temperatures makes it difficult to locate them exactly in the H—R diagram, still there may be some suspicion that light variability occurs generally for the coolest and reddest among giants, and this could connect its cause to the fact of being near the Hayashi limit. If this were the case, and if the evolutionary trend in the presupernova phase was still from left to right in the H—R plane, then the connection of red variability with such a phase could appear not too unlikely.

Not many reliable data on red variables of small amplitude are at hand. In order to partially supply for this lack of knowledge, we have collected the stars of this kind belonging to galactic clusters whose location in the H—R plane is determined. The data concerning them are given in Table II and their position in the H—R plane shown in Fig. 2; some Mira type stars belonging to the clusters are included for comparison.

At first sight, the red variables appear to be divided into two groups: an upper one of supergiants evolving from large mass clusters of early population I; and a lower one of giants belonging to low mass clusters of disk population. Mira variables are found only in this second group, so that small amplitude red variables of this group could be considered as transition stages leading to the Mira situation.

One would like to investigate whether the two groups outlined are in fact physically different, and separated by a real gap between them. Some indication on this question may be obtained from Table III, in which the clusters have been divided into three groups according to their different ages, and which contains the following data: number of clusters, number of red giants, number of red variables, ratio of the number of variables to the total number of giants for each group. The values into brackets for the third group include stars which have not been studied yet and which, therefore, are only suspected variables. It is seen that for the first or large mass group, the ratio of column 4 is much higher than for the third or low mass group.

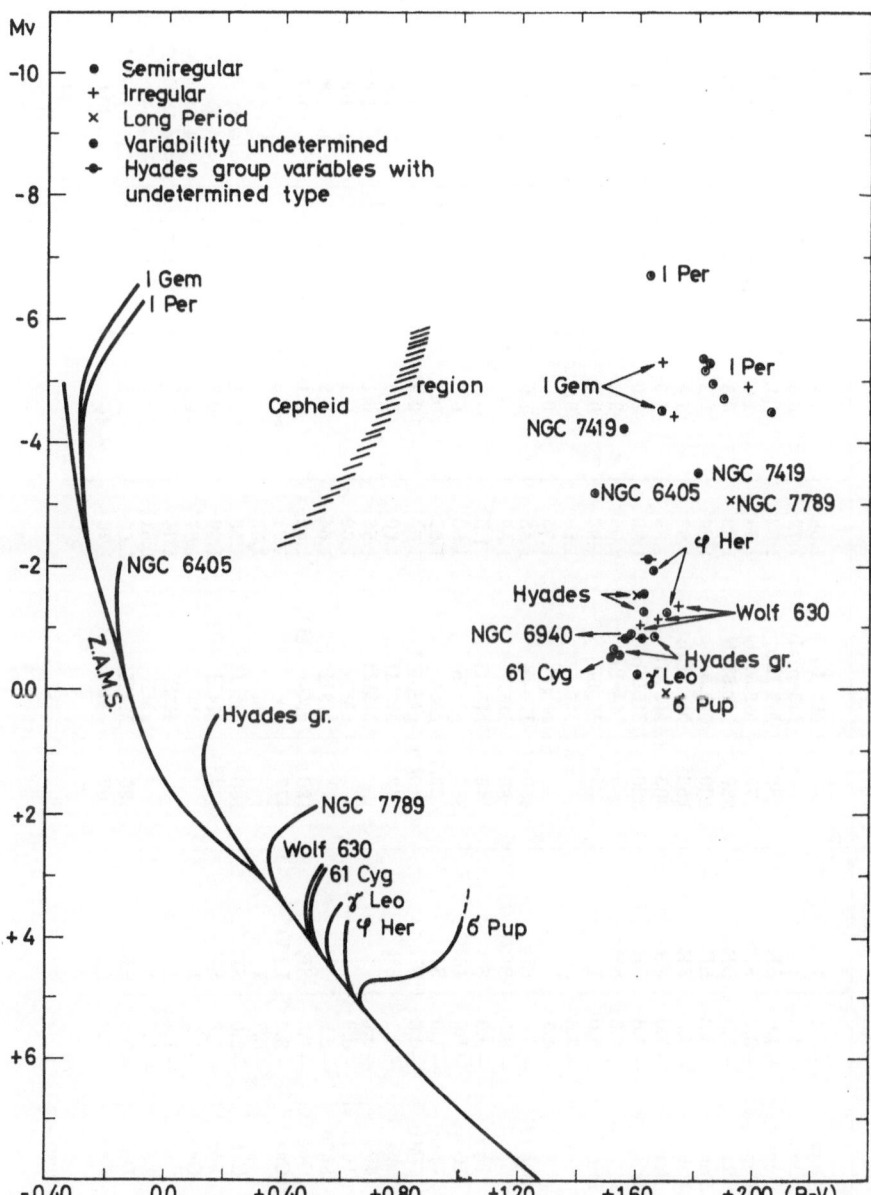

Fig. 2. H—R diagram of semiregular and irregular red variables. Schematic main
sequences of the corresponding clusters are also drawn.

So, either red variables are intrinsically more frequent in the first case, or
the stage of red variability is relatively longer for it. There is only one ascertain-
ed case belonging to the second group, BM Sco in NGC 6405; the period
assigned to this variable is 850 days, much longer in respect to all others in

Table II

Cluster	Star	M_v	kind of variability	Period (days)	Spect. type	$(B-V)_0$	ΔM_v	Mass $(H/M_\odot)$	Age (years)
I Gem	BU Gem	−5.4	I	—	M1 Ia	1.68	1.4	≤12	$1 \cdot 10^7$
	WY Gem	−4.5	I	—	M3epIab	1.72	0.6	9	$1 \cdot 10^7$
	TV Gem	−4.6	SR	182	M1 Iab	1.68	0.8(1.4)	9	$1 \cdot 10^7$
I Per (h, χ)	YZ Per	−6.8	SR	378	M2.5Iab	1.64	1.0	15+16	$1.2 \cdot 10^7$
	AD Per	−5.5	SR	320	M2.5Iab	1.82	0.8	13	$1.2 \cdot 10^7$
	SU Per	−5.2	SR	470	M3.5Iab	1.84	1.2	12	$1.2 \cdot 10^7$
	RS Per	−5.3	SR	152	M4.5Iab	1.83	1.6	12	$1.2 \cdot 10^7$
	BU Per	−5.0	SR	365	M3.5Ib	1.85	1.9	11	$1.2 \cdot 10^7$
	T Per	−4.8	SR	326	M2Iab	1.89	1.0	10	$1.2 \cdot 10^7$
	S Per	−4.6	SR	—	M4eIa	2.05	3.2	9	$1.2 \cdot 10^7$
	FZ Per	−5.0	I	—	M1Iab	1.97	0.7	11	$1.2 \cdot 10^7$
NGC 7419		−3.6			M7	1.80	9	8	$2 \cdot 10^7$
		−4.3			N (1)	1.55	9	9	$2 \cdot 10^7$
NGC 6405	BM Sco (2)	−3.2	SR	850	K—M	1.45	1.9	7	$7 \cdot 10^7$
NGC 6940		−0.9	SR(I) (3)	80	M5II	1.58		2-3	$4 \cdot 10^8$
Hyades (gr.)	R Lyr	−0.6	SR	46	M5III	1.52	~1	2-3	$4 \cdot 10^8$
	R Hya	−1.6	LPV	386	gM7e	1.60	6	2-3	$4 \cdot 10^8$
	VZ Cam	−1.6	SR	23.7	gM4	1.62	0.3	2-3	$4 \cdot 10^8$
	RR UMi	−0.6	SR	40(?)	gM5III	1.54	0.3	2-3	$4 \cdot 10^8$
	TV Psc	−0.1	SR	49	M3III	1.60	0.6	2-3	$4 \cdot 10^8$
	HR 46	−0.9			M3III	1.56	0.1	2-3	$4 \cdot 10^8$
	HR 1003	−0.9			gM3	1.62	0.1	2-3	$4 \cdot 10^8$
	HR 8636	−2.2			M3II	1.64	0.2	2-3	$4 \cdot 10^8$
NGC 7789	W Cyg	−1.3	SR	130	gM4e—M6	1.62	2.1	2	$4 \cdot 10^8$
61 Cyg (gr.)	WY Cas	−3.2	LPV	477	Se	1.91	>5.2		$1.2 \cdot 10^9$
ζ Herculis (group)	T Cet	−2.0	I	160	gM6	1.53	1.1	1.3	$3 \cdot 10^9$
	ϱ Per	−1.3	SR	33—55	M5eII	1.65	0.7	1.2	$\sim(4\text{-}5) \cdot 10^9$
Wolf 630 (gr.)	BQ Gem	−1.2	I		M4II—III	1.60	0.4	1.2	$\sim(4\text{-}5) \cdot 10^9$
		−1.4	I		M4	1.67		1.2	$\sim(4\text{-}5) \cdot 10^9$
		−1.1	I		M3S	1.74		1.2	$\sim(4\text{-}5) \cdot 10^9$
			I		gM1	1.61		1.2	$\sim(4\text{-}5) \cdot 10^9$
γ Leo (gr.)	R Dor	−0.3	SR	335	M7III	1.60	~1	1.2	$\sim(4\text{-}5) \cdot 10^9$
σ Pup	R Hor	0.0	LPV	402	gM8e	1.70	~10	≤1	$\sim 10^{10}$

(1) Probably non member
(2) Possible member
(3) Uncertain.

Table III

Group of clusters	Age limits (years)	Mass limits (solar unit)	Total number of clusters	Total number of red giants	Total number of red variables	Red variables / red giant stars
I	$5 \cdot 10^6 < t < 2 \cdot 10^7$	$M \geqslant 9$	16	37	13	0.35
II	$2 \cdot 10^7 < t < 2.5 \cdot 10^8$	$3 \leqslant M \leqslant 9$	47	184	1	0.005
III	$t > 2.5 \cdot 10$	$M < 3$	33	619	16 (22)	0.025 (0.035)

both groups; this fact suggests that this star, exceptional both for its period and its location, should be studied more accurately.

On the whole, the present data, although insufficient, seem to support the division of the red variables into two really different classes. The larger mass one, whose evolution towards more unstable states as are the Miras seems to be prevented by some other happening, might then perhaps be considered as a possible candidate to represent the presupernova type II stage, should all the present considerations correspond to some reality.

*

Our best thanks are due to Drs. G. Fabris and L. Nobili for their help in collecting and discussing the material related to red semiregular and irregular variables.

REFERENCES

Barbon, R., Dallaporta, N., Perinott, M. and Sussi, M. G. 1965, Mem. Soc. astr. ital. XXXVI, fasc. 1, 2.
Bertola, F., 1963, Contr. Oss. astrofis. Univ. Padova, N. 142 and 1965 N. 171.
Bertola, F. and Sussi, M. G., 1965, Contr. Oss. astrofis. Univ. Padova, N. 165.
Chiu, H. Y., 1961, Ann. of Phys. **15**, 1; **16**, 321.
Fraley, G. S., 1968, preprint.
Hayashi, C. and Cameron, R. C., 1962, Astrophys. J. **136**, 166.
Hayashi, C., Hoshi, R. and Sugimoto, D., 1962, Prog. Theor. Phys. Suppl. N. 22.
Hoyle, F. and Flower, W. A., 1960, Astrophys. J. **132**, 565.
Limber, D. N. 1960, Astrophys. J. **131**, 168.
Minkowski, R., 1964, A. Rev. Astr. Astrophys. **2**, 247.
Rakavy, G. and Shaviv, G., 1966, preprint.
Rakavy, G., Shaviv, G. and Zinamom, Z., 1967, Astrophys. J. **150**, 131.
Weymann, R., 1961, Mt. Wilson and Palomar Obs. Spec. Techn. Rept. No. 4.
Zwicky, F., 1964, Astrophys. J. **137**, 519.

SYNCHRONOUS THREE COLOUR STELLAR PHOTOMETRY AT THE CATANIA ASTROPHYSICAL OBSERVATORY

S. CRISTALDI and L. PATERNÒ

Astrophysical Observatory of Catania, Italy

SUMMARY

A synchronous three colour stellar photometer using a single photomultiplier has been constructed at Catania. In this communication the characteristics and the efficiency of this photometer are briefly described. At present the instrument is used for simultaneous UBV photometry of flare stars. A graph of simultaneous measurements in the UBV system of a flare of EV Lac is shown. A more detailed description of the instrument had been published elsewhere (Cristaldi. Paternò 1968).

INTRODUCTION

During the last few years various astronomers have emphasized the importance of simultaneous observations in different colours in stellar photometry. Besides, multicolour and simultaneous automatic photometry is indispensable in the study of fast phenomena, in general, and of stellar flares in particular. Finally, the use of this kind of photometry makes the observer's work easier and permits a uniform presentation of the data to equipments for digital measurements. For these reasons we constructed a synchronous photometer which executes simultaneous measurements within the UBV system. Unlike other multichannel photometers it uses only one photomultiplier. Fig. 1 shows the block diagram of the apparatus.

The light beam from the telescope, after having crossed the focal-plane diaphragm D is chopped by a filter-carrying disc, rotating at 900 revs. per minute. The disc has three round windows 120° apart, in which there are three filters: one UG1 (1 mm), one BG12 (1 mm) + GG13 (2 mm) and one OG4 (2 mm). The photomultiplier is of type EMI 6256 S.

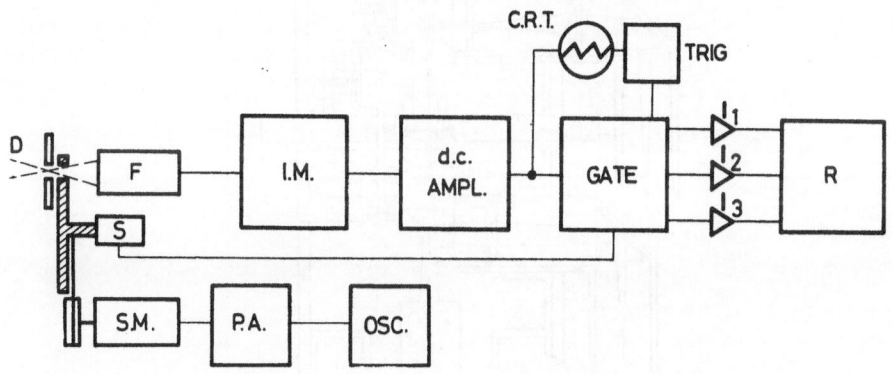

Fig. 1. Block diagram of the photometer.

The output of the photomulitiplier is connected by means of an impedance matcher (I.M.) to a d. c. amplifier which branches into two outputs; one is connected to an oscilloscope (CRT) for monitoring signals, the other is connected to a gate which acts in synchronism with three filters through the transducer S. The signals obtained through a given filter always exit through the same gate channel.

In our case we have three channels, the output signals of which are the UBV signals of our system. Each output gate is connected to an integrator circuit (I_1, I_2, I_3) whose output is measured in one of the three channels of a recording potentiometer (R) which records every six seconds.

Regularity in filter rotation and therefore in the synchronous signals for the gate is assured by a stepping motor (SM), powered by an amplifier (PA) and a regulating oscillator (OSC).

OPTICS OF THE PHOTOMETER

The optical part of the photometer is shown in Fig. 2. The focal-plane diaphragm D consists of a slide with three holes 1, 2, 3 mm in diameter.

The field lens is of fused quartz: diameter 20 mm and focus 70 mm. The lens is located approximately 70 mm from the photocathode of the photo-multiplier (F), so that with a f/10 reflector the diameter of the luminous disc which is formed on it is approximately 7 mm in size. The rotating disc with the three filters is located between the field lens and the photocathode.

Two small optical devices permit monitoring, respectively the pointing of the instrument and the inserted filter, if one desires to make continuous measurements using still filters.

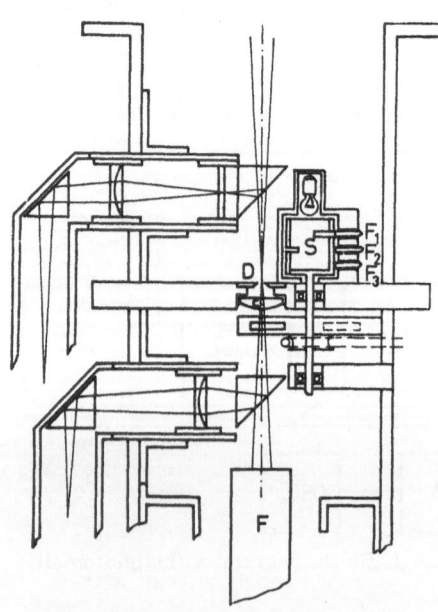

Fig. 2. Optics of the photometer.

The device (S) consists of a small cylinder mounted on the same shaft as the filters' carrier-disc. This cylinder is internally lighted and its inside walls are reflective; besides, it is cut transversely with three slots 0.7 mm wide and of such a length that each subtends, with respect to the axis of the cylinder, the same angle at the center as the corresponding filter in the disc. The light paths that emerge from the three slots are intercepted, at a fixed position, by three photodiodes F_1, F_2, F_3, one for each window opening for the entire time that the corresponding filter passes under the diaphragm opening. The signals from the photodiodes pilot the gate.

3. THE ELECTRONICS OF THE PHOTOMETER

The output of the photomultiplier is connected to an impedance matcher constructed with field effect transistors (Paternò 1967). At the matcher output, the voltage-pulses are amplified by a D.C. amplifier with variable gain from 1 to 1000. The amplifier consists of a cascade system of 5 identical operational amplifiers having strong negative feedback.

The UBV pulses at the amplifier output are selected and led to the proper channel by means of three photodiodes operated gates. The gates consist of a mercury relais type Clare HGSM 51111LOO.

All signals relative to each channel are detected and then, via integrator and differential amplifier, activate the recorder.

4. CHARACTERISTICS OF THE PHOTOMETER

It must be noted, above all, that the chopping of the radiation does not change the signal-to-noise ratio, so taking into consideration that the system of the amplification practically eliminates all the noise of the electronic apparatus, it is possible to reach, integrating on convenient time intervals, the same limit magnitude which is reached utilizing all the incident radiation.

The accuracy of the measurements, as known, is proportional to the total sum of the available radiation for each measure. Neglecting the radiation coming from the sky with respect to the signal, the average error (a.e.) of a measurement is given by the formula:

$$\text{a.e.} = \pm \left(\frac{1}{nqt} \right)^{1/2}$$

where n is the number of the photons due to irradiance of the star collected for a unit of time, q is the effective efficiency of the receiver and t is the exposure time.

At present the photometer is placed at the f/10 quasi-Cassegrain arrangement of the universal 61 cm reflector.

Considering a star of 10^m, the number of photoelectrons obtained from the collected photons is about 10^4 sec^{-1} (Allen 1963). Therefore, with an exposure of 1 sec the average error is about $\pm 0^m\!.01$ for one measurement; averaging 10 measurements the error is reduced to $\pm 0^m\!.001$. In our photometer the exposure time of each measurement in one band is 0.5 sec. However, as it was noted, the gate circuit is very prompt in action, it seems possible not only to increase the exposure time up to 1 sec, but even to increase it, gaining

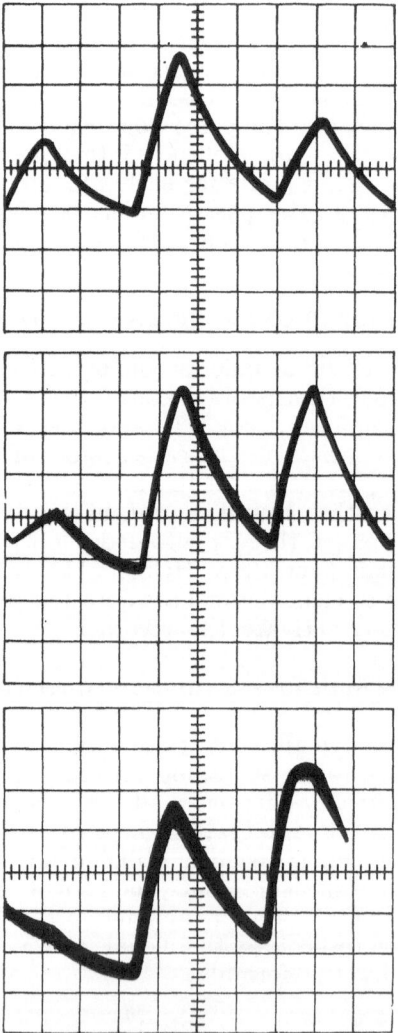

Fig. 3. The response of the photometer for B6, G2 and K5 spectral type stars (from the top to the bottom).

in the efficiency of the photometer which at present utilizes only 25 per cent of the incident radiation.

We have calculated that the magnitude limit which can be reached with our photometer, applied to a telescope of 61 cm, is 12.^{m}5; this result was verified by the observations. At present, the photometer works with the 61 cm reflector and is used for a research programme on flare stars.

Fig. 3 shows three photograms of oscilloscope tracings. They show the different responses of the photometer in the UBV for three stars of spectral types B6, G2 and K5 respectively (from the top to the bottom).

Finally, Fig. 4 shows a flare of the star EV Lacertae, observed with our apparatus on 8 August 1968 at 23^{h}23^m, Universal Time.

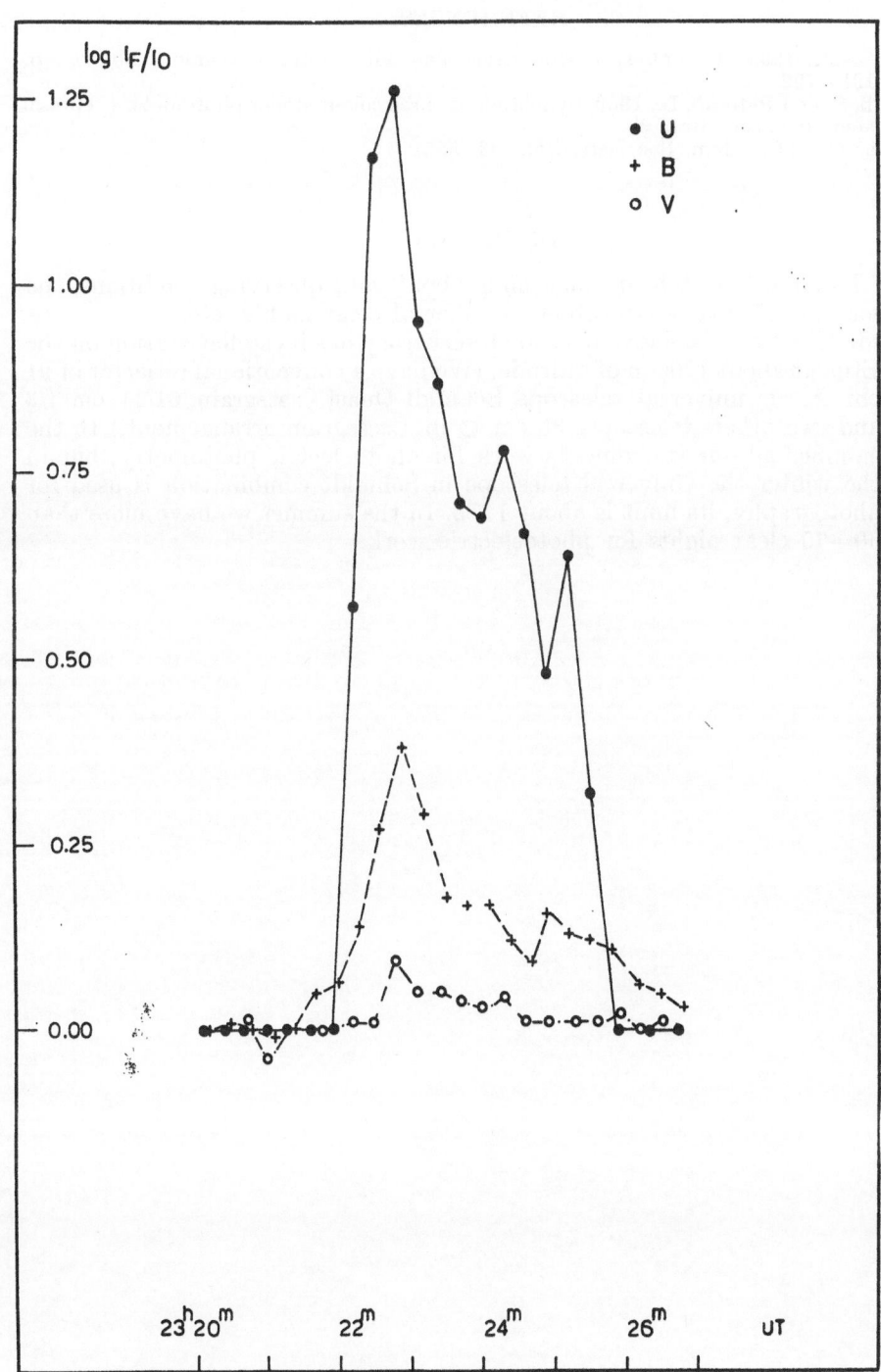

Fig. 4. Simultaneous light curves for a flare of EV Lac on 18th August. 1968 in the UBV system

REFERENCES

Allen, C. M., 1963, Astrophysical Quantities, 2nd edit. Univ. of London, London. p. 191—192.
Cristaldi, S. and Paternò, L., 1968, Synchronous three colour stellar photometer (in press) Mem. Soc. astr. ital., **39.**
Paternò, L., 1967, Mem. Soc. astr. ital., **38,** 555.

DISCUSSION

Detre: I should like to hear something about your observing conditions, the location of your observatory, number of clear nights, etc.

Cristaldi: The Catania Astrophysical Observatory has its stellar station on the Etna at about 1700 m of altitude. (We have a conventional reflector of 91 cm ∅; 1 universal telescope Schmidt-Quasi-Cassegrain 61/41 cm f/3 and two others telescopes 30 cm ∅ in Cassegrain arrangement.) In the summer all our instruments work for photoelectric photometry, but in the winter the Universal telescope in Schmidt combination is used for photography, its limit is about $17.^m5$. In the summer we have more than 60—70 clear nights for photoelectric work.

ABOUT THE T. V. PHOTOMETRY OF FAINT VARIABLE STARS

A. N. ABRAMENKO and V. V. PROKOFJEVA

Crimean Astrophysical Observatory, USSR

At the Crimean Astrophysical Observatory of the USSR a television equipment is employed for the measurement of stellar brightness using a high sensitive image-orthycon and a two cascade image tube. We can detect faint stars of the 20th and the 21st magnitude with the 0.5 meter telescope and with an exposure time of only 10—60 sec. (Abramenko et al. 1964.) Two methods of measuring T. V. image photos of stars are used. One of them is to measure the diameter of the images. This method is used for bright stars and has a range of 5—7 magnitudes (Abramenko and Prokofjeva, 1967). The other one is to measure the optical density or the transparency of the images of stars. This method is used for faint stars and has the range of 2—3 magnitudes. Observing variable stars we have an accuracy of 0.10—0.15 and 0.06—0.10 magnitude using the first and second method, respectively. Thus the full range of measurements in one T. V. photo is about 9 magnitudes and the accuracy is not worse than in the case of ordinary photography. Several variable stars have been observed. Fig. 1 shows the changes of brightness of

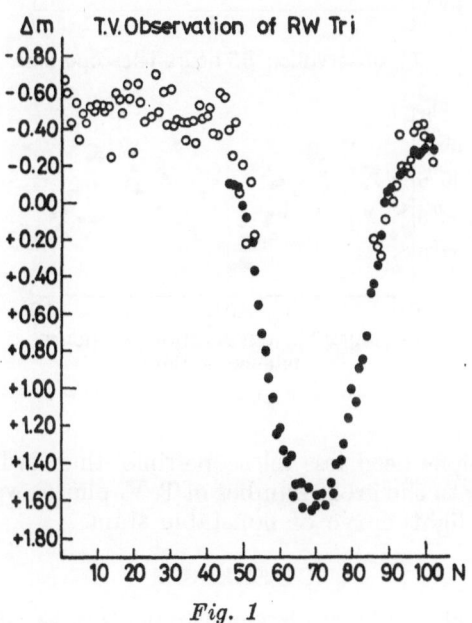

Fig. 1

RW Tri during one eclipse on 30 Dec. 1966 (Efimov and Prokofjeva, 1968).
The observations were made with the 0.5 meter coudé telescope. The exposure
time was 20 sec. The brightness of the star in eclipse was about 16 magnitude.
In 20 minutes the brightness of the star decreased two magnitudes. The measurements were made by two methods: by the measurement of diameters when
the star was bright (open circles) and by the measurement of optical density
when the star was faint (dots).

Fig. 2 shows the results of T. V. observations of S 8280 in NGC 188,
a variable of type W UMa (Istomin, 1967). During 5.5 hours in 3 different
nights about 400 photos were obtained, each with an exposure time of 40
sec. One point in Fig. 2. corresponds to the average value of 5 measurements.
Below there are results of ordinary photographic observations which
were made for the same star with a telescope having the same diameter.
About 50 measurements of the light were made during 40 hours of telescope-
time.

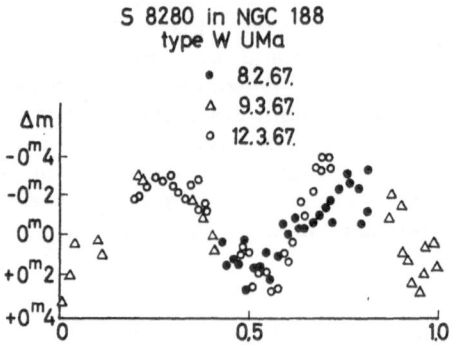

T.V. observation. 5.5 hours telescope time.

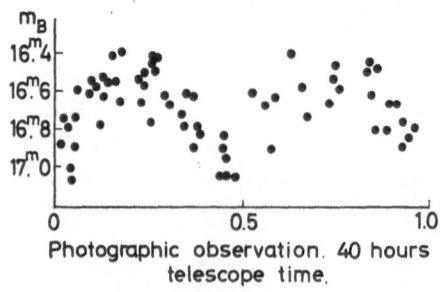

Photographic observation. 40 hours
telescope time.

Fig. 2

T. V. observations need less telescope-time than ordinary photographic
observations. Owing to the great number of T. V. photos we can easily register
the changes of the light curve of nonstable stars.

REFERENCES

Abramenko, A. N. et al., 1964, Izv. Krym. astrofiz. Obs. **33**, 315.
Abramenko, A. N. and Prokofjeva, V. V., 1967, Izv. Krym. astrofiz. Obs. **36**, 289.
Efimov, J. S. and Prokofjeva, V. V., 1968, Izv. Krym astrofiz. Obs. **39**.
Istomin, L. F., 1967, Astr. Cirk. Izdav. bjuro astr. Soobšč, Kazan, No. 448.

PART II

INTRINSIC IRREGULAR VARIABLES

A. VERY YOUNG STARS

B. FLARE STARS

C. Of, Be AND SHELL STARS

D. IRREGULAR MAGNETIC VARIABLES

E. R CrB VARIABLES

F. NOVA OUTBURSTS

EXTREMELY YOUNG STARS

Introductory Repdrt by

W. WENZEL

Institut für Sternphysik, Sternwarte Sonneberg, DDR

It is not so easy to give an Introductory Report on a topic like that, where during the last years new empirical material grew rapidly, a topic which is still too young for a consolidation of our knowledge. I must therefore concentrate on some partial questions which, according to my opinion, seem of some importance. When we speak now of extremely young variables we mean, of course, the evolutionary age (which is very short) and not the absolute age of the objects. In this way a B-type star of 10^6 years is no longer young, contrary to a K- or M-type star of the same age. Let us therefore call those stars very young ones which during their evolution have not yet reached the age zero main sequence, that is to say the gravitationally contracting stars.

The astronomers knew stars of that kind for a long time. They were hidden in the groups "T Tauri", "RW Aurigae" and "Nebular Variables". During the last fifteen years there was an increasing tendency to make sure that the above mentioned stellar types are, at least partially, such contracting stars.

Recently quite a few models of contracting stars have been computed. In this summary it is not our task to go into details as to these theoretical models. To-day the evolutionary tracks as well as the corresponding speeds in the HR-Diagram are known, at least in principle. We use to compare these evolutionary tracks with the HR-Diagrams of very young clusters or associations and in general we can notice a fairly good agreement. As always, difficulties will arise not before the conditions are more closely examined and the numerous details of the observations have to be explained.

Now the most important detail exists in the fact that a high percentage of the contracting stars in a young cluster is variable, but, as it seems, in a completely irregular fashion. This is the reason why we are engaged in this Colloquium also with "extremely young stars".

Another criterion of the variable contracting stars is their peculiar spectrum. At low dispersion the presence, for instance, of the emission line $H\alpha$ is used in a certain number of surveys for the search of very young stars. The question which percentage of the contracting objects is, in a given time interval, really constant in light (for example in a young cluster) and in which way these invariable stars differ physically from the variables is at the moment still open.

The same problem exists as to the presence of spectral peculiarities. And the matter becomes still more complicated by the fact that in a certain number of objects the amplitude of variability is subject to long-term variations whereby such a star may appear in constant light for some years. The time scale of this phenomenon and its frequency have still to be investigated.

Some research on young aggregates concerns the distribution of stars with respect to the masses or luminosities. This is important for us only in so far as the type of irregular light variation of a contracting star should undoubtedly depend on its mass (that is on its luminosity or on its spectral type). A detailed examination of this relation together with a statistical investigation of the various types of variables might therefore serve as a test for the duration of the state of variability or the initial mass-function, respectively. This must be emphasized because the variability of many faint objects is mostly easier to observe than abnormalities in the spectrum.

Knowledge of the bolometric corrections is, of course, necessary. Above all we must take into consideration the recent papers on the infra-red excess in T Tauri stars and related objects as well as the still insufficient knowledge of the intrinsic colours in other spectral regions.

In order to find the bolometric correction, especially in the infrared, and to explain the mechanisms of the variability we must take into account the influence of the surrounding interstellar material. In this respect the observation that the intensity of the Hα-emission in faint T Tauri stars is correlated with the strength of the interstellar extinction in the immediate surroundings of the object in question points to the existence of a rather extensive sphere of activity of the interstellar medium (Götz, 1967). On the other hand there are hints that the intensity of the Hα-emission is related to peculiarities of the irregular variations. But, it is true, the observations of different authors are in this case not all in good agreement and furthermore too few in number.

Nearly all well-examined contracting variables are in direct connection with clouds of interstellar matter. It must be emphasized that the investigators of variable stars should be very careful in classifying an object as "irregular" or "of RW Aurigae-type" if only insufficient observations are available.

The comparison of the variability of T Tauri stars and related objects inside and outside of interstellar clouds will supply another contribution towards finding the mechanism of variability. For there is no doubt that the light variation is partially determined by the circumstellar shells or clouds. These envelopes on the other hand should possess a physical connection to the above-mentioned spheres of activity of the interstellar matter. But we must not forget that in this case the real age of the respective variable plays a part, as one might expect that the connection with interstellar clouds on the average decreases with increasing age.

At present we know with some confidence that stellar formation in a certain region could last for some time (for instance Orion-associations). It is therefore an important task to analyse the light curves with respect to the different ages of the variables.

The analysis of the light curves must, of course, be accompanied by investigations on the spectral variations. It is known that the spectra of contracting stars are rich in pecularities, originating partly from the stars themselves, partly from the extensive atmospheres, shells and circumstellar clouds. But here we do not discuss the different components of these peculiarities. We have already mentioned the abnormal distribution of intensity in the continuous spectrum, brought about by the various additional superposed continua.

In this connection the question concerning the presence of solid particles in the circumstellar shells (especially in the variables of R Monocerotis-type) is of importance. The variable absorption effect, produced by such particles, is now and then taken for the interpretation of the variability in other types of variables (R Coronae and other carbon stars). If clouds of solid particles also play a rôle with T Tauri stars and related types, then, because of the short time-scale involved, we might have important hints as to the evolution of the above-mentioned shells by investigating the different kinds of the irregular variations.

So far we have mentioned the light variations of contracting young stars without giving details of these variations and their peculiarities. We will do this now more extensively.

At first we must notice that we are accustomed to describe the light variation as "completely irregular", although we know that there are in some objects temporary or permanent quasi-periodical phenomena which might be characterized by a certain length of the cycles. The search for other periodic components in the fluctuations is difficult because only very few accurate continuous series of observations are available.

As we are concerned with a large range of masses and evolutionary ages and as we are nevertheless inclined to consider the variable young stars as a whole, we must combine a great number of different forms of variability under one and the same aspect. We have a considerable number of classification schemes. Some of them are built up according to photometric characteristics, others partly according to spectral differences. The scheme, recommended in 1964 by Commission 27 of the IAU and included in the second Supplement of the General Catalogue of Variable Stars, represents a compromise in this respect. These classification principles of the irregular light variation (as far as presumably young stars are concerned) are the following: Spectrum early, intermediate to late, or similar to T Tauri; variation rapid, slow, or characterized by flares; with or without relation to diffuse nebulae.

I consider this scheme only as a tentative. Allow me to give some reasons.

1. The large difference between the light curves of the types T Orionis and, for instance, RW Aurigae is not properly expressed. In case the spectra of the respective stars were unknown, we would classify both objects as "Ins", that is "irregular observed in the region of diffuse nebulae and producing light variations of 0^m5 to 1^m0 in the course of several hours or days". It is in this connection without importance that RW Aurigae itself has just a distance of 2° from the nearest dark cloud. I quote this star only because it is well known.

To recall these types please remember the light curves of T Orionis (Parenago, 1955), DD Serpentis (Meinunger, 1967) and RW Aurigae (Kholopov, 1962), obtained from photographic and visual observations by several authors. Photoelectric observations will follow.

2. Another problematic case is presented by the so-called Is-stars (rapid irregular variables apparently not connected with diffuse nebulae). Extensive investigations have shown the number of these objects to be scarce in reality and it would be best to examine each newly-discovered Is-star meticulously whether it is correctly classified or not. In particular this is necessary for all those Is-stars, which lie within real T-associations or in their surroundings, for one had obviously to attribute a special astrophysical importance to these variables in case they were genuine.

An additional question of importance is the difference between Inb- or InT-stars on one side and the slowly, irregularly variable giants (abbreviated in the new catalogues by the symbol L) on the other side if, in routine work, the spectrum or the position in nebular regions is not properly investigated. This may be illustrated by the light curves of T Tauri (Ahnert, 1956) and the S-star AD Cygni (Beyer, 1948). In this field much observational work is still to be done.

There is also involved the question for the early phases of contraction, namely the configurations with very large radii and low temperatures at the beginning of those evolutionary tracks which are fairly well known at present. We must admit that for the time being we cannot identify these early phases with objects observed in the sky. Let me quote in this connection, without going into details, the Herbig—Haro-objects and the variable star FU Orionis. The outburst of that star (Wachmann, 1938) was interpreted as the last stage of the dynamical contraction of an opaque protostar before the quasi-hydro-static contraction begins, but there are wholly different models as well.

Let me add now a few words about our photoelectrically observed light curves.

For a rough empirical classification of variable stars in the contracting phase the light curves shown up till now might be sufficient (proto-types for instance RW Aurigae, T Orionis, BO Cephei or DD Serpentis, T Tauri, FU Orionis). However, in order to investigate the fluctuations in detail we need in addition photoelectric observations in several colours, simultaneously with spectral observations, if possible. The different kinds being manifold, a lot of work is waiting for the observers.

RW AURIGAE

The light variations are of an extremely complex nature. We find the following components (Wenzel 1966)

Waves of several hours' duration, amplitude some tenths of one magnitude;

Symmetrical outbursts, duration one to two hours, amplitude roughly $0^{m}.1$;

Unsymmetrical flares, presumably originating from the M-companion;

Fluctuations of some hundredths of one magnitude, apparently caused by variations of the emission line intensity;

Quasi-periodic fluctuations, cycle length roughly three days, amplitude 0.5 to 1 magnitude.

The colour-luminosity-diagram (l.c.) shows a large intrinsic scatter, as well as the two-colour-diagram (l.c.). Both these diagrams give the effect of the emission lines and the abnormal continua.

WW VULPECULAE

The star WW Vulpeculae seems to be of T Orionis-type (Fig. 1), charac-terized by unperiodic minima together with slow and short fluctuations of the normal light. It would be interesting to look for a T-association or an aggregate of faint Hα-stars in the neighbourhood of this object with the

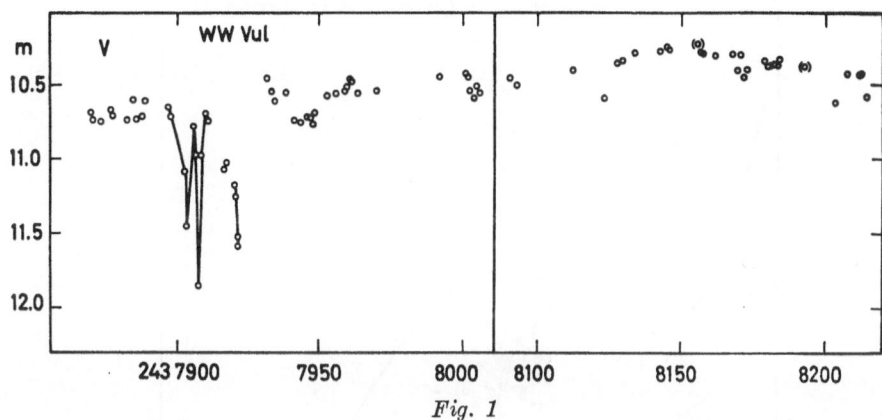

Fig. 1

spectral type A. The diagrams V/B—V (Fig. 2) and U—B/B—V (Fig. 3) show a much smaller scatter than it is the case with RW Aurigae.

In the diagrams V/(B—V) the direction of the main sequence in the respective interval of B—V is shown by the straight line; in the diagrams (U—B)/(B—V) the arrow indicates the interstellar extinction, which has not been applied to the observations.

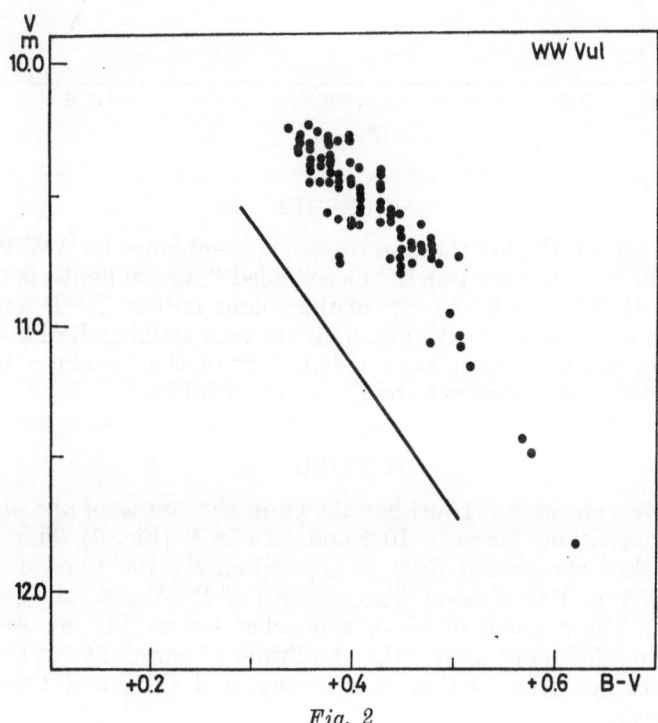

Fig. 2

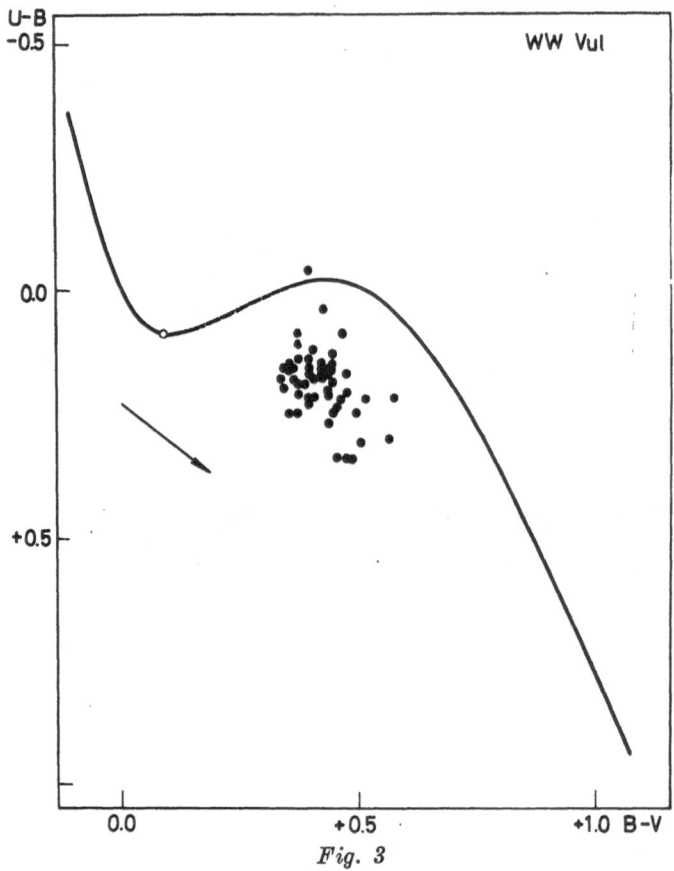

Fig. 3

SV CEPHEI

Apparently SV Cephei shows a certain resemblance to WW Vulpeculae, although the long-term variation of the so-called "normal light" is much more marked (Fig. 4). The small changes of the colour indices U—B and B—V in a range of two magnitudes for V (Fig. 5, 6) are very striking. Besides, we notice that slow fluctuations of the "normal light" or of the "medium brightness" are significant also for other extremely young variables.

T TAURI

The proto-type star T Tauri has shown in the course of our observations very small fluctuations between 10.2 and 10.5 in V (Fig. 7). This variability consists of a slow component (0^m2 in approximately 100^d) and a more rapid component (0^m1 in 1 to 5 days). The changes of B—V are very small (about 0^m05) (Fig. 8), the changes of U—B somewhat larger (Fig. 9). According to observations, made on old plates, the amplitude of variability in this star was in former times much larger than it is to-day, and amounted then to nearly four magnitudes.

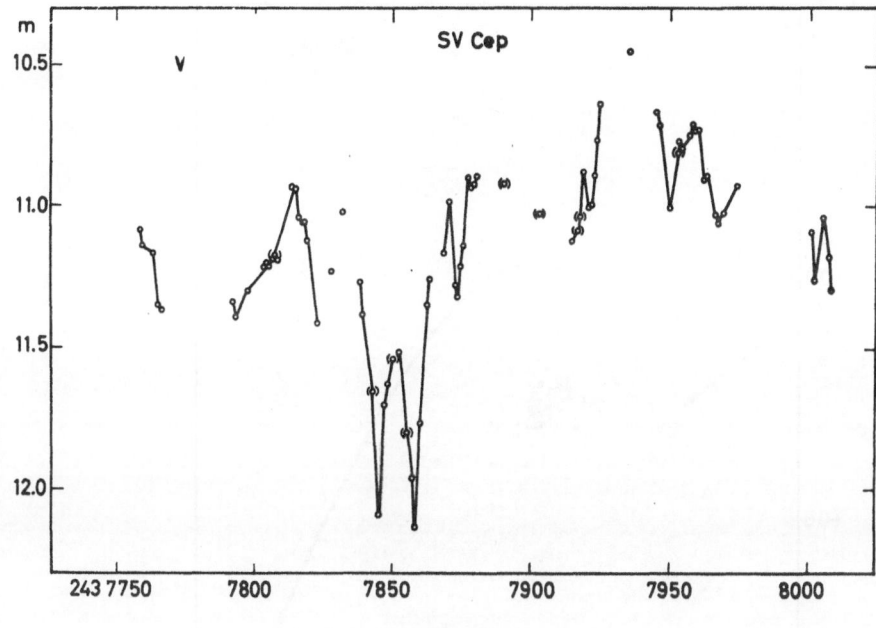

Fig. 4

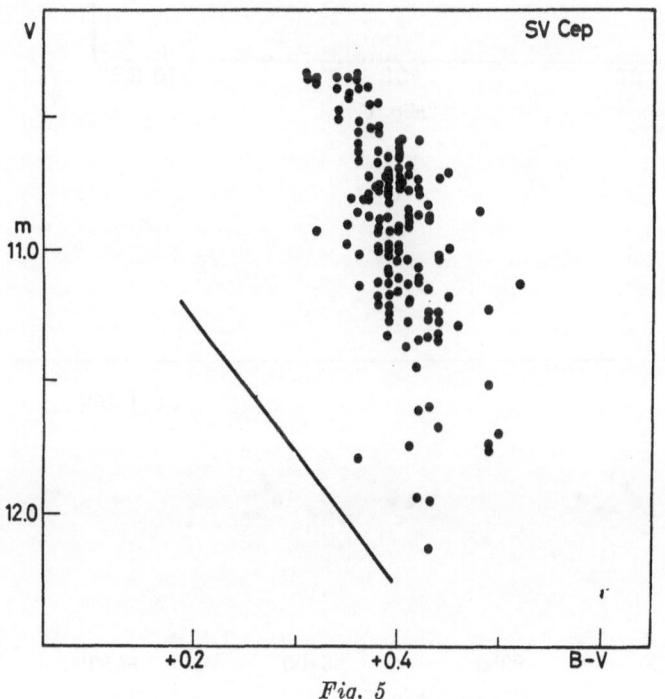

Fig. 5

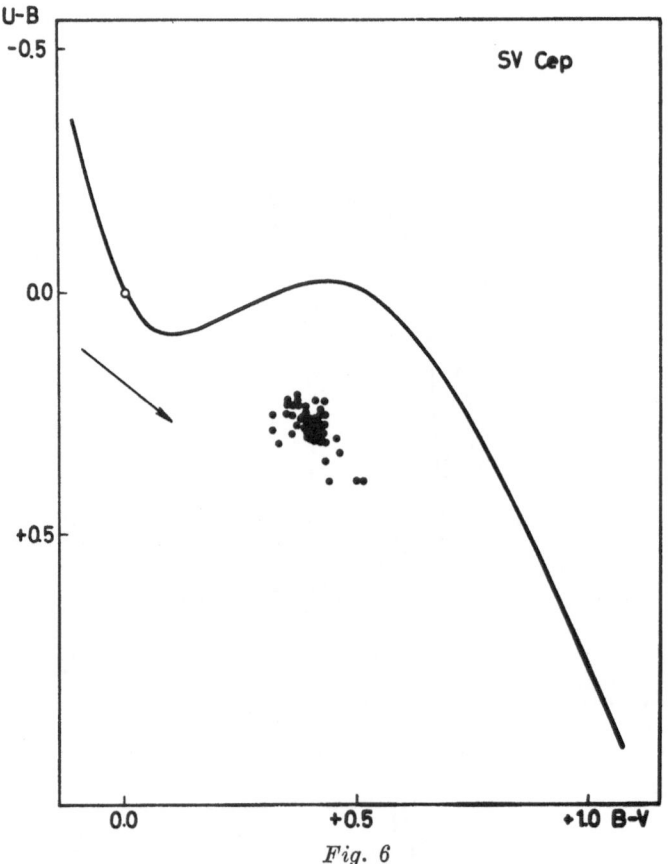

Fig. 6

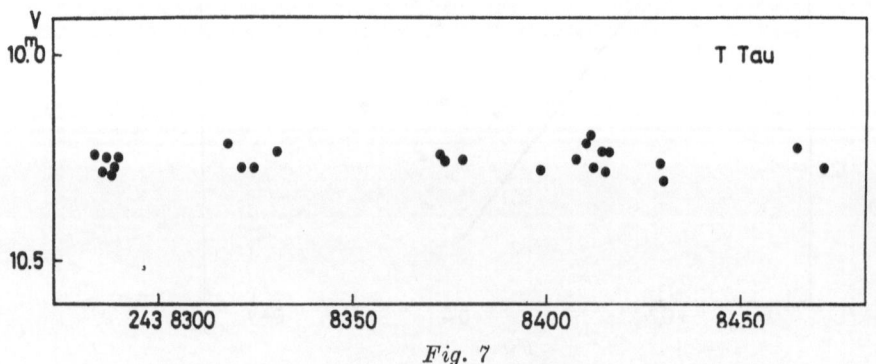

Fig. 7

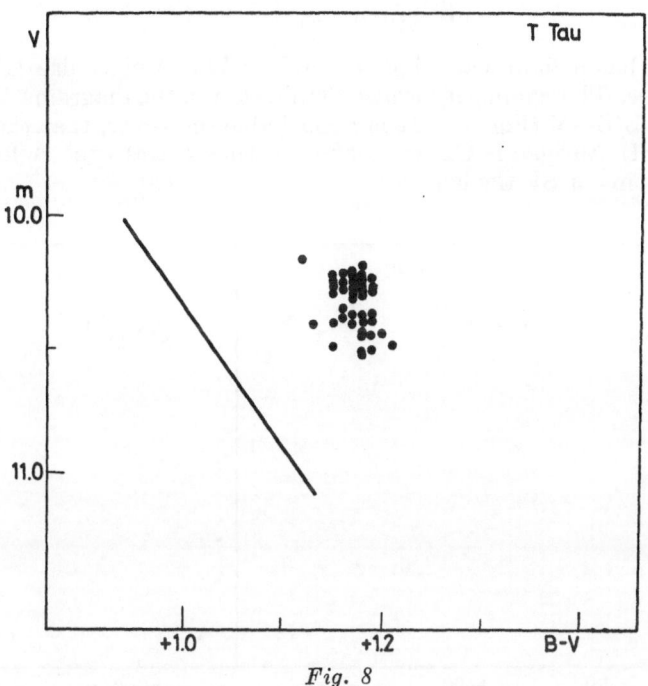

Fig. 8

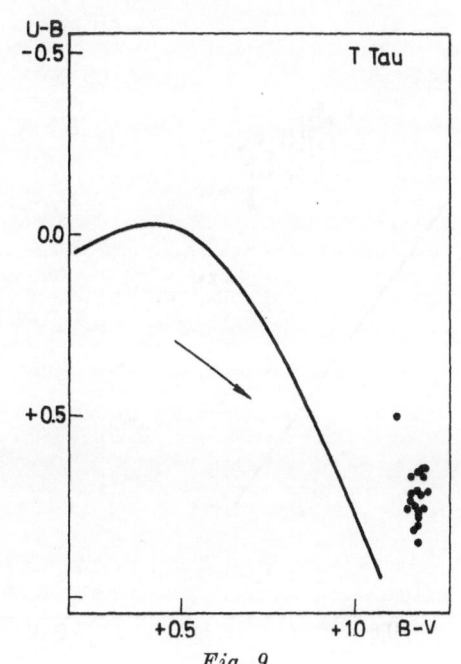

Fig. 9

SU AURIGAE

SU Aurigae has a light variation resembling WW Vulpeculae (Fig. 10) to a certain degree. This similarity is also manifested in the diagrams V/B—V (Fig. 11) and U—B/B—V (Fig. 12). It is remarkable, however, that the mean spectral type of SU Aurigae is G2, compared with spectral type A for WW Vulpeculae, T Orionis or SV Cephei.

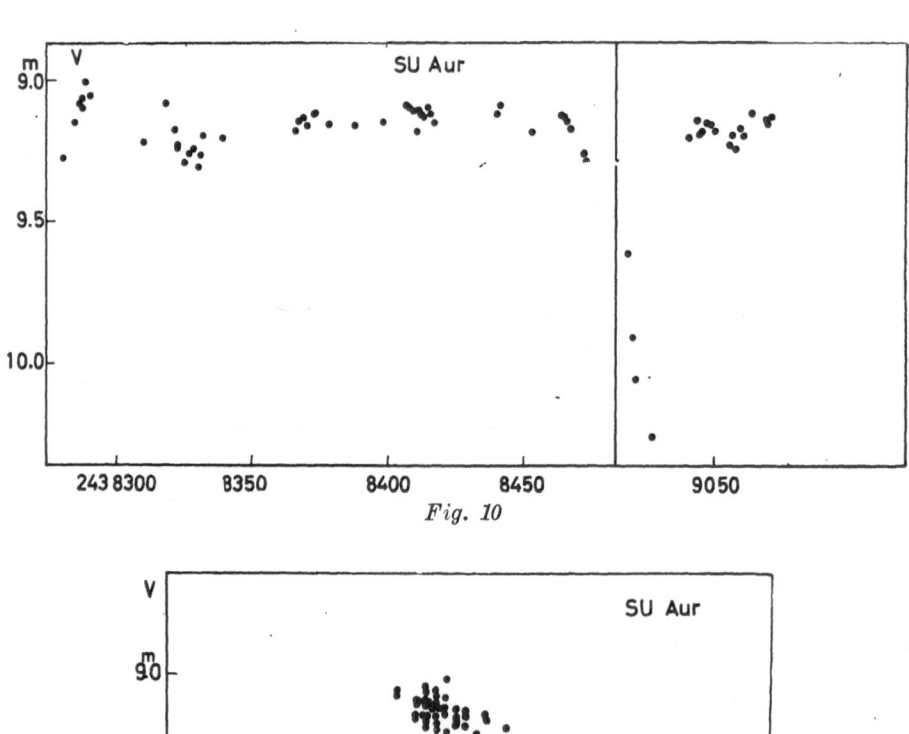

Fig. 10

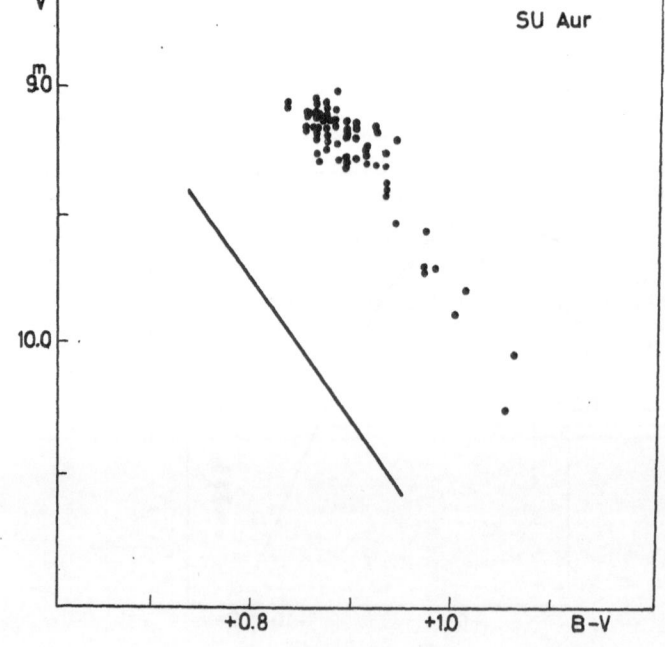

Fig. 11

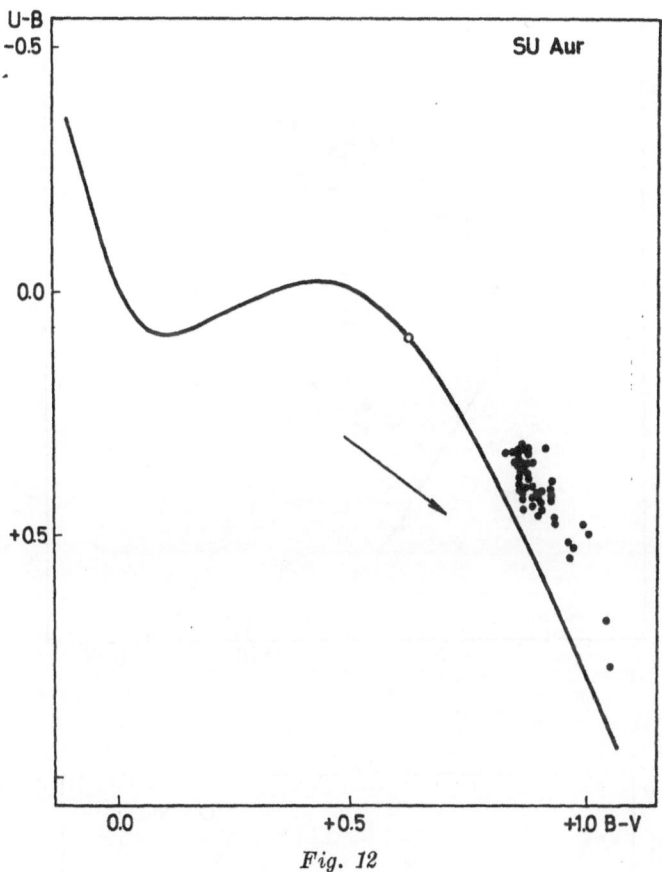

Fig. 12

RY TAURI AND CQ TAURI

Let me close now with two very strange diagrams V/B—V. RY Taur (Fig. 13) has two tendencies: the star moves vertically in the diagram (B—V remaining nearly constant), or, especially near the maximum, perpendicular to the direction of the main sequence.

With CQ Tauri (Fig. 14) the movement of the object across the diagram V/B—V is clearly curved: while getting brighter the star becomes at first redder, and then turns more blue.

I hope it was possible with these examples of our photoelectric work to underline in some way the variety of different forms which these extremely young variables show. In this discussion I have laid stress upon some aspects of the variability in brightness which might be observed most easily also in relatively faint objects. We must admit, of course, that remarkable spectral changes take place in these stars, too. To discuss these phenomena and to treat the theoretical mechanisms of variability I must leave, however, to my colleagues more versed in these fields.

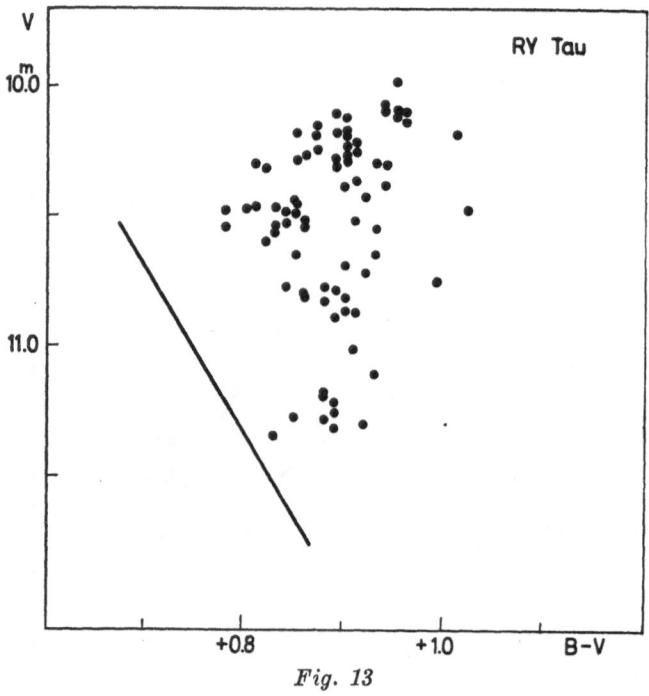

Fig. 13

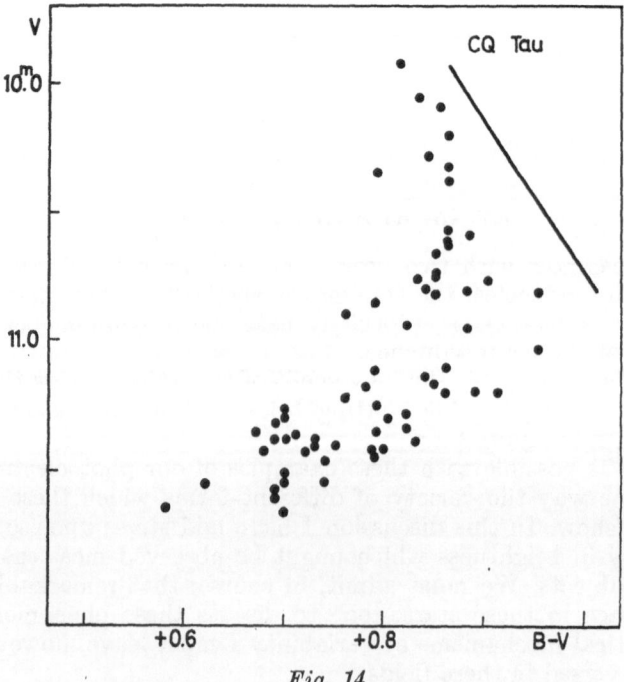

Fig. 14

REFERENCES

Ahnert, P., 1956, Budapest Mitt. **3,** No. 42.
Beyer, M., 1948, Erg. Astr. Nachr. **12,** No. 2.
Götz, W., 1967, Veröff. Sternw. Sonneberg **7,** No. 1.
Kholopov, P. N., 1962, Perem. Zvezdy **10,** No. 6.
Meinunger, L., 1967, Mitt. veränderl. Sterne **3,** 20.
Parenago, P. P., 1955, Trudy gos. astr. Inst. Sternberga **25,** 216.
Wachmann, A., 1938, Astrophys. J. **35,** p. 81.
Wenzel, W., 1966, Mitt. veränderl. Sterne **4,** No. 4.

THE LIGHT VARIATIONS OF THE NUCLEI IN HERBIG-HARO OBJECT NO. 2, 1946—1968*

G. H. HERBIG

Lick Observatory
University of California, Santa Cruz, California, USA

The first three of the peculiar semi-stellar emission nebulae now known as Herbig—Haro objects were detected by Herbig (1948, 1951) and by Haro (1950, 1952). The initial discoveries were made in the region south of the Orion Nebula, but a total of about 40 such Objects have now been found in Orion, Taurus, Perseus, and elsewhere. All known examples occur in heavily-obscured regions that are also rich in T Tauri stars. A number of other very small nebulous spots have been observed by Haro (1953, Table 2), Méndez (1967) and others, particularly near the Orion Nebula, but these are spectro-scopically distinct from the H—H Objects in having either a strong continuum especially in the near infrared, or a rather conventional emission spectrum. The H—H Objects have quite characteristic emission spectra: the H emission lines are strong, and [O I] and [S II] are unusually intense. The [N II] lines are also strong, and in those Objects not too heavily reddened, [O II] $\lambda\lambda 3726$—29 as well. These properties can be explained (Böhm 1956, Osterbrock 1958) if H and O are only partially ionized, and if $T_e \approx 7500°$ and $n_e \approx \approx 3 \times 10^3$ cm^{-3}. Only the brightest of the H—H Objects, Herbig No. 1 (= Haro 11a, near NGC 1999) has been observed in any detail. It shows a number of weaker emission lines that are not ordinarily found in appreciable strength in gaseous nebulae: H and K of Ca II, the infrared [Ca II] lines, Mg I $\lambda 4571$, and lines of [Fe II] and [Fe III]. It is significant that many T Tauri stars show, in integrated light, nebular lines very similar to those of the H—H Objects.

With only a few exceptions, most of the known H—H Objects have fairly simple structure: a small, bright, diffuse nucleus that may be elongated, often with a short fainter tail or extension attached. Herbig No. 2 (= Haro 10a) is one of the exceptions: at the present time it consists of a number (at least 8) of bright nuclei together with considerable fainter structure enclosed in an elliptical area about $25'' \times 40''$. This Object was first photographed at adequate scale in 1946—47, and when the plates were repeated in 1954—55 it was found that two new nuclei had appeared within this complex Object in the interim (Herbig 1957). Thereafter the region was photographed annually with the same telescope and emulsions (Crossley reflector, no filter, and Kodak 103a0 or IIa0) through 1959. A number of plates were obtained with the 120-inch reflector in 1959—63+, but the Crossley series was not resumed until 1968. Some further but rather minor changes were noted in the structure of

* *Contributions from the Lick Observatory*, No. 282.
+ A 120-inch photograph of this area appears in *Sky and Telescope*, 20, 338, 1960.

this Object after 1954 (Herbig 1966a). However, it was not until after the discovery by Magnan (1967) that by 1966 one of the nuclei had brightened still more, that all the Lick material was systematically re-examined and an effort made to fill gaps in the early record by inspection of other plate series.

Determination of the magnitudes of the individual nuclei in Object No. 2 is beset by problems of overlap and the background, the non-stellar nature of the condensations even at Crossley scale (39″/mm), the effects of differing seeing from plate to plate, and complications in the comparison of emission-line sources with stars having continuous spectra. A sequence of comparison stars was set up in the vicinity of Object No. 2 by direct comparison and extension, through objective-grating images on a plate taken with the 20-inch astrograph, with Special Selected Area 20 (Van Rhijn 1952). No high accuracy is claimed for these m_{pg}'s (listed in Table I) but they should be sufficient for this investigation, considering the other sources of uncertainty. Many of the fainter stars in this heavily-obscured region around NGC 1999 are variable, and several comparison stars had to be discarded for that reason, but the stars in Table 1 seemed to be constant over this time interval.

Table 1
Approximate m_{pg}'s and Relative Coordinates* of Comparison Stars

m_{pg}	$\Delta z \cos \delta$	$\Delta\delta$
16.0	$+11'.3$	$+1'.3$
16.5	$+0.7$	$+2.7$
17.0	-5.8	$+0.2$
17.6	$+6.2$	-1.8
18.0	-1.3	-3.4
18.5	$+4.8$	$+2.7$
19.5:	-3.4	$+1.5$

* With respect to Nucleus A in Object No. 2.

The principal nuclei in Object No. 2 are identified by letters in Fig. 1. The m_{pg}'s of the 7 brightest of these are listed in Table 2, estimated from all adequate plates available to me. It would have been desirable to use m_{pg}'s obtained only from the Crossley material in order to minimize the effect of the systematic errors already mentioned, but for the critical years 1947—1954 all available photographs were utilized. The 20-inch astrograph plates especially have to be employed with caution because of their small scale (55″/mm) and color-curve and low ultraviolet-transmission effects. Plates of the red region were used only to pronounce upon the presence or absence of a condensation. The magnitudes given for nuclei C, D and particularly E are of lower accuracy than for the others, because of their clearly non-stellar appearance.

None of the 120-inch plates were used for magnitude estimates, because at that scale (13″/mm) all the nuclei in Object No. 2 appear as complex, non-stellar structures. The best of the series of Crossley and 120-inch blue-ultraviolet plates of Object No. 2 are shown in Fig. 2.

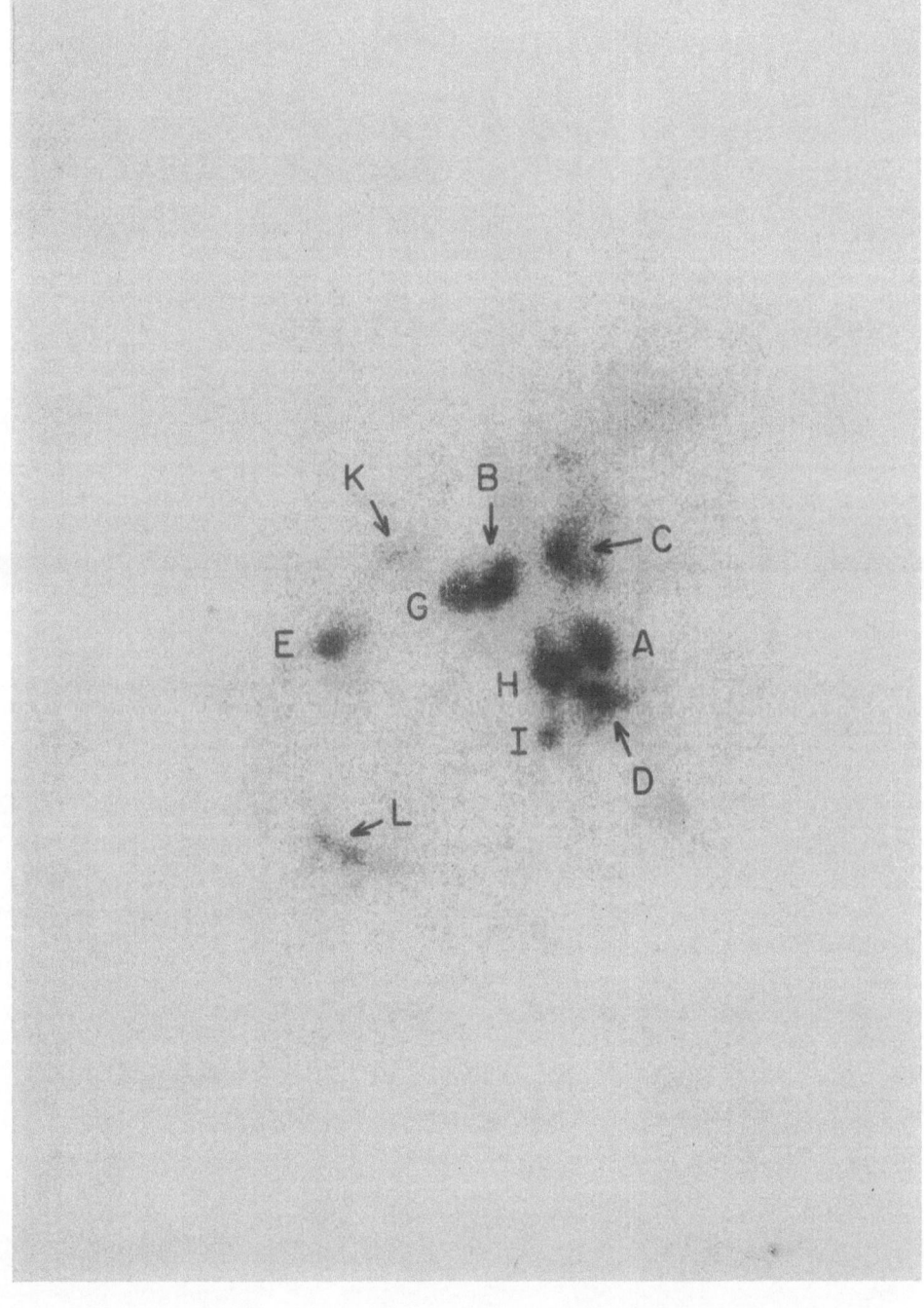

Fig. 1. Object Herbig No. 2, photographed with the 120-inch reflector in red light on 1959 Dec. 6. The letter designations of the individual nuclei are shown. The scale can be judged by the fact that the distance between Nuclei A and B is 8″.2.

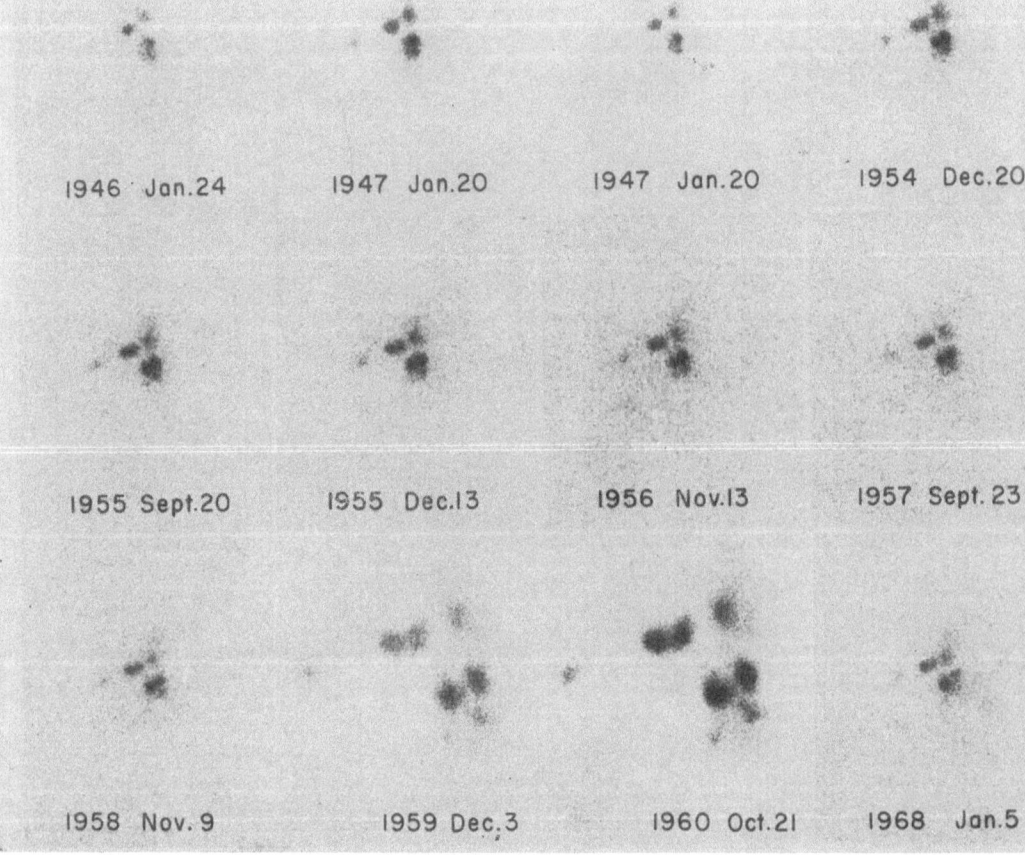

Fig. 2. Object No. 2, photographed with the Crossley and 120-inch reflectors in blue-ultraviole tween_1946_and 1968.

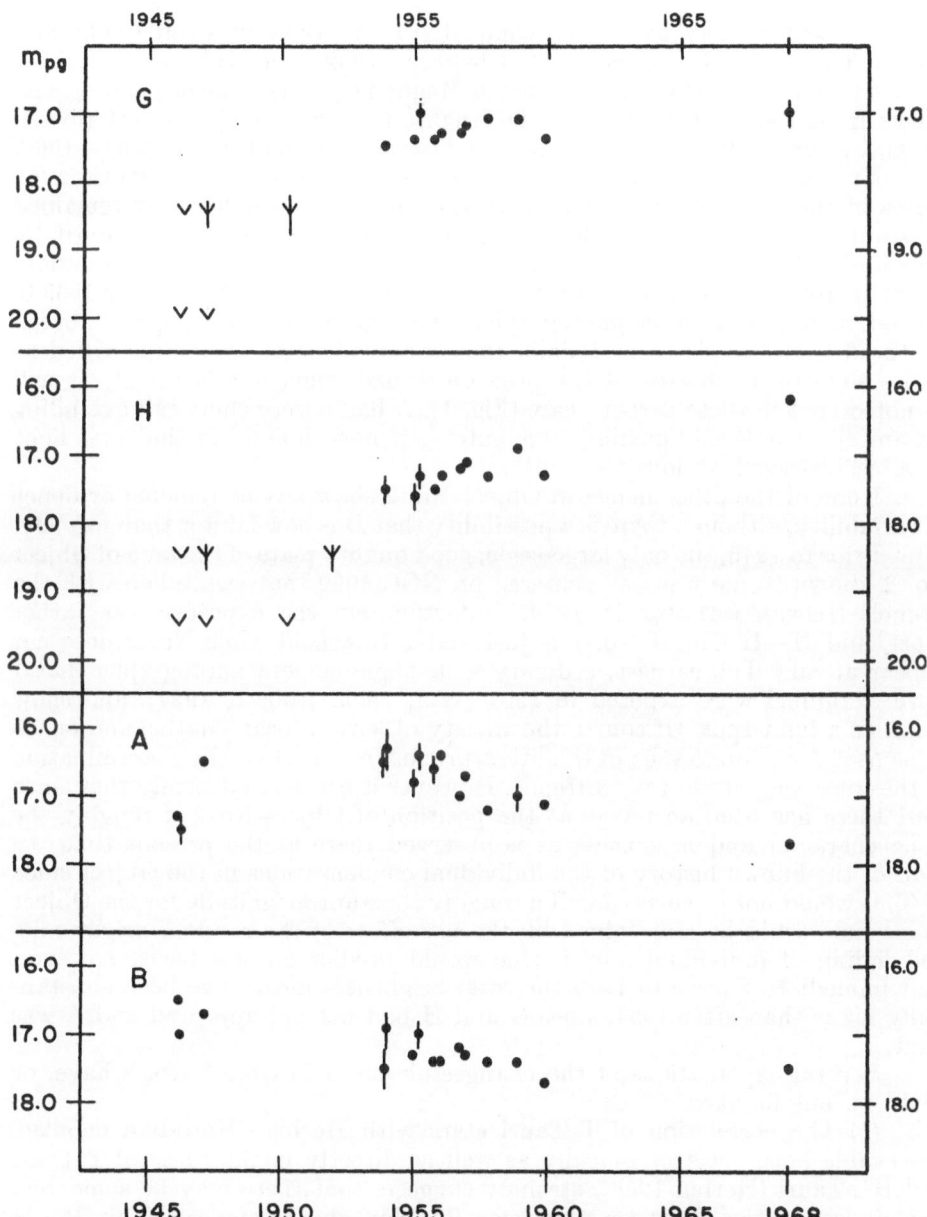

Fig. 3. Photographic light curves of nuclei G, H, A and B in Object No. 2. Vertical bars indicate uncertain points, and carets show plate limits.

The light-curves of nuclei A, B, G and H are plotted in Fig. 3. The rapid rise of nuclei G and H from fainter than about $m_{pg} = 20.0$ (only a rough estimate because of the lack of faint standards in SSA 20) to about $m_{pg} = 17.5$ is apparent. The data in Table 2 suggest that the brightening of G was certainly complete by late 1953, and if the evidence of the single 20-inch plate of 1951.9

can be accepted, may have been completed some time between 1950.2 and 1951.9. The rise of H was completed between 1952.8 and 1953.9. No further change took place in H until after 1959.9. Magnan's (1967) published photograph shows H in 1966.0 to have been the brightest nucleus in the Object (in red light); by 1968.0, H was fully 1 mag. brighter than it had been in 1954—1960.

Nucleus A is also variable. It was about 1 mag. fainter on the first Crossley plates of the area taken in early 1946 than it was a year later. It remained constant until about 1955. Beginning about 1956, A faded again until by 1968.0 it was fainter than at any other time during the 22 years of observation. Unfortunately no 120-inch plates have been taken of Object No. 2 since 1963.0. A good, modern large-scale photograph is required to check the suspicion, based on the 3 Crossley plates of 1968.0, that the condensation now identified as A lies slightly northward of the position A had when it was bright. It will be noted that in those earlier years (Fig. 1), A had a very short tail extending toward the northeast; possibly this extension now dominates the total light of A as observed at lower resolution.

None of the other nuclei in Object No. 2 show any convincing evidence of variability, although there is a possibility that D is now fainter than formely.

Prior to 1946, the only large-scale, good quality plate of the area of Object No. 2 known to me was one centered on NGC 1999 that was taken with the 30-inch Helwan reflector in 1915. Unfortunately the exposure was rather short, and H—H Object No. 1 is just above threshold while No. 2 does not appear at all.* The earliest, ordinary wide-angle camera photographs that I have examined were exposed in 1901 (Wolf 1903, Roberts 1927), and show No. 2 as a faint spot. Of course the variety of lenses, focal lengths, and emulsions used for photographs in the literature make a trustworthy determination of the total magnitude very difficult. However it can be said firmly that since 1901 there has been an image at the position of Object No. 2 of roughly the same character and brightness as is observed there at the present time. In view of the known history of the individual condensations in the Object since 1946, it would not be surprising if a roughly constant magnitude for the Object as a whole could be maintained by the net effect of the random brightening and fading of individual nuclei. One would predict on this basis, however, that immediately prior to 1946 the total brightness must have been substantially lower than after 1954, since G and H had not yet appeared and A was faint.

Several explanations of the changes observed in Object No. 2 have, or could be put forward.

(1) The correlation of T Tauri stars with Herbig—Haro-like nebulae, observable both spectroscopically as well as directly in the cases of T Tauri and HL Tauri (Herbig 1968), strongly suggests that there may be some real organic connection between the two. Possibly the existence of an H—H Object without an associated star marks the site of some electromagnetic activity preliminary to star formation, wherein is produced the flux of protons of about 100 KeV energy that Magnan and Schatzman (1965) calculate can explain the observed level of ionization. But the observed fading of nucleus A shows that an increase in brightness of an individual nucleus does not

---

* I am indebted to Dr. A. Samaha, Director of the Helwan Observatory, for very kindly sending me a print from this negative.

Table 2

Photographic Magnitudes of Nuclei, Object No. 2 between 1946 and 1968

Date (UT)	Telescope* (Number of plates)	A	B	O	Nucleus D	E	G	H
1946 Jan. 24	Cr (2)	17.3	16.5	17.6:	—	18.8±	<20.0	<19.5
1946 Feb. 23	Cr	17.5:	17.0	17.3	17.7:	—	<18.5	<18.5
1947 Jan. 20	Cr (3)	16.5	16.7	17.6:	—	20.0±	<20.0	<19.5
1947 Jan. 19	20 (2)	—	—	—	—	—	<18.5:	<18.5:
1949 Aug. 28	48, red	—	—	—	—	—	Invis	Invis
1950 Mar. 21	48	—	—	—	—	—	<18.5?	<19.5
1951 Nov. 30	20	—	—	—	—	—	17?[1]	<18.5:
1952 Oct. 26	Cr, red	—	—	—	—	—	Invis?[2]	Invis?
1953 Nov. 17	48	16.5:	17.5±	—	—	—	17.5	17.5:
1954 Jan. 2	Cr (2)	16.4±	16.9±	—	—	—	—	Present[3]
1954 Dec. 20	Cr	16.8:	17.3	18.0:	17.8	20.0±	17.4	17.6:
1955 Feb. 25	Cr	16.4:	17.0:	17.7	17.8	19.5±	17.0:	17.3:
1955 Sept. 20	Cr	16.6:	17.4	17.9:	18.5:	19.5±	17.4	17.5
1955 Dec. 13	Cr	16.4	17.4	17.8:	—	20.0±	17.3	17.3
1956 Sept. 3	Cr (2)	17.0	17.2	17.7	18.0:	20.0±	17.3	17.2
1956 Nov. 13	Cr	16.7	17.3	17.8	17.9	20.0±	17.2	17.1
1957 Sept. 23	Cr	17.2	17.4	17.8	17.8	—	17.1	17.3
1958 Nov. 9	Cr	17.0:	17.4	17.9	17.9	20.0±	17.1	16.9
1959 Nov. 10	Cr	17.1	17.7	18.0:	18.2:	20.0±	17.4	
1968 Jan. 5	Cr (2)	17.7[4]	17.5	17.6:	<19.0±	20.0±	17.0:	16.2

* Cr = Crossley reflector (Lick); 48 = 48-inch Schmidt (Palomar); 20 = 20-inch astrograph, blue lens (Lick).

Notes: [1] G seems to be present but at the scale and definition of the 20-inch, this result is marginal.
[2] Unwidened slitless spectrogram. The overlapping of many monochromatic images makes a decision difficult, but probably neither G nor H are present within a magnitude of their final brightness in Hα.
[3] Object No. 2 occurs near the corner of these plates, and the definition is very poor. The magnitudes given are only rough estimates.
[4] The nucleus A on the 3 plates of 1968 seems to be slightly north of its former position.

necessarily represent a progressive or a permanent change. Thus the spectacular brightening of G and H in the early 1950's must not be associated or
identified with the "birth" of two "new" stars. At best, one might expect
that at one of these sites an event like that of FU Orionis might eventually
take place (Herbig 1966).

(2) The appearance of G and H so close (about 3″) to A and B, respec-
tivaly, gave rise to one suggestion that this might be no more than the resolution, due to rapid orbital motion, of two binary pairs that had previously
been too close to be separated. This explanation can be dismissed quite convincingly for the following reasons.

(a) Prior to 1954, A and B, interpreted as A + H and B + G, would
have had to be about 0.75 mag. brighter than afterwards. This was not the case.

(b) The photocenters of A + H and B + G would have had to lie on
the line connecting the present positions of A and B, and of B and G. Careful
measurements were made of the position of the nuclei on Crossley plates, with
respect to a reference frame defined by nearby field stars. These show convincingly that A and B in 1947 were, within the errors of measurement (about
0.″2), at the same positions as they were in 1954 after G and H had appeared.

(c) The fact that A and H have subsequently been observed to change
in brightness demonstrates that the individual nuclei in Object No. 2 are
quite capable of intrinsic variation.

(3) The variations might be interpreted as due to the fluctuations of a
variable star within each nucleus. A strong objection to this is that, especially
in the red, essentially all the energy of the nuclei is contained in the emission
lines, yet the contribution of a variable star within — presumably a T Tauri
star — would largely be in the form of a continuous spectrum. Furthermore,
the best 120-inch direct plates exposed in the continuum between strong
emission lines (pass bands $\lambda\lambda 5200$—5800 and $\lambda\lambda 6800$—8800) show the nuclei
to be clearly non-stellar in these spectral regions. Thus there is no direct or
spectroscopic evidence of a star image within any of these condensations,
variable or not. This point has also been stressed by Haro. A further argument
to this same conclusion is based on the fact that T Tauri is known to be central
in a small complex emission-line nebula that appears itself to be a H—H
Object. There is strong evidence, although unfortunately most of it based on
early visual work, that this nebula varies in brightness by a substantial amount
(see Herbig 1950 for modern observations and references to the early work),
and that this variation is not obviously correlated with that of the star.
These facts support the belief that the presence of a faint T Tauri-like variable
could not explain the light variations of a H—H nucleus.

(4) It might be argued that the light variations of the nuclei are not
intrinsic, but rather are due to variable extinction by dense dust concentra-
tions moving across the line of sight in the foreground. This hypothesis meets
with the following difficulty. None of the nuclei in Object No. 2 are truly
stellar; diameters of 1″—2″ are measured on the 120-inch plates. Now, if
the motion of dust in the vicinity is typical of gas motion in H I regions,
a velocity of 1 km/sec for an element of the cloud would be reasonable. But
at 500 pc, the distance of Object No. 2, this corresponds to an angular cross-
motion of 1″ in 2000 years. Clearly, an opaque screen having such a slow motion
would not be able to uncover nuclei G and H quickly enough to explain their
observed brightening in an interval of 2—3 years or less.

(5) It has been suggested that the sudden appearance of new nuclei within Object No. 2 may have been the result of the sudden ionization of a small volume of a dense, neutral interstellar cloud by a blast of radiation or particles from some invisible source within. If this were the correct explanation, then the fading of nucleus A subsequent to 1959 has to be interpreted as recombination due to the withdrawal of excitation. At $T_\varepsilon = 7500°$ and $n_\varepsilon = 3 \times 10^3$ cm^{-3}, about 5 years are required to reduce n_ε to $0.8\ n_\varepsilon$ (Aller 1956) and thus to lower the surface brightness in the Balmer lines by the observed 0.5 mag. This is in fact approximately the observed time scale of the fading of A. On the other hand, if the H recombines on such a short time scale, it means that the hypothetical exciting sources must remain active in most H—H Objects. Otherwise, they all would quickly fade away. Furthermore, if this decay were going on in all Objects, one would expect a considerable range in their degrees of ionization, yet the "new" nuclei in Object No. 2 seemed to be spectroscopically indistinguishable from all the others, presumably much older.

The fact that H and O are partially ionized, apparently throughout the whole volume of the Object, seems to rule out radiative ionization from a single source within, as pointed out by Osterbrock (1958). Ionization by high-energy particles is a more acceptable explanation for this effect. Magnan and Schatzman (1965) have calculated the amount and energy (≈ 100 KeV) of a proton flux that would be required to maintain the ionization in Object No. 1 by collisional processes. But there is no explanation of why all H—H Objects are fed by particles of just these special properties.

(6) A very speculative interpretation is that the bright nuclei in Object No. 2 are only transient phenomena on the surface of a very dense, dark cloud, and are thus distantly analogous to surface phenomena on the sun. This presumably requires that the cloud be a coherent, gravitationally stable unit. The elliptical area of about 25″ × 40″ which contains all the bright structure of Object No. 2 would, in 3 dimensions and at a H density of $n \approx 10^4$ cm^{-3}, enclose a mass of about 0.05 ⊙. This mass would not be gravitationally bound because the surface escape velocity is only about 0.1 km/sec. If the mean density of the cloud were however raised to $n > 10^6$ H cm^{-3}, then $v_{escape} > v_{internal}$ and the cloud could remain intact. Such a mean density seems reasonable since the value of $n = 10^4$ cm^{-3} was inferred from the brightline spectra of the nuclei, which here are interpreted as surface phenomena on a body which could have strong central condensation. This hypothesis is appealing in that it provides at least intuitively, through analogy with the sun, an explanation for the light variations of the nuclei in Object No. 2. It also might explain why three nuclei brightened in the years 1946—1953: possibly the degree of surface activity on such an object would vary cyclically, as in the sun.

The foregoing outline has demonstrated that although some possibilities can be eliminated, there is still no completely straightforward interpretation of the light variations of the nuclei in Object No. 2. The most acute need in clarifying some of the questions raised here is a continuing series of high-resolution direct photographs, taken over a period of 5—10 years with a very large reflector. Such a series could demonstrate, for example, whether the position and the fine structure of a nucleus changes as its total magnitude varies. Possibly it would also help to answer also the question whether the same nucleus is able to reappear again, or whether each brightens and fades only once. It is not now clear whether variability is the exception or the rule among H—H Objects. The variability of the Object surrounding T Tauri has already been mentioned. Small changes are believed to have been observed in several other Objects as well, but to date, either the plate material is too limited or the Crossley scale too small to confirm this suspicion.

It is a pleasure to express my thanks to Dr. G. O. Abell for allowing me to examine some early 48-inch Schmidt negatives in his collection; to Mr. E. Harlan for taking the Crossley plates of Object No. 2 in January, 1968; to Dr. G. Haro for helpful and stimulating correspondence; to Dr. W. J. Luyten for sending me copies of some 48-inch Schmidt plates in his possession; to M. C. Magnan for interesting discussions as well as for communication of his results in advance of publication; and to Dr. S. Vasilevskis for taking several plates for me in 1959 with the 120-inch reflector.

REFERENCES

Aller, L. H., 1956, Gaseous Nebulae (New York: John Wiley and Sons), p. 66.
Böhm, K. H., 1956, Astrophys. J., **123**, 379.
Haro, G., 1950, Astr. J., **55**, 72.
Haro, G., 1952, Astrophys. J., **115**, 572.
Haro, G., 1953, Astrophys. J., **117**, 73.
Herbig, G. H., 1948, Thesis. University of California.
Herbig, G. H., 1950, Astrophys. J., **111**, 11.
Herbig, G. H., 1951, Astrophys. J., **113**, 697.
Herbig, G. H., 1957, Non-Stable Stars, ed. G. H. Herbig (I. A. U. Symposium No. 3. London and New York: Cambridge University Press), p. 3.
Herbig, G. H., 1962, in The Universe, by R. Bergamini (New York: Time, Inc.), p. 142.
Herbig, G. H., 1966, Vistas in Astronomy, **8**, 109.
Herbig, G. H., 1968, June: paper presented at Liège Astrophysical Symposium.
Magnan, C., 1967, l'Astronomie, **81**, 49.
Magnan, C. and Schatzman, E., 1965, C. r. hebd. Seanc. Acad. Sci., Paris, **260**, 6289.
Méndez, M. E., 1967, Bol. Tonantzintla y Tacubaya, **4**, 104.
Osterbrock, D. E., 1958, Publ. astr. Soc. Pacific, **70**, 399.
Roberts, Mrs. I., 1927, Isaac Roberts' Atlas of 52 Regions, Chart 22.
Van Rhijn, P. J., 1952, Durchmusterung of Selected Areas of the Special Plan, **1**. (Groningen: Kapteyn Astronomical Laboratory).

DISCUSSION

Seitter: Could the hypothesis of the object being a protostar with spot-like activity on its surface be checked from proper motions of the group of nuclei as a whole? If it is indeed a protostar one would expect it to be in rotation. The observation of a common proper motion of the expected order of magnitude could thus strengthen the hypothesis.

Herbig: Yes, in principle. But the cross-motion due to rotation at 1 km/sec would amount only to about 0″.0005/year.

Rosino: 1) I should like to know whether any radial velocity determination on H—H Object No. 2 has been made and, in the affirmative case, whether this radial velocity corresponds to that of the Orion Nebula.

2) Three H—H objects 1, 2, 3 are found in a peculiar region, near the small nebula NGC 1999, which is abnormal for the spectrum and the aspect. I am wondering whether there may be any connection between these objects and the three H—H objects.

Herbig: Observations of the radial velocities of the individual nuclei in No. 2 have been made at the prime focus of the 120-inch reflector. These results are beset by the well-known difficulties of velocity work with a spectrograph having a thick-mirror Schmidt and un-flattened field. Probably one should say only that there is no evidence for large internal motions greater than about 50 km/sec.

Anderson: Are there any infrared or radio observations?

Herbig: My own I-red photography extends only to 0.8 μ — these show the same structure in Objects No. 1 and 2 as in the emission lines in the red and blue. I understand that No. 1 has also been observed by one of the Arizona — Tonantzintla observers at 3 μ, but no detectable infrared radiation was found. Thus these H—H objects are not "infrared stars" in sense of the recent use of that word.

Feast: 1) Did I understand that there is no measurable relative proper motion of the various nuclei?

2) Is there any evidence for changes in size of the individual condensations?

Herbig: to 1)There was no relative motion between 1947 and 1954 greater than 0."2. This is based on Crossley material; the 120-inch plates have not been measured.

to 2) The scale of the Crossley plates is certainly too small to answer this question. The 120-inch material is perhaps adequate but has not been examined from this point of view.

PHOTOELECTRIC OBSERVATIONS OF 6 SOUTHERN RW AURIGAE VARIABLES

W. SEGGEWISS and E. H. GEYER

Sternwarte Bonn

A program for investigating the photometric behaviour of RW Aurigae variables has been started in 1962 by one of the authors (E.H.G.) at the Boyden Observatory in South Africa. The observations in the UBV colour system were carried out with the photoelectric equipment of the 60 inch Rockefeller reflector.

The following six stars classified as RW Aurigae type variables in the GCVS (1958) have been selected:

Star	l^{II}	(1950)	b^{II}
SY Phe	278° 5		—72.5
SZ Phe	277.5		—71.7
TT Phe	269.7		—71.5
BS Vel	260.0		— 1.5
DI Car	294.9		— 8.6
ES Car	290.7		+ 0.2

On the average these stars were measured three times per night during a total of 8 to 17 nights from 1962 to 1964.

The first four stars show only small changes in magnitude (Fig. 1) and colours confirming the observations by Hoffmeister (1958) and De Kort (1941). In Figure 2 the position of these stars in the two colour diagram is plotted. It seems that SY Phe is an F-Type star in agreement with the spectral type given in the HD-catalogue (F8) and the Potsdamer Spektraldurchmusterung (F4). For SZ Phe the PSD gives the spectral type K4. For the other two stars no spectral types are available. From their position in the diagram it is concluded that they are also K-type stars.

The star DI Car shows a brightness variation larger than $0^{m}7$ in V and $0^{m}3$ in B — V, resp. $0^{m}34$ in U — B (Fig. 3). The light curve resembles that of a cepheid star with a period of approximately 30 days. Yet Hoffmeister (1957) considered this star a member of the CN Ori variables, which are a subgroup of the U Gem stars. In the Henry Draper Extension the spectrum of DI Car is characterized as peculiar. We found a weak objective prism spectrum of this variable on a plate taken by one of us with the duplicate of the original Schmidt camera at the Boyden Observatory. Because of the faintness it is almost impossible to associate a correct spectral type. Three features, however, seem important assuming that they are not simulated by spots on the plate:

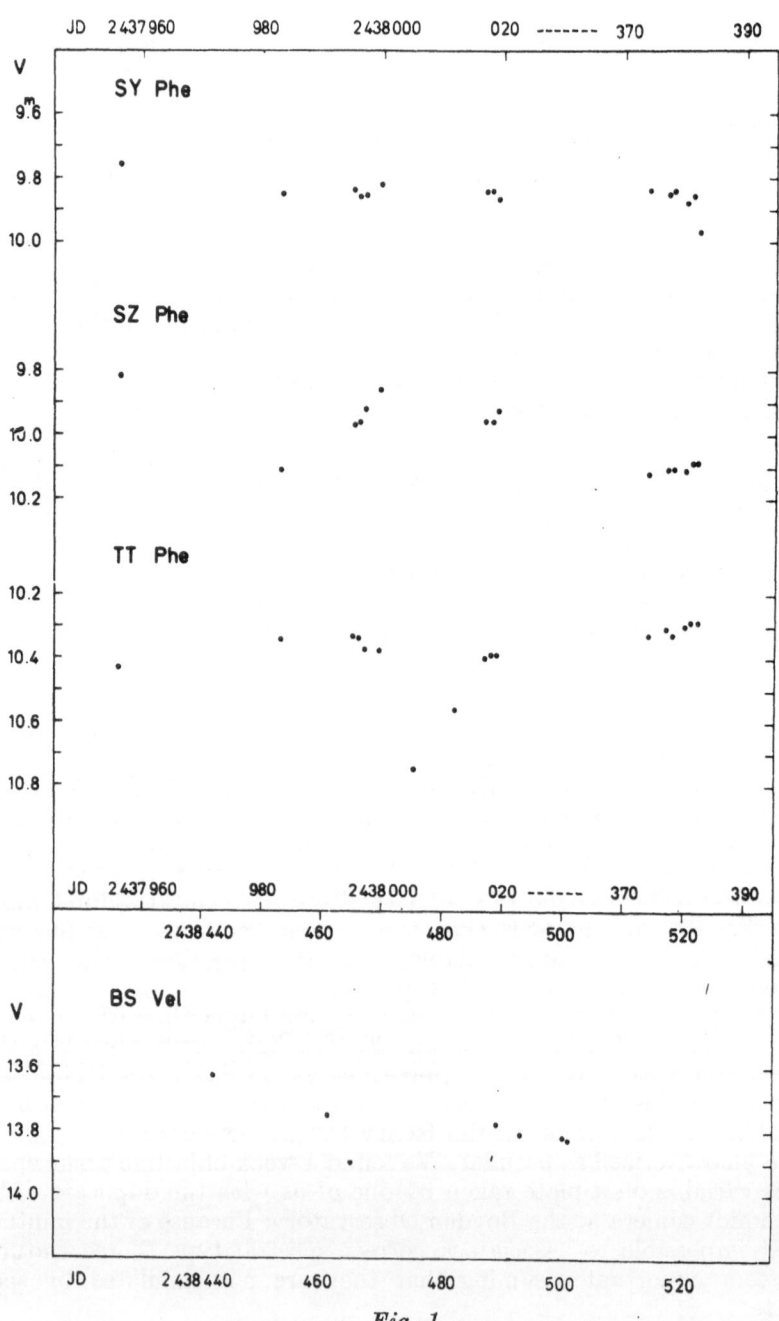

Fig. 1

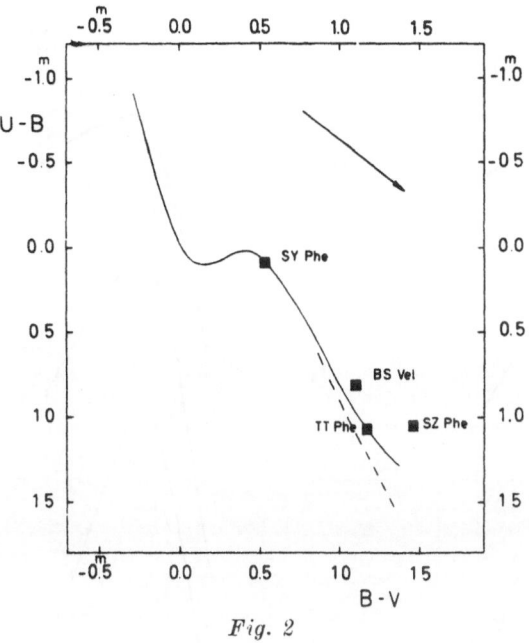

Fig. 2

(1) Sr II 4125 seems absent, as concluded from the undiminished strength of the continuum in this region.

(2) A strong absorption feature merges into the G-band from the red side; it may be interpreted as Hγ which is, however, too strong for the late type apparent from the H and K lines and Ca I.

(3) The continuum between Hβ and Hγ is extremely weak. Tentatively we might assume the star to be a K-dwarf, possibly composite.

During its variations the star moves along a straight line above the main sequence in the two colour diagram, in the colour magnitude diagrams the star describes loops (Fig. 4).

It is known through the photographic observations of Hertzsprung (1925) that the star ES Car shows besides isolated brightness variations up to 2 magnitudes small and rapid variations of $0.^m5$, which are confirmed by our measurements (Fig. 5). In Figure 5 the individual observations of each night are plotted. The magnitude and colour differences in the sense variable minus comparison star are given. The relevant data for the comparison star are: $V = 12.^m78$, $B — V = 0.^m28$, $U — B = 0.^m14$. This variable is situated close to the galactic plane in a very rich and complicated portion of the southern Milky Way studied by Schmidt and Diaz Santanilla (1964). They found in the NW edge of the open cluster NGC 3572 a second very young cluster located at a larger distance than NGC 3572 which they designated NGC 3572 b. ES Car lies close to the centre of this second cluster (star No. 33 in the paper of Schmidt and Diaz Santanilla). Figure 6 is taken from the above publication. The variable is shown as a square, filled circles are cluster members, open circles probable members, and crosses field stars. Assuming ES Car a cluster member, the following absolute magnitude and intrinsic colours can be derived

7*

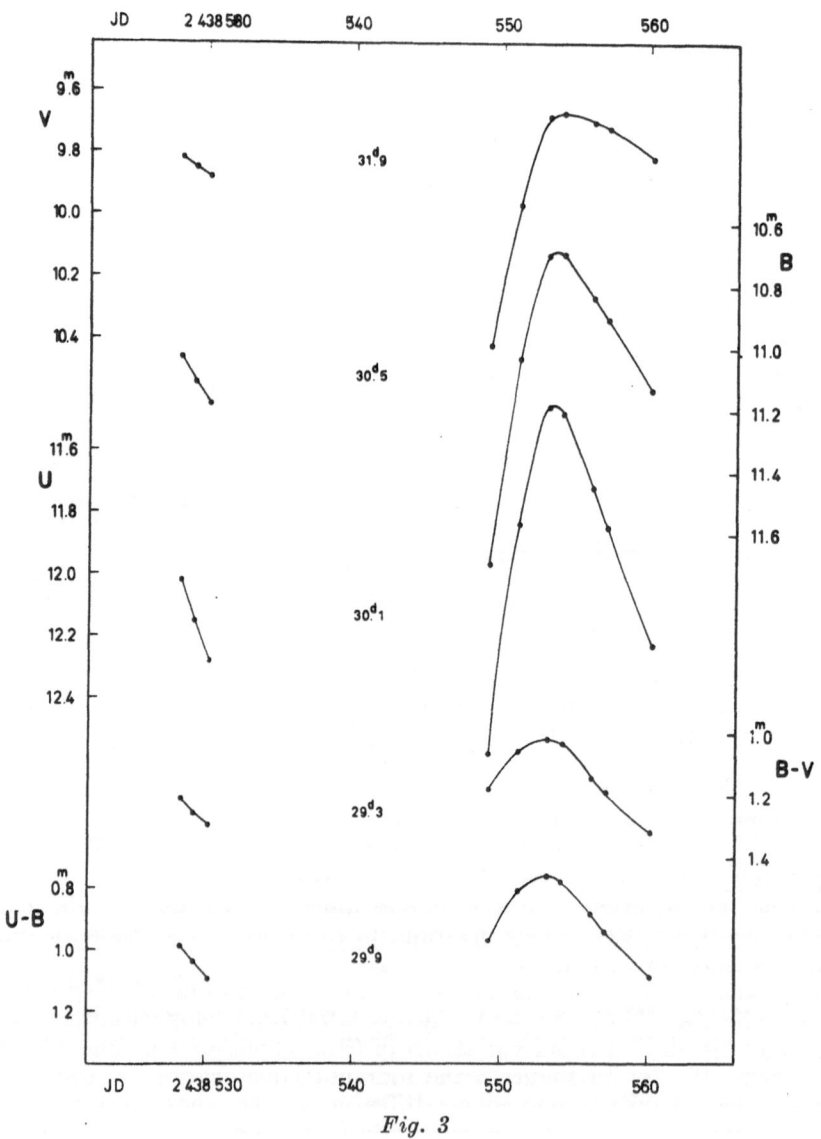

Fig. 3

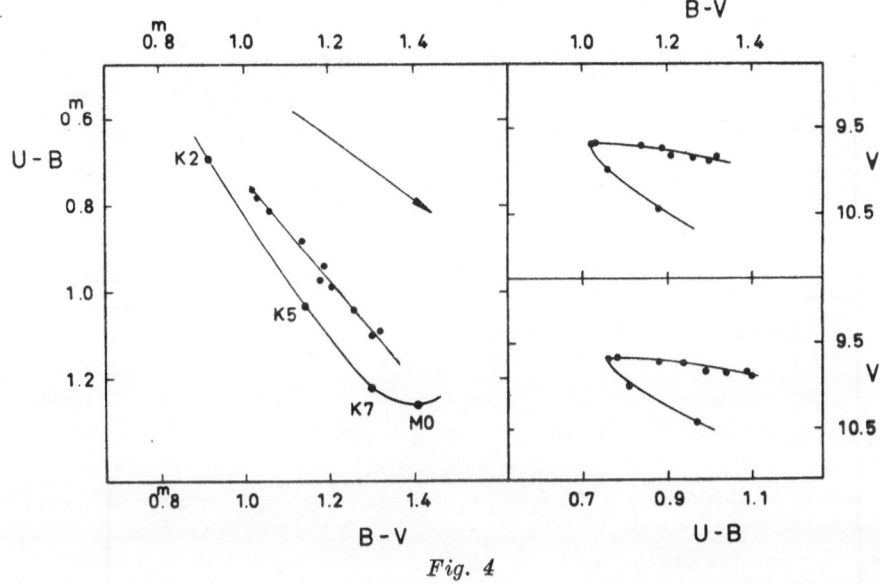

Fig. 4

from the known data of NGC 3572b: $M_v = -1.^m65$, $(B-V)_0 = -0.^m14$, $(U-B)_0 = -0.^m32$. This corresponds to a B7III star.

The position of the variable in the colour-magnitude diagrams is above the main sequence indicating the possibility that the star is still in the contracting phase of its evolutionary track.

In this case NGC 3572b would be the cluster with the earliest not yet to the main sequence contracted member.

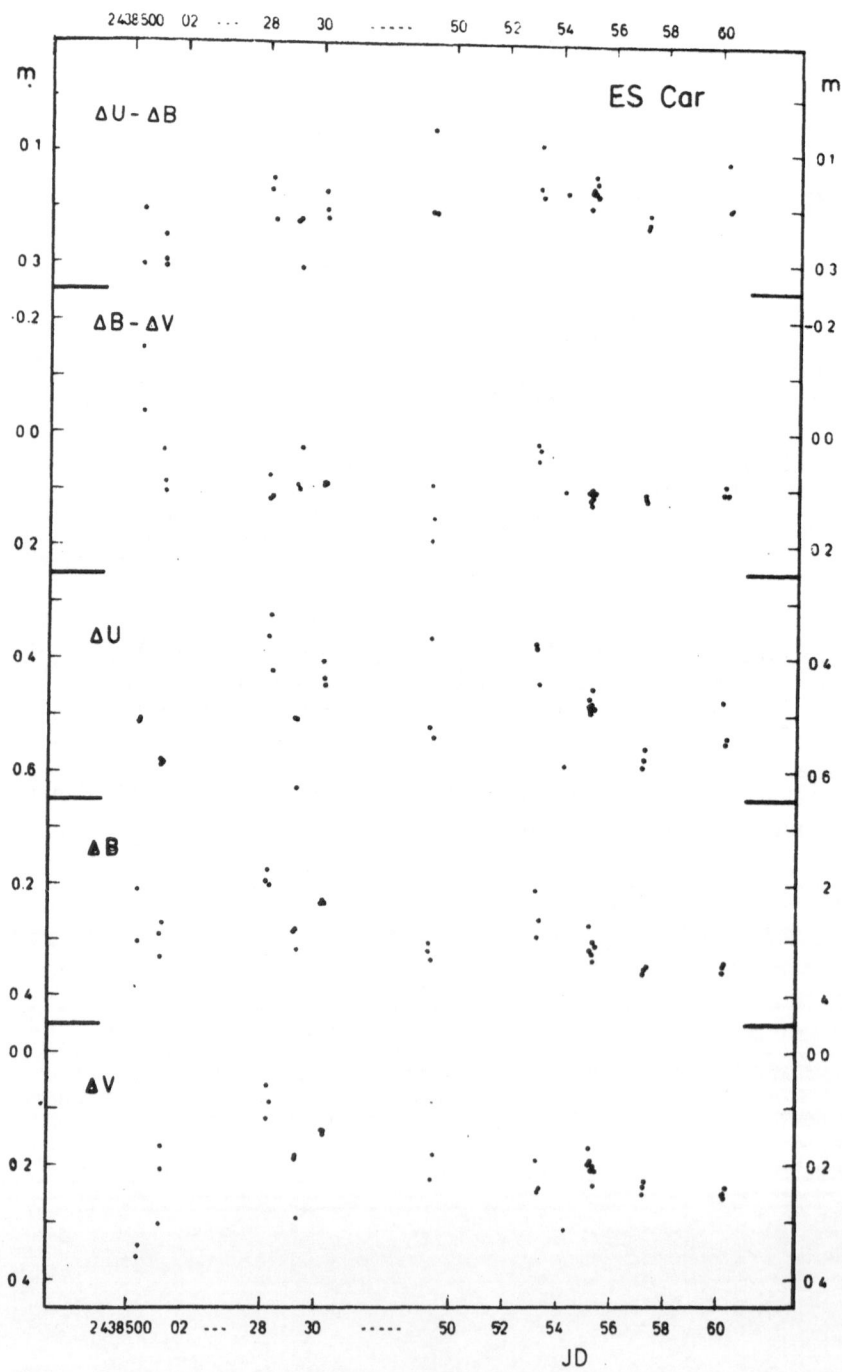

Fig. 5

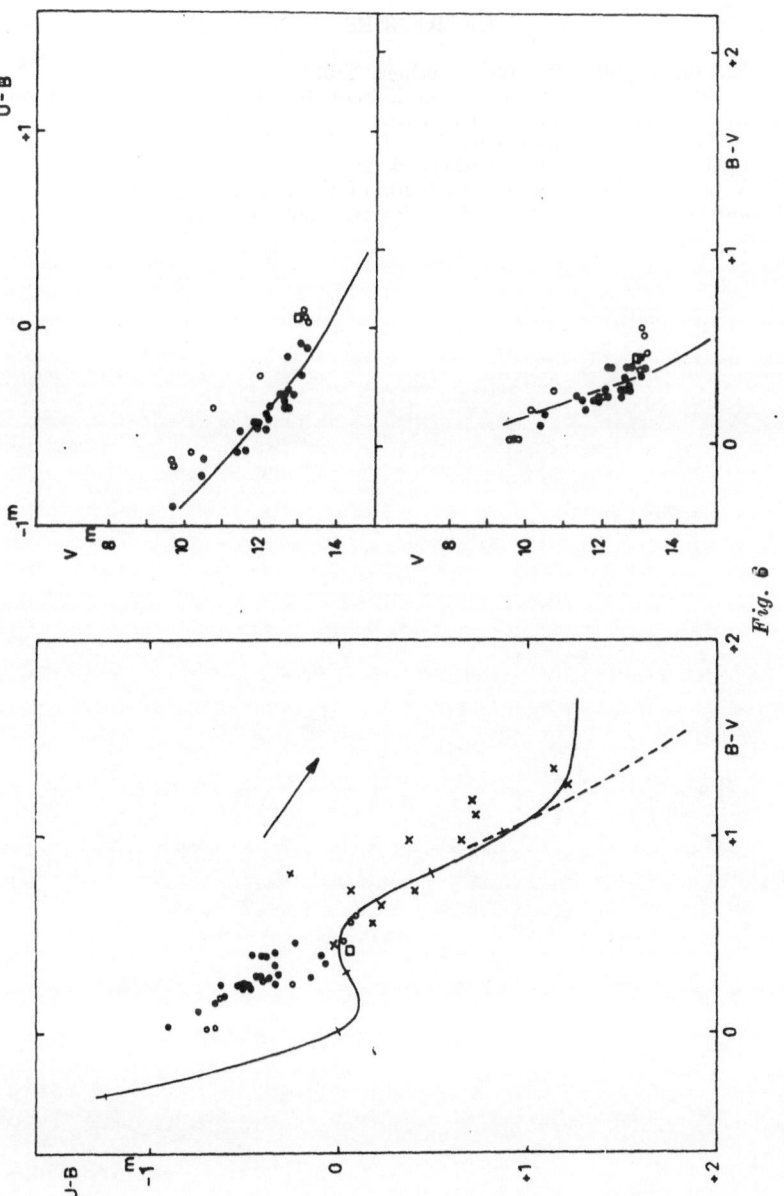

Fig. 6

We are indebted to W. S. Pretorius, former night assistant at Boyden Observatory, for carrying out most of the photoelectric observations.

REFERENCES

Hertzsprung, E., 1925, Bull. astr. Inst. Netherl. 210.
Hoffmeister, C., 1957, private communication, in Schneller, H., Geschichte und Literatur des Lichtwechsels der Veränderlichen Sterne, Vol. 4, Berlin.
Hoffmeister, C., 1958, Veröff. Sonneberg 3, 348.
De Kort, J., 1941, Bull. astr. Inst. Netherl. 9, 245.
Kukarkin, B. V. et al., 1958, General Catalogue of Variable Stars, Moscow.
Schmidt, H. and G. Diaz Santanilla, 1964, Veröff. Bonn No. 71.

HYDROGEN EMISSION PHENOMENA IN T TAURI STARS

L. ANDERSON and L. V. KUHI

University of California, Berkeley

INTRODUCTION

The sample studied includes the 25 or so T Tauri stars with apparent magnitude brighter than 13 visible from Lick Observatory. For photometric data, an FW130 with S20 response out to 7500 A was used, with exit slits of 48 A for $\lambda < 5500$ A and 64 A for $\lambda > 5500$ A. All observations were made on the 120-inch Lick telescope, except those marked 1965, which were made on the 200-inch telescope at Mt. Palomar. The reduction was carried out with a mean extinction curve for all nights, and using a new calibration of Vega (Hayes 1967). Data and reductions are most complete for the star AS209; this paper, as a preliminary report, shall deal exclusively with that star.

AS209 (1900: $16^h 43^m 6$, $-14° 13'$; $m_{pg} \sim 12$) is a T Tauri star which illuminates a bright nebulosity and has a strong emission spectrum including a UV excess shortward of $\sim \lambda 3700$ and the following lines:

The Balmer series out to $\sim$ H22 and Paschen to P14; very strong H and K and the infrared triplet of ionized Calcium;

the stronger multiplets of FeII and TiII;

λ 4063 and λ 4132 of FeI;

strong HeI λ 10830;

weak lines of MgI and other metals; and

very weak lines of [OII] λ 3727 and [SII] λ 4068

(these however do not appear on coudé plates obtained in 1967—68).

The absorption lines are all broadened and/or filled in by continuous emission so that no spectral type may be determined from them. Approximate UBV colors are V = 11.33, B—V = +1.2, U—B = —0.4 which indicates a type $\sim$ K5V but these are uncorrected for reddening.

THE HYDROGEN LINE SPECTRUM

Table 1 shows the Balmer decrement obtained from a 16A/mm Coudé spectrum (Apr. 3, 1961). The flatness of the decrement in the higher members and the steep rise to a very strong Hα implies that both collisional transitions and self absorption are important. This is further evidenced by Figure 1 which is a plot of Log_{10} Hβ/Hγ vs Log_{10} Hα/Hβ. Note the meandering over the diagram. The star was brighter in the visual in 1965 and 1966 than in 1967 and 1968.

The marks 5, 6, 7 on the reddening line show the location of the optical thickness line if the reddening is calculated from the ratio of Pγ/Hδ for the years 1965, '6, and '7 respectively. If the medium is optically thin in Pγ,

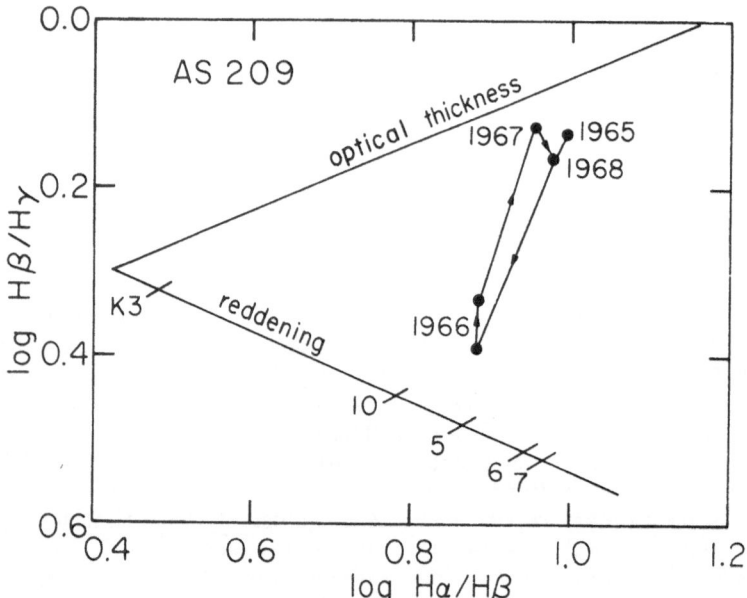

Fig. 1. Intensity ratios of Hα, β, and γ in AS209 from 1965 to 1968. The symbols along the reddening line are explained in the text.

Table 1

The Balmer Decrement (Correction factor for K5V)

Line	Hβ	γ	δ	8	9	10	12
Obs. Int.	45	90.4	60.3	32.8	35.0	32.9	25.4
Cor. Int.	34.9	21.3	9.1	3.3	3.2	3.0	2.3
1/2 width	6.5	5.07	5.27	5.27	4.09	3.45	3.70 A
Area	227	108	48	17	13	10	8.5
Decrement	1.000	.475	.211	.075	.057	.044	.037

Line	13	14	15	16	17	18	19
Obs. Int.	25.9	24.3	24.7	19.0	17.9	16.8	18.4
Cor. Int.	2.3	2.2	2.2	1.7	1.6	1.5	1.65
1/2 width	3.45	3.45	2.81	4.29	2.61	2.81	2.71 A
Area	7.9	7.6	6.2	7.3	4.2	4.2	4.5
Decrement	.035	.033	.027	.032	.018	.018	.020

Hδ, this ratio is useful since it is independent of such physical parameters as temperature, density, and gravity. However, it is obvious that using this ratio results in an estimation of the reddening which is far too large; therefore one may assume that the star is *not* optically thin in Hδ. The marks K0, K3 show the location of the optical thickness line when the 1966 data are corrected for reddening using the I(λ 4465) — I(λ 5556) color and assuming a K0V, K3V spectral type. If the spectral type remains constant, then some other agent, such as the presence of collisional transitions (Parker 1964) is responsible for the motion parallel to the reddening line.

SOME COMMENTS ON LINE BROADENING AND BLENDING AND THE SO CALLED "BLUE CONTINUUM"

I. M. Gordon (1957, 1958) has proposed that synchrotron emission in the infra red from selective "active zones" on the surface of the star may be responsible for the broadening and blending of the higher member Balmer lines into what has been referred to as the anomalous "blue continuum" of T Tauri objects. The polarized infrared emission induces transitions among the upper energy levels which decreases the mean life time and correspondingly increases the line width of emission resulting from transitions from those levels. This proposal is supported in AS209 by an observed infrared excess and a slight correlation between this excess and the variations in the ultraviolet

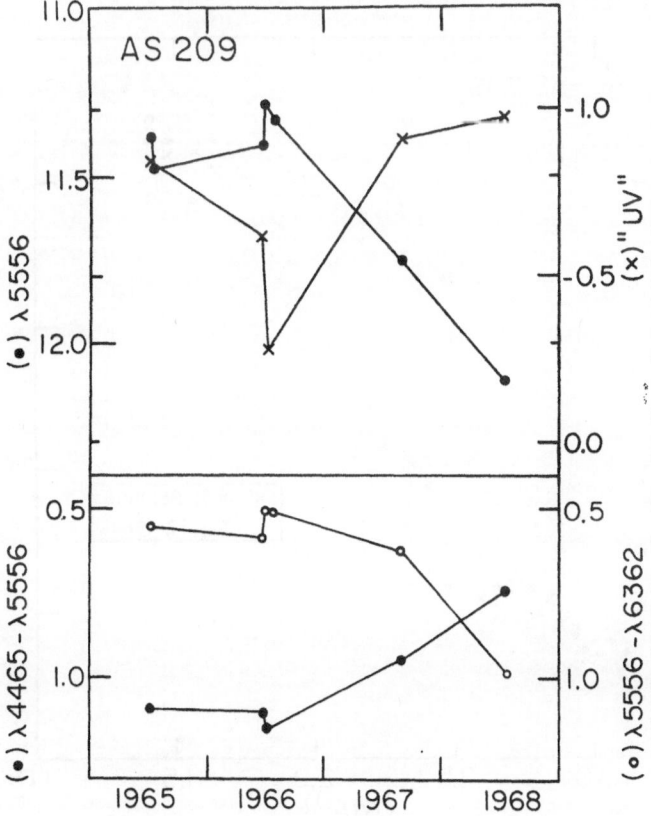

Fig. 2. Narrowband magnitudes and colors of AS209 from 1965 to 1968. "UV" is defined in the text and is a measure of the ultraviolet excess.

and blue continuum. However, one might expect to see radio emission as well as infrared from the synchrotron process, but this is not observed.

The authors prefer the simpler hypothesis of Böhm, that the blending is caused by turbulent motions, in combination with poor instrumental resolution. At the dispersion of 430 A/mm Böhm (1957) found the "blue continuum" shortward of λ 3760 and estimated that a turbulent velocity of 50 km/sec would be sufficient; with a dispersion of 16 A/mm, Kuhi has found that the "blue continuum" does not begin until λ 3690 and that the width of H8 indicates a turbulent velocity as large as ~ 100 km/sec (if no rotational velocity is present).

In reality, both Gordon's and Böhm's effects may be present, but the verification of the former must await polarization measurements in the infrared.

A NOTE ON THE Hα VS. UV EXCESS RELATION

T Tauri stars as a class seem to have a fairly well defined relation (of positive slope) between the intensity of Hα and the ultraviolet excess (Kuhi, 1966); however, individual examples, in particular AS209, have no such relation in their variations with time.

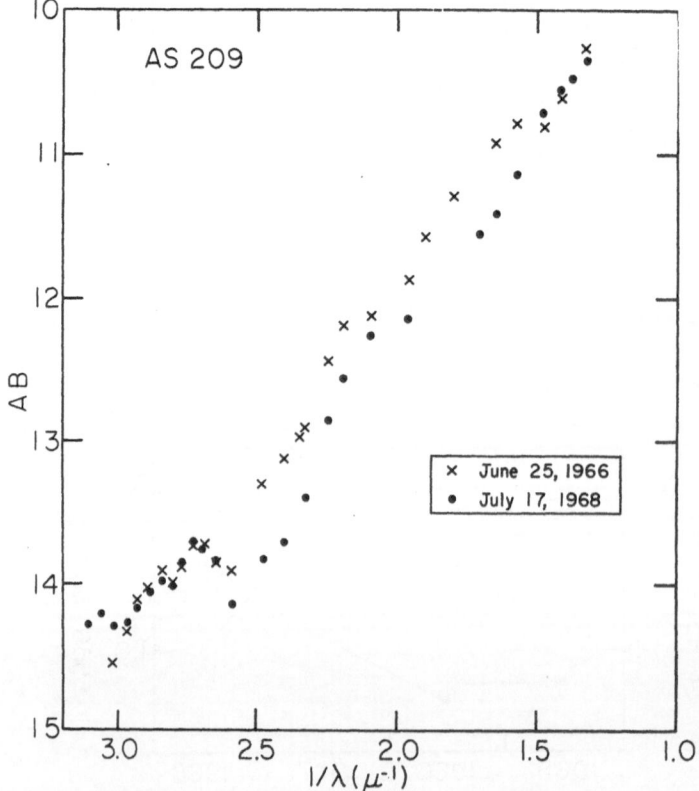

Fig. 3. Continuous energy distribution of AS209. AB gives the observed flux in mag. per unit frequency. No reddening corrections have been applied.

CONTINUUM OBSERVATIONS

Figure 2 is a plot of the variations in intensity (units: magnitudes/dν) for the following parts of the continuum:

    a) λ 5556 (approximately V; 64 A exit slit)

    b) λ 3620 — 2(λ 4032) + λ 4465 (a measure of the UV excess; 48 A exit slits)

    c) λ 4465 (48 A exit) — λ 5556 (64 A exit) (a measure of the blue color, $\sim$ B—V)

    d) λ 5556 —- λ 6362 (a measure of the red color, $\sim$ V—r; 64 A exit slits)

These plots show that the star is redder in I (λ 5556) — I(λ 6362) *and* bluer I(λ 4465) — I(λ 5556) and has a larger UV excess when it is fainter in the visual, i.e. I(λ 5556).

There are some peculiar apparent inconsistencies between figures 1 and 2. In 1965 and 1966 the I(λ 4465) — I(λ 5556) color is about the same, but the visual [I(λ 5556)] is higher in 1966 and the UV excess is less. If one then assumes that collisional excitation changes are responsible for these differences and the position changes on the log — log plot (figure 1), it becomes difficult to explain the observation that the star is *bluer* in 1967 and 1968 than in 1965 but has the same Hβ/Hγ, Hα/Hβ values. In 1968 I(λ 4465) — I(λ 5556) =

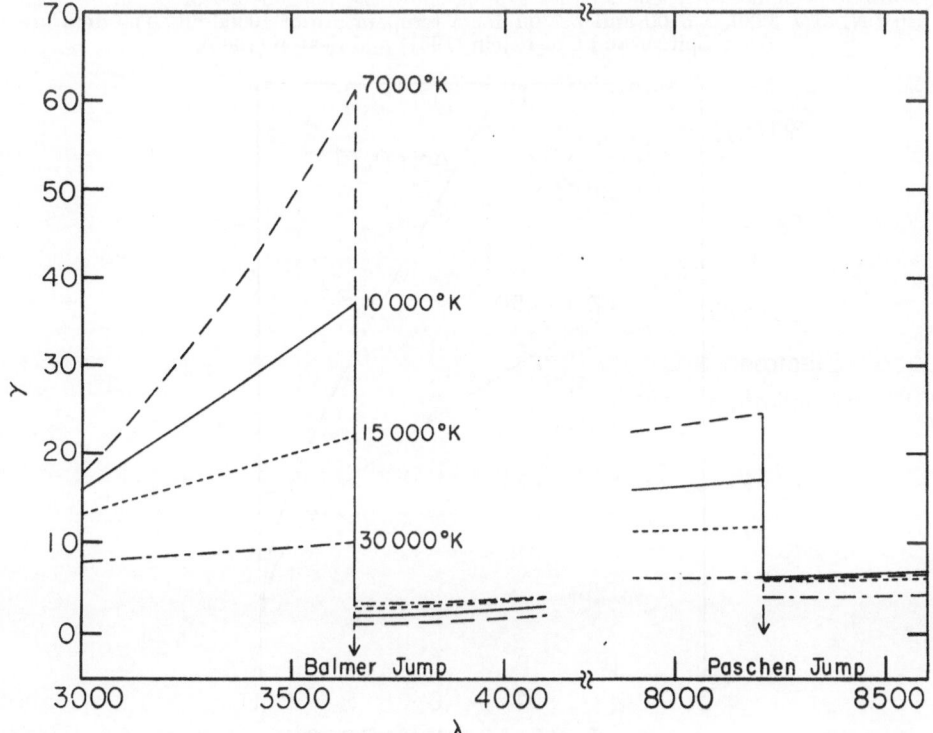

Fig. 4. Continuous emission of hydrogen (free-free and free-bound component) for Case B. The energy emitted is $N_p N_e \gamma d(h\nu)$. The units of γ are 10^{-14} cm^3—sec^{-1}; λ is in Angstroms

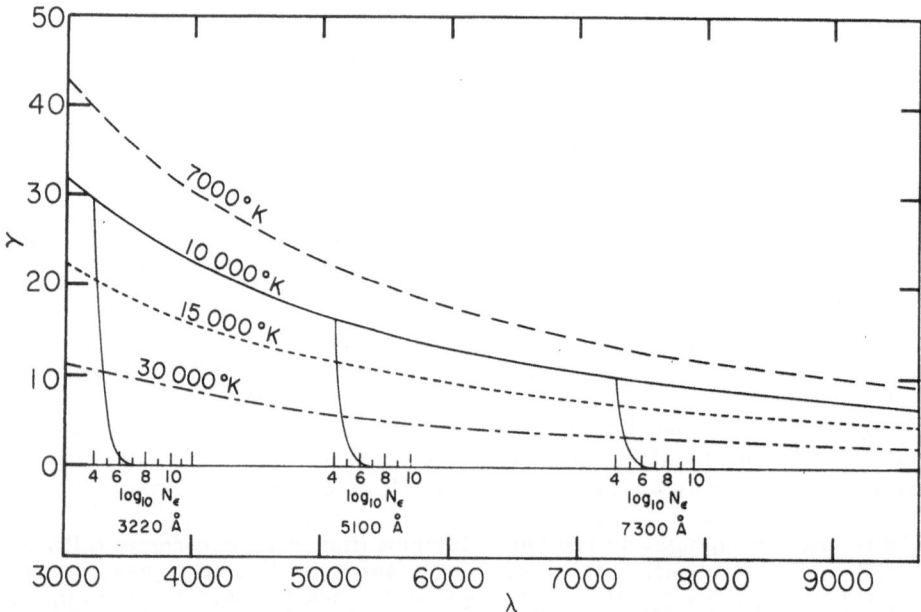

Fig. 5. Continuous emission of hydrogen (two-photon 2s → 1s component). The units of γ are 10^{-14} cm^3-sec^{-1}. The smaller graphs show the dependence of γ on electron density N_e at λ 3220, λ 5100 and λ 7300 for a temperature of 10000 °K. The data are from Spitzer and Greenstein (1951) and Seaton (1960).

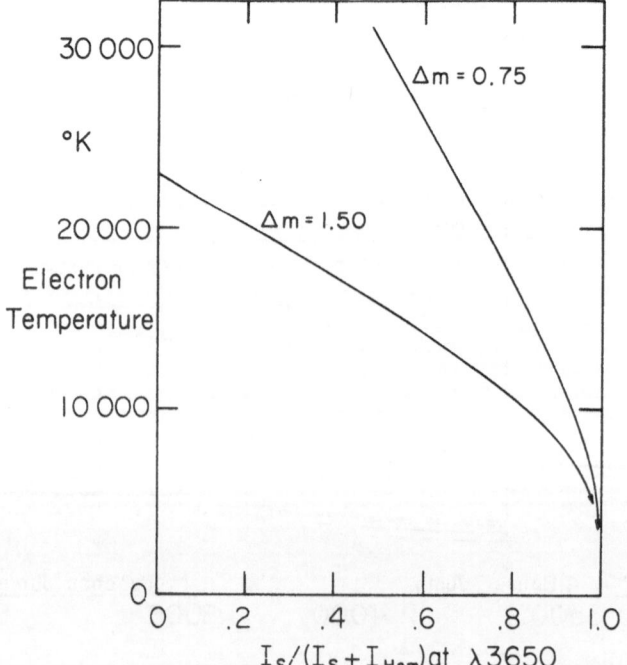

Fig. 6. Electron temperature versus the ratio of I_s to $I_s + I_{Hem}$ at λ 3650 for Balmer discontinuities of $\Delta m = 0.75$ and 1.50.

= .75 as opposed to 1.1 in 1966; if the star actually became hotter (to a K0), it would have required no reddening correction (which is unlikely), and would have become brighter, but I(λ 5556) is 0.6 magnitudes fainter. It should be noted that none of the continuum changes correlate with the changes in optical depth in figure 1. Note also that I(λ 5556) — I(λ 6362) is redder in 1968 than on any previous date which implies a later spectral type than when the star was brighter and reddening [from I(λ 4465) — I(λ 5556)] was greater.

THE ULTRAVIOLET EXCESS

Figure 4 shows theoretical curves for the free-bound and free-free continuous emission of hydrogen near the Balmer jump, in the case where the medium is optically thick in the Lyman region of the spectrum (Menzel and Baker: Case B). Figure 5 shows the hydrogen 2s → 1s two-photon transition continuum; shown are the wavelength dependence of the energy coefficient, γ, for the density $N_e = 10^4$ cm^{-3}, and the density dependence for the temperature $T_e = 10^4$ °K at three selected wavelengths. The formula for the two-photon emission contains a factor, X, which varies from .32 to 1.0 depending on what rôle 2s → 2p collisions play in depopulating level 2s; X is assumed equal to .32 in this paper.

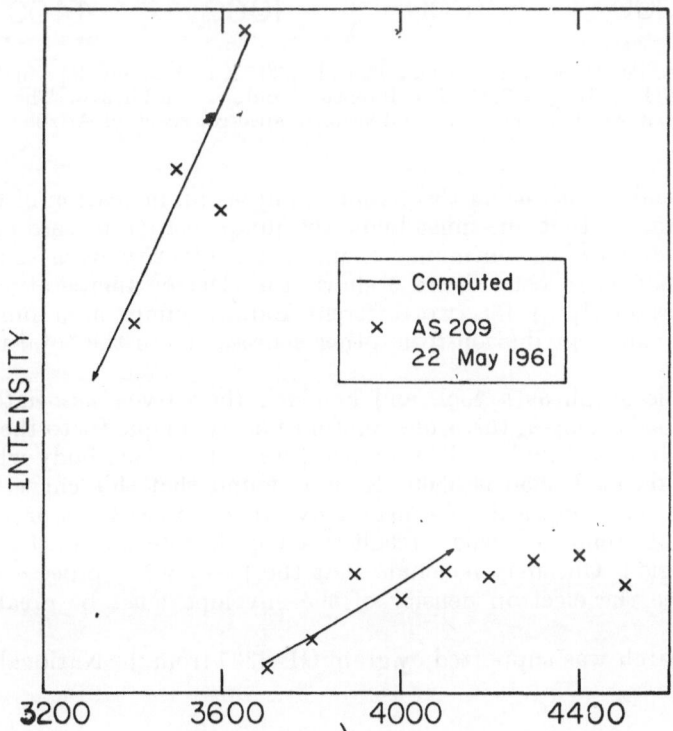

Fig. 7. Computed curve for an underlying star of 3500 °K and an envelope of 7000 °K. The ratio I_s to $I_s + I_{Hem} = 0.94$. The intensity is in arbitrary units. The observed points were obtained from 16A/mm coudé spectra of AS209.

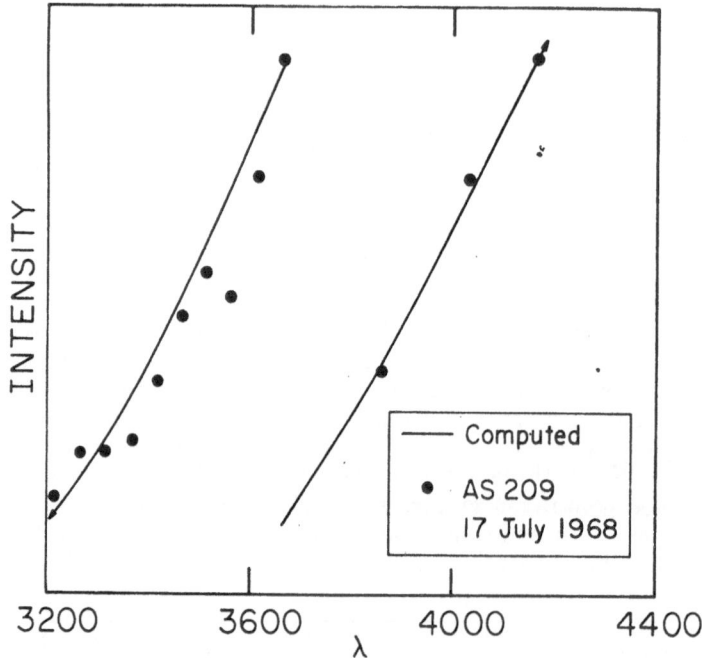

Fig. 8. Computed curve for an underlying star of 3500 °K and an envelope of 10000 °K. The ratio I_s to $I_s + I_{Hem} = 0.94$. The intensity units are arbitrary. The observed points were obtained from photoelectric spectral scans of AS209.

The difficulty with using the Balmer jump as an indication of the temperature for stars is that one must know the jump relative to zero intensity; if there is an underlying continuum of unknown intensity from some other source, this relation is lost. Figure 6 shows the electron temperature of the hydrogen emission (I_{Hem}) for two different Balmer jumps as a function of the fraction of the contribution from other sources (I_s) to the total intensity at λ 3650.

Using this graph as a tool, and knowing the curves associated with various black body temperatures, one can find a fairly unique fit to the observed continua (figures 7 and 8). The temperature of the black body which has the largest $dI/d\lambda$ at λ 3650 is 3500 °K; it is found that this curve best fits the continuum just longward of λ 3645. Any other temperature, or the inclusion of two-photon emission, would result in a slope less steep than that observed in this region. One may conclude that the two-photon process plays no rôle, and hence the electron density of the envelope must be greater than $5 \times 10^5 \, \mathrm{cm}^{-3}$.

This research was supported by grant GP-6337 from the National Science Foundation.

REFERENCES

Böhm, K. H., 1957, Z. Astrophys., **43,** 245.
Gordon, I. M., 1957, Astr. Zu., **34,** 739 (1957, Soviet Astr. **1,** 719).
Gordon, I. M., 1958, Astr. Zu., **35,** 458 (Soviet Astr. **2,** 420).
Hayes, D., 1957, Dissertation, UCLA.
Kuhi, L. V., 1966, Publ. astr. Soc. Pacific, **78,** 430.
Parker, R. A. R., 1964, Astrophys. J., **139,** 208.
Seaton, M. J., 1955, Mon. Not. R. astr. Soc., **115,** 279.
Seaton, M. J., 1960, Rep. Prog. Phys., **23,**
Spitzer, L. and Greenstein, J. L., 1951, Astrophys. J., **114,** 407.

THE EVIDENCE FOR VARIABLE INFALL OF MATERIAL IN THE ULTRAVIOLET EXCESS STARS*

MERLE F. WALKER

Lick Observatory, University of California+

In a previous paper (Walker 1966), the results of the writer's spectro-scopic study of the UV excess stars in the Orion Nebula and in NGC 2264 were outlined. These stars form a subgroup of the T Tau class of variables and like the T Tau stars are extremely young objects still in the process of contract-ing gravitationally from the pre-stellar medium. Apart from the UV excess itself, the spectra of the UV excess stars are similar to those of normal T Tau variables, having emission lines of hydrogen, Ca II, and sometimes helium, Fe I, and an underlying late-type absorption spectrum which is partially totally obscured by a blue continuum. However, unlike the regular T Tau variables which occasionally display a P Cyg spectrum with violet-displaced absorption lines, 10 out of a sample of 23 UV excess stars showed, at least at times, an inverse P Cyg or "YY Ori" spectrum. In these objects, the emission lines tend to have approximately the radial velocity of the cluster, while redward-displaced absorption lines of hydrogen and sometimes Ca II are observed having radial velocities of 150 to 400 km/sec more positive than the cluster velocity. Explanation in terms of binary motion appears ruled out by the fact that except for observations of YY Ori on two dates, the absorp-tion lines are always displaced to the red. Thus, it has been assumed that whereas the P Cyg spectrum in regular T Tau stars indicates ejection of mate-rial from the star, the inverse P Cyg spectrum in the UV excess stars indicates actual infall of material.

The fact that the inverse P Cyg spectrum is observed only among the UV excess stars suggests that the UV excess itself results in some way from the infall, while the fact that the phenomenon tends to occur among the intrin-sically brighter UV excess stars suggests that the interaction of the infalling material with the star causes the system to brighten, the stars with the larger amounts of infalling material being affected the most. If this interpretation is correct, then we might expect that if the rate of infall is variable, a correla-tion ought to exist between the brightness of the star and the intensity of the inverse P Cyg absorptions.

To investigate this question, spectra were obtained of SU Ori, one of the more rapid light-variables among the UV excess stars showing the inverse P Cyg lines. Table 1 lists the spectroscopic and photometric observations of this star. The spectra were obtained with the prime-focus spectrograph of the 120-inch reflector and a grating and camera giving a dispersion of 96 Å/mm.

Contributions from the Lick Observatory, No. 292

+ On leave, during 1968—1969, at Cerro Tololo Inter-American Observatory, La Serena, Chile.

The spectrum was recorded on baked Kodak IIa0 plates or film. The photometric observations were derived from photovisual plates taken simultaneously by the observer at the 20-inch Carnegie astrograph, or by Mr. Harlan using a yellow-corrected aerial camera lens. The magnitudes of the variable were obtained by measuring the plates in a Sartorius photometer, using as standards stars in the region for which photoelectric observations were available (Walker 1968).

Table 1

Spectroscopic and Photometric Observations of SU Ori

Plate No.	Date (UT)	Exp. (min)	V_r (km/sec) em.	V_r (km/sec) abs.	V (mag)
ES—384	Jan. 27, 1963	165	-40 ± 8	$+332 \pm 14$	14.5
ES—926	Nov. 30, 1964	206	$+ 3 \pm 2$	———[1]	15.3
ES—936*	Jan. 27, 1965	107	$+ 1 \pm 13$	———	15.3
ES—950	Jan. 30, 1965	330	$+53 \pm 3$	———[1]	15.0
ES—953	Jan. 31, 1965	285	$+39 \pm 5$	$+382 \pm 11$	14.5—15.0
ES—1206	Jan. 19, 1966	257	$+17 \pm 7$	———[1]	15.6

* Plate underexposed, only emission lines visible. Omitted from Figure 1.
[1] Absorption spectrum absent.

The five best spectra are reproduced in Figure 1, where they are arranged in order of decreasing brightness. The observations show that there is indeed a correlation between the brightness of the star and the presence and intensity of the redward-displaced absorption lines. A similar result is suggested by the existing observations of XX Ori, listed in Table 2. The data for this star are less satisfactory since only three plates are available and since the photometric observations consist merely of relatively crude visual estimates at the telescope, comparing the variable to other stars in the field. Nevertheless, it again appears that the inverse P Cyg spectrum disappears when the star becomes faint. Thus, the observations appear to confirm the hypothesis that material is falling into these stars at a variable rate and that an increase in infall causes the system to brighten.

Table 2

Spectroscopic and Photometric Observations of XX Ori

Plate No.	Date (UT)	Exp. (min)	V_r (km/sec) em.	V_r (km/sec) abs.	V (mag)
ECL—218*	Nov. 21, 1962	120	$+50 \pm 5$	———[1]	14.6
ES—312	Nov. 22, 1962	124	$+20 \pm 6$	$+333 \pm 9$	14.6
ES—402	Feb. 27, 1963	127	$- 6 \pm 3$	———[2]	15.1

* Plate taken with Lallemand electronic camera and coudé spectrograph; dispersion 48 A/mm.
[1] Absorption spectrum present, but too weak to measure.
[2] Absorption spectrum absent.

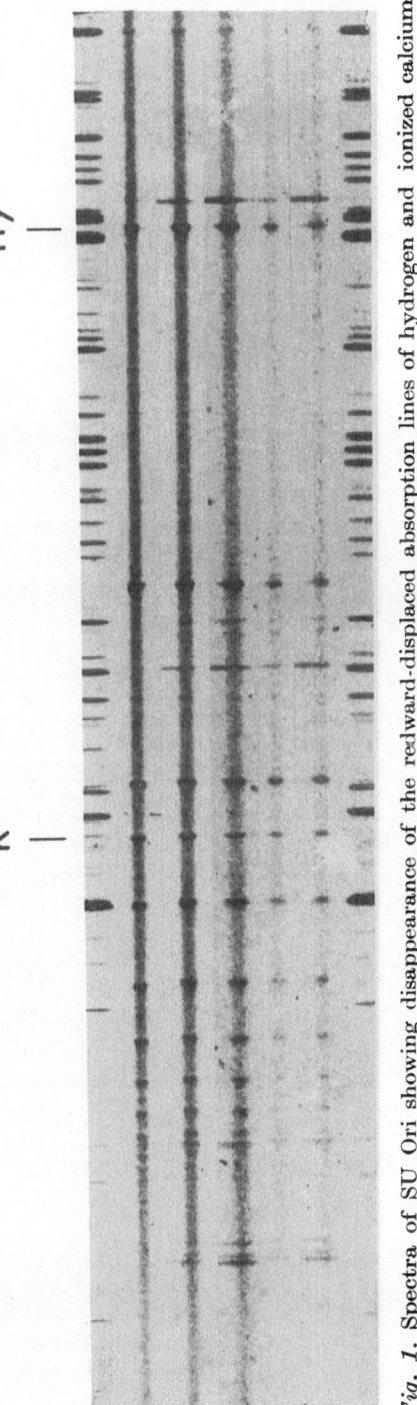

Fig. 1. Spectra of SU Ori showing disappearance of the reelward-displaced absorption lines of hydrogen and ionized calcium with decreasing brightness of the star. From top to bottom the plates and magnitudes are: ES−384, $V = 14.5$; ES−953, $V = 14.5−15.0$; ES−950, $V = 15.0$; ES−926, $V = 15.3$; ES−1206, $V = 15.6$.

That complications to this simple picture exist is shown by the radial velocity measurements. The large negative radial velocity of the emission lines of SU Ori on plate ES-384, when the "YY Ori" lines are strong, might be explained by encroachment of the absorption lines onto the red side of the emission features. However, Table 1 shows that considerable variation of the emission lines occurs which is *not* correlated with the brightness of the star or the intensity of the inverse P Cyg lines.

The nature of these stars is clearly very complex, and their observation is difficult owing both to their faintness (the brightest of them is about $B = 14$) and to the irregular nature of their variations; Table 1 shows that observations of SU Ori had to be continued over three observing seasons before plates covering a large range in magnitude could be obtained. Thus, a long period of study with large telescopes will be required before we will possess the necessary observational data for an understanding of this very interesting stage in the gravitational contraction phase of stellar evolution.

REFERENCES

Walker, M. F., 1966, Stellar Evolution. Ed. Cameron, A. G. W., and Stein, R. F., New York: Plenum Press, p. 405.
Walker, M. F., 1968, Astrophys. J., (in press).

EMISSION Hα-LINE PROFILES IN SEVERAL T TAU STARS

E. A. DIBAJ and V. F. ESIPOV

Sternberg State Astronomical Institute, Moscow

The emission Hα-line profiles in T Tau stars were observed by L. Kuhi, who proposed an expanding mass-loss envelope model (Astrophys. J. *140*, 1465, 1964; *143*, 991, 1966).

In October 1967 observations of eigth stars associated with nebulosities were carried out with the 50-inch reflector using an image-tube spectrograph (dispersion 25 A/mm, resolution 1 A).

Fig. 1—9 show the Hα-line profiles for the stars T Tau, RY Tau, FU Ori, V 380 Ori, R Mon, Z CMa, BD + 40°4124, Lk Hα 215 and for comparison the absorption Hα-line in an A0V type spectrum.

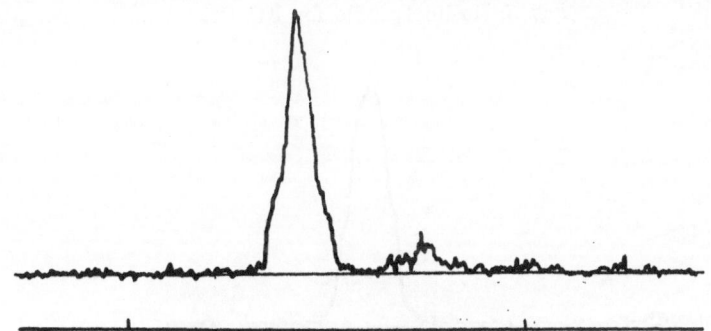

Fig. 1. Hα-line profile for T Tau

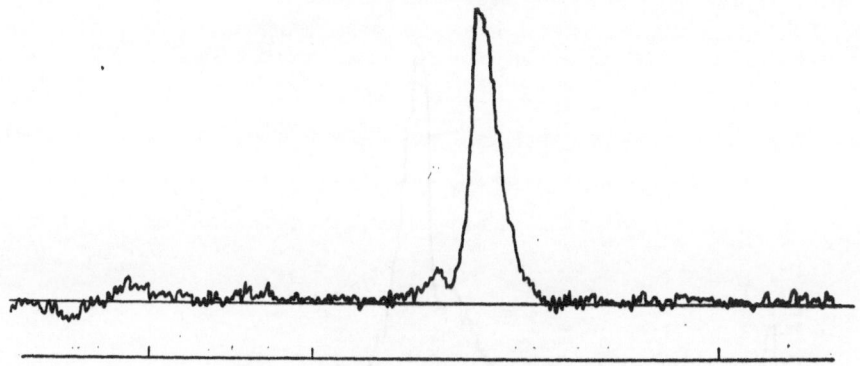

Fig. 2. Hα-line profile for RY Tau

It appears that motions in the mentioned stars are more complicated than pure expansion. We observe the expanding envelopes with self-absorption (FU Ori, Z CMa), the rotating envelopes ("pole-on-star" V 380 Ori and "shell-star" Lk Hα 215) and intermediate cases (T Tau, RY Tau, R Mon BD + 40° 4124).

The authors are planning further observations to detect possible spectral variability.

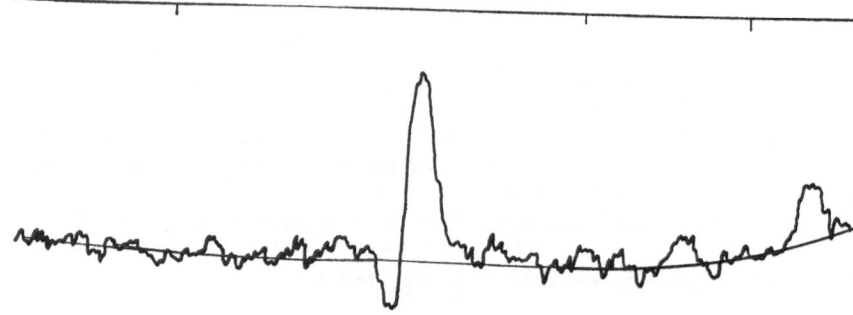

Fig. 3. Hα-line profile for FU Ori

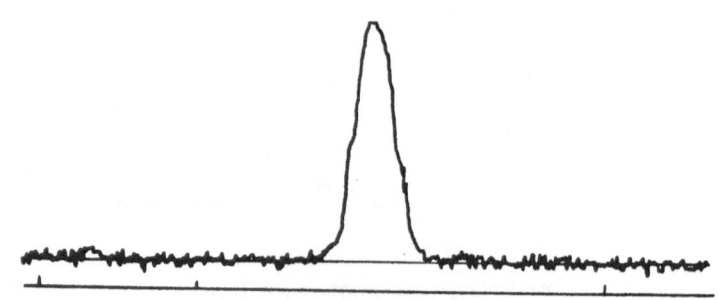

Fig. 4. Hα-line profile for V 380 Ori

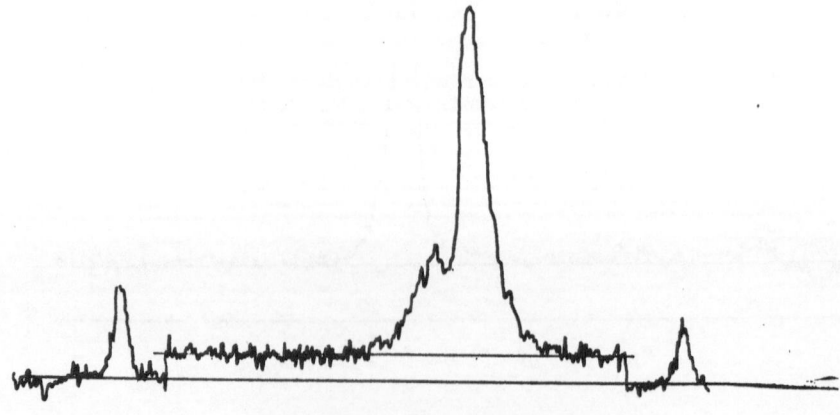

Fig. 5. Hα-line profile for R Mon

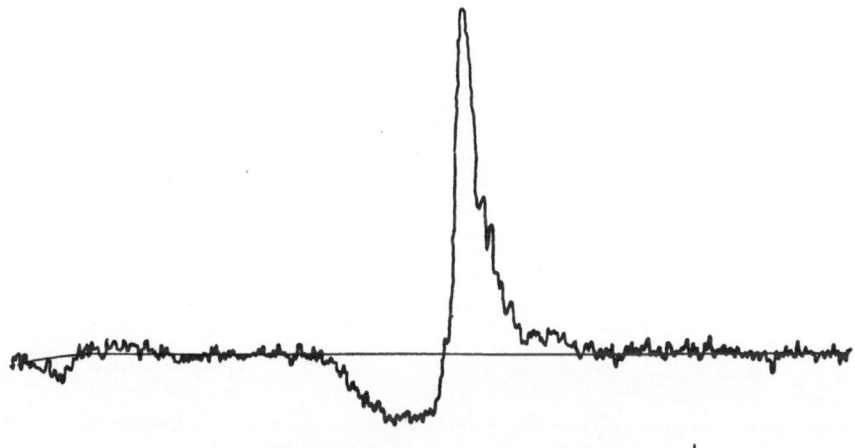

Fig. 6. Hα-line profile for Z CMa

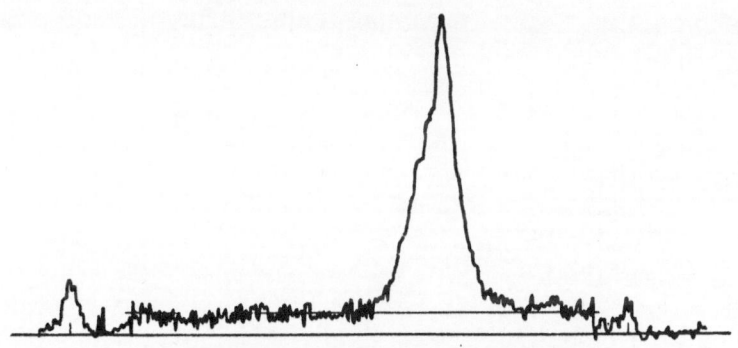

Fig. 7. Hα-line profile for BD+40°4124

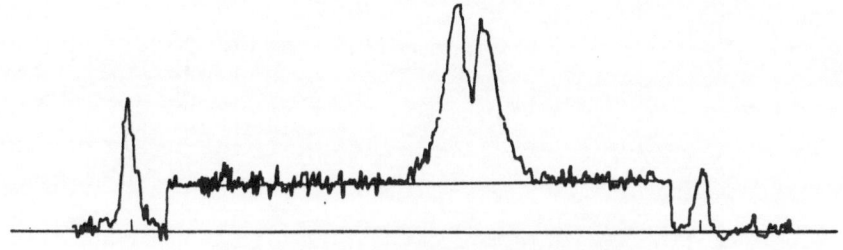

Fig. 8, Hα-line profile for Lk Hα 225

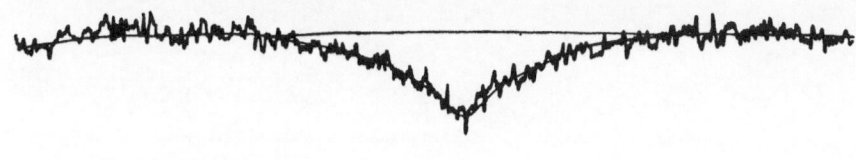

Fig. 9. Absorption Hα-line in an A0V type spectrum

FLARES OF UV CETI TYPE STARS

Introductory Paper by

R. E. GERSHBERG

Crimean Astrophysical Observatory, USSR

The term "flare stars" is used sometimes as a synonym to "eruptive stars" and in that case the term "flare" covers a wide range of phenomena of stellar variability. I intend to give a review of observational and theoretical results bearing on the classical flare stars of UV Ceti type only and I shall use the words "flare" and "flare stars" only in that limited sense. Owing to the restricted time, I have no possibility to give the detailed history of the investigations of the UV Ceti type stars. This history can be found in Joy's (1960), Oskanjan's (1964) and Haro's (1968) reviews — therefore I shall submit the state of the problem only for the present moment. That is why I shall not refer to a number of investigations which were important for their times but were surpassed by following studies.

The dMe-objects with quick flares of brightness are attributed to classical flare stars of UV Ceti type. Today no spectral or photometric criteria are known which would permit to establish the relation of a dMe to the UV Ceti type by observing it in a quiet state. About 25 UV Ceti type stars are at present known, and they make nearly a quarter of the known dMe stars and about 5 per cent of all the dM objects; because M dwarfs represent to about 80% per cent of galactic stellar population, one may suppose that the flares of UV Ceti-type stars are the most wide-spread kind of stellar variability.

Among the 25 UV Ceti type stars 19 are known as binaries. 3 of them are spectroscopic binaries, 2 of them have distances less than 1″ between the components; in the other 14 visual binaries the fainter components are flare stars. The masses of flare stars are small: the mass of UV Cet itself is equal to $0.04M_\odot$, that is less than a minimum mass of a main sequence star; the mass of EQ Peg is equal to $0.13M_\odot$, and that of DO Cep to $0.16M_\odot$ (Petit, 1961). The diameters of flare stars are about 3 times less than that of the Sun (Lippincott, 1953). The luminosities of these objects are low, and the absolutely faintest star, van Bisbroeck's object, BD $+4°4048$ B, $M_v = 18.^{m}6$, is a flare star. But we are not certain that there exist systematic differences in masses, sizes, luminosities and percentage of binaries between flare and normal M-dwarfs. The dispersion of the peculiar velocities of dMe and UV Ceti type stars is 2—2.5 times less than that of normal dM stars (Gliese, 1958).

After giving this short stellar statistical characterization of UV Ceti type stars, we may pass to discuss the flares themselves.

OBSERVATIONS

In accordance with the topic of our Colloquium it is necessary to begin with the time features of flares.

Time distribution of flares

For nearly 20 years there had been a belief that the flares of UV Ceti type stars occurred irregularly. But Andrews (1968) found some recurrence in the time distribution of 9 flares of YZ CMi: 2 intervals between flares were near to 122^h, 3 near to 73^h and 3 near to 47^h; later Andrews found the same effect with a characteristic interval near 48^h for flares of V 1216 Sgr. A closeness of all these quasiperiods to values wich are divisible by 24^h supposes a possible effect of observational selection.

The most detailed consideration of a possible periodicity of flares has been carried out by Chugainov: he has studied the time distribution of 28 flares which were registered during a cooperative observation of UV Cet organized by Lovell at several observatories. Chugainov has found as the best periodic representation of maximum flare moments:

$$T_{max} = \text{const} + 0\overset{d}{.}1821 \times E.$$

11 periods of this cycle are equal to $48\overset{h}{.}1$. But the deviations, O—C, are large: $\overline{O—C} = 43^m$ and $(O—C)_{max} = 99^m$. The registered flares have occurred not in all, but only in 70 per cent of "critical moments"; but that is not a contradiction to the hypothesis of periodicity of flares: the remaining 30 per cent of flares could have small amplitudes or occurred on the opposite side of the star. The arguments against the periodicity hypothesis are the large $\overline{O—C}$, a value which is close to 1/6 of the period proposed, and a possibility to represent the observable time distribution of flares as a Poisson distribution. This year the Working Group on Flare Stars organized several cooperative observations of UV Ceti type stars with attemps to realize a 24^h photometric patrol. We hope to receive an important information on the time-distribution of flares from these observations, but their discussions have not yet been finished.

Nearly 15 years ago Oskanjan (1964) has found variations of the flare activity level of UV Cet from season to season. A list of photoelectric observations of this star made by the end of 1967 is given in Table 1 (Gershberg and Chugainov, 1968). It is seen that the mean monitoring time per flare spent by different observers varies from $4\overset{h}{.}1$ to 47^h. But this table does not permit to reach a final conclusion: first, using different telescopes and different

Table 1.

List of photoelectric observations of UV Ceti

Observer	Season	Telescope	Spectral region	Total monitoring time (hours)	Number of flares registered	Mean monitoring time per flare (hours)
Roques	1952	12″ refractor	without filter	94	2	47
Chugainov	1963	64 cm meniscus	V	25	3	8.4
Chugainov	1964	telescope	V	47	4	12
Chugainov	1965		V	70	17	4.1
Chugainov	1966	70 cm reflector	H_β	49	12	4.1
Eksteen	1966	16″ reflector	V	24	3	8.0
Chugainov	1967	64 cm meniscus telescope	V	35	8	4.4

Table 2.

Different criteria of the flare activity level of UV Ceti

Season	Number of flares registered	Mean monitoring time per flare (hours)	Mean radiative energy of flare in V-region (ergs)	Ratio of radiative flare energy in V-region to the stellar radiation in V during monitoring
1963	3	8.4	9.3×10^{30}	0.0060
1964	4	12	3.2×10^{30}	0.0014
1965	17	4.1	9.0×10^{30}	0.012
1967	8	4.4	8.3×10^{30}	0.012

spectral bands we have different thresholds of flare detection; second, it is not clear whether a mean monitoring time per flare can characterize a flare activity level. In order to clear up these points, let us consider Chugainov's observations of UV Ceti which were carried out for 4 years with the same instrumental and photometric system. Three different criteria of the flare activity level are given in Table 2: the mean monitoring time per flare, the mean radiative energy of a flare and the ratio of the radiative energy of flares to the radiative energy of the star calculated by integrations over the monitoring time. These data show the reality of the flare activity level variations and detect some correlation between different criteria of this level.

Before finishing the discussion of time characteristics of UV Ceti type star flares and going to photometric characteristics, it is necessary to note, that observations carried out by different instrumental methods give us results which are difficult to compare. As seen from Table 3, even experienced visual observers overestimate systematically the amplitudes of flares registered and miss small flares. On the other hand, observations in UV region have a threshold of flare detection threetimes lower than those in blue and 9 times lower than those in visual region (Kunkel, 1967). This point complicates the statistical discussion of flare features.

Table 3.

Comparison of the results of simultaneous visual (Odessa) and photoelectric (Crimea) monitoring of the brightness of UV Ceti

Date	U. T.	M_{vis}	m_v
19.9.65	21^h03^m		0.9
20.9.	00 18	2.1	1.0
	00 52		0.35
22.9.	22 59		0.4
23.9	23 46	1.1	0.65
24.9	00 32	2.9	1.9
26.9	00 47	4.0	≥ 1.5
28.9	00 11		0.4
	21 09	2.1	1.15
1.10.	21 12	2.3	1.4
2.10.	21 57		0.4
	23 54	4.2	1.7

Table 4.

Comparison of the observed and calculated Balmer decrements according to Kunkel
(1967)
The observations of EV Lac flare on 11.12.1965

U. T.	Hβ	Hγ	Hδ	Hζ	Hη	H$_{10}$	H$_{11}$
3h55m	1.0	1.24	1.48	1.22	1.17	0.94	0.80
4 00	1.0	1.04	1.16	0.92	0.63	0.64	0.47
4 03		1.10	1.28	1.10	0.90	0.67	0.59
4 08	1.0	1.13	1.06	0.76	0.54	0.52	0.38
4 56	1.0	1.15	0.90				

Photometric characteristics and energetics of flares

Light curves of UV Ceti type star flares are very asymmetrical: as a
rule, after a very quick increase of brightness there is a sharp, momentary
maximum which is followed by a smoother decay (see Figs. 4 and 6). According
to statistics (Gershberg and Chugainov, 1968) which is based on the discussion
of about 100 photoelectric light curves, the time of flare growth is 10 to 30
sec for the half of the flares and 3 to 100 sec for 90 per cent of the flares. The
time of photometric decay of flares is 10 to 100 times as large as that of flare
growth, but, as a rule, the rate of increase of energy output just before the
maximum is only 2 to 3 times as large as the rate of decrease of energy output
immediately after the maximum. Then the flare decay slows down and such
details as secondary maxima and steps of constant brightness appear on the
light curve. Strong secondary maxima occur usually 5—10 min later than the
main maximum and the light curve of secondary maximum is more symmetrical; this photometric feature can be regarded as a criterion to distinguish two
close flares from a flare with a secondary maximum. As a rule, on the ascending branch of the light curve — in contrast to the descending branch — no
deviations from a monotonic growth of brightness are seen. Often, but not
always, a slow brightening appears some minutes before the sharp beginning
of the flare and the amplitude of such a slow brightening amounts to several
tenths of a stellar magnitude. Flares of UV Ceti type stars are known with
amplitudes up to 3—4 magn. Of course, the lower limit of flare amplitudes
is determined with the precision of photometric observations. The behavior
of UV Ceti type stars outside the flares is not clear up to now; the observers,
who were monitoring the brightness of these stars visually and photographically, sometimes noted small and slow variations of brightness with amplitudes up to 0.3—0.5 magn. and with a characteristic time close to half an
hour; but such secondary brightness variations were not confirmed by special
photoelectric observations.

According to Gershberg and Chugainov (1968) and Kunkel (1967) the
total radiation of flares of the most active UV Ceti type stars amounts to
0.1—1 per cent of the energy of the radiation of these stars outside the flares.

For the best studied 4 flare stars the distributions of flares according
to their energy of radiation (L) are given in Fig. 1. One sees that the total
energy of flare radiation in blue region amounts to $3 \times 10^{31 \pm 2}$ ergs and more
than half of the flares radiate $10^{31 \pm 1}$ ergs. Fig. 1 permits to conclude that an

absolutely brighter star shows stronger flares on the average and certainly this conclusion can not be due to the observational selection effect.

For the same 4 stars the distributions of flares according to their absolute rates of increase of energy output before maximum $\left(\dfrac{dl}{dt}\right)$ are given in Fig. 2. In all investigated cases these rates were within the limits 10^{27} and

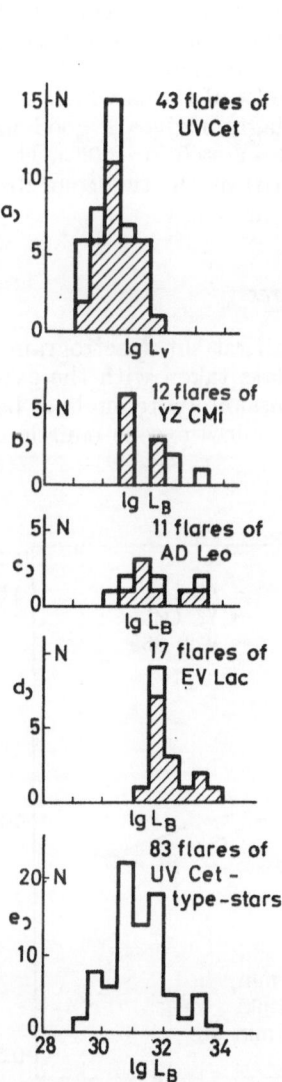

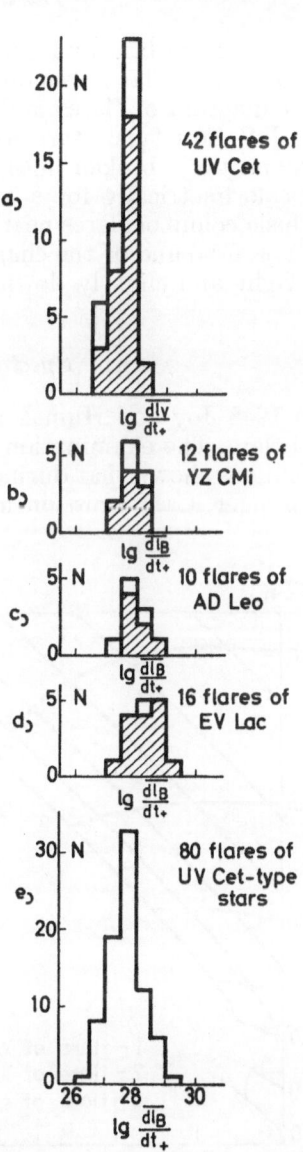

Fig. 1. Flare distributions according to their total radiative energy for 4 UV Ceti stars. Non-dashed districts are less certain data.

Fig. 2. Flare distributions according to their absolute rates of energy output increase before maximum for 4 UV Cet stars. Non-dashed districts are less certain data.

3×10^{28} ergs/sec². The narrowness of these hystograms should be noted, they
are 2—3 times narrower than the previous ones. It is suspected that the brighter the star is, the slower are the flares on an average, but we did not find any
correlation between the total radiative energy of individual flares and their
rate of increase.

Intrinsic colors of flares

The most certain and complete information on the intrinsic colors of
UV Ceti type star flares was obtained by Kunkel (1967). By using his data a
two-color diagram of flares is drawn in Fig. 3: the location of several flares
of three UV Ceti type stars near their maxima are marked with different
symbols and three broken lines represent the tracks of flares which could be
studied colorimetrically for a long time. This diagram gives a good idea of
the intrinsic colors of flares near their maxima (B—V $\approx 0.^m0 \pm 0.^m3$, U—B $\approx$
$\approx -1.^m1 \pm 0.^m2$) and of the character of a flare drift on the two-color diagram
(to the right and slightly downwards) during their decay.

Spectral features of flares

In 1948 Joy and Humason (1949) took the first slit spectrogram of an
UV Ceti flare. The examination of this unique plate taken with the exposure
of 144 min has shown that during the flare the emission hydrogen lines became
much stronger, CaII emission intensified, but to a less extent emission lines

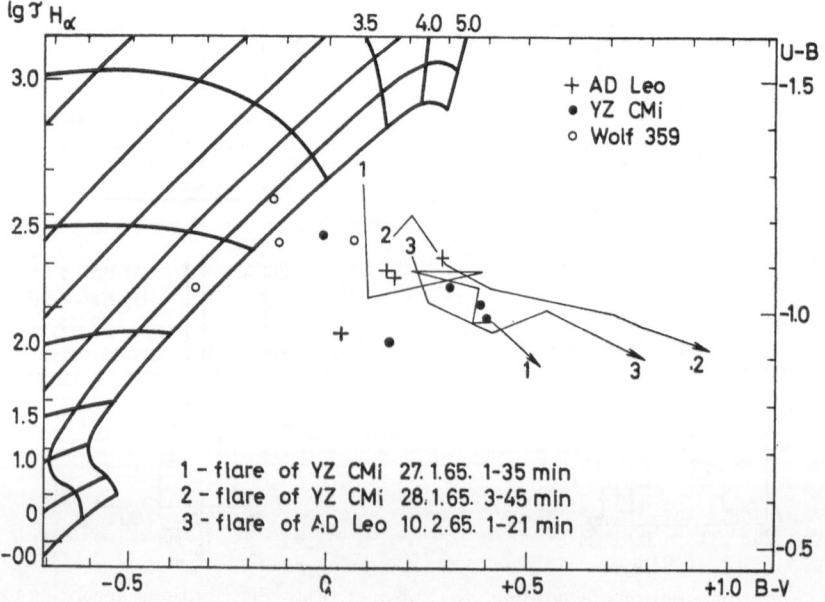

Fig. 3. Two-color diagram for UV Cet star flares. In the left and upper part of the plot
there are the colors of hot ionized hydrogen clouds of different temperatures and optical
thickness.

of HeI and HeII appeared which had not been seen in the quiet state star spectrum. Absorption lines almost disappeared, being veiled by a continuum which was very strong in UV spectral region; the spectrophotometric temperature of that continuum exceeded 10000°K, widths of the emission hydrogen lines amounted to 2 A, and the decrement was not steep.

Having used a high sensitive receiver (image tube), the spectral observations of flares were carried out in Crimea in 1965 and nearly 30 spectrograms of 10 flares of AD Leo and UV Cet were obtained with a time resolution from 20 sec up to 1—2 min (Gershberg and Chugainov, 1966, 1967); simultaneously the brightness of the star was being monitored photoelectrically. The light curve of the strongest AD Leo flare registered by us is given in Fig. 4, the time intervals of flare spectrographying are marked too. Five spectra of this flare are reproduced in Fig. 5. During the strong flare the stellar spectrum transformed beyond recognition in the photographic region, but the changes were not so striking in green and red. The most prominent feature in all flare spectra is an intensification of Balmer emission lines. The quantitative treatment of the spectrograms shows that the equivalent widths of the emission hydrogen lines during the flares approach tens of angstroms, the augmentations of the widths at a half intensity level amount to 3—5 A. With the flare decaying, the continuum radiation decreases first of all and line emission decreases more slowly; sometimes line emission is still visible when a wide-band photoelectric photometry does not find a trace of a flare. The widths of emission lines return to the normal state more quickly that the intensities of these lines do. The helium lines were found only near flare maxima, the maximum in CaII takes place later than that in hydrogen lines. The veiling of intensity jumps near the TiO-band limits and that of the absorption line λ 4227 A give a possibility to evaluate a part of flare continuum in the whole continuum radiation for moderate intense flares.

Later Chugainov (1968) carried out two sets of photoelectric observations of EV Lac and UV Cet flares; he used narrow-band interference filters and confirmed time variations; he determined absolute values of equivalent widths of the H_β-line spectrographically with the image tube device.

The same year important spectral studies of UV Ceti type star flares were carried out by Kunkel (1967). Kunkel's essential success is an investigation on the UV spectral region of flares and spectrophotometric measurements at a wide wave-length interval. Kunkel has found and measured the emission jump near the Balmer limit, he has confirmed the fact of a quick disappearance of the strong continuum radiation and the quick narrowing of emission lines after the flare maximum and found the Balmer decrement at several stages of the flare. He has shown that the relative rate of the flare decay in the lines is nearly one half of the rate in the continuum, and the decay in CaII is the slowest one.

At present we do not possess any information on flare line profiles, their Doppler shifts and possible anomalies in abundances of elements and isotopes. Nowadays such investigations are on the very limit or beyond the limit of instrumental power.

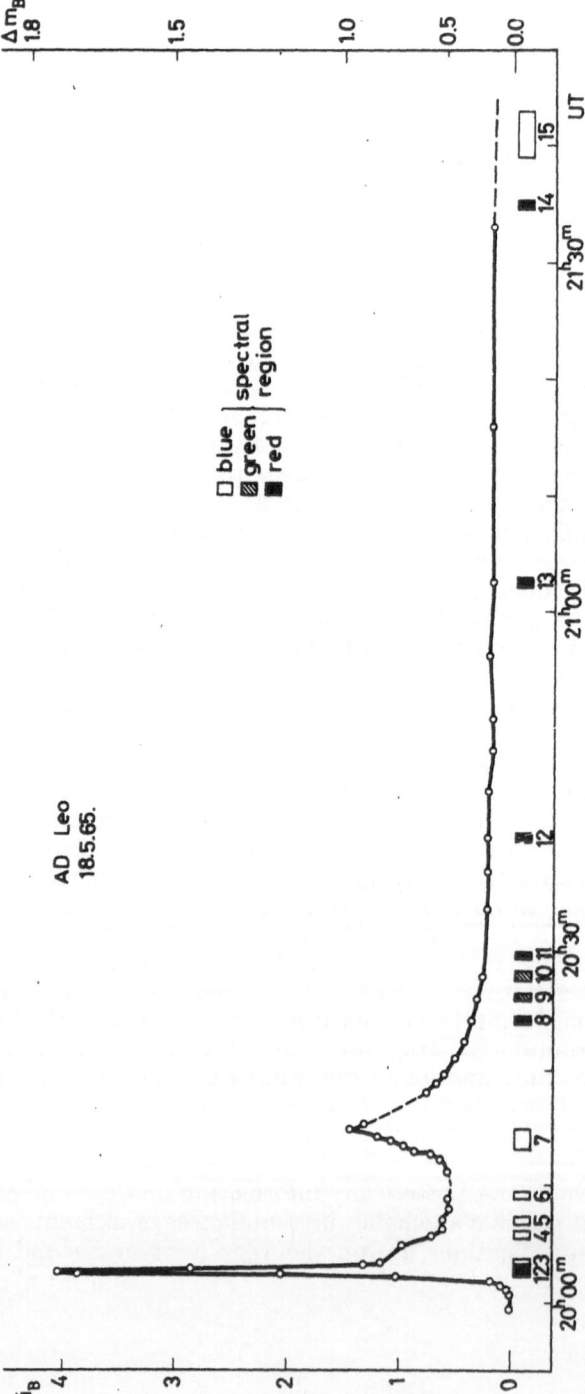

Fig. 4. Light curve of the AD Leo flare on 18.5.1965. Numbered rectangles mark time intervals of spectrographying the flare.

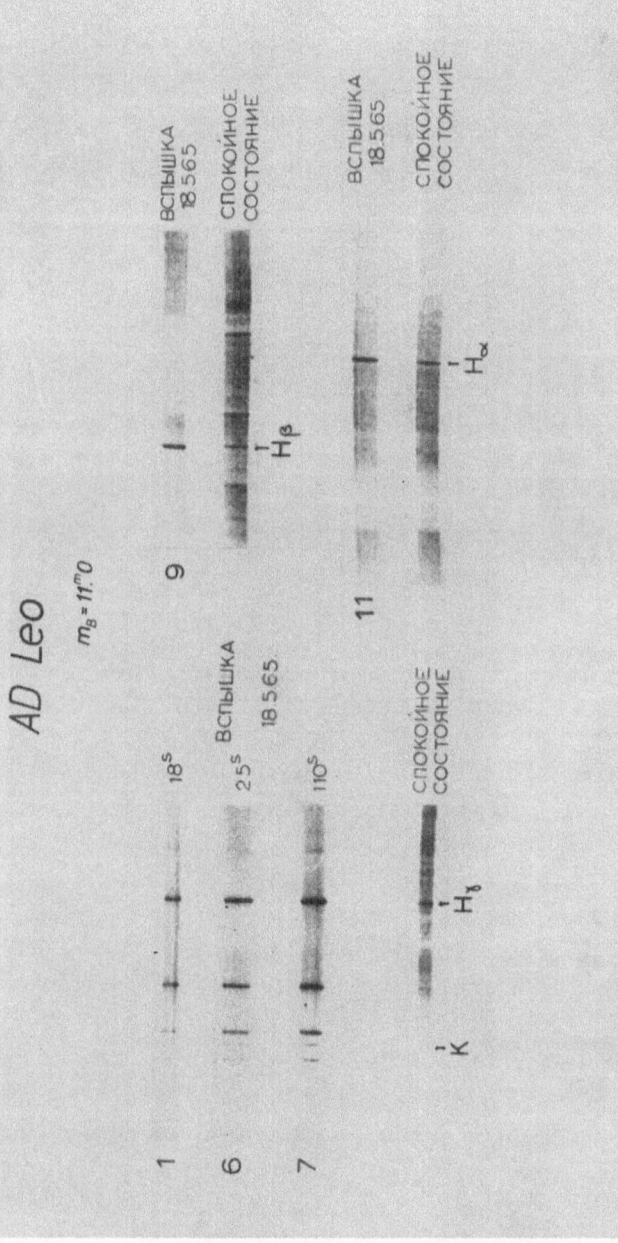

Fig. 5. Spectrograms of AD Leo during the strong flare on 18.5.1965 and in the quiet state on 4.6.1965. Numbers on the left correspond to the numeration in Fig. 4.

Polarimetric studies

Attempts to measure the polarization of UV Ceti type flare radiation were undertaken more than once. But as it was shown by Efimov (1968), all those observations had been made without due regard to the extreme rapidity of UV Ceti type flares. It is clear that when studying polarimetrically a variable source, the whole cycle of consecutive measurements must be made during the time which is small in comparison with the characteristic time of the source variations. But UV Ceti type flares are weak, therefore the quantum fluctuations of flare radiation flux turn out to be an essential obstacle when using small or moderate telescopes and small time averages for polarimetric measurements. It is necessary to use a large telescope and a special rapid-acting polarimeter (may by, similar to the device which was constructed by Oskanjan, Kubichela and Arsenijevich in Jugoslavia some years ago) in order to obtain reliable results on polarization features of flares.

Radio emission of flares

Excluding the sun, the UV Ceti stars are the only stellar bodies from which radio emission is certainly registered up to now. Radio emission of UV Ceti flares was found by Lovell (1964) with Jodrell Bank radio telescopes in 1958. First the radio emission of flares was found statistically by superposition of the radio records during 23 small optical flares (Lovell, 1963). But now we have more than two dozens of individual radio flare records at wave lengths in the range from 20 cm to 15 m.

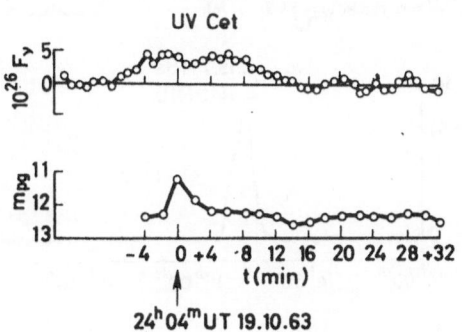

Fig. 6. Radio and optical flare of UV Cet on 19.10.1963.

The radio emission of flares varies as quickly as the optical radiation. The radio emission is characterized by a high brightness temperature: a moderate UV Cet radio flare distributed over the whole stellar disk corresponds to $T_b \approx 10^{15}$ °K. A typical radio flare record and its light curve are given in Fig. 6. Accordingly to a rough estimate, a flare radiates 100 times less energy in the radio wave-length range than in the optical flare energy. Lovell (1964) found a certain delay in radio emission at the lower frequencies when observing the UV Cet flare on 25.10.1963 at two frequences. (Fig. 7)

9*

    The radio emission of the V 371 Ori flare on 30.11.1962 was studied
in the fullest detail (Fig. 8): Australian investigators found the flare radio
emission at 3 frequencies and at 410 MHz sharp and deep fadings were observed
(Slee et al. 1963).

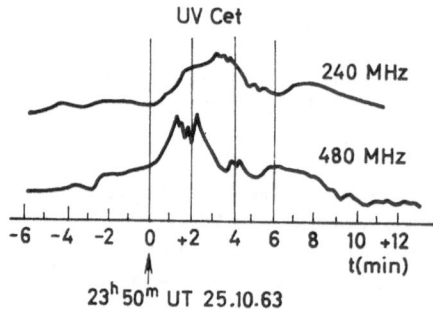

Fig. 7. Radio emission drift over frequencies during the UV Cet flare on 25.10.1963.

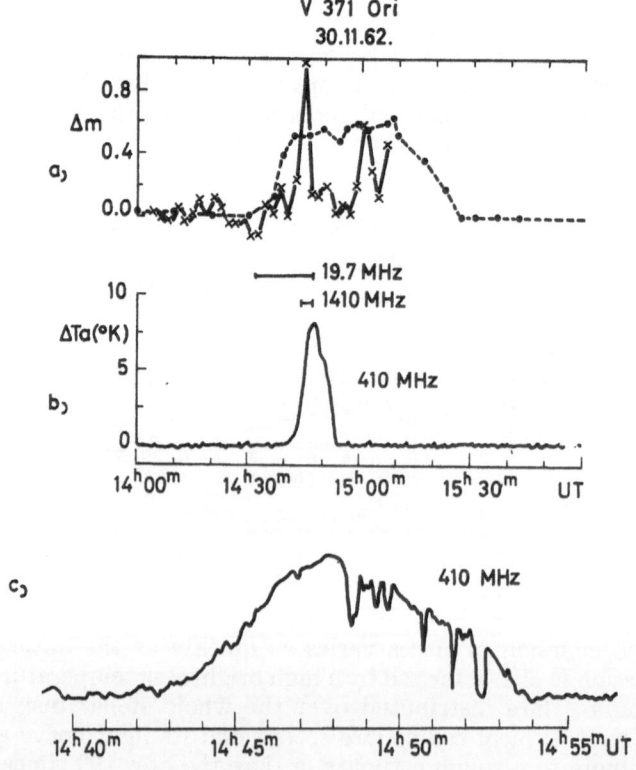

Fig. 8. V 371 Ori flare on 30.11.1962. *a)* optical observations: solid line — photo-
graphical, dashed line — visual monitoring of brightness; *b)* radio observations: solid
line is the smoothed record at 410 MHz, segments mark time intervals when the flare
is recorded at other frequencies; *c)* the flare-record at 410 MHz.

INTERPRETATION AND HYPOTHESES

Let us go to phenomenological interpretations and the physical hypotheses related to UV Ceti flares.

It is known that in 1924 Hertzsprung found for the first time a stellar flare, similar to UV Ceti flares, and in accordance with the spirit of the 20th years astronomy he supposed that the falling of an asteroid on the star could be regarded as a cause of the flare. Now this idea may be considered only as a historical curiosity. Since the beginning of intensive study of flare stars in 1948 nearly a dozen hypotheses have appeared. Today, the so-called nebular or choromospheric flare model has the closest contact with observations. Therefore, I shall give an account of this scheme and then shall describe other models and hypotheses in short.

Nebular or chromospheric model

The main supposition of the nebular model is that an optical flare is connected with a quick appearance of a hot ionized gasous cloud above the photosphere of a cold star; this cloud is deprived of external sources of ionization and radiates due to irreversible recombinations. Let us compare this scheme with the observations.

If during a flare the mass of the cloud is constant and its optical thickness is small, then it is not difficult to calculate the expected light curve. The comparison of 10 observed EV Lac flares with theoretical curves, which have been calculated for the simplest isothermal radiative process, are given in Fig. 9. (Gershberg, 1964). In half of the cases we have an agreement. Later calculations were carried out taking the cooling effects into account (Gershberg, 1967), and now we calculate the theoretical curves making allowance for self-absorption in the Balmer lines; as a result, the theoretical curve-family enriches and a possibility to fit the theory to the observations increases. But one ought not to undergo a delusion: on one hand, a rich theoretical curve-family makes the comparison of the theory with observations non-critical; on the other hand, no theoretical curves calculated for a homogeneous and uniformly expanded cloud are table to explain such details of light curves as secondary maxima and time intervals of constant brightness, and to interpret the ascending branches of flares. Therefore, the observed light curves are not in contradiction with the nebular model but this model is too primitive to give a complete theory of the observed light curves. It should be noted that many observers have represented the observed light curves of UV Ceti flares as one or two exponential curves, and this representation is not worse than the nebular one; but the exponential representation is not substantiated physically, it is an erroneous conclusion of a wrong hypothesis on a hot spot (see below).

The nebular interpretation of color features of flares is given in Fig. 3 according to Kunkel (1967). The colors of a hot ionized hydrogen cloud, which has the optical thickness $\tau_{H_\alpha} = 0$—10^5, are located in the left and upper part of the plot. From the relative positions of flares and nebular models on the two-color diagram one can conclude that the flare radiation at maxima has the same colors as the hot gas in the case of $T_e \geq 30000$ °K and $\tau_{H_\alpha} \approx 10^2$ to 10^3. The approximate character of Kunkel's calculations (a stationary gas

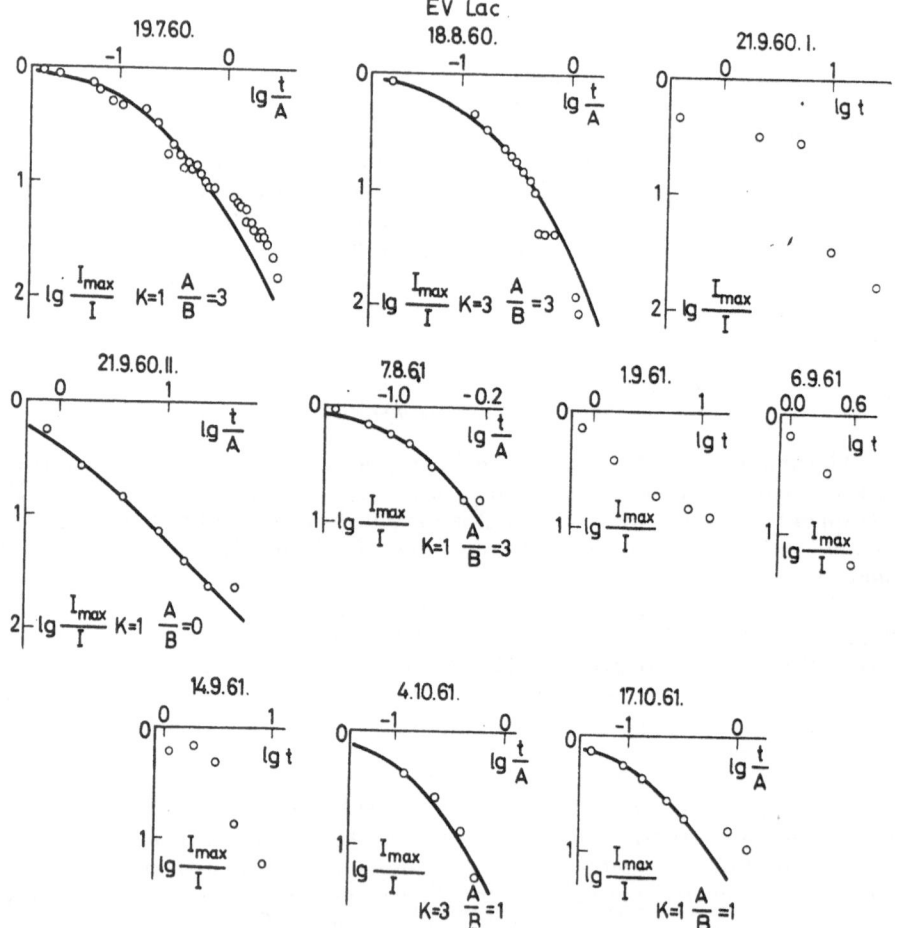

Fig. 9. Nebular representations of 10 light curves of EV Lac flares.

in LTE-conditions, self-absorption effects in coherent approximation) does not permit to insist on the values T_e and $\tau_{H\alpha}$ obtained, but the agreement between obsrevations and the nebular model is at hand. With the same certainty, Fig. 3 shows a flare drift according to the nebular model during the flare decay. Kunkel interprets this drift as an increasing contribution of an equilibrium hot photospheric spot — a photosphere's "burn" — to the surplus stellar radiation.

Qualitative spectral features of a UV Ceti type flare — the appearance of strong continuum emission and strong intensification of line emission — were known from previous observations by Wachmann (1939), Joy and Humason (1949), and Herbig (1956) and naturally they fit to the nebular model. Our observations (Gershberg and Chugainov 1966; 1967) and those of Kunkel (1967) permit to make a quantitative comparison.

The comparison of the equivalent widths of Balmer emission lines observed in the AD Leo flare on 18.5.1965 with the equivalent widths calculated in

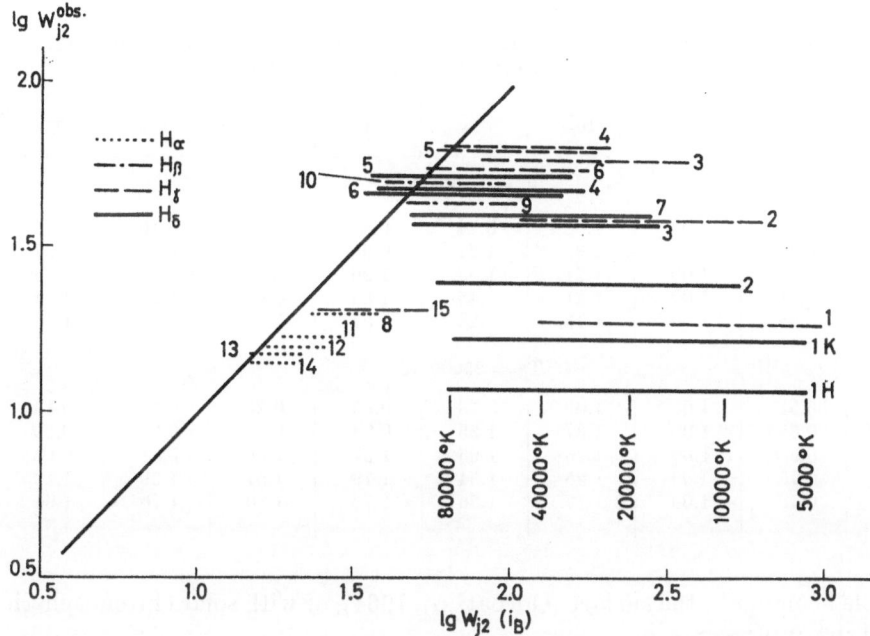

Fig. 10. Comparison of the observed equivalent widths of Balmer lines in the AD Leo flare on 18.5.1965. and corresponding equivalent widths calculated in accordance with the flare brightness at different times for the optically thin flare model.

accordance with the brightness of the flare and in supposition of an optically thin gas is given in Fig. 10. It is seen that at the beginning of the flare the equivalent widths observed are ten times less than the calculated ones. The same results follow from other spectrograms of AD Leo and UV Cet flares. Since our calculations have been carried out for $T_e \leq 80000$ °K, the obtained disagreement can be explained either by a higher temperature, or by self-absorption effects in the lines. Kunkel's observations decide this dilemma: the considerable magnitude of the emission Balmer jump found by him means $T_e < 25000$ °K. It should be noted that this upper limit of temparature may not be far from the real temperature because of the existing HeII emission. Comparing the observed and calculated Balmer decrements (see Table 5) Kunkel has found an independent and decisive argument for the chromospheric flare model: observed line intensity ratios from H_β to H_{11} correspond to radiation of the gas of the temperature $T_e = 20000$ to 25000 °K and of the same optical thickness what has been obtained from colorimetric studies.

Physical hypotheses

The phenomenological nebular model of flares permits several physical interpretations. The possible existence of a characteristic time between flares suggests the assumption of active regions on the stellar surface and of a rather quick rotation of the star. The origin of a hot cloud above the photosphere can be connected either with a shock wave appearance, or with a hot

Table 5

Calculated Balmer decrements for LTE-conditions, coherent re-emission in lines, $n_e = 30 \times 10^{13}$ cm^{-3} and $v_{turb.} = 20$ km/sec.

lg τ_{H_α}	H$_\alpha$	H$_\beta$	H$_\gamma$	H$_\delta$	H$_\zeta$	H$_\eta$	H$_{10}$	H$_{11}$
				$T_e = 20000°$K				
2.0	0.55	1.00	1.06	0.74	0.32	0.22	0.16	0.12
2.5	0.63	1.00	1.32	1.27	0.68	0.49	0.36	0.27
3.0	0.76	1.00	1.31	1.55	1.26	0.97	0.73	0.56
3.5	0.81	1.00	1.21	1.45	1.65	1.49	1.26	1.02
4.0	0.84	1.00	1.13	1.29	1.57	1.64	1.60	1.46
				$T_e = 25000°$K				
2.0	0.51	1.00	1.09	0.78	0.35	0.24	0.17	0.13
2.5	0.58	1.00	1.37	1.35	0.74	0.53	0.39	0.29
3.0	0.68	1.00	1.36	1.65	1.37	1.06	0.81	0.62
3.5	0.75	1.00	1.25	1.54	1.79	1.64	1.39	1.12
4.0	0.78	1.00	1.17	1.36	1.70	1.80	1.76	1.61

"bubble" coming to the surface (Gorbatzkij, 1964), or with solar chromospheric flare type processes.

An analogy between solar and UV Cet type flares was suspected by Greenstein and Whipple nearly 20 years ago, then it was spoken about by the Burbidges, Schatzman, Struve and al., but now this analogy becomes clearer and deeper. Indeed, both types of events have a strongly pronounced explosive behavior, they have no clear periodicity, but there are epochs of different activity level; the same emission lines appear in the spectra of both types of flares and there is similarity in the sequence of the flaring of different lines and in the character of intensity and line width variations during the flare decay; both types of optical flares are accompanied by strong radio flares, and the physical parameters of the hot gas responsible for the optical flares are similar. Therefore we have a good reason to suspect the similarity in intrinsic physical causes of both types of flares.

As it is known, the energy source of chromospheric flares is the magnetic field, and the solar activity is determined by magneto-hydrodynamic phenomena which are mostly caused by convective motions. As UV Ceti stars are bodies of small mass and their inner structure is completely convective, it is natural to expect strong magneto-hydrodynamic motions and, as a result, a flare activity in these stars. The decisive confirmation of this conception has been obtained recently. Poveda (1965) showed that the convection must be very strong in stars of low luminosity up to K1-stars and Haro (1968) found a high flare activity of stars (in clusters of different age) up to the same spectral class. The main but the only parameter of the modern theory of stellar evolution — the mass — is used in this conception; therefore, in order to explain the differences between dMe and UV Ceti type stars and between dMe and normal M dwarfs, it is necessary to appeal either to evolutionary considerations or to additional parameters, as rotation, anomaly of element abundance, binary systems' features etc. We may note that the approximate characteristic time of flare activity level variations is a few months for UV Cet and the orbital motion characteristic time is tens of years for the L726—8AB-system (UV Cet = L726—8B), therefore a close connection between these phenomena can not exist.

Alternative models and hypotheses

Finally we submit some critical comments on other phenomenological models and physical hypotheses.

Recent results on UV Ceti-type star flares permit to reject some previous models of flares. For instance, the ideas about a flare as an appearance of a hot equilibrium photospheric spot, and hypotheses on pure synchrotronic or pure Compton nature of flare radiation must be rejected in view of the discovery of the Balmer emission jump and the strong line emission in flares. It is also necessary to reject different hypotheses of external excitation of flares as the flare frequency and the drift of radio emission to long wavelength range indicate inner causes of these events.

Ambarzumjan (1954) supposed that the UV Ceti type flares were connected with ejections of an unknown "pre-stellar matter" from the stellar interior. Recent experimental data do not confirm this hypothesis but they do not disprove it either.

Since 1956 Schatzman (1967) has developed a stellar vibrational instability theory and applied it to different stellar flares; but this mechanism cannot be responsible for UV Ceti type flares as its basis is a strong dependence of the thermonuclear reaction rate on temperature and density of matter and such a reaction is not maintained in small mass M dwarfs.

Kolesnik (1966) has used a maser effect in nonequilibrium plasma to interpret the flares. The necessity of such a hypothesis is not obvious today.

For the last years Gurzadjan (1965, 1966) has been developing a theory of UV Ceti type flares; all nebular features of flares are regarded as secondary effects while the primary one is supposed to be a fast electron ejection and their Compton interaction with the photospheric radiation. According to this hypothesis the electron energy must exceed 10^5 to 10^7 times the optical energy of the flare and does not excite any observable consequences on the star; this is the main difficulty in Gurzadjan's theory.

It is my pleasure to thank Dr. W. E. Kunkel, who courteously sent me his Dissertation. Prof. Haro for making available a preprint of his review on flare stars and Dr. P. F. Chugainov, Chairman of Working Group on Flare Stars, for the information on premilinary results of cooperative observations. All these data were very useful while preparing this report.

REFERENCES

Ambarzumjan, V. A., 1954, Sobbsch. Bjurak. Obs. **13.**
Andrews, A. D., 1966, Publ. astr. Soc. Pacific **78,** 324; 542.
"Non stable stars". Ed. by. Arakeljan, M. A., Erevan 1957.
Chugainov, P. F., 1966, Astr. J. **43,** 1168.
Chugainov, P. F., 1968, Izv. Krym. astrofiz. Obs. **38,** 200, **40.**
Efimov, YU. S. 1968, Krym. astrofiz. Obs. **41.**
Gershberg, R. E., 1964, Izv. Krym. astrofiz. Obs. **32,** 133
Gershberg, R. E., 1967, Izv. Krym. astrofiz. Obs. **36,** 216.

Gershberg, R. E., and Chugainov, P. F., 1966, Astr. Zu. **43.** 1168.
Gershberg, R. E., and Chugainov, P. F., 1967. Astr. Zu. **44,** 260.
Gershberg, R. E., and Chugainov, P. F., 1968. Izv. Krym. astrofiz. Obs. **40.**
Gliese, W., 1958, Z. Astrophys. **45,** 293.
Gorbatzkij, V. G., 1964, Astr. Zu. **41,** 53.
Gurzadjan, G. A., 1965, Astrofizika **1,** 319, C. R. Ac. Sci. USSR.
Gurzadjan, G. A., 1966a, Astrofizika **2,** 217.
Gurzadjan, G. A., 1966b, C. R. Ac. Sci. USSR, **166,** 53.
Haro, G., 1968, Flare stars. "Clusters and binaries", Ed. by B. Middlehurst, Chicago.
Non-stable stars. Symposium No. 3. IAU. Ed by Herbig, G. H. Cambridge, 1957.
Herbig, G. H., 1956, Publ. astr. Soc. Pacific **68,** 531.
Joy, A. H. and Humason, M. L., 1949, Publ. astr. Soc. Pacific **61,** 133.
Joy, A. H., 1960, "Stellar Atmospheres". Ed. by Greenstein, J. L. Chicago, p. 666.
Kolesnik, I. G., 1966, Dissertation, Kiev.
Kunkel, W. E., 1967, An optical study of stellar flares. Austin, Texas.
Lippincott, S. L., 1953, J. astr. Soc. Can. **47,** 24.
Lovell, B., Whipple, F. L., and Solomon, L. H., 1963. Nature **198,** 228.
Lovell, B., Whipple, F. L., and Solomon, L. H., 1964. Nature **201,** 1013.
Lovell, B., 1964. Scient. Am. **211,** No. 2, 13.
Oskanjan, V., 1964. Publ. Obs. astr. Beograd No. 10.
Petit, M., 1961, J. Observateurs **44,** 11.
Poveda, A., 1964, Nature **202,** 1319.
Slee, O. B., Solomon, L. H., and Patston, G. E., 1963, Nature, **199,** 991.
Wachmann, A. A., 1939, Beob. Zirk. **21,** 25.

COMMENT

Lortet-Zuckerman: I did not understand whether the nebular model may account
for the high observed ratio of radio to optical energy. The ratio radio
to optical energy is two or three orders greater for flare stars than for
the sun.

COOPERATIVE 24-HOUR OBSERVATIONS OF UV CETI-TYPE STARS

P. F. CHUGAINOV

Crimean Astrophysical Observatory, USSR

Since 1967 the Working Group on UV Cet-type stars has organized and carried out runs of 24-hour photometric observations of these stars. Our aim is to study the time distribution of flares.

The following observers took part in the observations:

Australia: members of the Astronomical Society of N.S.W., coordinators C. S. Higgins and G. E. Patston.

Italy: Catania Observatory, G. Godoli.

Japan: Tokyo Astronomical Observatory, K. Osawa et al.

New Zealand: Mt. John Observatory and amateur astronomers, coordinator F. M. Bateson.

South Africa: Boyden Observatory, J. P. Eksteen.

U.S.A.: Smithsonian Astrophysical Observatory, L. H. Solomon. Steward Observatory, B. Westerlund.

U.S.S.R.: Abastumani Astrophysical Observatory, V. S. Oskanjan. Crimean Astrophysical Observatory P. F. Chugainov.

At present, several other observatories have agreed to take part in future programmes.

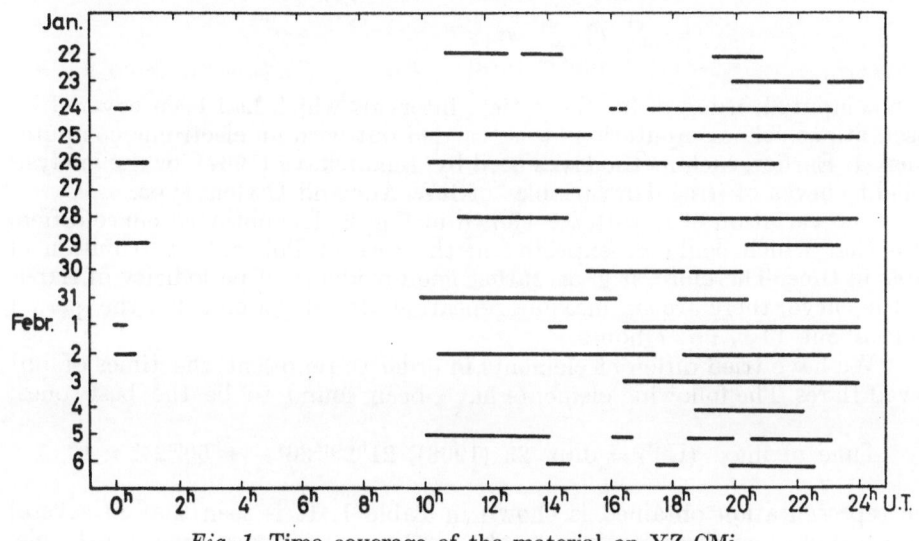

Fig. 1. Time coverage of the material on YZ CMi

Up to now two observational campaignes have been carried out: September 26—October 10, 1967, UV Cet and January 22—February 6, 1968, YZ CMi. During the first period 50.5 per cent of total time was covered by observations. 17.9 per cent being photoelectric and 32.6 per cent photographic and visual observations. Only photoelectric observations were made during the second period, the coverage being 29.5 per cent of the total time.

Most of the observational results have been published in I.B.V.S. (1968). The study of the time distribution of flares has been carried out by K. Osawa et al. (1968). At the Crimean Astrophysical Observatory A. A. Korovyakovskaya and the Author have analyzed the observations from both periods. Now we have completed the study of the material on YZ CMi.

We used the autocorrelation analysis in order to study the time distribution of flares. The observational period was divided into equal time intervals τ. $\tau = 30$ minutes was adopted. For the periods of 15 days the total number of time intervals was equal to 720. Energies I emitted for each interval were computed according to observational data. One can see that $I = I_{norm}$, if the flares are absent, and $I = I_{norm} + I_{flare}$ in the presence of a flare, I_{norm}, I_{flare} being the integrated energies emitted by the star and flare, respectively. For each time interval i, the quantity

$$u_i = I_i - \overline{I}$$

was determined where

$$\overline{I} = \frac{\Sigma I_i}{N}.$$

Autocorrelation functions

$$r_k = \frac{\sum_{i=1}^{720-k} u_i u_{i+k}}{\sqrt{\sum_{i=1}^{720-k} u_i^2 \sum_{i=1}^{720-k} u_{i+k}^2}}, \quad k = 0, 1, 2, \ldots, 100$$

were computed, using u_i for those time intervals which had been covered by observations. The computations were carried out with an electronic computer Minsk-1. Earlier, such method was used by Lukatskaya (1967) for the analysis of light curves of irregular variables of RW Aur and U Gem-type.

The variation of r_k with k is shown in Fig. 2. The obtained curve differs from that which could be expected in the case of Poisson's distribution of flares in time. Therefore, it gives rather good evidence of periodicity of flares. On the curve, there are six maxima repeating almost periodically, the period being about 14 τ, i.e. 7 hours.

We have tried different elements in order to represent the times of observed flares. The following elements have been found to be the best ones:

Time of max. (UT) = Jan. 23 (1968), $21^h29^m36^s + 6^h50^m24^s \times E$.

The representation obtained is shown in Table 1. It is seen that in several cases two or even three flares were observed near the times given by the elements. In such cases, if the energies of flares are nearly equal, the mean time of maxima was adopted as the observed time. If the energy of one flare of the

Table 1. Flares of YZ CMi

Observer	Time of max. of flare, U. T.	Integral energy of flare, minutes	O—C
Eksteen	Jan. 23, 21^{h}29^{m}6 22^{h}09^{m}0 23^{h}23^{m}6	19.2 1.3 0.4	0^{h}00^m
Osawa et al.	Jan. 24, 17^{h}22^{m}5 17^{h}34^{m}2 17^{h}56^{m}7	0.35 0.25 3.5	—0^{h}04^m
Osawa et al.	Jan. 25, 12^{h}53^{m}2 13^{h}42^{m}7	0.7 } 0.25 }	—1^{h}13^m
Oskanjan	Jan. 26, 19^{h}17^{m}0	1.5	+1^{h}24^m
Eksteen	Jan. 29, 0^{h}29^{m}7	1.4	—0^{h}07^m
Eksteen	Jan. 29, 20^{h}29^{m}7 21^{h}39^{m}7	1.4 14.0	+0^{h}32^m
Osawa et al.	Jan. 31, 12^{h}23^{m}4 12^{h}45^{m}5	0.3 } 0.3 }	—1^{h}36^m
Chugainov	Jan. 31, 19^{h}58^{m}0	0.23	—1^{h}03^m
Chugianov	Feb. 1, 18^{h}21^{m}0	0.32	+0^{h}49^m
Osawa et al.	Feb. 2, 15^{h}58^{m}9	0.15	+1^{h}55^m
Eksteen	Feb. 2, 19^{h}14^{m}3	0.6	—1^{h}41^m
Chugainov	Feb. 3, 22^{h}49^{m}0	0.68	—1^{h}26^m
Oskanjan	Feb. 4, 20^{h}36^{m}2 21^{h}13^{m}4	5.0 15.0	—0^{h}26^m
Osawa et al.	Feb. 5, 10^{h}45^{m}0 11^{h}20^{m}0 12^{h}14^{m}9	7.0 } 5.0 } 0.35	—0^{h}35^m

group is essentially greater than that of the others, then O—C was computed only for this flare and the others were neglected. The mean O—C was obtained to be 1^{h}06^m. The tendency of flares to form groups was noticed by Osawa et al. (1968).

130 P. F. CHUGAINOV

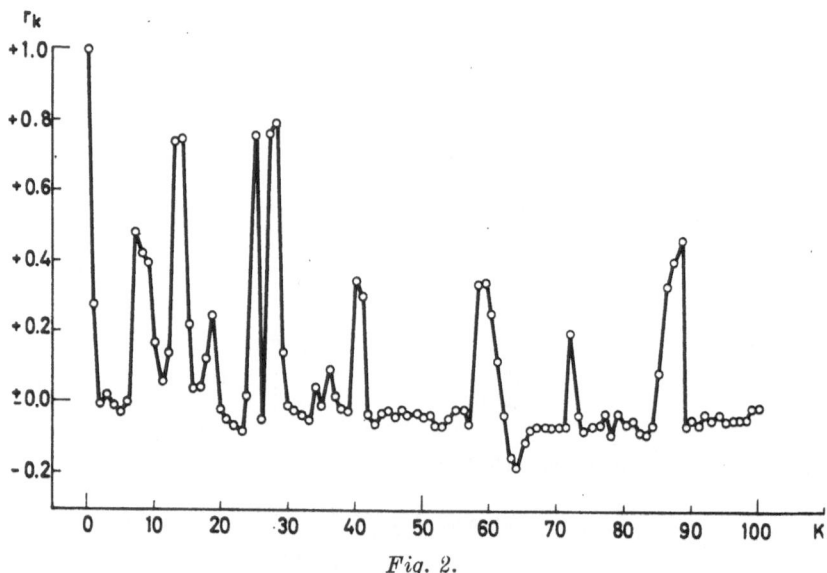

Fig. 2.

The period of repetition of flares was found to be 20^h2 by Osawa et al. (1968). Andrews (1966) found from observations of YZ CMi in 1966 that the intervals between flares were equal to 47^h, 73^h or 122^h. These data do not contradict our result because $6^h50^m24^s \times 3 = 20^h5$, $6^h50^m24^s \times 7 = 47^h9$, $6^h50^m24^s \times 11 = 75^h2$ and $6^h50^m24^s \times 18 = 123^h1$. It is obvious that Andrews and Osawa have found periods which are divisible by our period. Our result shows the importance of making observations at several observatories located at different longitudes.

We have tried to represent the times of flares of YZ CMi observed by Andrews in 1966. It was found that the elements:

$$\text{Time of max (UT)} = \text{Feb. 21 (1966)}, 18^h32^m + 6^h38^m9 \times E,$$

represent all the flares observed, with a mean deviation of 1^h22^m.

REFERENCES

Andrews, A. D., 1966, Publ. astron. Soc. Pacific **78**, 324.
Information Bulletin on Variable Stars nos. 264, 265, 267, 268. 1968.
Lukatskaya, F. I., 1967, Perem. Zvezdy, **16**, 168.
Osawa, K., Ichimura, K., Noguchi, T., and Watanabe, E., 1968, Tokyo astr. Bull., Second Series no. 180.

THE CLASSIFICATION OF PHOTOMETRIC
LIGHT-CURVES OF FLARES OF UV CETI STARS

V. S. OSKANIAN

Byurakan Astrophysical Observatory, Armenia, USSR

One of the means to investigate the flare phenomenon of the UV Ceti stars is to study its photometric light-curve characteristics, for it can be supposed that in principle the form of the light-curve depends on the type of process taking place on the star. Such studies have already been made (Chugainov, 1962; Oskanian, 1957; Roques, 1961). It is noteworthy, that all of them were not dealing with the light-curve as a whole but with its decreasing branch only. Moreover, the number of curves studied in this way has been very limited.

The number of precise photoelectric observations increased significantly during the las few years resulting in the discovery of a great variety of forms of flare light-curves. This fact called for a rather different approach to the study of light-curves. (An example of such a new approach is W. Kunkel's (1967) attempt to treat the curves as composed from two — slow and fast — superposed components). Now it seems more correct to substitute the detailed study of single curves by a classification of them, based on such parameters which can be determined for all kinds of flare curves. In this case, the study of details of light-curves should be renounced as it can be supposed that their general features are more substantial than the small amplitude details.

The aim of the present paper is to propose one of the possible kinds of classification of flare light-curves. The proposed classification is based on the following two parameters of the light-curve:

a) the rate of brightness-increase;

b) the character of the decreasing branch of the light-curve.

Four types of photometric light-curves can be defined by means of these two parameters. Two of them may be considered as extremal in the sense that they possess extremal values of the parameter "a" and substantially different parameters "b", while the other two types should be regarded as intermediate.

These four types of flare light-curves could be denoted by I, II, III, IV and defined as follows:[*]

---

[*] A third parameter t/T (t the time of rise from normal state to maximum and T the duration of the whole flare) could be used as well, but it seems that it is less precise than the proposed two parameters for the following reasons:

1. It is difficult to determine the precise moment of ending of the flare, i.e. the right value of T.

2. A prolonged "tail" of the flare, that appears sometimes, can make this parameter rather indefinite.

Nevertheless, it is quite obvious that on the average the value of this parameter should increase with the increasing number of curve types, attaining a value of about 0.5 for Type IV curves.

Type I. This type is characterized by a great rate of brightness-increase (in the studied cases this rate was between 5 and 1 magnitudes per minute). The brightness-decrease starts immediately after the maximum and takes place with the same rapidity as the increase, producing thus a very sharp maximum. The rate of decrease slows down only in the final phase of the flare. In the stellar magnitude scale the decreasing branch of the curve can be approximated by two straight lines. In other words, if one tries to represent the decreasing branch — expressed in intensity scale — by the formula

$$I = I_M . e^{-\alpha(t - t_M)}, \tag{1}$$

it would be necessary to choose two different values of α in order to approximate the curve of this type (Chugainov, 1962).

Examples of Type 1 curves are given on Fig. 1.

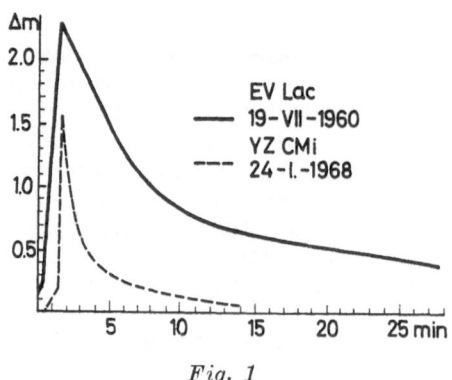

Fig. 1

Type IV. Being the other extremal case, this type of curves is characterized by a very small rate of brightness-increase (not greater than some tenths of magnitude per minute), a flat maximum and a small rate of brightness-decrease. In the stellar magnitude scale the decreasing branch can be approximated by a single straight line. The approximation by formula (1) can be realized by using a single value of α.

Examples of Type IV curves are given on Fig. 2.

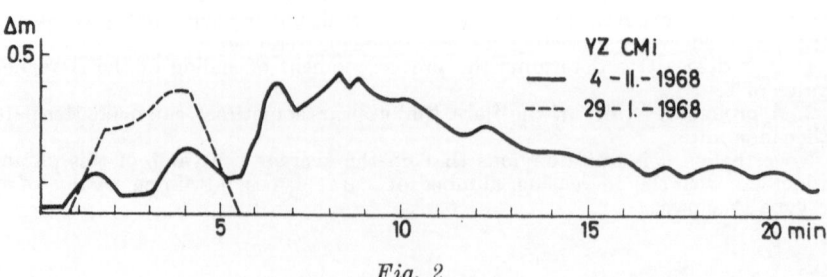

Fig. 2

Type II. The curves of this type seem to be composed of curves of Types I and IV, the first part of the curve having Type I and the second part Type IV characteristics. The transition from Type I to Type IV is very rapid and takes place at different phases of the decreasing branch.

Examples of Type II curves are given on Fig. 3.

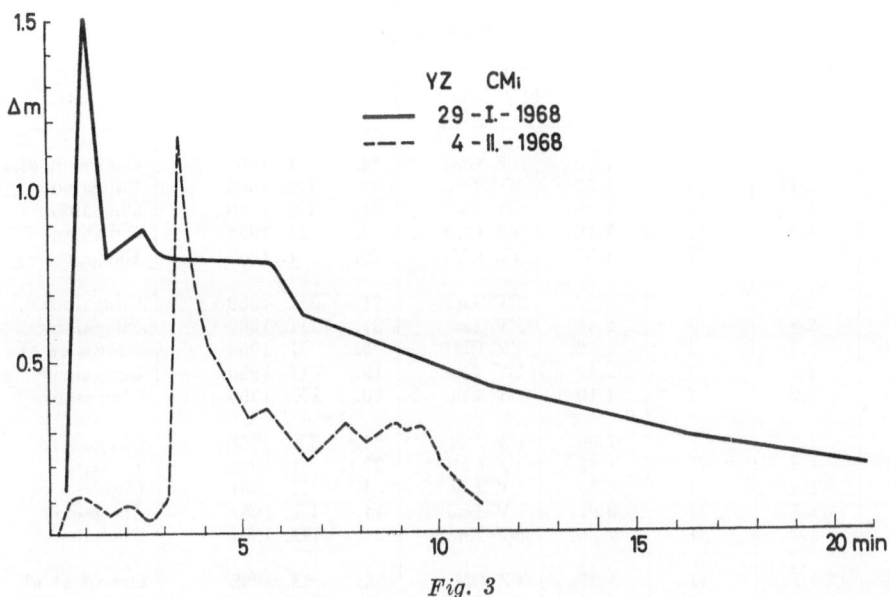

Fig. 3

Type III. By its appearance — especially by its relatively sharp maximum — the curves of this type are very similar to those of Type I, but they differ from the last ones by their smaller rate of brightness-increase (nearly always smaller than one magnitude per minute). In the stellar magnitude scale the decreasing branch can be approximated by a single straight line, i.e., by one value of α in formula 1.

Examples of Type III curves are given on Fig. 4.

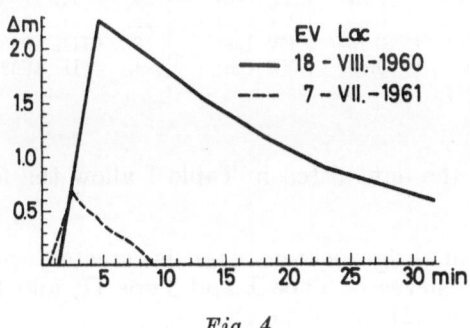

Fig. 4

A list of 30 flares classified by means of the two mentioned parameters is given in Table I. It should be noted that the accuracy of data used to prepare this Table was not sufficiently homogeneous, so that some unessential changes — specially in the values of $\Delta m/\Delta t$ — could be allowed. The uncertainty in the classification of flare No. 21 should also be ascribed to the impossibility of getting more accurate data from the published curves.

Table I

No.	$\Delta m/\Delta t$	Type	Δm_B	Star	Date		Author
1	5.2	I	1.56	YZ CMi	24.	I. 1968	Osawa et al.
2	5.1	I	1.52	UV Cet	26.	IX. 1965	Chugainov
3	4.9	I	1.23	EV Lac	21.	IX. 1960	Chugainov
4	4.8	II	1.12	YZ CMi	4.	II. 1968	Oskanian
5	3.0	II	1.51	YZ CMi	29.	I. 1968	Eksteen
6	3.0	I	0.90	EV Lac	21.	IX. 1960	Chugainov
7	2.9	I	2.88	EV Lac	27.	VIII. 1962	Chugainov
8	2.6	I	1.32	YZ CMi	5.	II. 1968	Osawa et al.
9	2.3	I	2.29	EV Lac	19.	VII. 1960	Chugainov
10	2.2	I	1.10	UV Cet	20.	IX. 1965	Chugainov
11	1.8	I	1.90	UV Cet	24.	IX. 1965	Chugainov
12	1.8	II	1.83	YZ CMi	23.	I. 1968	Eksteen
13	1.8	I	0.54	EV Lac	4.	X. 1961	Chugainov
14	1.5	II	0.81	EV Lac	14.	IX. 1961	Chugainov
15	1.4	I	2.86	EV Lac	27.	VIII. 1962	Chugainov
16	1.4	II	1.39	YZ CMi	5.	II. 1968	Osawa et al.
17	1.3	I	1.52	EV Lac	31.	VII. 1962	Chugainov
18	1.2	I	0.81	EV Lac	17.	X. 1961	Chugainov
19	1.1	I	0.56	YZ CMi	1.	III. 1968	Cristaldi
20	0.7	III	0.72	EV Lac	6.	IX. 1961	Chugainov
21	0.6	III I	0.63	EV Lac	1.	IX. 1961	Chugainov
22	0.6	III	0.36	YZ CMi	26.	I. 1968	Oskanian
23	0.5	III	0.33	YZ CMi	23.	II. 1968	Oskanian
24	0.5	III	2.30	EV Lac	18.	VIII. 1960	Chugainov
25	0.4	III	0.70	EV Lac	7.	VIII. 1961	Chugainov
26	0.4	III	0.75	V 1216 Sgr	28.	VI. 1961	Grigorian, Vardanian
27	0.2	III	3.21	YZ CMi	24.	II. 1968	Oskanian
28	0.14	IV	0.39	YZ CMi	29.	I. 1968	Eksteen
29	0.1	III	0.75	EV Lac	18.	VIII. 1963	Chugainov
30	0.06	IV	0.46	YZ CMi	4.	II. 1968	Oskanian

Nevertheless, the data listed in Table I allow the following qualitative conclusions:

a) The rate of brightness-increase is greater than one magnitude per minute for the curves of Type I and Type II, and less than this value for the curves of Type III and Type IV.

b) There are some reasons to suppose that curves of Type **IV** appear really more rarely than those of other types. As to the frequency distribution of curves of different types (Table II) resulting from Table I, it can not pretend to be a real one, owing to the sampling effect caused by the suppression of a number of small amplitude flares.

Table II

Type	Number of flares	Mean values of Δm_B
I	14	1.42
II	5	1.33
III	9	1.08
IV	2	0.43

c) There is no obvious correlation between the amplitude of light-variation and curve-type. Nevertheless, it seems that the mean values of Δm_B for different types of curves show tendency to diminish from Type I to Type IV. But, because of the above mentioned sampling effect, this conclusion too must be accepted with some precaution.

It should be noted, at last, that in some rare cases the light-curve can not be classified according to this classification. But in these cases too the proposed classification does not lose its value, as the mentioned curves are nearly always a combination of two or more curves of the types defined by this classification.

So, for instance, the curve represented on Figure 5 can be interpreted as a superposition of two Type I curves.

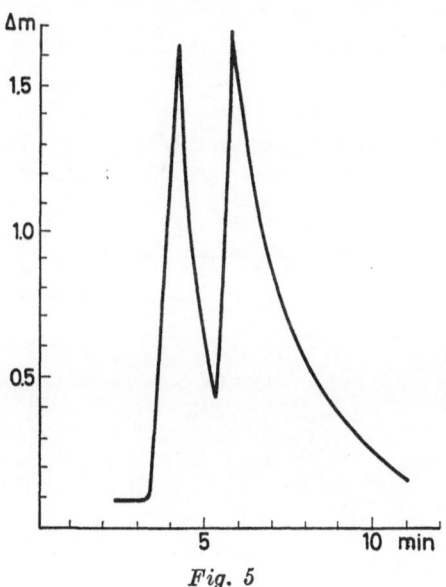

Fig. 5

10*

REFERENCES

Chugainov, P. 1962. Izv. Krym. astr. Obs. XXVIII., 150.
Kunkel, W. 1967. Thesis, Univ. of Texas, Austin.
Oskanian, V. 1957. Nestacinarnie zvjozdi, Ac. Sc. of Armenia, Erevan.
Roques, P. 1961. Astrophys. J. **133,** 914.

DISCUSSION

Godoli: Could it be possible, to approximate the decreasing part of *your type I* flare light curves by an exponential function instead of two linear functions?

Oskanian: In intensity scale you need two exponential functions, in stellar magnitude scale two straight lines,

MULTI-COLOUR PHOTOMETRY OF ORION FLARE STARS

A. D. ANDREWS

Armagh Observatory, Northern Ireland
(read by J. D. FERNIE)

ABSTRACT

The first results of a photometric study of the Orion flare stars is presented using material from the Boyden Observatory. A 60-inch photoelectric
sequence, in U, B, V and R, and a photographic reduction technique developed
for ADH Baker—Schmidt plates by C. J. Butler, are utilized to construct
colour-magnitude and colour-colour diagrams for flare stars to $V = 16^{m}$.
The scattre of the flare stars about the main sequence, pointed out by Haro,
is confirmed in the B—V/V diagram. However, a fairly well-defined band in
the V—R/V diagram is evident, extending from V—R = $1^{m}0$, V = $12^{m}5$
to V—R = $1^{m}7$, V = $16^{m}0$. The classical flare stars appear to fall within the
same region of the B—V/V—R diagram but to the red of the majority of
Orion flare stars.

INTRODUCTION

The wealth of material on flare stars in stellar aggregates of differing
age, systematically accumulated since the early fifties mainly by Haro, has
inspired many fresh inquiries into the early evolution of stars. The list of 176
flare stars in the vicinity of the Orion Nebula published by Haro (1968)
summarizes the discoveries of Haro and Chavira at Tonantzintla, Rosino at
Asiago, and their collaborators up until 1965. These flare stars, in common
with many T Tauri stars, are strongly concentrated towards the centre of the
Orion Nebula, and are, almost beyond doubt, T-association members of Orion
T2 (Kholopov, 1959). From the evidence of extensive photographic work,
Haro has stated that the Orion flare stars appear to lie both above and below,
as well as on the main sequence. In view of the lack of spectra any but the
brightest stars, obtained by Herbig (1962), and the paucity of reliable magnitude determinations, it is worthwhile to attempt, as far as possible traditional
UBV photometry even in this difficult, nebulous region to examine this peculiar feature of the Orion flare stars in the H—R diagram. Mendoza (1968)
has already made multicolour photometry for seven of these stars and has
emphasized large infrared colour-excesses in these and other related, T Tauri-
like stars. The primary questions asked today concerning flare stars are:
a) What is the true extent of their scatter about the lower main sequence?
b) What is the observed leftward limit, following Poveda (1964), of the flare
stars in the H—R diagram? *c)* Is there evidence for broad-band colour changes
in flare stars such as found in the RW Aurigae stars (Broglia, Lenouvel 1960,
Mosidze 1967) and *d)* What is the relation between flare stars of the Orion
type, for example, and the classical UV Ceti variables?

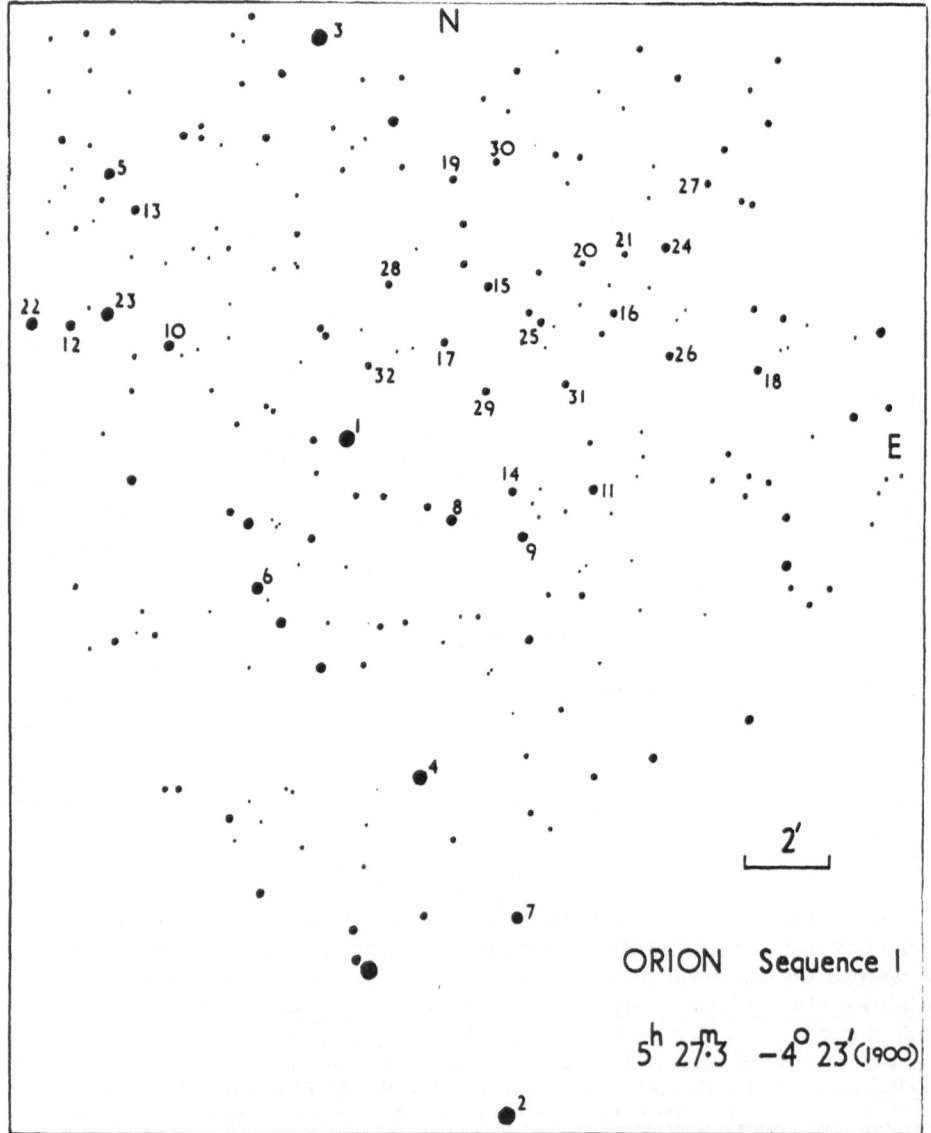

Fig. 1. Sequence stars (Centre R. A. 5^h 27^{m}3, Dec.—4°23', 1900)

  My original intention was to extend Mendoza's photoelectric work in Orion to flare stars with V = 16^m in four bands, U, B, V and R, using the 60-inch reflector of the Boyden Observatory. This initial work was abandoned as being far too time-consuming for a substantial number of stars to be measured with sufficient accuracy. Instead, photographic photometry, based on a new photoelectric sequence, was attempted for those stars only slightly affected by nebulosity. Correction for small variations in nebulous fog has

Table 1a
Photoelectric Sequence

No.	P	V	V—R	B—V	U—B	n	V'	(V—R)'	(B—V)'	(U—B)'
1	866	8^{m}769	0^{m}887	1^{m}219	1^{m}263	6	primary standard			
2	908	8.806	0.038	−0.034	−0.199	6	primary standard			
3	857	8.25	1.38	1.78	1.94	2	8^{m}21	1^{m}44	1^{m}78	2^{m}16
4	878	10.12	0.50	0.51	−0.02	1	10.10	0.49	0.57	—0.14
5	784	10.88	0.61	0.48	0.16	2	10.77	0.52	0.52	0.22
6	824	10.91	0.58	0.64	0.09	1	10.92	0.56	0.69	—0,02
7	917	11.36	0.54	0.59	0.01	1	11.29	0.47	0.73	—0.09
8	895	11.71	1.18	1.42	1.14	4	11.73	1.17	1.43	1.09
9	930	11.88	0.92	1.06	0.70	4	11.84	0.87	1.10	0.66
10	804	12.05	0.58	0.56	0.08	1	11.96	0.56	0.65	0.07
11	952	12.30	0.79	0.95	0.62	3	12.40	0.85	0.93	0.56
12	767	12.57	0.64	0.67	0.11	1	12.69	0.61	0.57	0.03
13	793	12.86		0.66		2	12.77	0.72	0.68	0.11
14	924	13.94	0.80	0.90	0.16	2	13.99	1.02	0.83	0.26
15		14.29	1.30	1.48	1.07	1	14.50	1.48	1.51	1.34
16		14.67	0.84	0.86	0.10	1	14.85	0.99	0.71	0.01
17		15.50	1.19	1.31	1.11	1	15.53	1.21	1.18	0.79
18		15.50	1.14	0.79	0.49	1	15.53	0.90	0.95	0.23
19		15.66		0.87		1	15.71	0.86	0.85	0.08
20		16.03	0.54	1.23		1	16.18	0.67	1.14	0.17
21		16.11	0.89	0.53	0.39	1	15.79	0.74	1.25	0.24

Table 1b
Photographic Sequence

No.	P	V'	(V—R)'	(B—V)'	(U—B)'
22	748	10^{m}19	1^{m}52	1^{m}78	1^{m}69
23	775	10.22	0.20	0.20	0.10
24	979	14.41	0.98	0.82	0.13
25	936	14.56	1.21	1.05	0.33
26	980	14.75	1.05	0.80	0.15
27		15.75	1.13	1.22	0.54
28		15.77	1.00	1.11	0.31
29		15.90	0.38	1.44	0.38
30		16.01	0.92	1.12	0.43
31		16.05	0.76	1.09	0.20
32		16.12	0.69	1.15	0.32

been successfully applied using an empirical technique developed for ADH Baker—Schmidt plates by C. J. Butler at Dunsink Observatory (private communication). In this report is presented the first part of the reduced material only, with a brief discussion of the photographic accuracy, and the applications and limitations of broad-band photometry applied to flare stars.

The Boyden 60-inch reflector, freshly aluminized, was equipped with an E.M.I. 9558 QA photomultiplier, magnetically-shielded and cooled to 0° C, with sensitivity ranging from the ultraviolet to the near-infrared. The

following filter combinations were used which allow reproduction of the standard system of Johnson et al. (1966), and remove the red leak:

U 1 mm UG 2 + 2.5 mm 80% saturated $CuSO_4$ soln. at 15 °C
B 1 mm BG 12 + 2   mm GG 13 + 1 mm BG 18
V 2 mm GG 14 + 2   mm BG 18.
R 2 mm RG 5

The effective wavelength of the red filter-tube combination at about 7150 A
is 150 A longward of Johnson's value, cutting off the H-alpha line.

The magnitude sequence extends to $17^{m}9$, $17^{m}2$, $16^{m}1$ and $15^{m}5$ in
U, B, V and R, respectively. Colour transformations were studied for dwarf
and giant stars to a redward limit of $1^{m}8$, $1^{m}6$ and $1^{m}5$, in U—B, B—V and
V—R, respectively. The effects of reddening and peculiar spectra have not
yet been examined. Nightly-determined zero-points and extinction coefficients
(with second-order colour dependence in V and B—V only) were used together
with colour equations derived during the same observing period. For the faint
R-scale extension, a 4-magnitude perforated aperture-screen was employed
at the 60-inch to first establish V—R colours for primary standards at about
$V = 9^{m}$, using the Arizona Tonantzintla Catalogue stars (Johnson, 1966).
The sequence is given in Table 1a. The columns give 1) Reference number as
per Finding Chart (Fig. 1), 2) Parenago's (1954) designation, 3) to 6) Photoelectric magnitudes and colours, 7) Number of observations, 8) Photographi-
cally-smoothed magnitudes and colours (not used in the present work). A
photographic interpolation to a number of other stars in the field of the
sequence is added in Table 1b. The probable errors for the primary standards
in V, U—B, B—V and V—R are $\pm 0^{m}014$, $\pm 0^{m}035$, $\pm 0^{m}007$ and $\pm 0^{m}018$,
respectively, and for the faint end of the sequence, about five times these
values.

A large number of scattered photoelectric standards were also set up
in the Orion region for the study of photographic colour-corrections, field-
errors etc. The agreement with the work of Johnson (1957), Sharpless (1952,
1954, 1962) and Lee (1968), for a number of common stars, was within the
errors of measurement. A few additional stars taken from their work were
consequently used in the photographic reductions but given one-third weight to
reduce possible systematic effects across our plates.

PHOTOGRAPHIC MATERIAL

Over a period of two months, 76 plates were taken at the ADH 32/36-inch
Baker—Schmidt telescope, with the following plate-filter combinations as
frequently as possible on the same night:

U   103a-O   +   2 mm UG 2.
B   IIa-O   +   1 mm BG 12 + 2 mm GG 18
V   103-aD   +   2 mm GG 11
      IIa-D
R   103a-U   +   RG 1.

The present discussion is limited to three sets of UBVR plates and another single V plate, with the centre, R.A. $5^h30^m.0$, Dec. —5.0, Equinox 1900. Several different exposure times, ranging from 7 to 59 minutes, were taken on each emulsion in order to study the effects of nebulous fog. Fuller details are given later in Table 2.

PLATE MEASUREMENT AND REDUCTION

A maximum number of 303 stars, depending on the emulsion and exposure time, has been measured on the above plates using the Sartorius iris-photometer of Armagh Obervatory. This number comprises 113 flare stars, 43 other Orion variables, 99 photoelectric standards and 48 control stars of unknown magnitude. For each star a mean measure of the neighbouring fog-density (W) on an arbitrary scales was made, after completion of the iris measurement ($\emptyset$) for the whole plate, by wedge photometry of the densest plates. Differences due to sky fog from plate to plate of the same emulsion were shown to be negligible compared with the nebulous-fog. An X and Y measure for each star was also made with an arbitrary centre of co-ordinates.

Basically, Butler's ADH plate reduction technique involves a least-square solution for the coefficients in normal equations of the form:

$$\text{p.e. mag} + F(\emptyset) + G(X, Y) + K(Col) + L(Den) + \text{const} = 0$$

where the functions, F, G, K and L, are carried to as high an order as required for a satisfactory solution. See Remarks under Table 2. In fact, the solution is built up step by step, with attention to stars with exceptional residuals, δm (p.g. — p.e.), with the solution for position, colour-and density-dependence limited to stars with magnitude brighter than $15^m.5$, and with a system of weighting to ensure that the final calibration curve is dominated by stars of the p.e. sequence. The reduction for a set of UBVR plates is illustrated in Fig. 2 showing the accuracy attainable with the densest of our plates. Table 2 summarizes the colour and density coefficients used in calculating photographic magnitudes, and gives r.m.s. errors in the fitting of the final calibration curve. All least-square solutions were performed on the I.B.M. 1620 computer of Dunsink Observatory.

From a comparison of the derived magnitudes for non-standard stars from each series of plates a general empirical limit for an acceptable background-density variation was set. This was found to correspond to a maximum magnitude-correction of $0^m.5$ (for density alone). By contrast, final magnitudes of these stars agreed, then, to better than $\pm 0^m.1$ from one series to another. The fog-image interaction is, thereby, not directly studied. Practically the full field of an ADH plate (16 × 16 sq. cms.) within an area of 6 sq. degs., could be utilized to yield magnitudes and colours with probable errors less than $\pm 0^m.1$ to V = 16^m, indicating that field errors were well corrected. The value of these reults to this order of accuracy is evident in the colour-magnitude diagrams for the Orion flare stars.

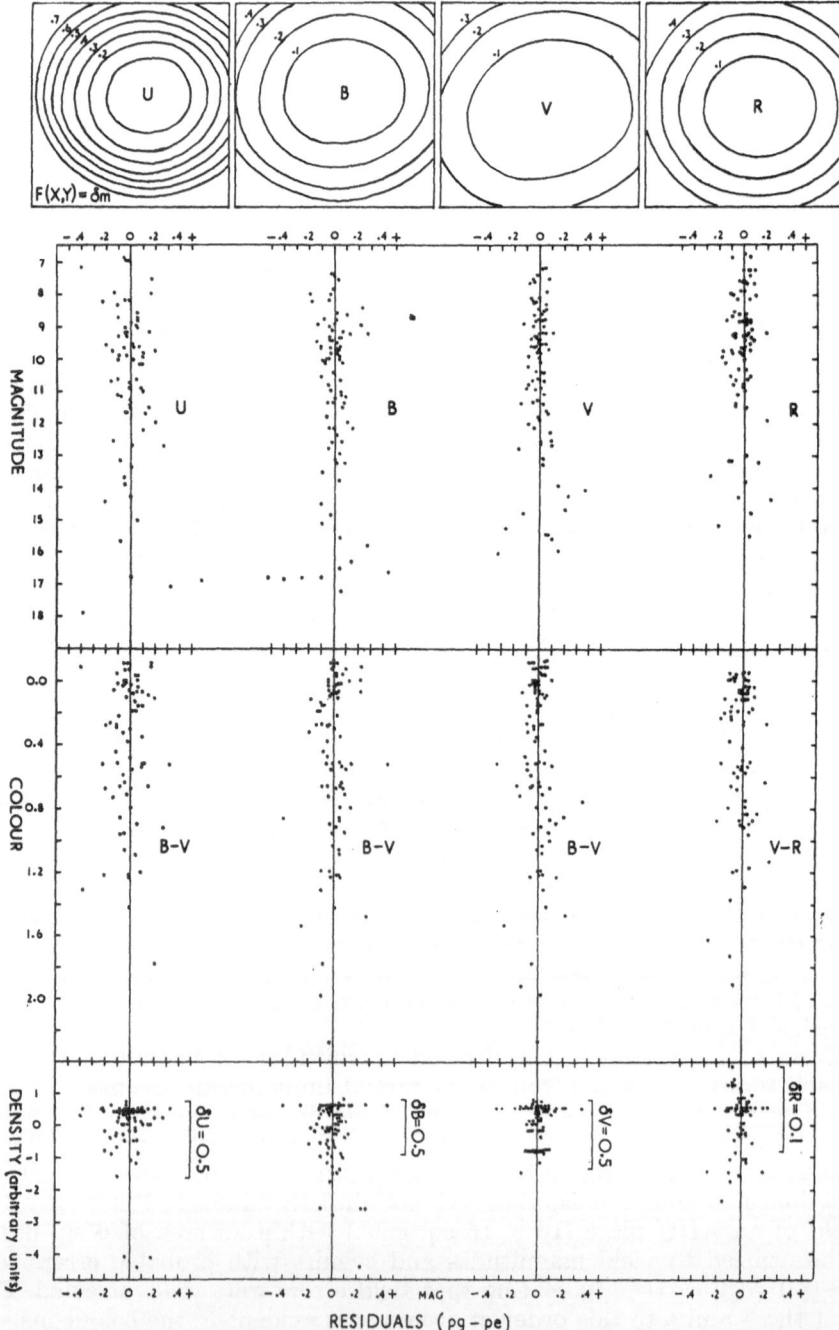

Fig. 2. Photographic reductions of ADH Baker—Schmidt plates (Nos. 8282, 8279, 8278 and 8280) showing field-correction contours in tenths of a magnitude over an area of 2.9 × 2.9 sq. degs., and photographic minus photoelectric magnitude residuals as a function of magnitude, colour and background density. See Table 2.

Table 2

Summary of Photographic Data: Colour and Density Coefficients and r. m. s. Errors of Calibration Curves

ADH Plate No.	8314	8345	8282	8313	8343	8279	8311	8278	8341	8364	8312	8346	8280
J. D. − 2439400	$52^{d}.072$	$68^{d}.957$	$43^{d}.067$	$52^{d}.058$	$68^{d}.897$	$42^{d}.946$	$52^{d}.028$	$42^{d}.916$	$68^{d}.863$	$90^{d}.792$	$52^{d}.043$	$68^{d}.997$	$42^{d}.997$
Exp. (filter)	15^{m}(U)	40^{m}(U)	59^{m}(U)	7^{m}(B)	20^{m}(B)	31^{m}(B)	7^{m}(V)	15^{m}(V)	21^{m}(V)	30^{m}(V)*	15^{m}(R)	40^{m}(R)	59^{m}(R)
k = Col. coeff.	−0.013	−0.047	−0.037	−0.157	−0.116	−0.149	0.109	−0.169	0.166	0.166	−0.071	−0.038	−0.033
l = Den. coeff.	−0.055	−0.157	−0.211	−0.019	−0.209	−0.291	−0.083	−0.232	−0.288	−0.304	−0.004	−0.046	−0.037
r . m. s. error	$\pm0^{m}.075$	$\pm0^{m}.094$	$\pm0^{m}.163$	$\pm0^{m}.112$	$\pm0^{m}.064$	$\pm0^{m}.082$	$\pm0^{m}.070$	$\pm0^{m}.081$	$\pm0^{m}.096$	$\pm0^{m}.085$	$\pm0^{m}.086$	$\pm0^{m}.090$	$\pm0^{m}.084$

Remarks to Table 2

Photographic magnitudes (U, B and V) are derived from the following equation:

$$\text{p.g. mag.} = a^5 + b^4 + c^3 + d^2 + e + fX^2 + gY^2 + hXY + iX + jY + k(B-V) + l(W) + \text{const.}$$

A similar equation is used for R except that the colour, $V-R$, is substituted. Since the nebulous background is most serious in the ultraviolet, affecting both the photoelectric and photographic measures, a $U-B$ colour dependence was not used. After correction on the U plates for $B-V$ dependence, however, no further dependence on $U-B$ was evident. In Table 2, the colour coefficient, k, may be defined as the magnitude correction at $B-V$ (or $V-R$) $= 1^{m}.0$, and the relative magnitude corrections for density may be derived by applying the coefficient, l, to the density scale in Fig. 2. It may be seen that the density coefficients for a given emulsion are fairly smooth (almost linear) functions of the exposure times. N. B. The asterisk indicates use of 103a−D instead of the usual IIa−D emulsion employed. Also, we note that the colour coefficients are consistent within photographic accuracies. The form of the field corrections, $G(X, Y)$, are, however, considerably different from plate to plate and from emulsion to emulsion (not tabulated). This may be due to differences in image quality, focus, plate-tilt, etc. From the r. m. s. errors for the fitting of the final calibration curves over the whole plate (2.9×2.9 sq. degs.), the plate reductions appear most satisfactory.

COLOUR-MAGNITUDE AND COLOUR-COLOUR DIAGRAMS

In B, V and R, 68 flare stars satisfy the following conditions, *a)* within 85' of the plate centre, *b)* background-density corrections less than 0.^{m}5 and *c)* all magnitudes within the limits of the photoelectric sequence. In U, the

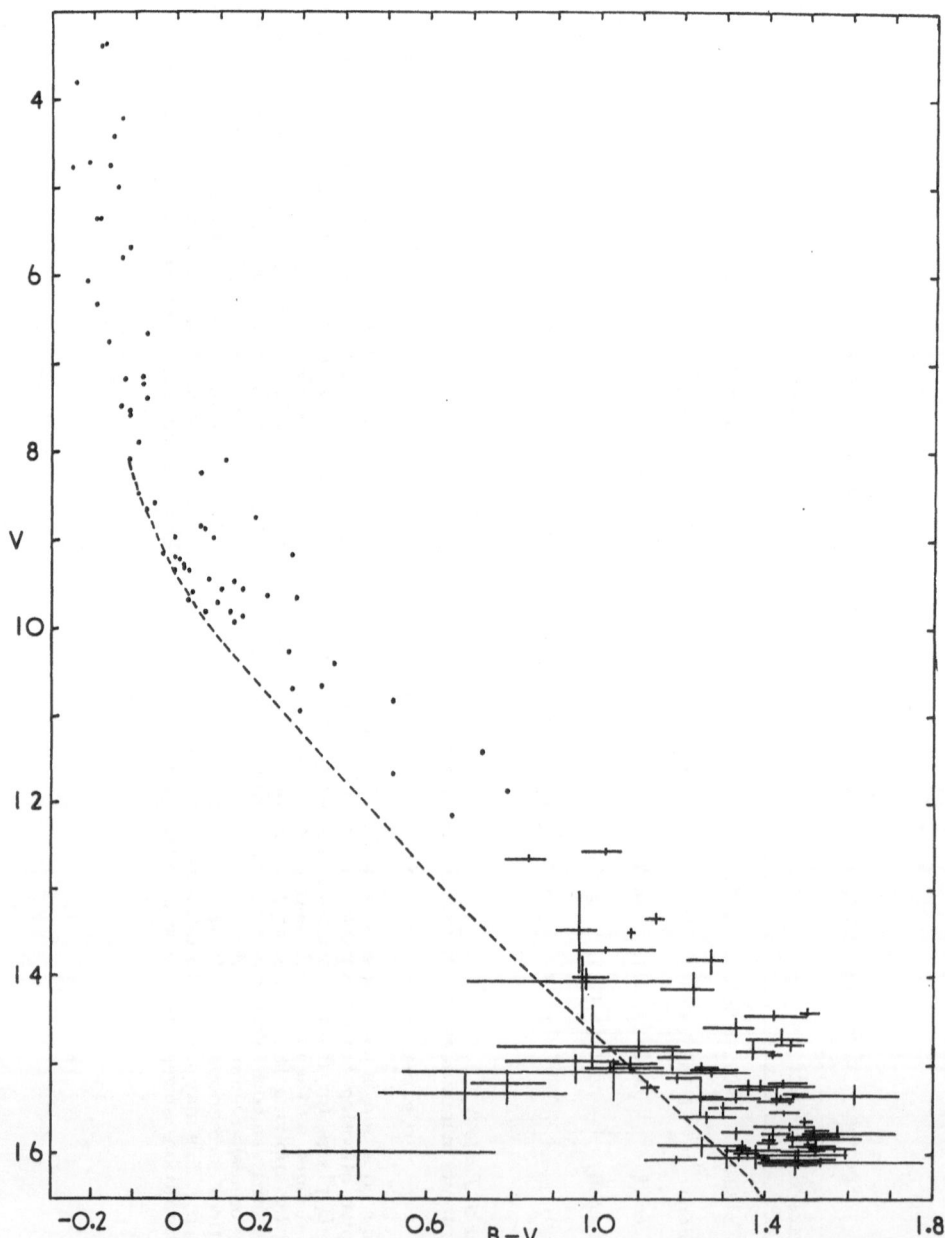

Fig. 3. Colour-magnitude diagram for Orion flare stars showing the scatter in the observed values. The dots indicate bright members of the Orion association and the dashed line is the zero-age main sequence using Mendoza's (1967) distance modulus of 7.^{m}9.

number is smaller, only 37. These selected flare stars havé been plotted in colour-magnitude and colour-colour diagrams (Figs. 3, 4, 5 and 6), showing at the same time the scatter in the values for the 15 plates. The zero-age main-sequence and the standard colour relations for normal, unreddened main-sequence stars are indicated (Johnson 1963, Mendoza 1967). Also, the position

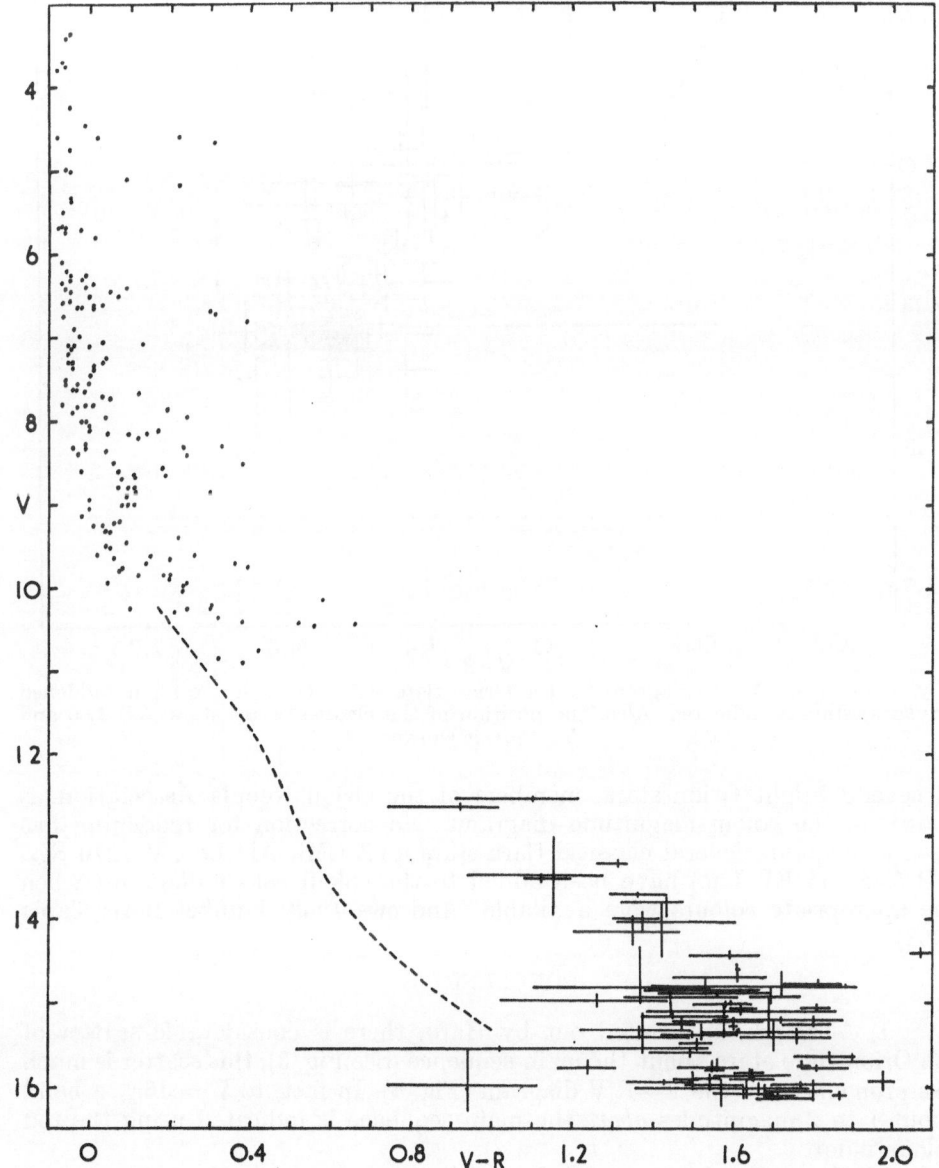

Fig. 4. As in Fig. 3 except using the V—R colour. The zero-age main sequence is derived using Mendoza's (1967) relation between B—V and V—R for the Hyades cluster stars. The bright stars are taken from the work of Lee (1968).

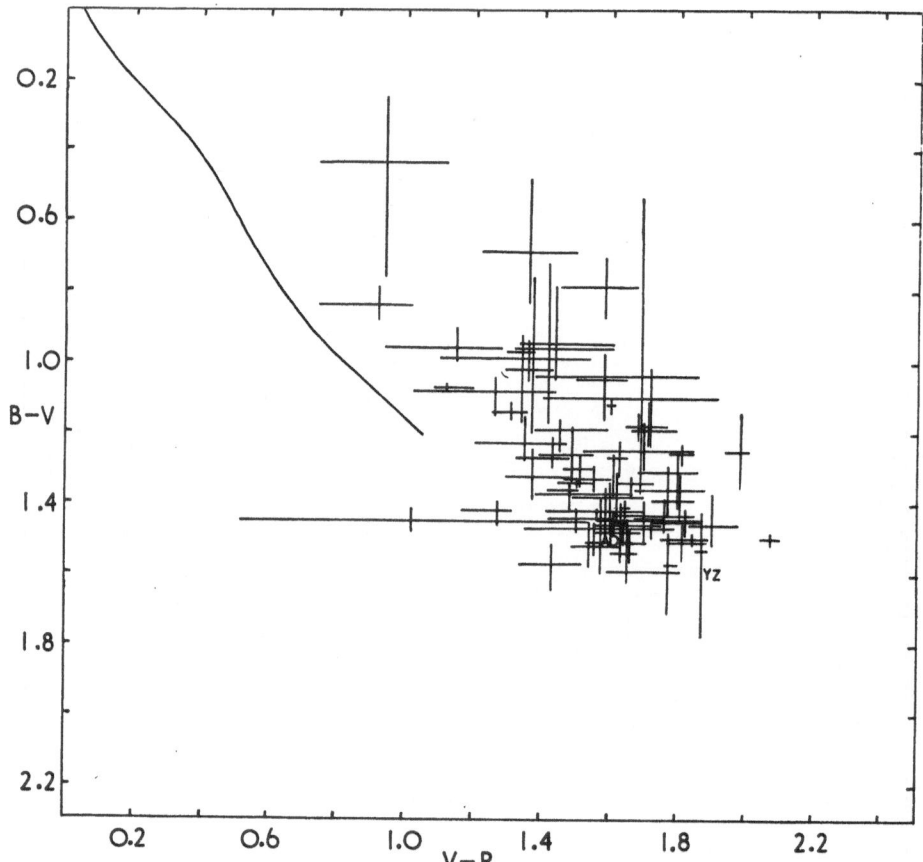

Fig. 5. The B—V/V—R diagram for the Orion flare stars. The relation for unreddened (Hyades) stars is indicated. Also, the position of the classical flare stars, AD Leo and YZ CMi, is shown.

of several bright Orion stars, members of the Orion Nebula Association, is shown in the colour-magnitude diagrams. No correction for reddening has been attempted. Several classical flare stars (YZ CMi, AD Leo, V 1216 Sgr, DH Car and EV Lac) have been added to the colour-colour diagrams when the appropriate colours were available (Andrews 1968; Kunkel 1967; Tapia 1968).

CONCLUSIONS

I) Although, as pointed out by Haro, there is considerable scatter of the Orion flare stars about the main-sequence (See Fig. 3), this scatter is much less-pronounced in the V—R/V diagram (Fig. 4). In fact, to V = 16^m, a band about 1 to 2 magnitudes *above* the main sequence is indicated using the red colour index.

II) There is a fairly clear leftward limit of the Orion flare stars in the V—R/V diagram at about V—R = 1^{m}0, defined by the brighter flare stars of spectral type K0 to K1.

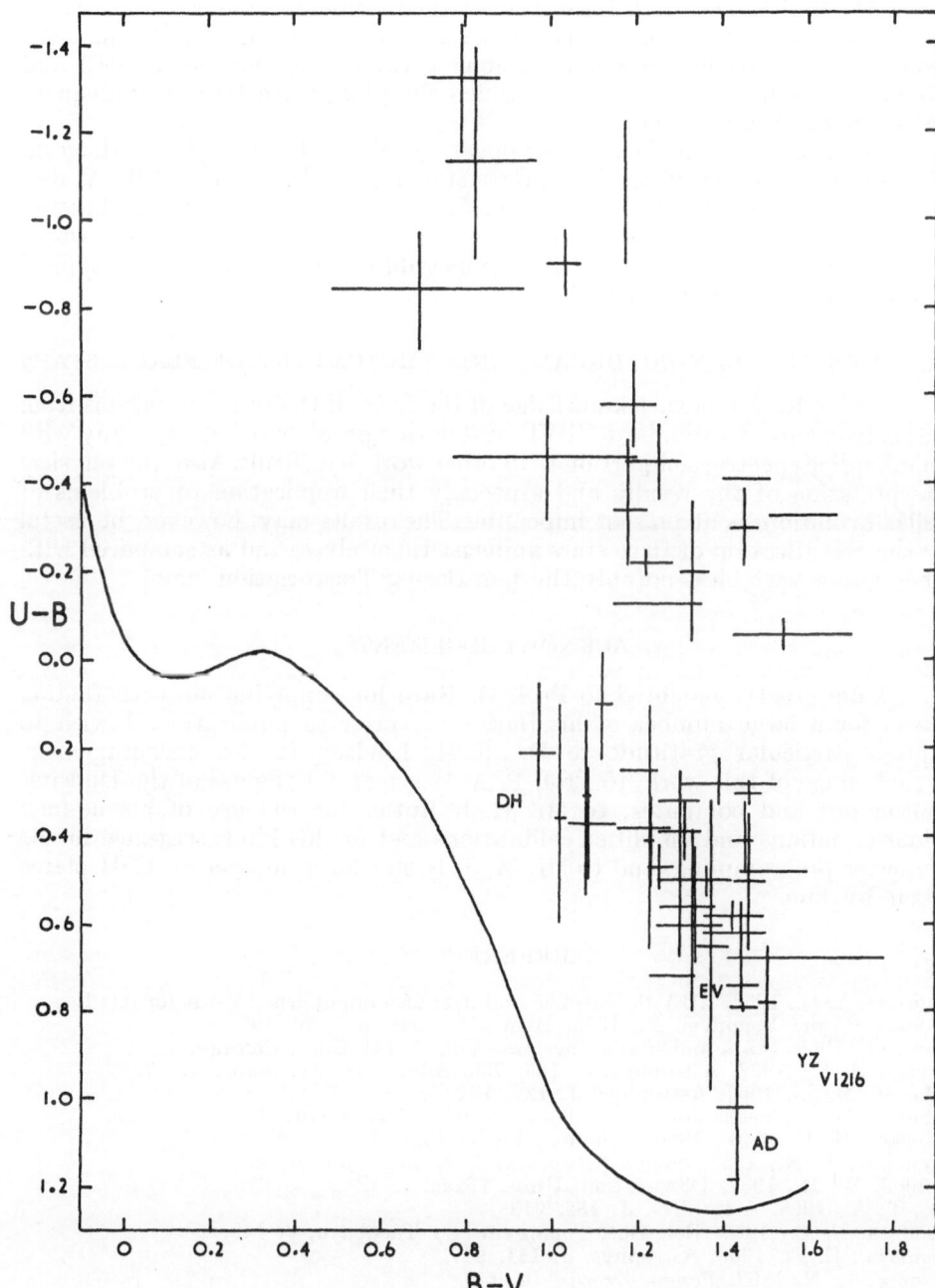

Fig. 6. The U−B/B−V diagram for the Orion flare stars. The standard colour relation for normal, unreddened main-sequence stars is indicated. The classical flare stars, AD Leo. YZ CMi, V 1216 Sgr, EV Lac and DH Car, are shown.

III) Half the selected flare stars showed no magnitude or colour variations during the 2-month observing period, certainly not greater than the
probable errors of measurement. Colour variations amounting to 0.^{m}3 and
more are evident in some stars for which the plate corrections are adequate
(in the sense stated above).

IV) The classical flare stars appear to fall in the B—V/V—R diagram
within but to the red of the Orion flare-star region. In the U—B/B—V diagram, all but DH Car fall somewhat below the reddest Orion stars. There is,
however, some doubt attached to the enormous ultraviolet colour-excesses
in some Orion flare stars, and it is impossible to decide what is a typical
U—B colour for these stars.

REMARKS CONCERNING BROAD-BAND PHOTOMETRY OF FLARE STARS

Owing to the inexact knowledge of the form of the transformations from
the instrumental to standard UBVR system for peculiar red stars, many with
emission-line spectra, comparison with other work is difficult. Also, the physical
interpretation of the results and especially their application to problems of
stellar evolution, is at present impossible. The results may, however, be useful
for the classification of flare stars amongst themselves, and as compared with
other Orion variables, notably the non-flaring T-association stars.

ACKNOWLEDGEMENTS

I am greatly indebted to Prof. G. Haro for supplying me with finding
charts for a large number of his flare stars prior to publication. I wish to
express particular gratitude to Dr. E. M. Lindsay for his encouragement
at each stage of this work, to Prof. P. A. Wayman for the use of the Dunsink
photometer and computer, to Mr. G. J. Butler for the use of his 60-inch
colour-equations and amplifier calibrations and for his kind assistance in the
computer programming, and to Mr. M. J. Bester for a number of ADH plates
taken by him.

REFERENCES

Andrews, A. D., 1968, I.B.V.S. Nos. 265 and 273; also unpublished V—R for AD Leo.
Broglia, P. and Lenouvel, F., 1960, Mém. Soc. astr. ital. **30**. 199.
Haro, G., 1968. Stars and Stellar Systems Vol. 7, 141 Univ. Chicago.
Herbig, G. H., 1962. Astrophys. J. **135**, 736, Adv. Astr. Astrophys. **1**, 47.
Johnson, H. L., 1957, Astrophys. J. **126**, 134.
Johnson, H. L., 1963, Stars and Stellar Systems. Vol. 3. 204. Univ. Chicago.
Johnson, H. L. et al., 1966, Commun. lunar planet Lab. **4**. 99.
Kholopov, P. N., 1959, Soviet Astron. A. J. **3**. 291. translation.
Kunkel, W. E., 1967, Dissertation. Univ. Texas.
Lee, T. A., 1968, Astrophys. J. **152**, 913.
Mendoza, E. E., 1967, Bol. Obs. Tonantzintla y Tacubaya, **4**. 149.
Mendoza,, E. E., 1968, Astrophys. J. **151**, 977.
Mosidze, L. N. 1967, Perem. Zvezdy **16**, 149.
Parenago, P., 1954, Trudy Gos. astr. Inst. Sternberga, **25**.
Poveda, A., 1964, Nature **202**, 1319.
Sharpless, S., 1952, Astrophys. J. **116**. 251.
Sharpless, S., 1954, Astrophys. J. **119**, 200.
Sharpless, S., 1962, Astrophys. J. **136**, 767.
Tapia, S., 1968 .I.B.V.S. No. 286.

PHOTOELECTRIC RESEARCH ON FLARE STARS AT THE CATANIA ASTROPHYSICAL OBSERVATORY

S. CRISTALDI, G. GODOLI, M. NARBONE and M. RODONÒ

Catania Astrophysical Observatory, Italy

SUMMARY

In this paper the results recently obtained at Catania from the observations of the flare stars PZ Mon, YZ CMi, AD Leo, BD +55°1823, BD +51°2402 and EV Lac are summarized.

INTRODUCTION

In 1967 systematic photoelectric observations of flare stars were started at the Catania Astrophysical Observatory.

These observations have been carried out in the course of the Catania research programme on stellar activity of the solar type (Godoli, 1967, 1968).

Some results have already been published (Cristaldi, Rodonò, 1968a; 1968b; Cristaldi, Narbone, Rodonò, 1968; Godoli, 1968). Here the observations and the results obtained until August 15, 1968 will be summarized.

OBSERVATIONS

At Catania four main instruments are available for stellar work (Fracastoro, 1967).

1. A Newtonian-Cassegrain reflector of 91 cm aperture, 400 cm and 1400 cm focal length respectively.

2. A Schmidt quasi-Cassegrain universal reflector (41 cm aperture and 119 cm focal length in the Schmidt combitation, 61 cm aperture and 600 cm focal length in the quasi-Cassegrain combination).

3. Two Cassegrain telescopes of 31 cm aperture and 500 cm focal length, which we discriminate by their position (N or S) in their common dome.

All these instruments are equipped for photoelectric observations with EMI 6256 photomultiplier tubes.

The characteristics of the stars observed, the instruments and the filters used are reported in Table 1.

For the study of the physics of the flares it is very important to carry out three-colour photometry which due to the abrupt nature of the phenomena is not feasible with traditional photometers. For this reason at Catania a suitable photometer for simultaneous multi-colour photometry has recently been completed (Cristaldi, Paterno, 1968a; 1968b). By this new apparatus and the universal reflector used in the quasi-Cassegrain combination observations of EV Lac and BD +55°1823 have recently been carried out.

The star YZ CMi was observed on occasion of an international optical and radio flare patrol programme organized by Prof. Chugainov and Prof.

Table 1

Characteristics of the observed flare stars (August, 1968) and instruments and filters used.

Flare star	Coordinates (1950.0)		m_v	Sp.	Instrument (cm aperture)	Filters	λ equ.
	R. A.	D.					
PZ Mon	06h43m2	+01°13′	9m5	K2e	30N	V	5150 Å
YZ CMi	07 39 .5	+03 48	11 .6	M4.5e	91	B	4300
AD Leo	10 14 .2	+20 22	9 .5	M4e	91	V	5150
BD+55°1823	16 14 .9	+55 32	10 .1	M1.5e	30S	V	5150
BD+55°1823					61	B V	4300, 5150
BD+51°2402	18 31 .6	+51 39	8 .3	K6e	30S	V	5150
BD+51°2402					30N	V	5150
EV Lac	22 42 .6	+43 49	10 .2	M4.5e	30S	V	5150
EV Lac					61	B V	4300, 5150
EV Lac					91	V	5150

Table 2

Comparison stars.

Flare star	Comparison star		
	Star	m_v	Sp.
PZ Mon	BD +1°1495	9.2	K
YZ CMi	—	—	—
AD Leo	BD +20°2475	9.0	K
BD +55°1823	BD +55°1834	9.0	K
BD +51°2402	BD +51°2410	7.7	K2
EV Lac	BD +42°4527	8.6	M

Lovell in two observing periods: January 22—February 6 and February 21—March 7, 1968.

As we were interested not only in the light curves of flares themselves but also in the brightness fluctuations at minimum, sky and comparison stars measurements have been performed about every 20 minutes. The comparison stars are listed in Table 2.

The observations $(m_v - m_c)$ versus JD_{hel} are corrected for atmospheric extinction using a statistically determined mean absorption coefficient. Also $\log I_{flare}/I_{normal}$ versus Universal Time are considered for the periods in which flares occur. The data are reduced by the IBM 1620 computer of the Science Faculty of Catania University.

RESULTS

The duration of patrolling in hours, the number and the characteristics of the flares observed for each star are given in Table 3.

Up to now (August 15, 1968) during about 278 hours of the flare patrol programme 12 flares have been observed.

Table 3

Flares observed until August 15, 1968.

Flare star	Hours of patrol (August 15, 1968)	Number of flares for 100 hours	Flare observed				
			Epoch of max.	Δm (magn.)	rise time (minutes)	total duration (minutes)	rise speed (magn/sec)
PZ Mon	38^h.2	2.6	67/09/21 03^{h}40^m	0^{m}11	0^{m}7	1^{m}7	0.003
YZ CMi	16 .5	24.2	68/02/26 21 26	0 . 24	0 . 9	7 . 2	0.004
			68/02/26 22 26	0 . 35	?	?	?
			68/03/01 20 00	0 . 62	1 . 4	?	0.007
			68/03/03 20 01	0 . 23	1 . 4	23 . 3	0.003
AD Leo	36 .3	5.5	68/03/04 02 27	0 . 08	0 . 4	10 . 0	0.003
			68/04/24 19 28	0 . 46	1 . 4	22 . 0	0.005
BD +55°1823	25 .2	0.0					
BD +51°2402	96 .9	1.0	67/07/25 01 59	0 . 07	5 . 0	12 . 2	0.001
EV Lac	64 .5	6.2	67/09/27 22 08	0 . 14	0 . 6	5 . 5	0.004
				0 . 16			(first rise)
			67/09/30 00 27	0 . 32	0 . 6	7 . 5	0.009
			67/12/29 19 19	0 . 92	1 . 3	6 . 7	0.012
			68/07/21 02 24	1 . 06 (V)	0 . 5	22 . 0	0.035
			02 24	1 . 12 (B)	0 . 7	?	0.027
TOTAL	277 .6	4.3					

The light curves of the observed flares are plotted in Figs. 1—3. In Fig. 4 the simultaneous B and V measurements of the EV Lac flare observed on July 21, 1968 are plotted together with the (B—V) colour indices. The star appears bluer just before the flare maximum. A strong reddening of 0^{m}5 took place during the decreasing phase.

We should notice that the two stars PZ Mon and BD +51°2402 of relatively early type were less active than the normal M type flare stars. The most rewarding stars have been YZ CMi and EV Lac. Naturally it is not possible to draw a decisive conclusion on this matter from our data alone.

We notice that the first observed flare of EV Lac shows two peaks. The structure of this flare recalls the existence of the symphathetic solar flares.

From the light curves of the flares one can see that often before a flare starts, the star's brightness decreases gradually or abruptly by the order of 0^{m}01. This light variation appears to be clearly beyond the range of error.

For EV Lac, brightness fluctuations at minimum of light have been observed. These fluctuations have an amplitude of about 0^{m}1 and seem to be cyclic, although we cannot give a definite period because of the discontinuity in the observations. It is of interest pointing out that all the observed flares of EV Lac took place during the minima of these fluctuations.

Also for BD +51°2402 small brightness fluctuations, which in this case seem to be not periodical, have been found at the minimum.

In Fig. 5 for each flare the maximum intensity versus the ratio d/D has been plotted, where d indicates the rising time and D indicates the total duration of a flare. Catania and other observations have been used. It is evident that the ratio d/D is smaller for larger flares.

One might suppose that this ratio is overestimated for flares of small intensity, D in this case being underestimated. In order to avoid this criticism,

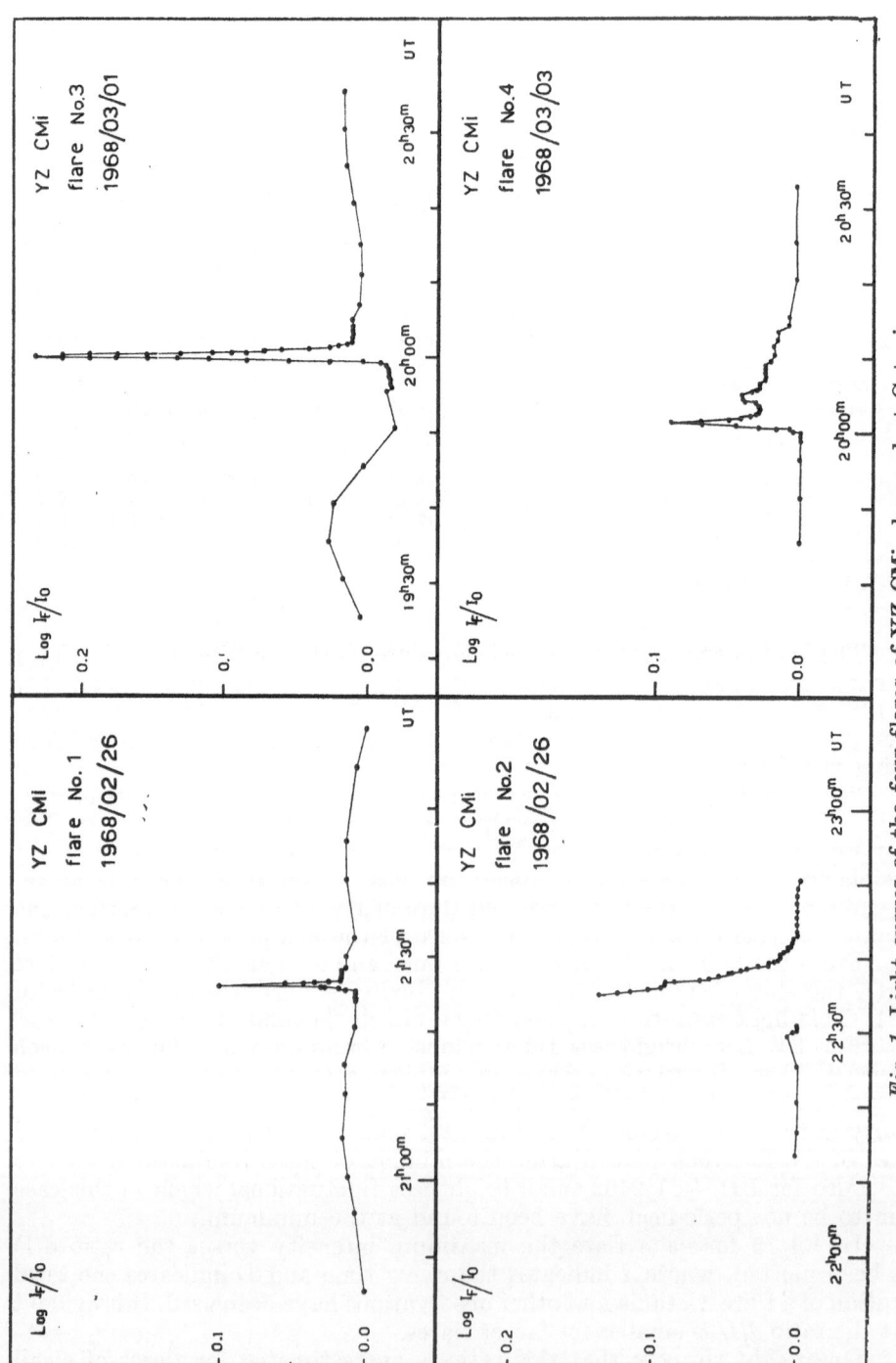

Fig. 1. Light curves of the four flares of YZ CMi observed at Catania.

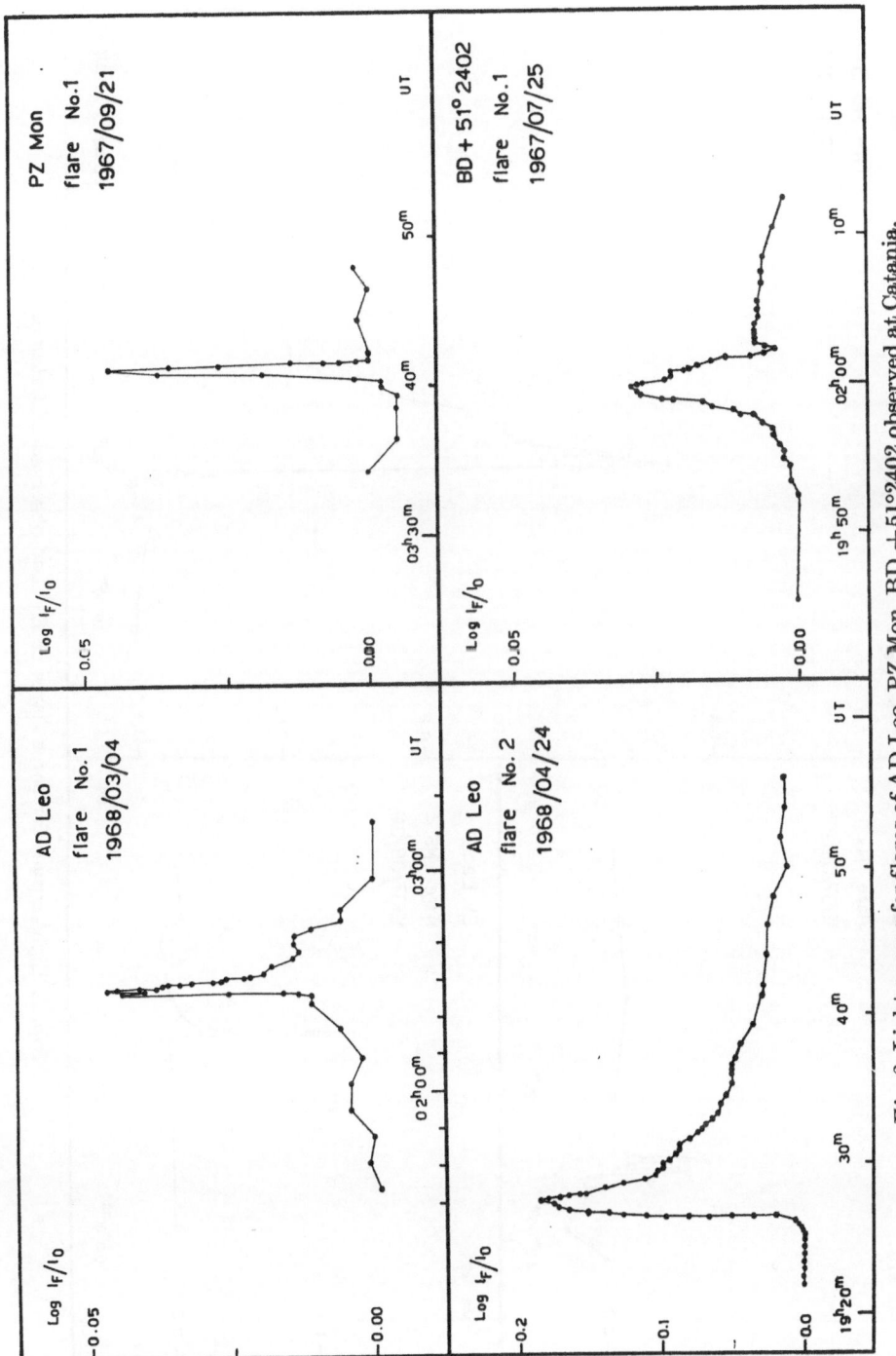

Fig. 2. Light curves for flares of AD Leo, PZ Mon, BD +51°2402 observed at Catania.

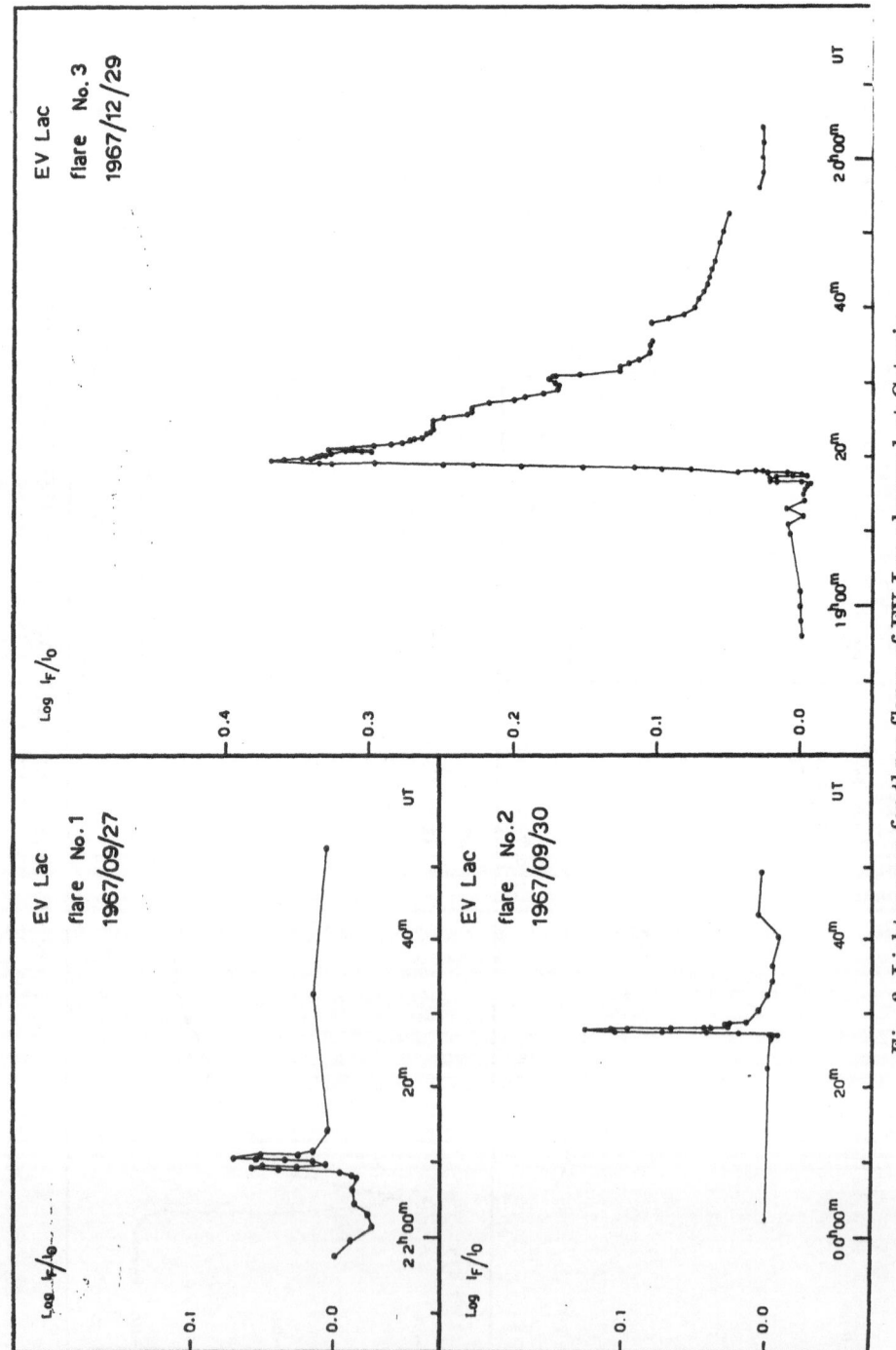

Fig. 3. Light curves for three flares of EV Lac observed at Catania.

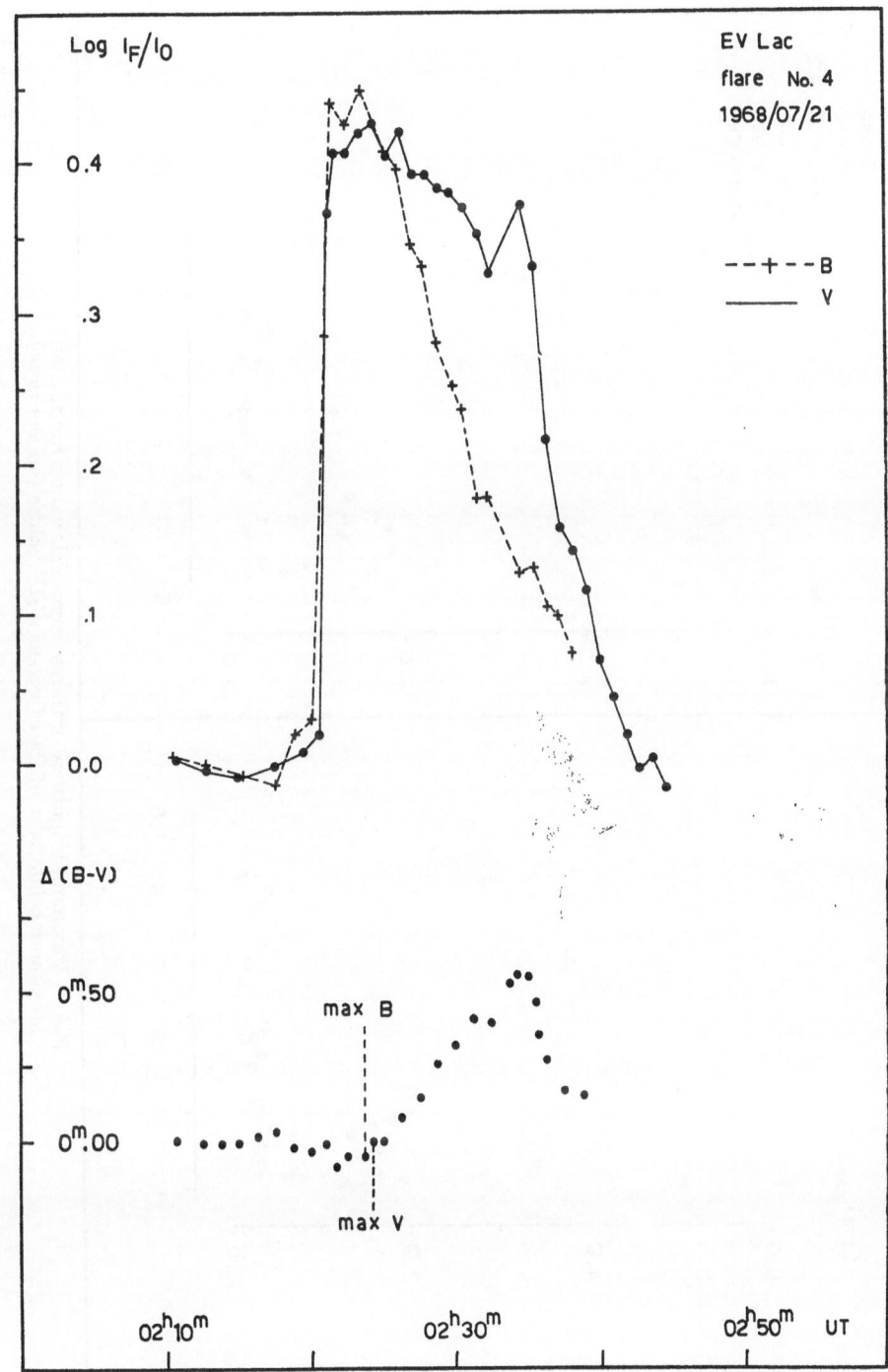

Fig. 4. B and V light curves for flare No. 4 of EV Lac.

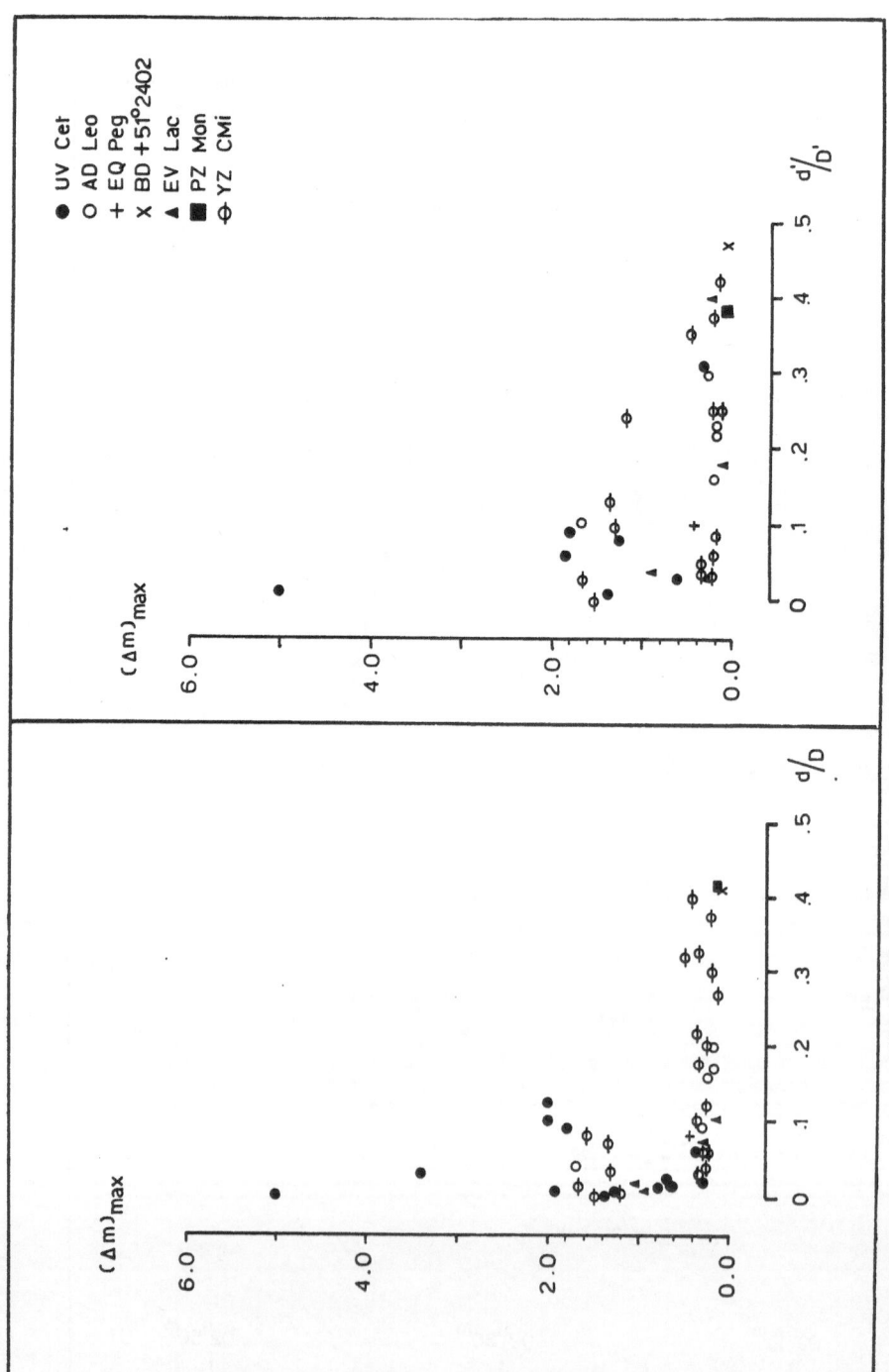

Fig. 5. Maximum brightness of flares versus d/D values (on the left). Maximum brightness of flares versus d'/D' values (on the right).

in Fig. 5. (on the right) the ratio d'/D' has been plotted, where the durations now refer to the part of the flares with log $I_f/I_n \geq 20\%$ $(I_f/I_n)_{max}$. Also in this case the relationship is confirmed.

Notes

1. The photoelectric data of each night are too numerous to be published in this paper. The interested researchers may have a copy by request directly from the authors.

2. The individual observations will be sent to the IAU Comm. 27 depository of unpublished photoelectric observations of variable stars as recommended in IAU Transactions XIII A, p. 510.

REFERENCES

Cristaldi, S., Paternò, L., 1968a, Mem. Soc. astr. ital., **39.**
Cristaldi, S., Paternò, L., 1968b, These Proceedings.
Cristaldi, S., Rodonò, M., 1968a, Comm. 27 of IAU, Inf. Bull. Var. Stars No. 252.
Cristaldi, S., Rodonò, M., 1968b, Comm. 27 of IAU, Inf. Bull. Var. Stars No. 274.
Cristaldi, S., Narbone, M., Rodonò, M., 1968, Mem. Soc. astr. ital., **39,** 339.
Fracastoro, M. G., 1967, Oss. astrofis. Catania, Pubbl. No. 107.
Godoli, G., 1967, Oss. astrofis. Catania, Pubbl. No. 115.
Godoli, G., 1968, in "Mass Motion in Solar Flares and Related Phenomena" IX Nobel Symposium, Anacapri.

COMMENT

Wood: I am much interested in the light fluctuations shown just before some of the flares. In observing SX Phe — quite a different type of variable but one which also shows a rapid increase of brightness — it was found that on the one night quite similar fluctuations appeared just before the increase. This was the only night observation made in this particular region, and I have never known whether or not to believe that the fluctuations were really in the star, although the observing conditions were excellent. If flare stars show such changes preceding the outbursts, it seems more likely that other stars showing sudden increases may do the same. This suggests that in stars such as SX Phe, where the intervals between maxima are predictable, observers should concentrate on the interval just preceding the beginning of the rise to maximum.

HD 160202 — AN EARLY-TYPE FLARE STAR

GUSTAV A. BAKOS

University of Waterloo
Waterloo, Ontario, Canada

In the course of a search for variable stars in the galactic cluster NGC 6405 (M6) it was found that the brightest star of the cluster displayed what might be described as a flare.

On July 3, 1965, in the course of a five-hour continuous observing with the image orthicon system of the Organ Mountain Station, about 20 minutes after the start the star brightened by about two magnitudes followed by another outburst of seven magnitudes 30 minutes later. After the flare subsided the star remained at its pre-flare brightness for the rest of the observing period. A graphical representation of the phenomenon is shown in Fig. 1. In this diagram a straight line was drawn between points of observations taken at five-minute intervals.

There might be some uncertainty in the brightness of the second flare since it represents an extrapolation of the calibration curve over a range of five magnitudes. The photometric properties of the image orthicon tube indicate, however, that this uncertainty should not exceed $\pm$ 1 magnitude.

After the discovery of the flare an investigation showed that the following facts are known about the star:

The cluster, including the flare star, was observed photoelectrically by Rohlfs, Schrick and Stock in 1957 and again in 1960 and 1961 by Eggen and Talbert respectively. During this period the star showed a small variation in brightness as seen from the following tabulation:

1957, Rohlfs, etc.	V = 6.76 mag	B — V =	0.04 mag
1960, Eggen	6.63		—0.03
1961, Talbert	6.75		—0.01

In 1962 the cluster was observed by means of the image orthicon tube at Organ on two nights, namely on May 10 and August 13. On both nights the star appeared at a brightness of V = 6.75 mag. On the other hand, in 1963 fluctuations were observed as follows:

June 18, 1963	V = 7.25 mag
June 24	9.50
July 2	6.75
July 14	8.00

The estimated mean error of these observations is 0.25 mag. During 1964 and 1965 the brightness of the star fluctuated between 8th and 9th magnitude except for the brief flare-up on July 3, 1965. It is remarkable, however, that

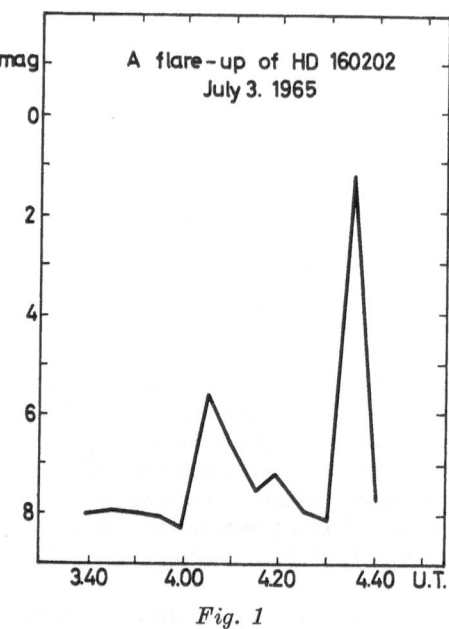

Fig. 1

after the flare the star remained at a constant brightness of 8th mag for $4^1/_2$ hours of observing time.

Recent photoelectric observations of the star by the writer made in July and August of 1968 showed a constant brightness of V = 6.75 on 14 nights.

The spectral type of HD 160202 is found in the literature to vary from B1ne to B5 and B8. Apparently there are no high dispersion spectrograms available.

Although flares of the type described are not common among early type stars, as an hypothesis it is suggested that the star, because of its *ne* characteristic, might be a shell-producing star whose shell, or its part, gets occasionally blown off. If this process is periodic at all, the periodicity might be of the order of 5—6 years.

ON THE GROUP OF YOUNG STARS IN THE SOLAR VICINITY

M. A. ARAKELIAN

Byurakan Astrophysical Observatory, Armenia, USSR

The problem of the origin of flare stars in the immediate vicinity of the sun is rather interesting, since the flare phenomenon is usually associated with stars which are almost certainly young. Two points of view have been expressed concerning the origin of these stars. The first one suggests that flare stars in the solar vicinity are members of a stellar association which once existed in the region where at present the sun is located [Ambarzumian (1957), Haro and Chavira (1965)]. The other point of view suggests that UV Cet stars are outcomers from the nearest stellar associations which exist now (Herbig 1962). If the latter is the case these stars should be distributed more or less uniformly in some volume exceeding considerably that of a single association.

In this report some data are presented, which seem to be in favour of the first of the mentioned possibilities.

I. *The Space Density of the Flare Stars in the Solar Vicinity.* The method of the stellar space density determination was proposed earlier by the author (Arakelian, 1968), which permits to receive some rather reliable estimates using the distances of a comparatively small number of the nearest stars. This method was applied to the determination of the luminosity function of K and M dwarfs with hydrogen emission lines (Arakelian, in press). The total density of dwarfs with hydrogen emission is found to be 0.055 ps^{-3}, while the value of 0.118 ps^{-3} was obtained for the total stellar space density (Arakelian, 1968b). Thus, the space density of dwarfs with hydrogen emission is about the half of that of all stars.

The dependence of the ratio of the number of known flare stars to the number of dwarfs with hydrogen emission upon the absolute visual magnitude was obtained and used for the determination of the luminosity function of stars known as flaring variables. This relation is presented by the following linear function

$$n/N = (0.04 \pm 0.01)\, M_v - (0.12 \pm 0.09), \tag{1}$$

where n is the number of flare stars and N is the number of dwarfs with hydrogen emission. The luminosity function of flare stars and the luminosity function of stars with hydrogen emission has been determined using relation (1). The space density of flare stars in the solar vicinity is found to be 0.027 ps^{-3}. This value must obviously be considered as a lower limit, since it may increase with the accumulation of observational data on flare stars. Nevertheless, even the data we have at present demonstrate high density of flare stars in the solar vicinity.

Supposing that the density of flare stars is constant up to sufficiently large distances and using the obtained luminosity function we may estimate the number of flare stars brighter than the arbitrary apparent magnitude m_v in a given solid angle with its summit in the sun. Let us assume this angle to be 16 square degrees in order to make some comparisons of these results with that obtained with the 40"-Schmidt telescope of the Byurakan Observatory in the course of searching for flare stars in the region of NGC 7023 (Mirzoyan and Parsamian 1969). We obtained by numerical integration

$$N_{16}(m_v) = 3.5 \cdot 10^{-9} \cdot 10^{0.6 m_v} \tag{2}$$

The Byurakan observations show that the number of discovered flare stars is much less, than it should be, provided that the space density of these stars is constant and equal to its value in the solar vicinity. Indeed, we have from (2) that $N_{16}(17) = 56$.

The mentioned searches were made by means of consecutive exposures used firstly by G. Haro and his collaborators (1965). Five minute exposures were used at Byurakan and their limiting apparent photographic magnitude was $m_{pg} = 18.5$. Thus, the limiting apparent visual magnitude of flare stars at Byurakan plates may be supposed to be 17^m0.

On the other hand, only 9 flare stars were discovered at Byurakan during 40 hours of observation in the region of NGC 7023. One flare star in this region was discovered earlier by L. Rosino and G. Romano. It is very probable, that most of these flare stars are actually members of NGC 7023, and not field stars. Nevertheless, even neglecting this possibility one can see that the number of discovered flare stars is much less than their expected number.

One may suppose that this deficiency in discovered flare stars is due to the small duration of observations. Therefore we discuss the mathematical expectation of the number of the flares themselves. Let us assume the mean value of the time interval between those flares which are accessible for photographic observations to be 100 hours (this value is apparently overestimated). Then the mathematical expectation of the number of flares of 56 stars during 40 hours will be of the expected value. This deviation may be explained only by the assumption that the value of the mean space density of the flare stars at large distances, used in the calculation of the expected number of flares, has been highly overestimated.

Thus, the value of the space density of flare stars previously referred to must be attributed to the immediate vicinity of the sun only. Their density at larger distances is much lower.

2. *Kinematical Properties of Dwarfs with Hydrogen Emission.* Ten years ago A. Vyssotsky and Dyer (1957) and W. Gliese (1958) showed that the space velocity dispersion of dwarfs with hydrogen emission was nearly one half of that of the dwarfs of the same type without emission in their spectra. Intending to use a larger number of stars, we studied the kinematics of these stars on the basis of radial velocities (Arakelian, 1958). The conclusion of the mentioned authors has been confirmed. It has been shown, thereby, that a K-term was present in the velocity of these stars amounting to several km/sec. This property makes the group of dwarfs with hydrogen emission similar to known stellar associations, since the expansion of the latter was predicted by

Ambarzumian (1949) and verified for two associations by Blaauw (1952) and E. Markarian (1953).

3. *The Stellar Luminosity Function in the Immediate Vicinity of the Sun.* The stellar luminosity function was determined on the basis of stars nearer than 20 ps (Arakelian, 1968). The behaviour of this function is somewhat unusual in the region of faint stars, since two maxima are present at $M_v = 13$ and $M_v = 15$. This property has apparently been noted by other authors, but it has been attributed to random fluctuations. However, it has been shown (Arakelian, 1968) that the probability of such fluctuations is rather small.

The maximum at $M_v = 15$ and the partial densities for $M_v > 15$ are almost entirely due to stars with hydrogen emission. Therefore, one can suggest that the luminosity function obtained in my above mentioned paper, is a superposition of that of the field stars and that of the members of some group.

Thus, the data presented above seem to suggest the presence of a group of genetically related young stars in the solar vicinity.

REFERENCE

1) Ambazsumian, V. A., 1949, Astr. Zu., **26,** 3.
2) Ambazsumian, V. A., 1957, Non-Stable Stars, p. 9. Yerevan.
3) Arakelian, M. A., 1958, Izvestia of Academy of Sciences of Armenian SSR, 11. No. 5. 79.
4) Arakelian, M. A., 1968, Doklady of Academy of Sciences of USSR, **179,** 555.
5) Arakelian, M. A., 1968a in press.
6) Arakelian, M. A., 1968b, Astrophysika, **4.**
7) Blaauw, A., 1952, No. 433. Bull. astr. Inst. Netherl.
8) Gliese, W., 1958, Z. Astrophys. **45.** 293.
9) Haro, G. and Chavira, E., 1965. Vistas in Astronomy, **8,** 89.
10) Herbig, G. H., 1962. Symposium on Stellar Evolution, p. 45, La Plata.
11) Markarian, B. E., 1953. Soobsch. Byurakan Observatory, No. 11, 3.
12) Mirzoyan, L. V., 1969. Parsamian, E. S., 1969. Report on the present conference.
13) Vyssotski, A. and Dyer, E., 1957. Astrophys. J. **125.,** 297.

DISCUSSION

Herbig: What is the kinematic age of this cloud of Me stars around the Sun as inferred from the velocity dispersion?

Arakelian: The age of this group is rather low, of the order of 10^6 years. This fact is indeed the difficulty of the proposed point of view, since there are no T Tauri stars in the vicinity of the Sun.

FLARE STARS NEAR NGC 7023

L. V. MIRZOYAN and E. S. PARSAMIAN

Byurakan Astrophysical Observatory, Armenia, USSR

The distribution of flare stars in T-associations can give some definite information about the early stages of stellar evolution. In order to determine this distribution the small T-association near NGC 7023 has been selected. Being very compact and comparatively close to us it is suitable for the mentioned purpose. The photographic observations of the NGC 7023 region began in 1962. The 40″ Schmidt telescope of the Byurakan Astrophysical Observatory of the Academy of Sciences of Armenian SSR was used. 92 plates with 479 five-minutes serial exposures embracing an effective observational time of 40^h were obtained. The limiting magnitude on our plates is 18^m5, consequently flare stars with absolute magnitude brighter than $+11^m$ (for the distance of the T-association equal to 280 pc, Weston 1953) can be detected on these plates. Nine flare stars (Mirzoyan et al. 1968 a, b), two of which flared twice during our exposures, have been found. Another possible flare star in this region has been announced by Rosino and Romano (1962). Thus, the number of known flare stars in this region is equal to 10.

This number can be used for a rough estimation of the total number of flare stars in this region brighter than $M \sim +11^m$.

Assuming, that the probability P_K of flare appearance obeys the Poissons law:

$$P_K = \frac{e^{-vt}\,(vt)^K}{K!}$$

where t denotes the duration of the observations, v the frequency of flares, K the number of flares during t, respectively, and using for the mathematical expectation of the number of stars flared K times, the expression

$$N_K = N P_K,$$

one can calculate the number N of all probable flare stars in this region. With our small statistics ($N_1 = 8$, $N_2 = 2$) we obtain $N = 27$. From this data also follows that the mean time interval between flares is 3^d3. It is difficult to discover photographically flares with amplitudes less than 0^m5. Therefore, these numbers correspond to flare stars with amplitudes larger than 0^m5. Since the duration of the flare-maximum can be shorter than 5 min, i.e., shorter than our exposure time and the starting point of the exposures may correspond to different moments of the flares, the estimated amplitudes are in fact always less than the real ones. Therefore, the real number of flares with amplitudes exceeding 0^m5 must be larger and the mean time interval

between flares correspondingly smaller than estimated. Naturally, our calculations refer only to stars brighter than $M = +11$. The number of corresponding stars in the Pleiades ($r = 126$ pc), is equal to 20 (Haro 1968). The
total duration of the corresponding observations in Pleiades is about 189^h.
The calculations similar to that brought above give $N = 112$ for the total
number of flare stars in the Pleiades. Therefore, assuming that the luminosity
function of flare stars in the region of NGC 7023 is close to that of the Pleiades,
one can conclude that the number of all flare stars in the region of NGC 7023
differs not much from the corresponding number in the Pleiades.

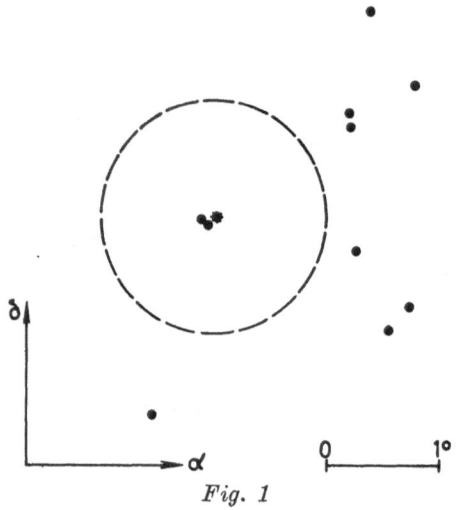

Fig. 1

Of particular interest is the distribution of the known flare stars in the
region of NGC 7023 presented in Fig. 1. It shows that eight flare stars out
of ten are situated far from the centre of the T-association, probably farther
than the limits of the association showed by the dotted circle having a radius
of 1° (Kholopov, 1959). The remaining two stars have probably come to the
central part of the cluster through the effect of projection, in reality they
may be far from the centre in space. Thus, it may be suggested that the flare
stars surround the T-association. It must be noted, we suppose, that all
detected flare stars in the region of NGC 7023 are in a distance equal to the
distance of the T-association. The opposite assumption, i.e. that they are
field stars, is less probable, because of the very high frequency of flares of
field stars needed in this case.

If one assumes that all of them have been originated in the central
region of the association, then the distribution of flare stars confirms Haro's
(1965) hypothesis, according to which the flare stage of a star follows the
T Tauri stage.

On the other hand if we suppose Haro's hypothesis to be correct, the
observed distribution of flare stars arround the T-association NGC 7023 can be
interpreted as a confirmation of the expansion of this association. However,
our statistical sample is not rich enough for a final conclusion.

The authors thank to Prof. Ambarzumian for valuable advices.

REFERENCES

Haro, G., Chavira, E., 1965. Vistas in Astronomy, vol. **8,** 89.
Haro, G., Stars and Stellar Systems, vol. VII. 147.
Kholopov, P. P., Astron. Zu. **36,** 295.
Mirzoyan, L. V., Parsamian, E. S., Chavushian, O. S., 1968. Soobshch. Byurakan Obs.
    39, 3.
Mirzoyan, L. V., Parsamian, E. S., Kalloglian, N. L. 1968b, unpublished.
Rosino, L., Romano, G., 1962. Asiago Contr. N. 127.
Weston, E. B. 1953. Astr. J. **58,** 48. 1953.

DISCUSSION

Rosino: I should like to know whether flares have been observed in known
    nebular variables of NGC 7023 or wether the flares are usually ob-
    served in stars which are *not* nebular variables.

Parsamian: No one of these flare stars coincide with known nebular variable
    stars in this region.

ON THE POLARIMETRIC STUDY OF UV CETI TYPE STAR FLARES

Yu. S. EFIMOV

Crimean Astrophysical Observatory, USSR

A study of the polarization of UV Ceti stars during an outburst, especially in its rising part, is very important for better understanding of the physical mechanism of the outburst. These observations are connected with the measurement of the small variable polarization of the radiation. Of course, the time of one observation must be less than the duration of the outburst, that is, the measurement must be taken at a rather high speed and with good accuracy. But the reduction of time for which information on the Stokes parameters might be obtained is limited by statistical fluctuations of the signal. These fluctuations consist mainly of the light flux quantum fluctuations, the seeing, and the dark current noise of the photomultiplier. On the other hand, a sufficient number of quanta must be registered in order to guarantee an acceptable accuracy of the measurements. Taking into account only the accidental errors of measurements, the dependence of the accuracy of a measurement upon the observational conditions is shown in Fig. 1. This plot refers to observations with a one-channel integrating polarimeter containing a continually rotating analyser. The initial stellar magnitude was chosen equal to 10^m. The time of observations is on the ordinate-axis, the amplitude of the brightness variations on the abscissa-axis.

Tilted straight lines correspond to different rates of brightness variations for the time of observation. The curve lines are lines of constant accuracy (for different absolute errors σ_p). There is a possibility to find the error of a measurement of the degree of polarization from Fig. 1, if the amplitude of brightness-change and its rate for the time of observation are known. So, it is possible to measure the polarization of the light of some UV Ceti stars using the 2.6 meter telescope. For a well known star of this type, AD Leo (10 magn. in its normal state) one can suppose that its brightness increases approximately by one magnitude per 10—12 seconds (that is at a mean rate of $0^m1/\text{sec}$). As it follows from the Figure, the light polarization of this star in the rising part of the outburst might be measured in 8—10 seconds using the 2.6 m telescope, with an absolute error of 0.2 per cent. Making observations in the descending part of the outburst, it is possible to get the same accuracy. Using the 70 cm telescope, such observations would be practically useless, because the error of the measurement will be more or equal to 0.7—1.0 per cent.

In this way, using a special method and a large telescope, it is possible to study the changes of the polarization during the outburst in detail. A special polarimeter for this aim has recently been built by Drs. Arsenijević and Kubichela (Yugoslavia). Such an equipment attached to a 100″ or larger

D=2.6 m
D=0.7 m
t
160
140
120
100
80
60
40
20
-4
-3
-2
-1
0
1
2
3
4
Δm

Fig. 1.

lgσp
σp %
dm/dt = 0
t = 100 sec
d = 10"
ε = 0.1
D=0.7 m
D=2.6 m
mag

Fig. 2.

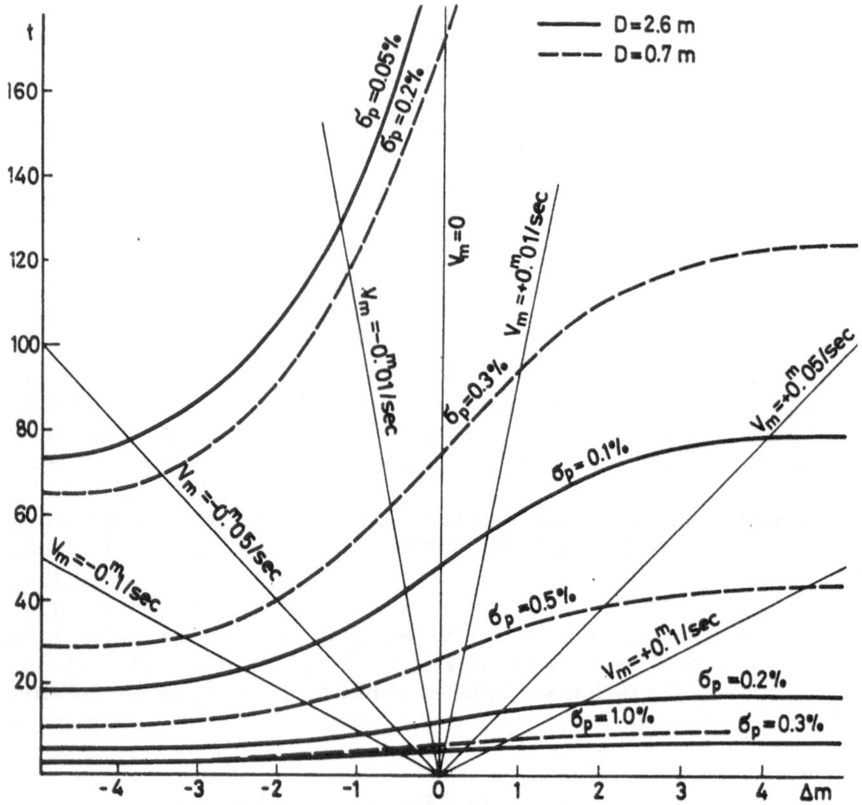

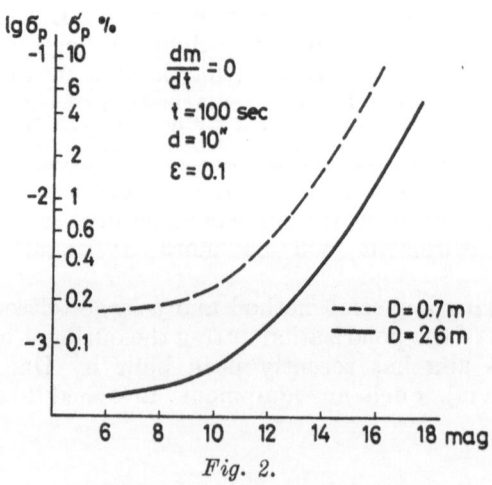

telescope will be able to yield valuable information on the polarization during the outburst.

It is interesting to consider the practically important case of observations of stars of different brightness when one can suppose the stellar magnitude to be constant (dm/dt = 0). The dependence of the error of the measurement upon stellar magnitude is shown in Fig. 2 (for the 2.6 m and the 70 cm tele-scopes solid and dashed lines, respectively). The duration of an observation is equal to 100 seconds, the quantum efficiency is equal to 0.1. One can see that the error of the measurements rises rather fast for fainter stars, and the polarization of stars fainter than 16 magn. cannot be measured with an error less than 1 per cent even if the 2.6 meter telescope is used.

The results obtained in this theoretical investigation are rather general. Therefore, any polarimetric observation given with an accuracy greater than shown here must be considered as incorrect.

PRELIMINARY RESULTS OF A SURVEY OF NEBULAR VARIABLES AND FLARE STARS

L. ROSINO

Astrophysical Observatory of Asiago, Italy

ABSTRACT

Preliminary results of the survey of nebular variables and flare stars carried out at Asiago with the 92—67 cm Schmidt telescope are reported.

An extended survey of nebular variables in stellar aggregates of different age and distance has been carried out in the last years at the Astrophysical Observatory of Asiago, partly with the 122 cm parabolic telescope and partly with the new Schmidt of 92—67 cm, F : 3.3, covering a field of 25 square degrees. Kodak blue and infrared material has been mostly used. Although only a part of the material have been reduced, it is possible to give preliminary results at least for some of the fields which are under control.

1. *Orion aggregate.* Two fields have been extensively observed: *a)* The Trapezium area including the nebulae NGC 1976, 1977, 1982 and 1999. *b)* The Horsehead nebula including also NGC 2024 near ζ Orionis. Results of previous surveys in the Trapezium area have already been published by the writer (1946, 1956, 1962). After 1962, forty-two new variable stars have been found. Twenty-two are concentrated near the Trapezium; 25 are visible only in infrared ($m_{pg} > 18.5$). Their positions and identification charts will be published in a forthcoming paper. The total number of known variables in the field (flare stars *not* included) is now 456. The density is particularly high near the Trapezium, where 70% of the stars are found to be variable, and along a strip going from NGC 1977 to NGC 1999 and continuing towards the Horsehead Nebula. As observed elsewhere (Rosino, 1962) in heavily obscured regions the variables are mostly found in the fringes of dark nebulosities.

The Asiago observations after 1962 have shown another interesting property of the nebular variables, particularly of those having a strong ultra-violet excess: they present at times rapid fluctuations of brightness with amplitudes of one magnitude or more, which, however, are not flares. A photoelectric survey of such stars (YZ, YY, HS, XX, SY Ori) should be of the greatest interest.

The mean amplitude of nebular variables in Orion is $1^{m}60$, with a fairly large dispersion. The frequency distribution of the apparent magnitudes has a maximum near 17 pg, corresponding to an absolute magnitude of about $+9$. It is likely, however, as indicated by the infrared survey, that the number of variables may be still higher for fainter luminosities, below 18.

Thirty new variables have been found in the region of the Horsehead nebula. These variables and those already known have the same characteristics of the variables observed in the Trapezium region. Some of them show at

L. ROSINO

Table I

Flare stars in Orion recently discovered at Asiago

No.	P	1900 R. A.	1900 D.	m	Date		Dur.	Notes
1		5h 22m00s	—6°04′5	15.2— 17.4	11 Dec,	1966	40m	—
2		5 22 54	—4 25.6	15.8— 18.5	8 Dec,	1967	30	—
3		5 24 34	—6 37.6	14.5— (18	2 Dec,	1967	—	—
4		5 25 00	—7 09,0	16.6— 17.8	30 Jan,	1968	—	—
5		5 25 09	—4 27.7	15.3— 17.5	23 Jan,	1966	>20	Haro 38
6	981	5 27 47	—5 03.9	16.0— 16.8	30 Jan,	1968	10	—
7		5 28 13	—5 23.8	16.2— 17.5	4 Jan,	1968	20	—
8	1333	5 29 08	—5 40.5	16.2— (17.1	30 Jan,	1968	15	I I Ori
9		5 29 09	—4 11.7	15.8— (18.5	18 Jan,	1966	10?	—
10		5 29 27	—6 22.0	16.2— (18.5	12 Jan,	1967	37	—
11		5 29 35	—6 03.3	16.4— 17.2	4 Jan,	1968	20?	—
12		5 29 38	—0 31.0	15.0— 17.5	7 Dec,	1967	>30	—
13		5 29 42	—6 12.3	15.0— (18.5	1 Feb,	1968	>30	—
14		5 29 55	—1 50.0	15.6— 17.1	27 Feb,	1968	30	—
15	1625	5 29 55	—5 50.0	15.5— 16.8	19 Jan,	1964	>60	—
16		5 30 29	—6 51.2	15.5— 17.2	23 Jan,	1966	25	—
17	2039	5 30 32	—6 05.4	15.0— 16.5	25 Feb,	1963	—	NS Ori
18	2112	5 30 40	—5 33.2	15.0— (16.8	1 Feb,	1964	35	—
19		5 30 40	—7 06.1	15.8— 17.3	30 Jan,	1968	30?	—
20		5 30 54	—5 33.8	15.2— (17	19 Jan,	1966	—	—
21	2210	5 30 52	—5 44.9	15.1— 16.8	30 Jan,	1968	25?	V 378 Ori
22	2235	5 30 55	—5 39.3	15.8— 16.8	27 Dec,	1967	10	—
23	2245	5 30 56	—5 19.0	16.4— (17	12 Dec,	1966	10	V 379 Ori
24	2246	5 30 56	—5 20.4	16.0— 16.9	30 Jan,	1968	>20	OT Ori
25		5 30 56	—6 21.8	15.8— 18	1 Feb,	1968	>20	—
26		5 31 05	—4 22.3	16.4— (17	19 Jan,	1966	—	—
27	2295	5 31 06	—5 27.3	16.6— (17.2	21 Jan,	1968	—	V 365 Ori
28		5 31 12	—6 29.2	15.4— 16.8	8 Dec,	1967	—	—
29		5 31 32	—5 34.4	16.0— 17.5	27 Feb,	1965	40	—
30		5 31 40	—6 43.6	16.6—(17.2	15 Jan,	1966	15	—
31		5 31 41	—6 46.0	15.5— 16.9	19 Jan,	1963	10	—
32		5 31 51	—6 37.6	16.2— 17.2	6 Jan,	1968	10	—
33		5 32 10	—2 55.3	15.0— (18	9 Dec,	1966	—	—
34		5 32 18	—6 49.3	16.2— (17.2	19 Jan,	1966	—	—
35		5 32 52	—0 49.5	15.5— 17.2	31 Jan,	1968	20	—
36		5 33 40	—3 50.0	15.9— 16.7	27 Jan,	1968	20	—
37		5 34 29	—2 53.2	16.2— 18.2	24 Jan,	1966	—	—
38		5 36 31	—1 45.0	15.9— (17.2	27 Jan,	1968	20	—
39		5 40 02	—1 11.9	14.8— 16.7	8 Feb,	1959	—	—

times a sort of periodicity which disappears after a few months. The possibility that among the nebular variables there may also be eclipsing binaries should not be disregarded. However, the erratic variation would tend to mask the eclipses, unless an harmonic analysis of the light curves over a long period of time could be made.

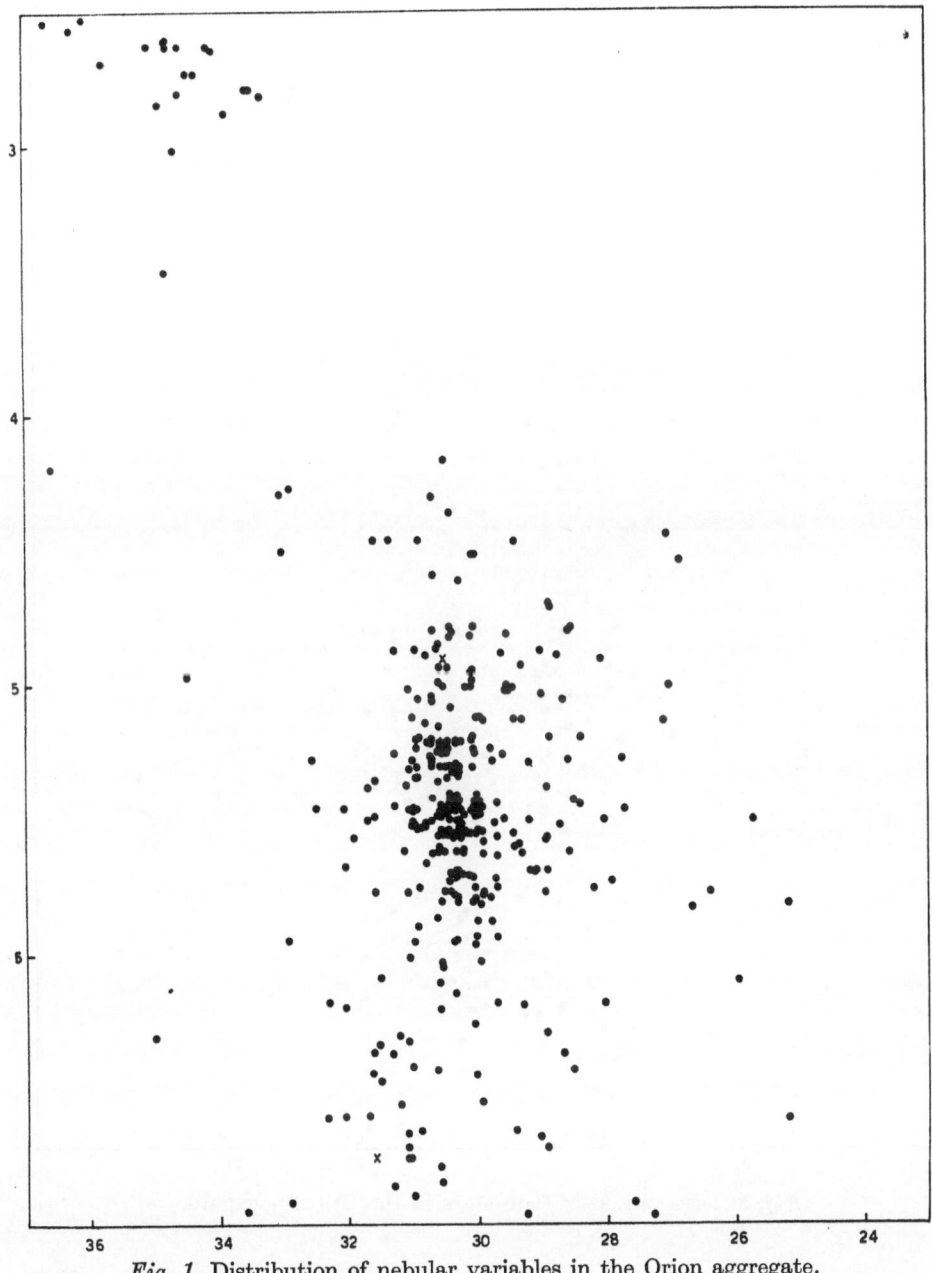

Fig. 1. Distribution of nebular variables in the Orion aggregate.

2. *Flare stars.* After the new 67 cm Schmidt telescope has entered in operation at Asiago, a great deal of time has been dedicated to the search of flares in the Orion aggregate and other fields. The highest frequency of flares (1 flare every 100 minutes of effective observation) has been found in the 25

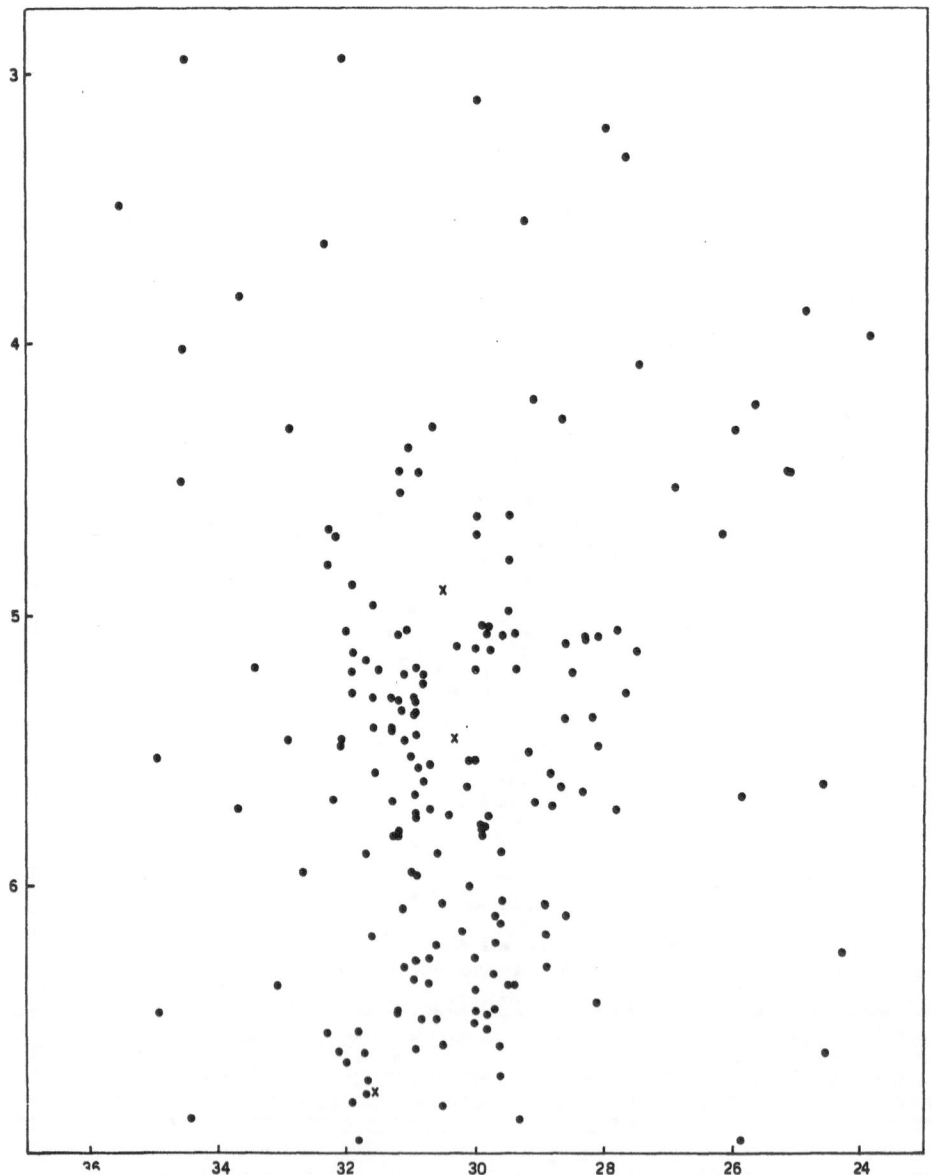

Fig. 2. Distribution of flare stars in the Orion aggregate.

square degrees area centered in the Trapezium, while in the area around ζ Orionis the frequency has been of one flare every 4 hours.
In total, the flares discovered at Asiago in the Orion aggregate (Trapezium plus ζ Orionis) from 1962 to March 1968 have been 39; they are reported in Table I. Details on the flares, light curves and identification charts will be published in a forthcoming paper. In their normal condition of minimum

the flare stars discovered at Asiago are all fainter than 16.5 pg. Their mean photographic magnitude is 17.2. By a comparison with the flare stars found in the same region by Haro and Chavira (1965) the mean P—V color index for a flare star in the aggregate is of the order of $+1.5$, with a V mean absolute magnitude ~ 7.6, so that the representative point is slightly to the right of the zero age main sequence. However, it is apparent from the Asiago observations that some of the flare stars lie on the main sequence and even to the left of the main sequence, although the large majority is on the right.

Figs. 1 and 2 illustrate the distribution of nebular and flare stars in the Orion aggregate. Although they are obviously correlated, the dependence is not so strong as it was believed. The effects of selection in the discovery of flares, due to the presence of bright nebulosity, should, however, be carefully considered.

Flares are mostly observed in stars which show no variations outside the flares. Very seldom, they are also found in typical nebular variables and in this case the amplitude of the flare is smaller ($1^{m}1$) than normal ($2^{m}0$). The frequency of flares in the same star is in general rather low in the Orion aggregate. Of 222 flare stars in Orion hitherto published only 23 have been caught in flare more than once, and only 5 have had three flares or more. This means that the mean interval between two successive flares in Orion is higher than 10 days and in fact the occurrence of a flare in a given star is really a very rare event. As it was pointed out by Haro, the rate of flares in a star depends from the nature of the aggregate and increases with age. In the Pleiades and Hyades it is considerably higher than in Orion.

BIBLIOGRAPHY

Haro, G. and Chavira, E. 1965, Vistas in Astronomy, Vol. **8**, 89.
Rosino, L., 1946, Pubbl. Bologna Vol. V, No. 1.
Rosino, L., 1956, Mem. Soc. astr. it. **27**, 3.
Rosino, L., 1962, Mem. Soc. astr. it. **32**, 4.

Non-Periodic Phenomena in Variable Stars
IAU Colloquium, Budapest, 1968

Of AND Be STARS

Introductory Report by

ARNE SLETTEBAK

Perkins Observatory, Ohio State and Ohio Wesleyan Universities, Delaware, Ohio, USA

I. THE Of STARS

The Of stars are characterized by broad emission lines of N III 4634, 4641, and He II 4686, in addition to absorption lines shown by normal O-type stars of corresponding spectral type. The presence of these emission features can be understood in terms of selective fluorescence processes proposed by Bowen (1935). Emission can also usually be found at Hα and C III 5696 in most Of stars, and at Si IV 4089 and 4116, Ni IV 4057, N V 4603 and 4619, and He I 5876 in some Of stars (Underhill 1966).

Some 13 per cent of the O-stars listed in recent catalogues are of type Of, according to Underhill (1966). No sharp distinction between the absorption-line O-type stars and the Of stars exists however, as has been shown by Wilson (1958), Underhill (1958a), and Kumajgorodskaya (1962).

The extent of the shell producing the emission in Of stars has been estimated by Struve and Swings (1940), Swings (1942), Oke (1954), Underhill (1958b), Hutchings (1968a), and others. Generally, the emission appears to arise from a shell not more than one or two stellar radii in extent. The existence of P-Cygni line profiles with very broad emission features in the spectra of Of stars suggests that these objects have expanding atmospheres with considerable internal motions. Wilson (1958) found for the Of star λ Cephei band widths for He II 4686 and N III 4634—41 of the same order as observed in the spectra of Wolf-Rayet stars, indicating Doppler velocities in excess of 1000 km/sec. Radial velocity measurements by Hutchings (1968a) of two Of stars suggest extended accelerating atmospheres with velocities reaching 600 km/sec and beyond.

Struve and Swings (1940, 1941), Oke (1954), Mannino and Humblet (1955), Underhill (1959), Kumajgorodskaya (1960), Hutchings (1967a, 1968a), and others have commented on the variability of the spectra of Of stars. The emission features show pronounced variations, while variability of the absorption lines has also been reported. Evidence of extremely rapid variations was first given by Oke (1954), who found emission-line intensities in the Of stars HD 34656 and HD 190429 N varying in a matter of hours. Oke pointed out that the variations of the N III and He II line profiles take two forms: (a) the main part of the line varies considerably in shape; (b) the main lines are accompanied by emission and absorption satellites on the red and violet sides which come and go from plate to plate. More recently, Hutchings (1967a, 1968a) has found similar variations.

In October of 1966, an attempt was made to observe rapid variations, of the kind reported by Oke, with the 72-inch Perkins reflector of the Ohio State and Ohio Wesleyan Universities at the Lowell Observatory in Flagstaff,

Arizona. Spectrograms of dispersion 40 A/mm at Hγ on Kodak IIa-O emulsion were taken during a series of consecutive nights of the Of stars λ Cephei, 9 Sagittae, HD 34656, and HD 190429 N. The spectrograms were considerably widened in order to aid the detection and measurement of the diffuse and shallow emission features. Line profiles of the N III 4634—41 and He II 4686 features for the four stars are illustrated in Figures 1—4. These are intended only to show observed line profile changes; no correction for underlying absorption has been made, nor are radial velocity differences shown.

Although the spectrograms are of rather low dispersion, rapid variations in the line profiles are evident. These may occur from night to night, as in the spectra of λ Cephei (Fig. 1) taken on the nights of Oct. 22nd and Oct. 23rd, or within a few hours, as shown in the λ Cephei spectra of Oct. 24th. In addition to the rapid variations, the great widths of the emission bands noted by Wilson (1958) is confirmed by the present material. Brief comments about the aforementioned Of stars plus two additional Of stars follow:

λ *Cep.* Spectrograms taken of this O6f star with the Perkins 69-inch telescope (28 A/mm at Hγ) in Delaware, Ohio prior to those illustrated in Fig. 1 also show interesting changes. On Sep. 6, 1952, N III 4641 was stronger in emission than N III 4634 and He II 4686 had a violet satellite emission feature. Eight nights later, on Sep. 14, 1952, the two N III lines were more equal in intensity, while the He II satellite line was much weaker than previously. Again, on Nov. 6, 1956, the emission spectrum was similar to that just described for Sep. 6, 1952, while by Nov. 24, 1956, the N III emission was much stronger and broader and the violet He II satellite line had disappeared.

In addition to the changes in the emission lines illustrated in Fig. 1, the absorption lines also show rapid changes at times. On Oct. 25.28, 1966, for example, the He I 4471 and He II 4541 lines appeared rather sharp relative to their usual diffuse appearance.

9 Sag. The N III 4641 emission was generally stronger than the 4634 component in this O8f star, as shown in Fig. 2, but on occasion (i.e., Oct. 26.19) they appeared nearly equal. Meanwhile, the He II 4686 emission also showed rapid intensity variations and actually disappeared on the night of Oct. 25.19.

HD 34656. This O7f star showed essentially equal N III 4634 and 4641 components except on the night of Oct. 24th, when the 4641 feature was the stronger of the two as illustrated in Fig. 3. On the night of Oct. 27th, the N III emission was markedly weaker than on the other nights. The He II 4686 line appears in absorption in this star and was much sharper on the nights of Oct. 24th and 27th than on the other three nights. No corresponding pronounced changes in other absorption lines were detected.

HD 190429 N. Variations in the emission line profiles of this O5f star are shown in Fig. 4. Rapid changes in the absorption lines also occured: the He II 4541 and 4200 lines appeared more diffuse on the spectrogram of Oct. 27.08 than on the other plates.

λ Cep   (O6f)

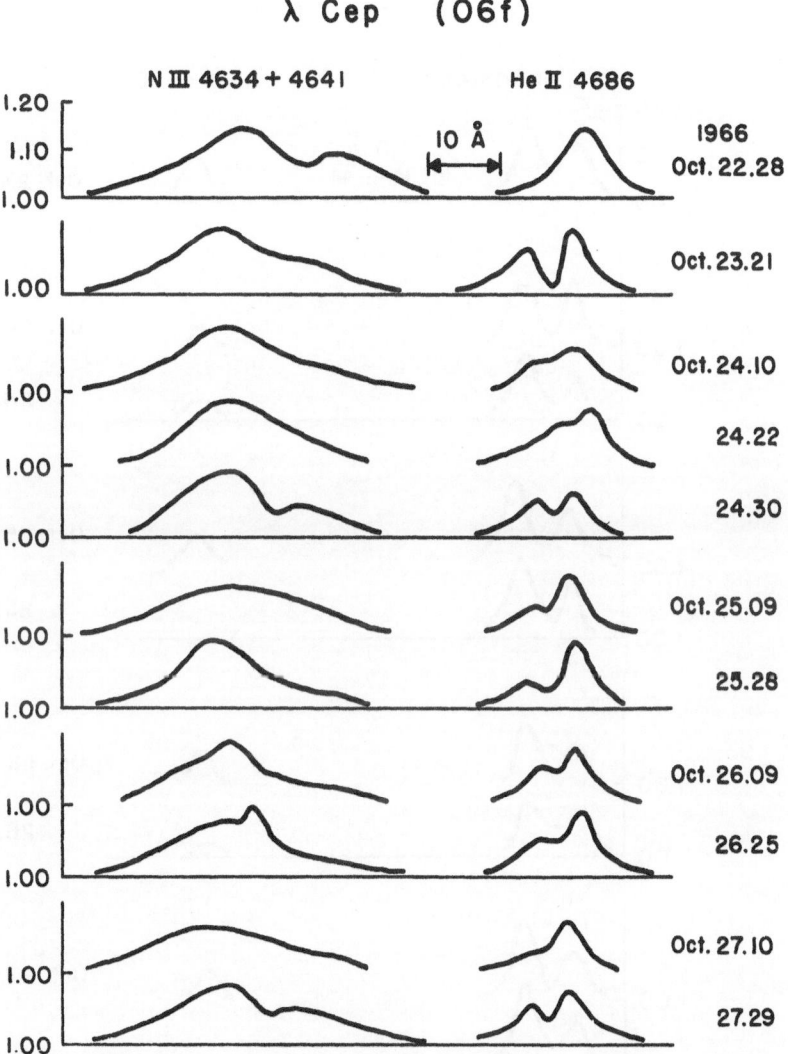

Fig. 1. Variations in the profiles of the N III 4634—41 and He II 4686 emission features in the spectrum of the O6f star λ Cephei during six consecutive nights.

29 CMa. Spectrograms of 20 A/mm dispersion taken with the Perkins 72-inch telescope on the nights of March 18 and 21, 1968 show a violet satellite emission line to the He II 4686 emission feature on the 21st, which was not present on the 18th.

ζ Pup. Both N III 4634—41 and He II 4686 show a complex emission structure on 20 A/mm spectrograms taken with the Perkins 72-inch telescope on the nights of March 18 and 21, 1968, but no obvious differences were visible.

13

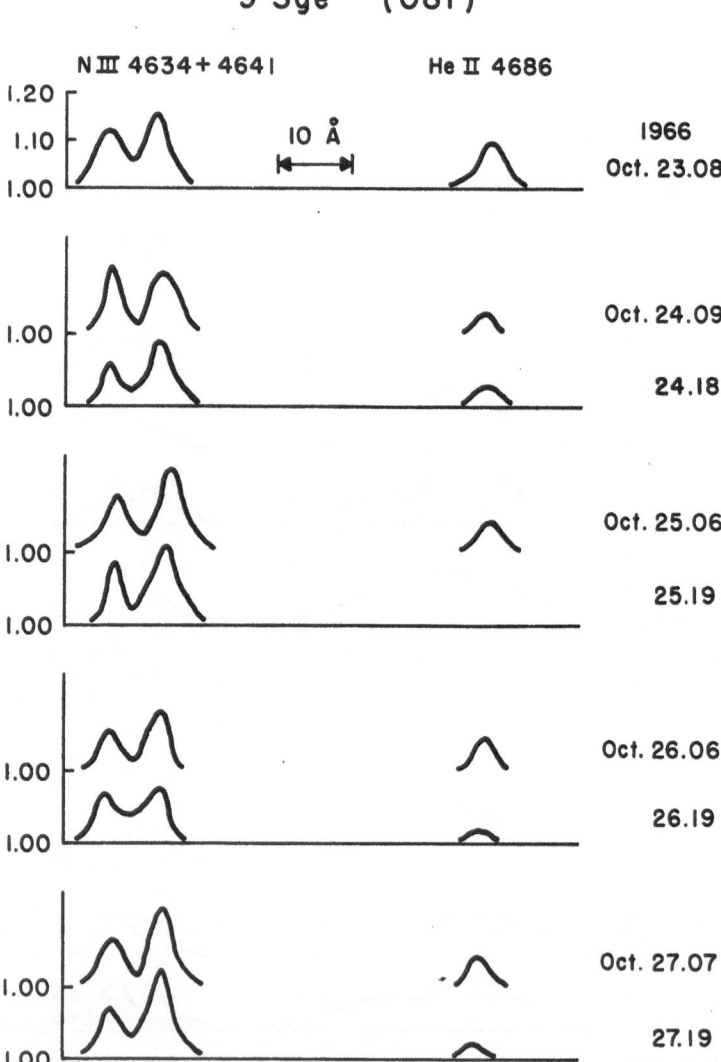

Fig. 2. Variations in the profiles of the N III 4634—41 and He II 4686 emission features in the spectrum of the O8f star 9 Sagittae during five consecutive nights.

Rapid variations of the kind illustrated in Figures 1—4 should be studied with high-dispersion spectra over continuous time periods and include both intensity and radial velocity measurements. If, as seems probable at present, the variations are non-periodic, the picture brought to mind is one of large-scale, turbulent atmospheric motions with velocities of hundreds of kilometers per second. Whether or not axial rotation plays a role is not clear but it probably does not play the dominant role since sharp-lined Of stars (as would be expected from those seen pole-on) are non-existent or very rare

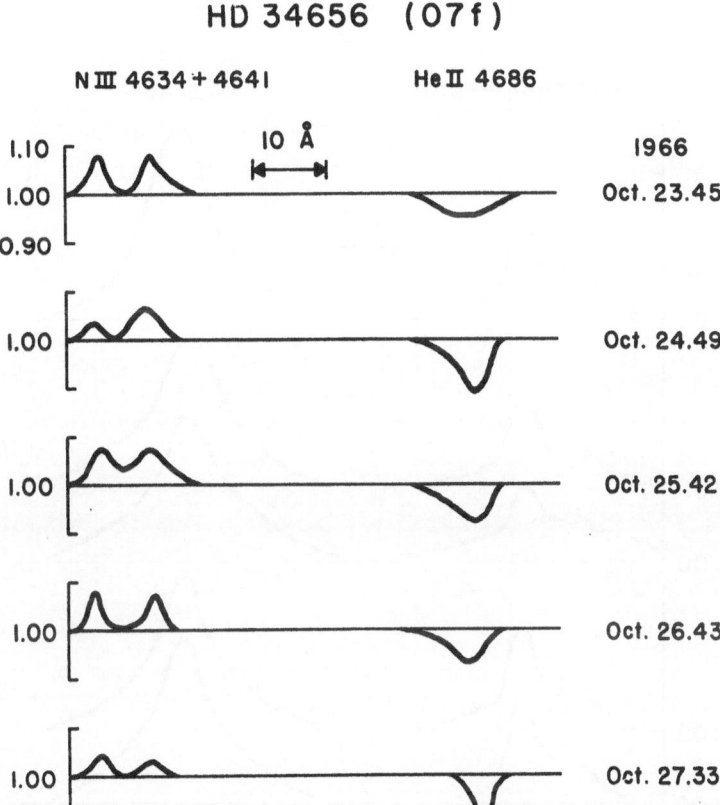

HD 34656 (O7f)

N III 4634 + 4641   He II 4686

Fig. 3. Variations in the profiles of the N III 4634—41 emission feature and He II 4686 absorption line in the spectrum of the O7f star HD 34656 during five consecutive nights.

(Slettebak 1956). The analogy between Of stars and Be stars (where rotation is the dominant line-broadening agent) which is sometimes made should therefore probably not be carried too far.

What then is the nature of the Of stars and how are they related to normal O-type stars? The picture is not yet clear. There is evidence in favor of these objects being somewhat more luminous than absorption-line O-type stars of corresponding type (Roman 1951; Slettebak 1956; Kumajgorodskaya 1962; van den Bergh 1968) but no general agreement (Underhill 1955). The recent suggestion by van den Bergh (1968) that Of stars of a given spectral type are older and hence presumably more highly evolved than are absorption-line O stars of the same spectral type is interesting in this connection and should be investigated further.

II. THE Be STARS

Some 10 per cent of the brighter B-type stars show emission lines of hydrogen (Merrill and Burwell 1933) and are designated as Be stars. Of these, about 10 per cent are Ia supergiant stars (Abt and Golson 1966). The latter

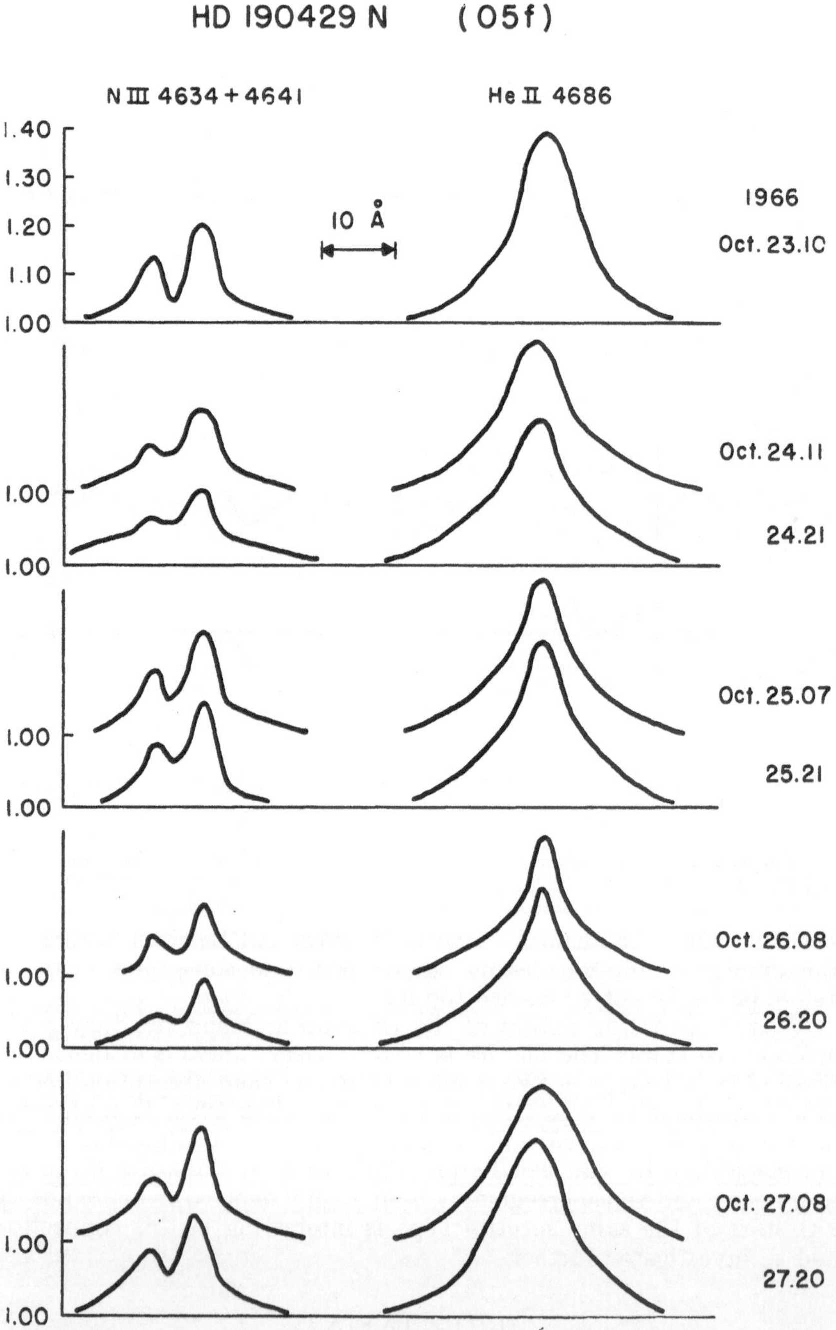

HD 190429 N   (O5f)

Fig. 4. Variations in the profiles of the N III 4634—41 and He II 4686 emission features in the spectrum of the O5f star HD 190429 N during five consecutive nights.

have emission at Hα (sometimes also at Hβ), which usually appears as a P Cygni-type line profile, and which is often variable. A detailed study of the variations for three B-type supergiants was made by Underhill (1961), who found non-periodic radial velocity changes and suggested that the variations are due to atmospheric motions.

The remaining 90 per cent of Be stars are of much lower luminosity, apparently lying a magnitude or two above the main sequence, and are characterized by very large line broadening. The use of the term "Be stars" in the remainder of this paper shall refer only to these objects. Struve (1931) was the first to suggest that the emission lines in Be stars arise in gaseous rings ejected from rapidly-rotating stars at the limit of instability, and this interpretation still seems valid today.

The spectra of Be stars typically show variations. Among a group of 40 of the brightest Be stars observed by McLaughlin (1961) for many years, only 8 failed to show convincing evidence of spectral changes. In the terminology of McLaughlin, the spectrum variations are of three kinds: (1) appearance and disappearance of a shell absorption spectrum; (2) E/C variation; (3) V/R variation. An enormous literature regarding spectrum variation in Be stars exists and only a few authors and papers can be cited in the following paragraphs.

Absorption shell spectrum variation. When the axis of rotation of a Be star is oriented such that the equatorial ring is in or near the plane including the observer, an absorption-line spectrum arising from the ring or shell is visible in addition to the emission lines. Be stars with this orientation are called "shell stars", and the shell spectra strengthen and weaken as their shells come and go. Such changes in some of the brighter shell stars have been studied intensively by Struve, Merrill, McLaughlin, Underhill and others. There is evidence for an "oscillation" of 8—10 years during which the equatorial shell appears and disappears (Merrill, 1956), but some shell stars (γ Cas, ζ Tau) behave much more erratically. The mechanism which triggers the formation of a shell remains an unsolved problem.

E/C Variation. Changes in the ratio of intensity of the emission lines to the neighboring continuous spectrum are designated as E/C variations. Such changes are usually associated with the appearance and disappearance of the absorption shell spectra for shell stars and also reflect the coming and going of the shell. For stars seen nearly pole-on, these E/C variations are the only spectral evidence of the shell phase. In a study of 8 Be stars, Lacoarret (1965) found E/C variations ranging between 3 and about 15 years.

Spectrograms taken with the Perkins 69-inch telescope of the shell star Pleione in 1949 show a relatively strong absorption shell spectrum with weak emission at Hβ. Two years later, the absorption shell spectrum was much weaker, while the emission was somewhat stronger. In 1957 the absorption shell spectrum had disappeared, except for the cores of the Balmer lines, while the emission at Hβ was very strong. The appearance of the spectrum in 1968 was again similar to that in 1957.

V/R Variation. In a typical Balmer-line profile shown by a Be star which is oriented such that the equatorial shell is seen projected against at least a portion of the photosphere of the star, a narrow central absorption divides the wider emission into a violet and a red component. These two emission components frequently show variations in relative intensity which have been designated as V/R variations. In a study of 54 Be stars for which spectrograms were available over a 24-year period, Copeland and Heard (1963) found that two-thirds showed V/R variation.

The V/R variations often show apparent periodicities of several years. McLaughlin (1963, 1966) found a period of about 10 years for the V/R variation in 105 Tauri, and about 4.5 years for HD 20336 in the interval 1916—31. The V/R variation behaved much more erratically for the latter star outside the aforementioned time interval, however, at times stopping altogether. A similar erratic behaviour was found for π Aquarii, while β^1 Monocerotis started V/R variation with a period of about 12.5 years in 1924 after showing no variation for 20 years prior to that year (McLaughlin 1958). Copeland and Heard (1963) found periodic V/R variations in nearly half of their V/R variables, with a mean period of 6.8 years. Although several have been proposed, no model exists as yet which can satisfactorily explain these V/R variations (McLaughlin 1961).

In addition to such long-period variations in V/R, much shorter variations have been observed. Recently, Hutchings (1967b) has observed very rapid profile variations in the spectrum of γ Cas using photoelectric scanning techniques. He finds significant changes in the Hγ profile in intervals of one hour or less and suggests that a continuously changing velocity of expansion covering most of the stellar surface is responsible (Hutchings 1968b).

An attempt to observe short-period variations in the spectra of two bright Be stars was made by the writer in October of 1966. Using spectrograms of dispersion 40 A/mm at Hγ taken on Kodak IIa—O emulsion with the Perkins 72-inch reflector, 28 Tau and γ Cas were each observed once and sometimes twice during seven consecutive nights. No convincing changes were observed in the spectrum of 28 Tau, which showed R sligthly stronger than V at Hβ during the entire period of observation. The spectrum of γ Cas was also well behaved, for the most part, but a definite and sudden change occurred during the night of Oct. 25th, as shown in Fig. 5. Although V was usually somewhat stronger than R at Hβ during the observation period, the two emission components became nearly equal for a short time on Oct. 25th before returning to normal again on the following night.

It should be emphasized that such relatively low-dispersion spectrograms will only show the more obvious variations. Scanning techniques of the type employed by Hutching (1967b) are probably required to bring out with certainty the small-amplitude rapid variations. The existence of such sudden and sporadic changes suggests again, as in the case of the Of stars, rapid atmospheric motions. The Perkins observations suggest a difference in the degree of atmospheric activity, however: the Be stars seem to act up only on occasion whereas the Of atmospheres are apparently in a state of constant turmoil.

Another way of attacking the problem of variability in the Be stars is by looking for changes in the total light and colors of these objects. Although γ Cas showed changes of one magnitude or more during its shell phase, most Be stars vary by much smaller amounts. Schmidt (1959) found variations in light of slightly more than 0.1 mag. for the shell star Omicron Andromedae and concluded that it is a contact binary. Jackisch (1963) made UBV obser-

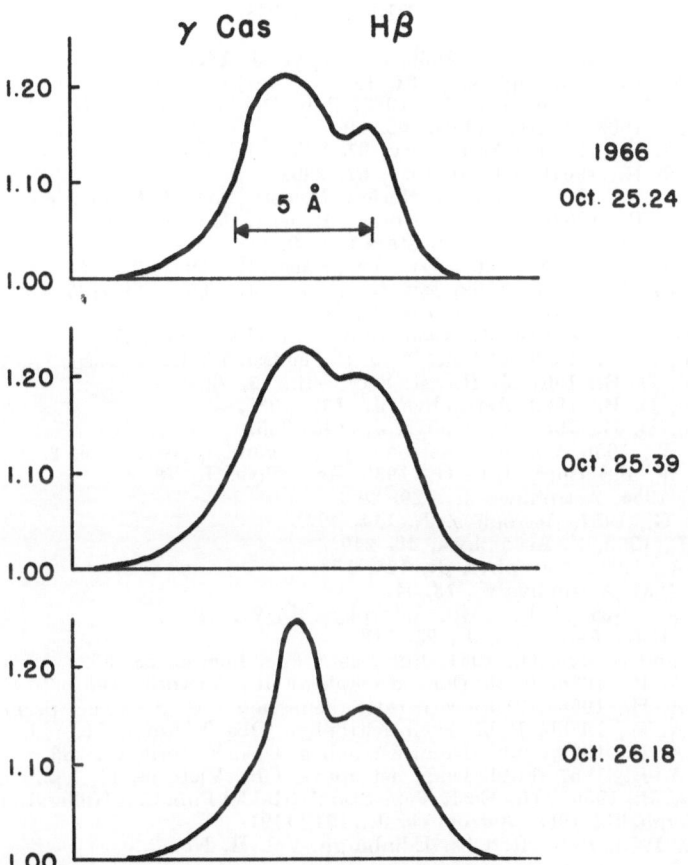

Fig. 5. Rapid V/R variations in Hβ in the spectrum of the Be star γ Cassiopeiae.

vations of the Be stars 48 Per and 53 Per and found magnitude variations
of 0.10 and 0.07, respectively, for the two stars. Feinstein (1968) observed
72 bright southern Be stars in the UBV system over a three-year period.
He found that 33 stars displayed variations in V larger than 0.06 mag. and
that 21 changed in U—B by more than 0.06 mag. The variations were found
to be either progressive or irregular, with some changes as large as 0.3 mag.

Recently, T. P. Roark has started a series of photometric observations
of selected Be stars with the Perkins 72-inch reflector to look for short-period
variations. He is measuring uvby colors on the Strömgren system, plus Hα
spectrum scans with the Boyce scanner. Preliminary results show night to
night variations in Hα and the Strömgren c_1 index for several of the stars,
with possible variations in $(b—y)$.

The need for more observations is obvious. The most valuable information
would include simultaneous spectroscopic and photometric observations over
continuous periods of time. Only in this way can any periodicities be sorted
out from the random changes and the true causes of the variations finally be
understood.

REFERENCES

Abt, H. A. and Golson, J. C., 1966, Astrophys. J. **143**, 306.
Bowen, I. S., 1935, Astrophys. J. **81**, 1.
Copeland, J. A. and Heard, J. F., 1963, Pub. David Dunlap Obs., **11**, 317.
Feinstein, A., 1968, Z. Astrophys., **68**, 29.
Hutchings, J. B., 1967a, Observatory, **87**, 273.
Hutchings, J. B., 1967b, Observatory, **87**, 289.
Hutchings, J. B., 1968a, Monthly Notices R. astr. Soc., **141**, (in press).
Hutchings, J. B., 1968b, Monthly Notices R. astr. Soc., **141** (in press).
Jackisch, G., 1963, Inf. Bul. Var. Stars no. 40.
Kumajgorodskaya, R. N., 1960, Izv. Krym. astrofiz. Obs., **24**, 91.
Kumajgorodskava, R. N., 1962, Izv. Krym. astrofiz. Obs., **28,** 135.
Lacoarret, M., 1965, Ann. Astrophys. **28**, 321.
Mannino, G. and Humblet, J., 1955, Ann. Astrophys., **18**, 237.
McLaughlin, D. B., 1958, "Etoiles à raies d'émission", U. de Liège, p. 231.
McLaughlin, D. B., 1961, J. R. ast. Soc. Can., **55**, 73.
McLaughlin, D. B., 1963, Astrophys. J., **137**, 1085.
McLaughlin, D. B., 1966, Astrophys. J., **143**, 285.
Merrill, P. W., 1956, Vistas in Astronomy. (Pergamon Press), Vol. **2**, 1375.
Merill, P. W. and Burwell, C. G., 1933, Astrophys. J., **78**, 87.
Oke, J. B., 1954, Astrophys. J., **120**, 22.
Roman, N. G., 1951, Astrophys. J., **114**, 492.
Schmidt, H., 1959, Z. Astrophys., **48**, 249.
Slettebak, A., 1956, Astrophys. J., **124**, 173.
Struve, O., 1931, Astrophys. J., **73**, 94.
Struve, O. and Swings, P., 1940, Astrophys. J., **91,** 546.
Swings, P., 1942, Astrophys. J., **95**, 112.
Swings, P. and Struve, O., 1941, Publ. astr. Soc. Pacific, **53**, 35.
Underhill, A. B., 1955, Publ. Dom. astrophys. Obs. Victoria, **10**, 169.
Underhill, A. B., 1958a, "Etoiles à raies d'émission", U. de Liège, p. 17.
Underhill, A. B., 1958b, Publ. Dom. astrophys. Obs. Victoria, **11**, 143.
Underhill, A. B., 1959, Publ. Dom. astrophys. Obs. Victoria **11**, 283.
Underhill, A. B., 1961, Publ. Dom. astrophys. Obs. Victoria, **11**, 353.
Underhill, A. B., 1966, "The Early Type Stars" (Reidel Pub. Co., Holland), pp. 242−245.
van den Bergh, S., 1968, Astrophys. J., **151**, 1191.
Wilson, R., 1958, Publ. R. Obs. Edinburgh. Vol. **II**, No. 3.

DISCUSSION

Bakos: How does rotational broadening effect line profiles?

Slettebak: At the present time, rotational broadening and line broadening due to large-scale turbulent motions cannot be distinguished from a study of individual line profiles. — Both mechanisms broaden lines in the same way. Therefore it is impossible to asses the relative importance of these two broadening agents in the Of stars. Although large-scale turbulence probably dominates, rotation may also play a role in the line broadening of the Of stars.

Detre: We have some long runs of photoelectric measures of o Andromedae. We have obtained only some random fluctuations smaller in amplitude than 0.02 which are not correlated with Schmidt's period. The star seems to be *not* an eclipsing binary.

As I have mentioned in my Introductory Report, there are some recent theoretical papers considering magnetic effects in Of and Be stars, e.g., the paper by Hazlehurst on the magnetic release of a ring. Have we some observational evidence about the shape of the shell?

Slettebak: In his study of Of stars (1954) J. B. Oke concluded from the spectroscopic data that the emission lines probably arise in regions where the
fairly strong absorption lines are formed and not more than one or two
stellar radii above the star. I do not know of observational evidence
regarding the actual shape of the emitting shell region, although it
seems unlikely that this is ring-shaped, as in the Be stars.

Bakos: γ Cas is a visual binary with separation of about 20″. Would be of
interest to obtain UBV magnitudes of this secondary?

Sahade: You mentioned the appearance of satellite lines in the spectrum of 29
CMa. Perhaps it is in order to remember that 29 CMa is a spectroscopic
binary — and that there are variations in the absorption line intensities
probably due to the opacity of the gaseous streams in the system.

Milone: Are any of the Of stars magnetic variables?

Slettebak: I do not know of any measurement of magnetic fields in these objects.

Feast: There appears to be a much higher ratio of Of stars to normal O stars
amongst runaway stars than amongst non-runaway stars. This could point
to higher than average masses for the Of stars though the difficulties —
determining velocities of Of stars must be borne in mind.

Hutchings: Line radial velocity measurements of the Of stars HD 151804,
152408 show a range in velocities, correlated with the excitation potential
of the lines. This suggests a spherically symmetrical expanding envelope
and strong line profiles computed with such a model match the observations, with a small rotation — $v \sin i$ of the order of 50 km/sec.

Almár: Are there any curve of growth analyses of Of stars showing large turbulent velocity?

Slettebak: I do not know.

PHOTOELECTRIC OBSERVATION OF LINE PROFILES WITH HIGH TIME RESOLUTION IN B AND Be STARS

J. B. HUTCHINGS

Dominion Astrophysical Observatory, Victoria, B. C. Canada

INTRODUCTION

One of the most important problems connected with the observation of irregular variable stars is that of spectroscopic time resolution. In a star whose variable properties are not repeated in a predictable manner it is essential to make continuous observations until the time scale and magnitude of its variations become apparent. In addition, high time resolution observation may reveal short time variation in properties of stars previously thought to be stable. It is known that many early type stars for instance, especially those having emission lines (and hence extended atmospheres), show changes in their spectra, but until we can watch such changes taking place we cannot expect to explain their occurrence.

High dispersion high time resolution spectrophotography is limited to the few brightest stars or the few largest telescopes in the world. Both of these limitations are unsatisfactory, but as the photomultiplier is some 10 to 20 times more efficient than the photographic plate it is the obvious choice for such work. In addition the photomultiplier's linear response to light enables one to obtain direct intensity profiles immediately, saving hours of daytime drudgery, and allowing on-the-spot monitoring of the results.

Several systems for photoelectric line scanning are being developed now, so it is important to discuss the techniques being used in order to achieve reliable cross-comparison of results and to overcome design and performance problems. It is also important to bring to the attention of astronomers the sort of work which can be done with this technique. I propose therefore to describe the Victoria scanner and the problems in its operation and then show the results which early runs on it have produced. These results should at this stage be regarded as an introduction to the astrophysical problems brought to light by the observation.

THE SCANNER

Basically the scanner is a photomultiplier, Fabry lens, and slit which are moved in a straight line tangential to the focal plane of the 96″ focal length coude spectrograph of the Victoria 48″ telescope, using an accurate screw and speed stabilised motor. The dispersion in the spectrum is about 2.3 A/mm and with the present resolution scans of up to 50 A are possible before defocussing is significant.

The motor speed is continuously variable by remote control and is normally used in the range between 0.5 and 50 A/minute. The speed is constant to within 1% at any setting. The screw trips a micro-switch every revolution, which provides a fiducial mark on the tracing every 2.3 A. Thus the wavelength scale on the tracing is accurate to within .02 A, which is far below the resolution of the instrument.

The zero point of the wavelength scale is determined by scanning the comparison spectrum, which is an Fe—A discharge tube. This comparison exactly replaces the stellar spectrum by means of a swinging quartz prism. Initially a grating setting is made and the exact position of the scan identified by making a scan of the comparison spectrum. Once the position and length of the scan are decided, they can be fixed by means of adjustable limit switches on the screw.

The cooling mechanism is thermoelectric, with the hot junction cooled by water circulation. Temperatures accurate to 1°C, down to —30°C are remotely controlled. The best compromise between thermal noise and light response is obtained at about —20°C with the EMI tube used in the blue.

The recorded tracing is the ratio of the output of two channels — the scanner and a monitor channel. The monitor system measures light reflected off a diagonal quartz plate in the beam immediately behind the objective slit. The output from each channel is amplified and impedances matched. Further optional adjustments allow the scale of the tracing and the time constant to suit the speed of scan and brightness of the star. A filter is used in the main beam to isolate the spectral region being scanned for the monitor response. Using this system, variations in seeing and extinction during the time of scan are largely overcome. A 25% change in light transmitted through the slit results in a 2% change in the recorded output. In good conditions the seeing variation is less than 5% over the time for a scan so that final accuracy can be well within 1%.

In order to achieve this stability however it is necessary to pass most of the starlight through the objective slit, which requires a width of about 7 mm. The resolution of the scanner is then some 0.5 A. While this is not very high it is sufficient to indicate changes which occur is spectral features of almost all early type stars.

Finally I should mention the disadvantages of the system and possible ways of overcoming them. The monitoring system used requires a filter which cuts down light or covers too wide a range in wavelength. It also requires a wide slit. Unevenness or dirt on the quartz flat can lead to inaccuracy in the monitoring system, and defocussing and decollimation effects are not monitored. A stationary monitor channel covering some 10—20 A in the spectrum would overcome all these drawbacks. It would also allow the limiting magnitude to be increased by one, or the resolution to be improved for brighter stars. Efficiency and resolution could also be improved by replacing both slits with image slicers of the type now used for spectrography in Victoria. This would require redesign of the Fabry optics, but is a possibility. Lastly, a recorder which showed both monitor and ratio outputs would assist in the assessment of the profile accuracy.

At present the scanner can produce profiles in the blue region of early type spectra of 2% accuracy in a 4th m. star over 20 A in some 5 minutes, or 3% accuracy in a 6th m. star over 6 A in about the same time.

RESULTS

The figures below show selected profiles of lines in some early type stars which show irregular and rapid spectral changes.

1. γ *Cas*. Figure 1 shows series of profiles of Hγ and Hβ taken on two consecutive nights. Each shows rapid change in the double emission structure found in this Be star. The star is well known for its longer term changes and similar short term changes have been obtained previously (Hutchings 1967, 1968). The separation of the Hγ peaks here falls fairly steadily during the period of observation, increases sharply at the second last profile and then starts to fall again. This is shown by the table below.

time	4.03	4.07	4.10	4.13	4.16	4.21	4.24	4.27	4.31
separation (arbitrary units)	3.9	3.7	3.6	3.5	3.6	3.4	3.3	3.9	3.6

Whether this is a regular phenomenon or not must be decided by further systematic observation.

2. ϰ *Dra*. This is another Be star, with much weaker emission components. Figure 2 shows a series of scans of the emission components in the bottom of the Hγ line. Here again there is rapid activity, especially in the shortward peak, whose sharpness appears to fall off steadily throughout the time of observation. Further observation is needed to confirm and study this type of activity, which may be connected with the rapid rotation of the star. This star also shows emission peaks in other lines, often very weakly, but whose mean separations are quite different. Again, evidence so far is fragmentary

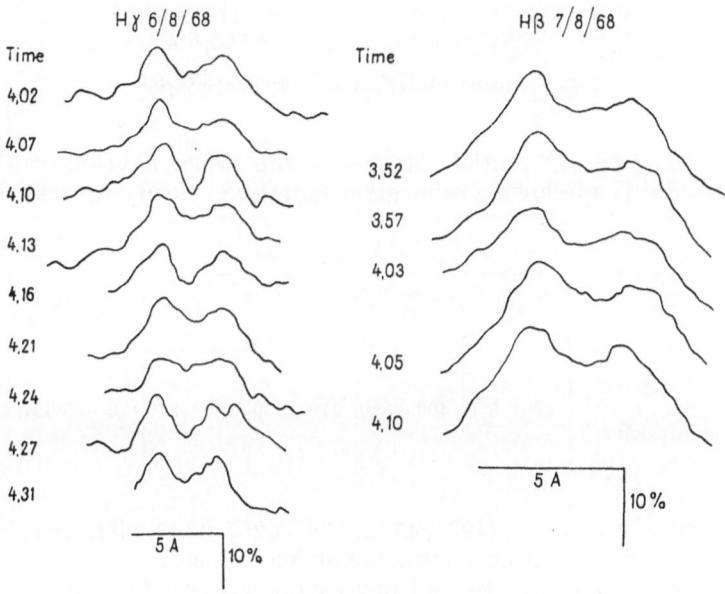

Fig. 1. Line profiles in γ *Cas*

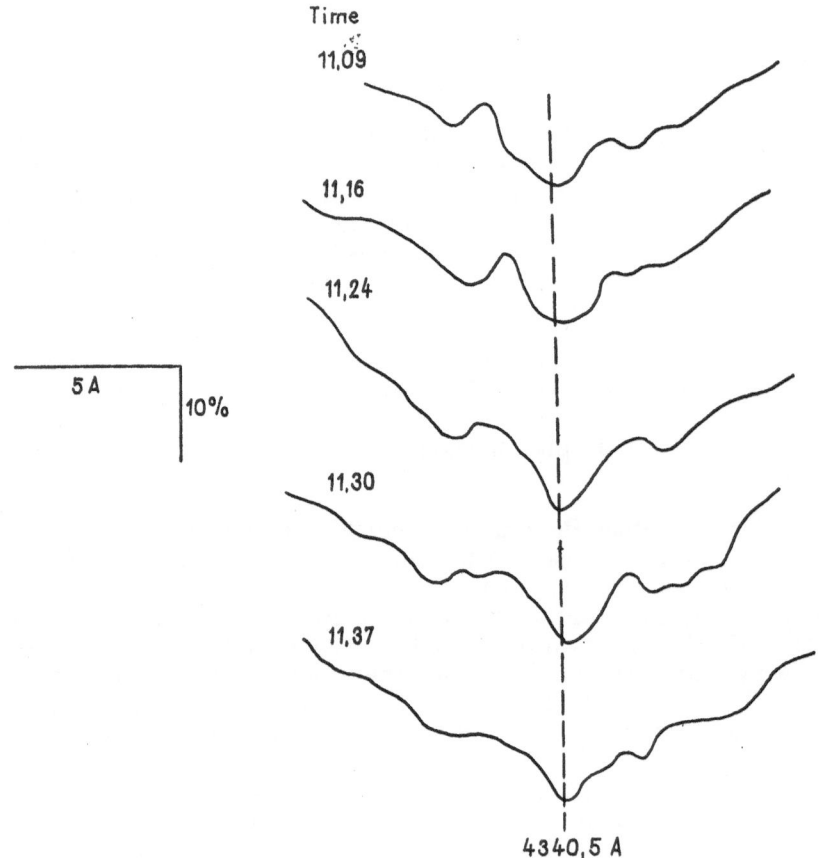

Fig. 2. Scans of Hγ in ϰ Draconis 30/5/68

but it may indicate differential rotation of the stellar envelope with height or even latitude. The following table gives tentative velocity ranges for various lines.

Line	4471	4481	Hγ	Hβ	Hα (triple peaked)
Vel. indicated by peak separation	490—650	400—520	210—300	160—240	170—650km/s

Other lines without emission (4921, 4713, 4387, 4267, 4143) all have rotationally broadened profiles indicating *v* sin *i* about 350 km/sec.

The star is being investigated further observationally and by computation of line profiles for various geometrical models.

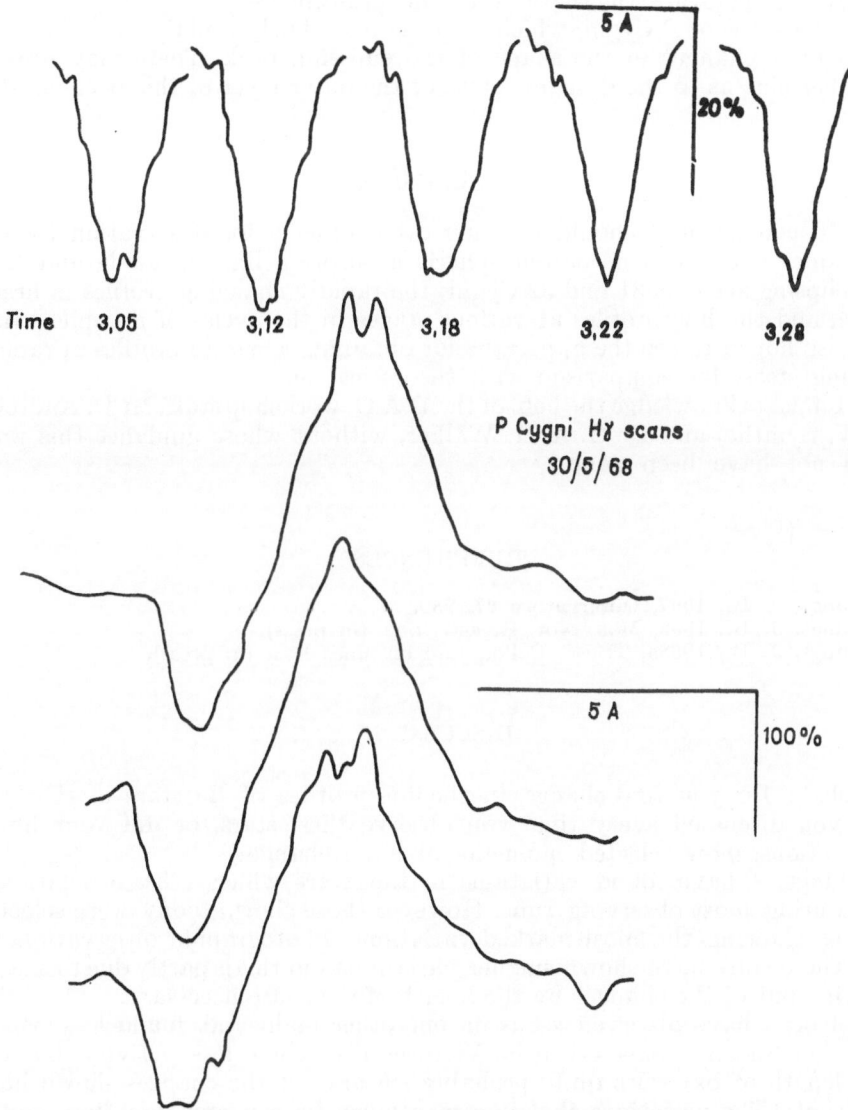

Fig. 3. Scans of Hγ in B-type supergiants

3. *B Supergiants.* Figure 3 shows profiles of Hγ for the B3 supergiant 55 Cygni, and for P Cygni. The core of the Hγ absorption in 55 Cygni is seen to be in rapid and irregular activity. As the line core is formed in the outer layers of the atmosphere and the star is probably undergoing mass loss (Hutchings 1968a), this type of observation may be a valuable method of extracting information about the atmospheres of such stars. This sort of activity is seen in

other supergiants (Hutchings 1967). Scans of the B 1.5 supergiant HD 190603 have shown rapid changes in strength in the O II line at 4351 A, which may indicate temperature fluctuations in the atmosphere.

The scans of P Cygni, which is losing mass fairly rapidly show evidence of irregular changes in the shape of the emission peak. These may provide a further clue as to the dynamic state of the outer layer of this peculiar star.

CONCLUSION

In conclusion I should mention other objects for observation by this technique. There are the chromospheric absorption lines in Ca H and K of the eclipsing systems 31 and 32 Cygni; the rapidly changing profiles in bright novae, and the line profiles at various stages in the cycles of β Cephei stars. It is also hoped to use the apparatus for obtaining accurate profiles of rapidly rotating stars, for comparison with theoretical ones.

I must acknowledge the help of the D.A.O. workshop staff, Mr D. Andrews, Mr W. Symthe, and Dr G. A. H. Walker, without whose guidance this work could not have been done.

REFERENCES

Hutchings, J. B., 1967, Observatory **87**, 289.
Hutchings, J. B., 1968, Mon. Not. R. astr. Soc. (in press).
Hutchings, J. B., 1968a, Trieste Colloquium on Mass Loss (in press).

DISCUSSION

Slettebak: Do you find changes in the line profiles of Be stars of the type you discussed every time you observe these stars, or did your illust-rations show selected moments of large change?

Hutchings: I have found variations in the stars which I have mentioned during most observing runs. However those shown today were selected as showing the most marked variations. Photographic observations of these stars have shown smaller variations but this is partly due to smear-ing out of the changes by the length of exposure necessary.

Slettebak: I have observed γ Cas on one other night and found less varia-tion. Spectrograms taken in Victoria also show less activity but the length of exposure quite probably smears out the changes shown here.

De Groot: The variations that you mentioned in the emission line profile of Hγ are also indicated in our material; but there they are more dif-ficult to detect because our profiles are from photographic observa-tions. The emission peaks sometimes are quite black and difficult to re-duce, but there are indications for the same variations.

SOME REMARKS ON THE SPECTRAL AND LIGHT VARIABILITY OF P CYGNI

L. LUUD

Estonian Academy of Sciences, Tartu

The unusual profiles of the spectral lines and its peculiar variability make P Cygni one of the most interesting stars in the sky. This interest is enhanced by the fact that each observational study adds as many problems as it resolves.

P Cygni was observed spectroscopically at the Tartu Observatory during the years 1961 to 1963. 44 spectrograms secured with a one-prism spectrograph giving dispersion 160 A/mm at H_γ were used to determine spectrophotometric gradients in the Greenwich system. The gradients show according to the sequential test no variations exceeding the accuracy of measurement. In Table 1 the mean absolute gradients for different years are given. It must be mentioned that Dolidze (1958) found variations of the spectrophotometric gradient, but the spectral range used by her was smaller.

Table 1

Observations, year	Greenwich Catalogue	1961	1962	1963
Gradient	1.03 ± 0.02	1.02 ± 0.04	1.00 ± 0.04	1.03 ± 0.07

For spectral line investigations 32 spectrograms with dispersions from 1.5 A/mm to 36 A/mm were taken at the Crimean Astrophysical Observatory in 1964—1966, every year during some weeks. On the spectrograms obtained in the same year the differences in spectral line contours do not exceed the accuracy of measurement. The mean contours determined by spectrograms taken in different years differed systematically (Fig. 1). The detailed review of spectral line contours is given in the Publications of Tartu Observatory (Luud et al. 1968).

The Balmer decrements observed are in good agreement with the calculations by A. Boyarchuk (1966), if we assume that $T_e = 10000°$, $T^* = 20000°$ and $W = 10^{-2}$. The probabilities for the exit of L_α-quanta (β_{12}) are given in Table 2.

Supposing that in a rough approximation the P Cygni-type contours can be divided into components as shown in Fig. 2, we should be able to carry out the coarse analysis of the P Cyg atmosphere. From the curve-of-growth analysis we get the data given in Table 2.

Table 2

Year	log n_e	log N_2H	τ_{H_β}	β_{12}	V_T	T_i
1964	12.36	15.98	30	$2.5 \cdot 10^{-4}$	23	19400°
1965	12.26	15.68	21	$5 \cdot 10^{-4}$	40	21800°
1966	12.80:	15.37	12	$1 \cdot 10^{-3}$	24	21400°

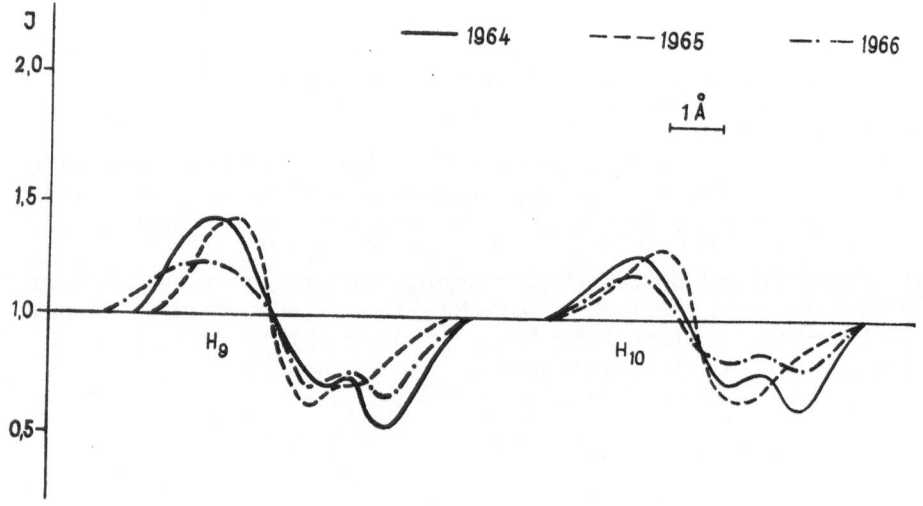

Fig. 1

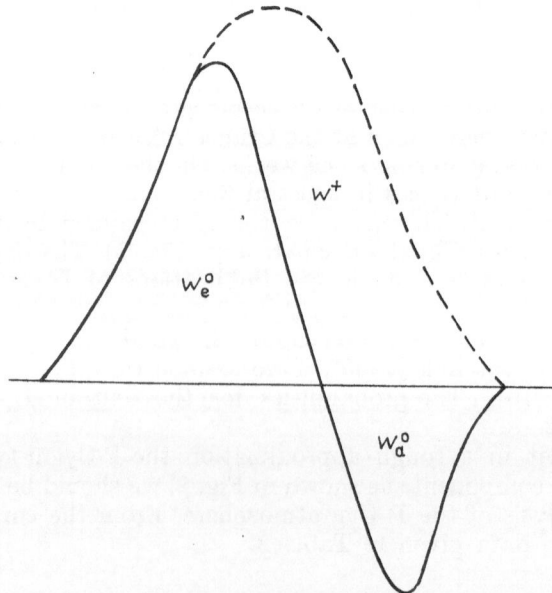

Fig. 2

From more extended data published in our (Luud 1967a, b) papers we concluded that spectral variations are due mainly to variable amount of atoms in the volume in which the spectrum is formed. The variations of physical conditions seem to have less effect.

If we assume that the luminosity of P Cyg is $M_v = $ -8 and $T^* = 29000°$ (Luud 1967 b, c) then we get for the photosphere $R^* = 62\ R_\odot$. From the formula

$$\varDelta\mathfrak{M} = 4\pi R^2 \varrho(R)\ V(R)dt$$

we obtain, assuming $\varrho(R^*) = n_e\,m_H, n_e = 10^{12}\,\text{cm}^{-3}$ and $V\ (R^*) \approx 150$ km/sec, that the mass loss of P Cygni is $9 \cdot 10^{-5}\ \mathfrak{M}_\odot$/year. The result $\varDelta\mathfrak{M} = 2 \cdot 10^{-6}$ $\mathfrak{M}_\odot$/year for giant B stars (Jenkins and Morton, 1967) suggests that this may be a real value.

The mass of the atmosphere can be checked by the formula

$$\mathfrak{M}_\text{atm} = N_+\,H \cdot 4\pi\,R_\text{atm}^2 \cdot m_H$$

where N_+H are available from N_2H using the Boltzman and Saha equations. From $W = 10^{-2}$ we get $R_\text{atm} = 5\ R^*$, and finally $\mathfrak{M}_\text{atm} = 6 \cdot 10^{-6}\ \mathfrak{M}_\odot$.

These figures show that as a result of mass-loss atoms of the atmosphere replace rapidly — during some tens of days. Therefore the spectral line variations are probably due to the nonstationary outflow of matter.

It should be mentioned, that the variability of P Cygni spectral line contours is described by a number of observers (Herman 1968; Lacroute, 1938; Wilson, 1936) No periodicity has been found.

The light variability of P Cygni was at first discovered in 1600. Here we shall only provide some remarks about contemporary photoelectric observations and will not touch earlier history of light variations.

The most extensive data are published by Magalashvili and Kharadze (1956). By these and unpublished data they have found that P Cygni was variable with a W UMa-type light curve and with a period of $0^d500656$. The observations published after the paper of the Georgian scientists do not seem to confirm regular variability (Alexander and Wallerstein, 1967). Here we shall briefly discuss these results.

According to the mass-luminosity relation by Paranego and Masjeviĉ (1951) the luminosity of P Cyg, $M_v = -8^m$, corresponds to 100 $\mathfrak{M}_\odot$. If we could detect the W UMa type eclipse, the secondary component would have a brightness of $M_v \approx -7^m$ and the minimum mass of 30 $\mathfrak{M}_\odot$, if it is of the spectral class A0. If we now assume that $\mathfrak{M}_1 + \mathfrak{M}_2 = 100\ \mathfrak{M}_\odot$, we get from the orbital period $a_1 + a_2 = 12\ R_\odot$. Consequently the radii of the stars turn out to be greater than their separation and the assumption of eclipsing type variability must be dropped.

If P Cygni has very peculiar components and they are actually main sequence stars with $\mathfrak{M}_1 + \mathfrak{M}_2 = 30\mathfrak{M}_\odot$ and $\mathfrak{M}_1 : \mathfrak{M}_2 = 2$, the orbital velocity of the primary should be 600 km/sec. If the light varies with $\varDelta m = = 0^m1$, we should have $i \gtrsim 45°$ and an observable radial velocity of 420 km/sec. Line displacements of 14 A, that correspond to this velocity, had never been observed. For example in Table 3 we give H_{10} displacements from the spectra taken in July 24/25 with a dispersion of 15 A/mm. If we had observed in the most unfavourable time, the displacement would have a value of ~ 60 km/sec. It follows that orbital movements are excluded.

Table 3

Time (UT)	V_e	V_{a1}	V_{a2}
20^h39m	−36	−142	−218
21^h00m	−36	−142	−224
21^h42m	−38	−141	−227

On Fig. 3 we plotted the published Abastumani observations with the period $0^d500656$. We see no serious arguments why this periodicity is not valid. Near the phases 0.8—1.2 there is a very great scatter that seems to suggest that P Cygni had at that time irregular light outbursts.

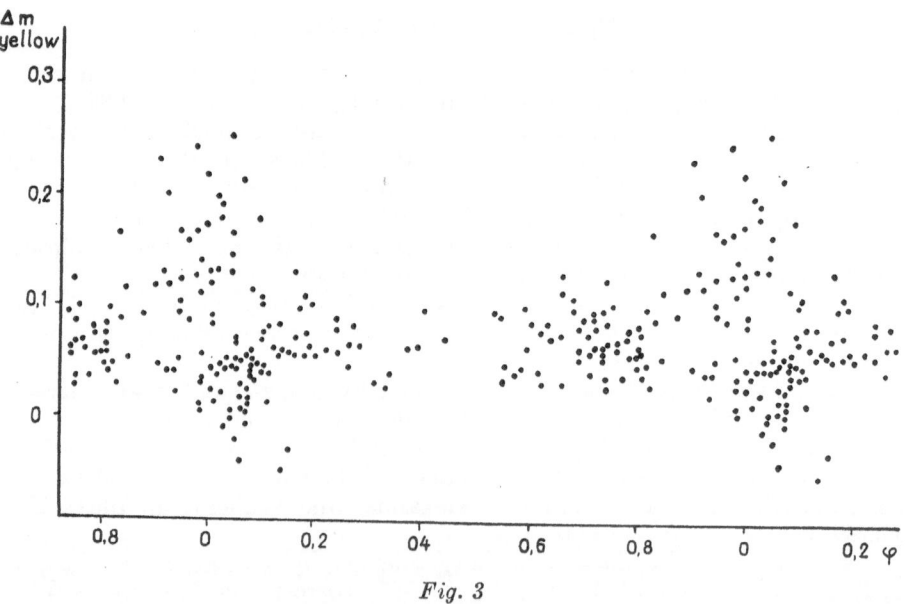

Fig. 3

There is a serious suggestion that the observations of the Georgian scientists may be interpreted in another way. Schwarzschild and Härm (1959) have shown that very massive stars must be pulsationally non-stable and must have little light variations with periods of approximately half a day and amplitudes near to 0^m05. Taking into account the possible differences of the main structural parameters of P Cygni from those used in this stellar structure calculations the agreement seems to be quite reasonable.

We may guess that in the case of P Cygni we have two kinds of irregular variations generated respectively by non-stationary mass loss and by pulsational instability.

To close our brief discussion, we must say that in spite of the quite favourable brightness of P Cygni we have no systematic simultaneous photo-electric and spectral observations of it, but this seems to be the only way for receiving new important data on the subject treated above.

REFERENCES

Alexander, T., Wallerstein, G., 1967, Publ. astr. Soc. Pacific **79,** 500.
Boyarchuk, A. A., 1966, Izv. Krym. astrofiz. Obs. **35,** 45.
Dolidze, M. V., 1958, Abastumanski astrofiz. Obs. Gora Kanobili Bjull. No. **23.** 69.
Herman, R., 1964, Ann. Astrophys. **27,** 507.
Jenkins, E. B., Morton, D. C., 1967, Report on XIII IAU Gen. Assembly.
Lacroute, P., 1938, C. R. hebd. Seanc. Acad. Sci. Paris **206,** 1091.
Luud, L., 1967a, Astron. Zu. **44,** 267.
Luud, L., 1967b, Astrophysics, **3,** 379.
Luud, L., 1967c, Publ. Estonian Acad. Sci. Phys. Math. **16,** 319.
Luud, L., Pôldmets, A., Leesmäe, H., 1968, Tartu Publ. **36,** in press.
Magalashvili, N. L., Kharadze, E. K., 1956, Abastumanski astrofiz. Obs. Gora Kanobili
    Bjull. No. **20,** 3.
Magalashvili, N. L., Kharadze, E. K., 1967, Astr. Cirk. Izdov bjuro astr. Soobšč. Kazan
    No. 467; Inf. Bull. Var. Stars, No. 210.
Paranego, P. P., Masjevič, A. H., 1951, Publ. Sternberg astr. Inst. No. 20.
Schwarzschild, M., Härm, R., 1959, Astrophys. J., **129,** 637.

SPECTRAL VARIATIONS OF P CYGNI

MART DE GROOT

Sonnenborgh Observatory, University of Utrecht, The Netherlands

ABSTRACT

From a careful study of 35 high-dispersion spectrograms of P Cygni it is concluded that the spectroscopic data do not confirm the conclusion of Magalashvili and Kharadze that P Cygni is a W UMa type system. It is found that many of the absorption lines are double, the hydrogen absorption lines even triple. This is attributed to line formation in different shells. In the outer shell variations with a period of 114 days lead to observed radial velocity variations between −180 and −240 km/sec. A preliminary conclusion about the velocity field in the atmosphere of P Cygni is drawn.

P Cygni, in the Henry Draper Catalogue classified as B1p, has been known as a variable star from the year 1600 when it was discovered as a third magnitude nova by the Dutch chartmaker, geographer and mathematician Willem Janszoon Blaeu. The early history of the light variation of this star is nearly unequaled and rather puzzling. However, since 1880 P Cygni has been of nearly constant brightness. Some observers have reported irregular light variations with an amplitude up to 0.2 magnitude (e.g. Nikonov 1936, 1937). About one year ago Magalashvili and Kharadze reported some interesting two and three color observations of P Cygni. From their observations made during the period 1951—1960 they concluded that P Cygni is a W UMa system with a period of 0.500656 days and with amplitudes of $0^{m}.10$ and $0^{m}.08$ for the primary and secondary minimum respectively. (Magalashvili and Kharadze 1967a, b.)

When these results were first reported in the Information Bulletin on Variable Stars (no. 210) P Cygni was put on a constant observation program for 5 nights by Alexander and Wallerstein (1967) who reported that their observations did not reveal any variations of the brightness of P Cygni and thus did not confirm the observations made by Magalashvili and Kharadze.

In this paper some facts pertaining to the character of these light variations are presented from a different point of view. We have been working upon a collection of high-dispersion spectrograms of P Cygni, covering the period 1942—1964. From the study of some 35 spectrograms the following facts have been established.

1. On most of the spectrograms the lines of hydrogen, many lines of He I and the strongest lines of Fe III show besides the nearly undisplaced emission line two shortward displaced absorption components. In the case of the hydrogen lines with Balmer number $n \geq 9$ there often are even three components with velocities of about −95, −125 and −210 km/sec (cf. Figure 1).

2. The radial velocity of the most shortward displaced component of the hydrogen lines is not constant but shows variations which after a closer inspection have a period of 114 days. Other lines do not show this periodicity.

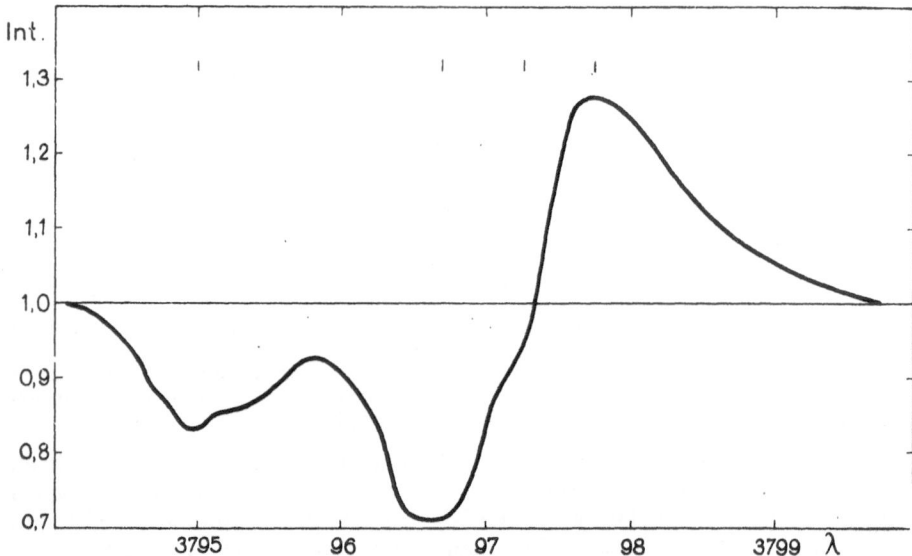

Fig. 1. Profile of H10 λ 3797 showing three absorption components.

3. There are variations in the relative intensities of different absorption components. These variations seem to be rather irregular.

With this information let us consider again the conclusion of Magalashvili and Kharadze about the binary nature of P Cygni. Should the fact that the spectral lines often are double be regarded as a proof that P Cygni is a binary? The two absorption components which appear at the positions of the hydrogen and helium are of comparable strength. This means that a companion star should not be more than one magnitude fainter than the main star. If this statement were true, then also other spectral lines of the companion should be visible in the spectrum of P Cygni, this providing more double absorption lines. This is not the case. Only the hydrogen and some of the helium lines are double. One might think of a late B type companion with few strong spectral lines except those of hydrogen. But then the Si II spectrum and the line of Mg II at λ 4481 should be more prominent than the lines actually observed in the spectrum of P Cygni.

Furthermore, the mean velocity of approach, as derived from the two absorption components of the hydrogen lines equals about — 170 km/sec. If the duplicity of the lines were a proof of the binary nature of P Cygni this figure would mean either that the system as a whole has a velocity of —170 km/sec with respect to the sun, or that the W UMa binary is surrounded by a large expanding atmosphere. The first suggestion is not acceptable because it leaves unexplained the fact that all the emission lines lie at an average displacement of about —15 km/sec. Also a velocity of —170 km/sec is impossible to combine with the membership of P Cygni of the galactic cluster NGC 6871. The second suggestion is difficult to maintain because the two components always fall in the same limited radial-velocity intervals between —180 and —240 km/sec and between —120 and —160 km/sec respectively, but their relative intensities change. If these were two lines from the spectra of different stars their radial velocities should pass through all values between say —120 and —240 km/sec.

In order to find out if the intensity ratio between the two absorption components or their radial velocities show any correlation with the phase of the light variations given by Magalashvili and Kharadze (1967a) the phases of all the plates of this study were determined and in Figure 2 are shown plotted against the intensity ratio of the two absorption components at —210 and at —125 km/sec. The intensity ratios for the Balmer lines, H9, H10, H11 and H12 were used, since these lines are essentially free from blends and nearly always show the two components concerned. The same phases are also shown plotted against the radial velocities of the components of Hγ and H9 at about —210 km/sec and of H9 at about —125 km/sec (see Figure 3).

In both figures there is much scatter. In Figure 2 this is caused by the roughness of the visual intensity estimates that were made on the spectrograms while measuring them for their radial velocity. In Figure 3 much of the scatter is introduced by unresolved double or triple absorptions. No convincing evidence appears of a change either in the intensity ratio or the radial velocity in a period of 0.500656 day.

One must conclude that the result of Magalashvili and Kharadze, that P Cygni is a W UMa system, though very interesting from the points of view of stellar evolution and of explaining nova outbursts, is not supported by the spectroscopic information.

As is indicated above the radial velocity of the most shortward displaced component of the hydrogen lines shows variations with a 114 day period. This results is more fully illustrated in Figure 4 which shows the radial velocities of Hβ, Hγ, Hδ, H9, H10 and H11 against their phases in the 114 day period. It is found that all these lines show very much the same variations with corresponding phases and amplitudes. In evaluating Figure 4 one should keep in mind that many of the points in the lower part of the diagram at small and at large phases are from dates on which the H-lines did not show all three components. These points then are either the results of blends between the third and second component, or they are only the second components the third being absent. In both cases these points give lower limits to the radial velocity of the third component.

Not only are the phases and amplitudes of these variations about the same, but also the mean value around which the radial velocity varies is strikingly similar for the various lines studied. If one assumes a unique relation between radial velocity and level in the stellar atmosphere, which in fact is a unique relation between radial velocity with respect to the star and the distance from the stellar surface, Figure 4 could be explained in either of two ways:

1. At some high level in the atmospheres of P Cygni there is a layer which shows periodic velocity fluctuations. The velocity of that particular part of the atmosphere varies with a 114 day period between —180 and —240 km/sec.

2. The velocity field in the stellar atmosphere is fixed. The variations are introduced by variations in the opacity of the atmosphere. Sometimes we can only see as deep as the layer with a velocity of —240 km/sec and half a period later we see a deeper layer with a velocity of —180 km/sec.

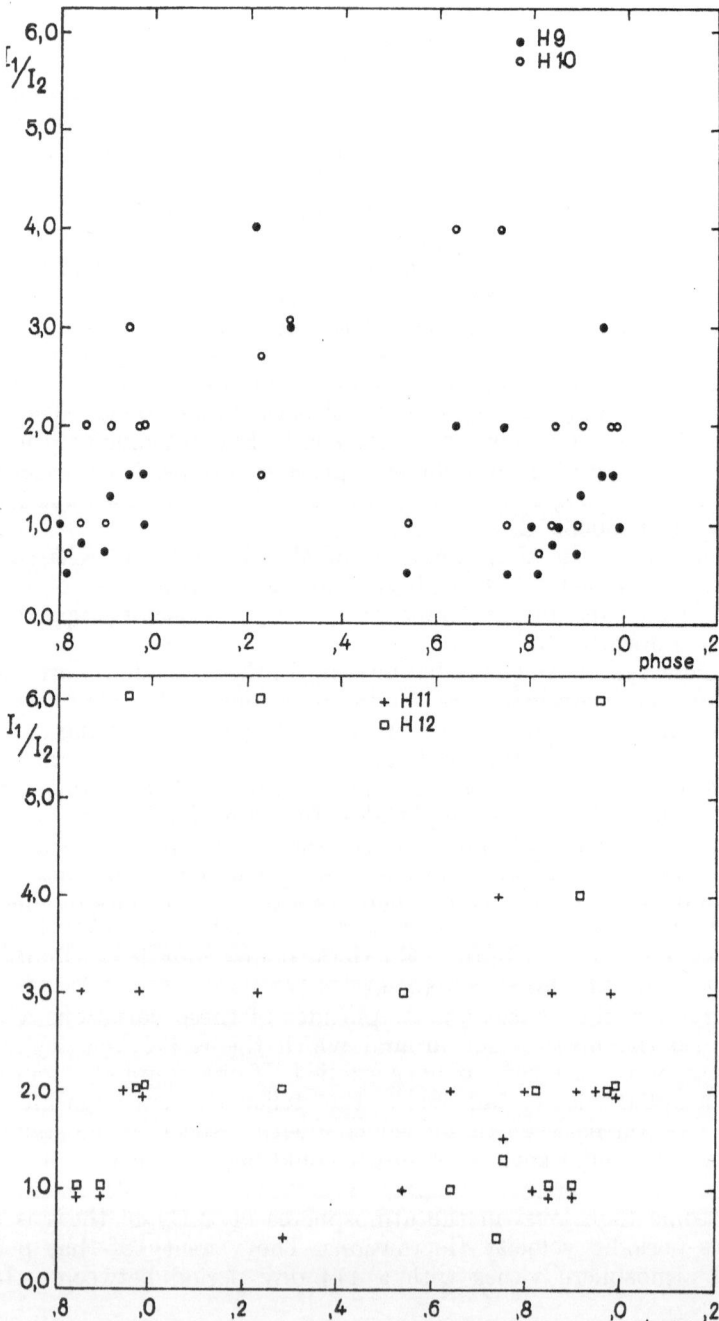

Fig. 2. Intensity ratio of second and third components of hydrogen lines against phase
of Magalashvili and Kharadze.

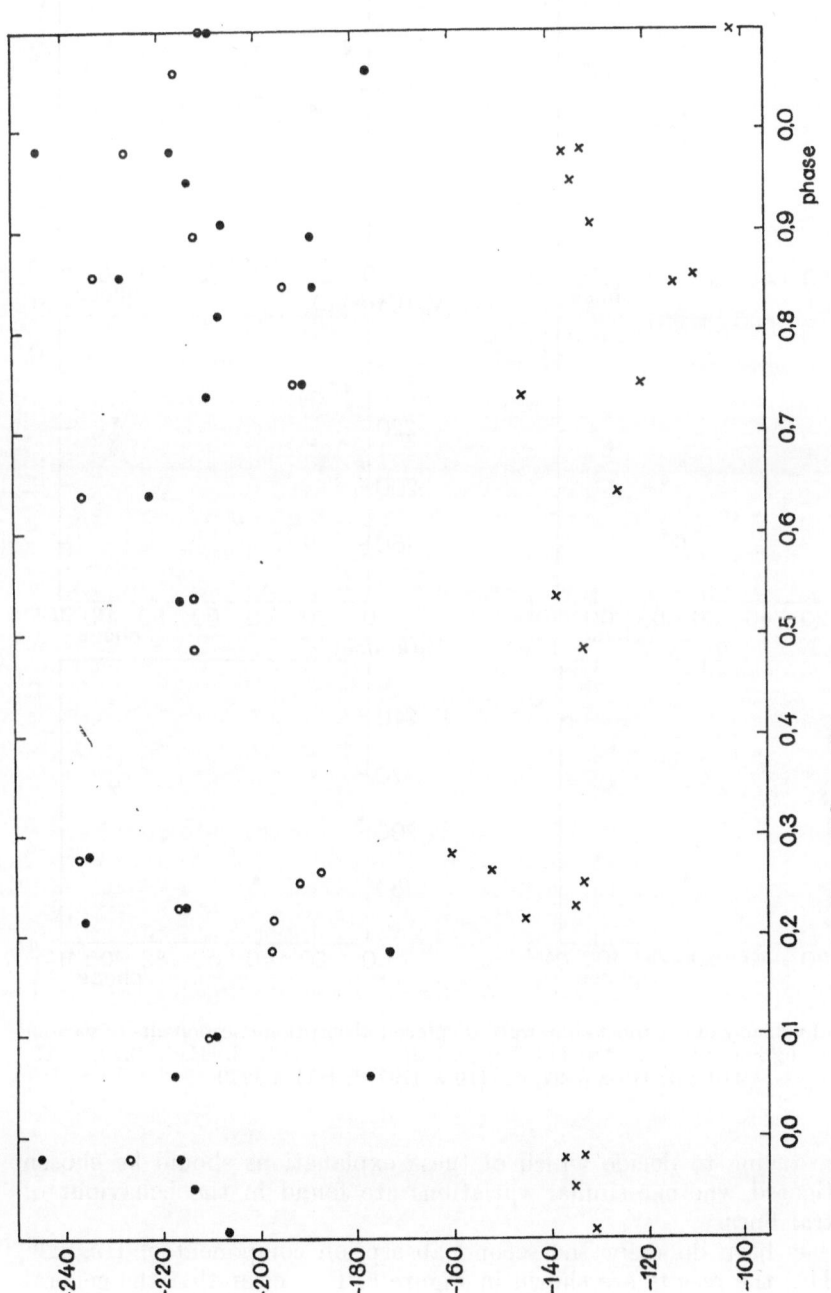

Fig. 3. Radial velocity of some hydrogen absorption lines against phase of Magalashvili and Kharadze; open circles: second component of Hγ; dots: third component of Hγ; crosses: second component of H9.

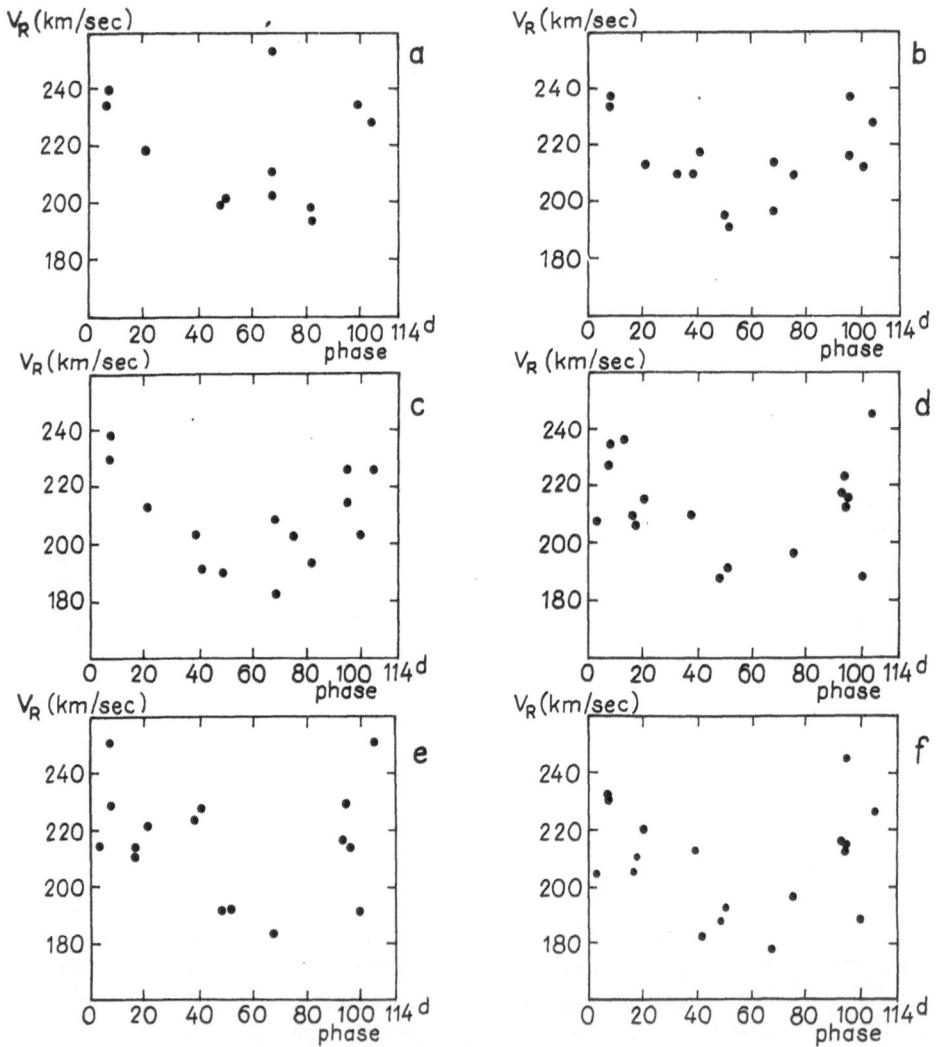

Fig. 4. Radial velocity of the most shortward displaced absorption components of various hydrogen lines against phase in the 114-day period; a: Hβ λ 4861; b: Hγ λ 4340; c: Hδ λ 4101; d: H9 λ 3797; e: H10 λ 3797; f: H11 λ 3770.

Before trying to decide which of these explanations should be chosen it is investigated whether similar variations are found in the behaviour of other spectral lines.

This has been done for the second absorption component of Hδ, H9, H10 and H11; the results are shown in Figure 5. It is clear that the general pattern of Figure 4 is not retained. The variations are more at random. This means that second absorption are formed in a layer where no radial-velocity fluctuations or opacity variations of the stellar atmosphere occur.

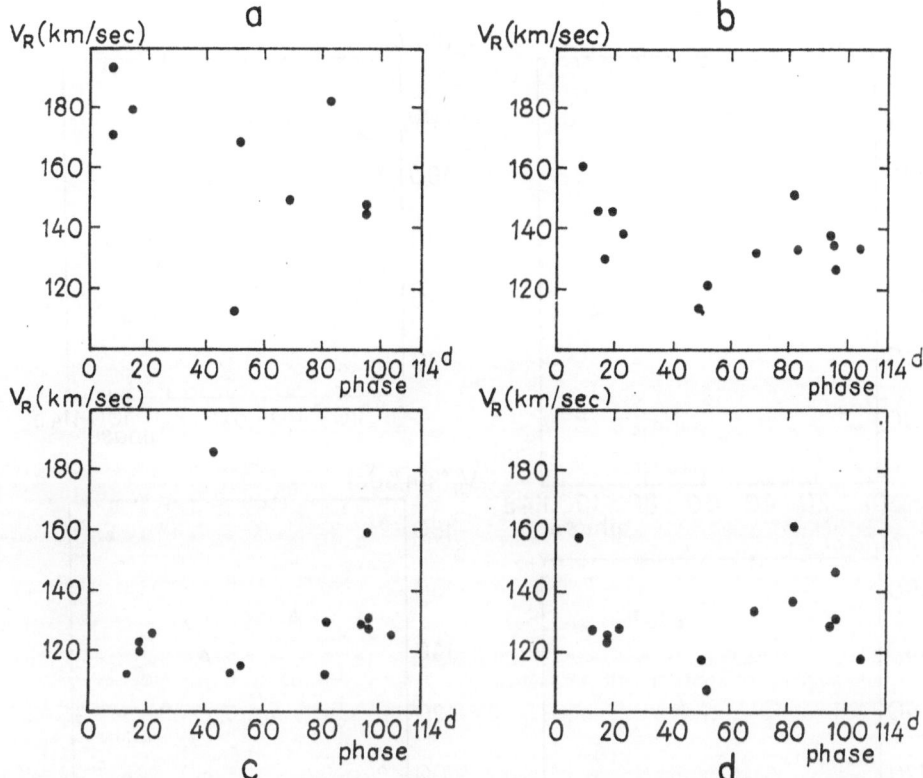

Fig. 5. Radial velocity of the second absorption component of various hydrogen lines against phase in the 114-day period; a: Hδ λ 4101; b: H9 λ 3835; c: H10 λ 3797; d: H11 λ 3770.

The same results are obtained for the radial velocities of the helium lines. From different series the best measured lines were selected and their radial velocities plotted against the phase in the 114 day period in Figure 6. The lines at λλ 3964, 4471, 4387 and 4120 are used for this purpose. There are no indications of variations in that part of the atmosphere where these helium lines are formed. The second components of the lines at λ 4387 and at λ 4120 have radial velocities of about —180 km/sec and this value is well below the value found in the case of the varying velocity of the third components of the hydrogen lines. For the two other lines, λλ 3964 and 4471, the second components have radial velocities of nearly —200 km/sec. This value is about equal to the velocity minima of the third components of the hydrogen lines. That no variations are found in the case of λ 4471 may be due to the small number of measured second components. For λ 3964 the mean velocity of the second component is —193 km/sec whereas the third hydrogen absorption component with smallest radial velocity, H11, still gives —208 km/sec. The conclusion is that even the radial velocity of λ 3964 is not subject to variations because this line is formed just below the layer of the atmosphere in which the variations occur.

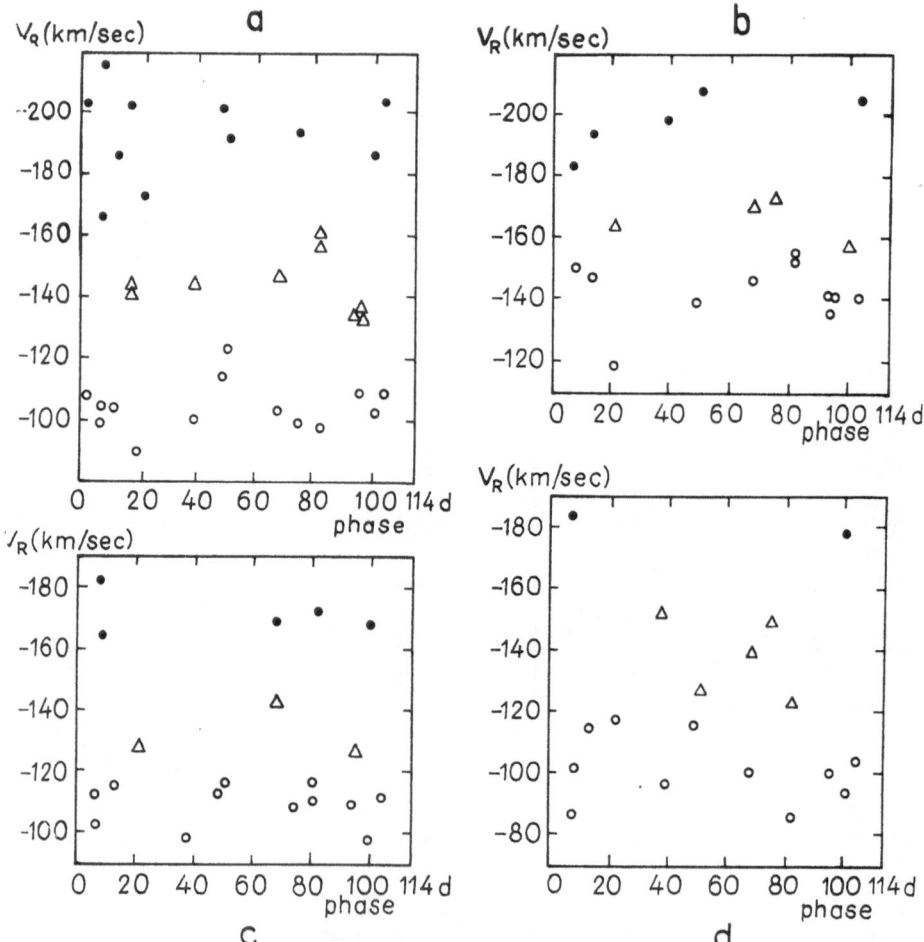

Fig. 6. Radial velocity of all components of some He I lines; a: λ 3964; b: λ 4471; c: λ 4387; d: λ 4120. First components are indicated by open circles, second components by dots, and unresolved pairs by triangles.

The influence of the emission lines upon the measured radial velocities has been investigated also. The tendency is that a strong line fills in a larger part of the adjacent absorption and thus will cause a larger absorption velocity to be measured. By studying the radial velocities and the line profiles simultaneously it is possible to separate this "emission-line effect" from the influence upon the radial velocity of the velocity gradient of the atmosphere. It appears that the corrections to be applied in correcting for the emission-line effect are always smaller than 15 km/sec. The effect of the stratification of the atmosphere which can be determined from a study of the radial velocities of lines from ions with different ionization potentials is much larger than the emission-line effect.

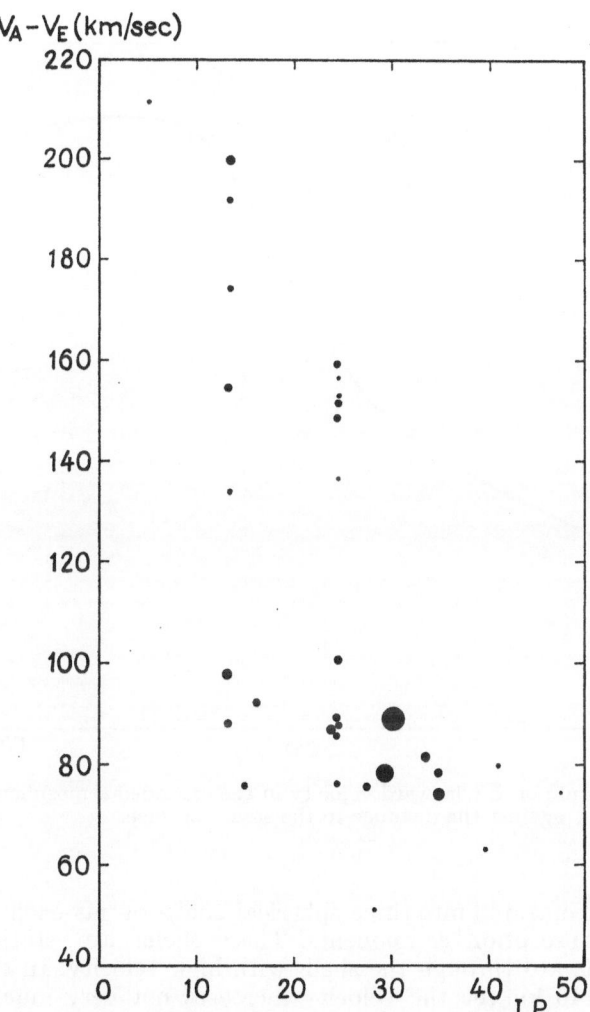

Fig. 7. Absorption minus emission radial velocity against Ionization Potential. The size of the dots is a measure for their weight.

The results obtained by previous investigators (Struve 1935, Kharadze 1936) about the dependence of the radial velocity upon the ionization potential are confirmed. This dependence is found best to show up if the ionization potential is plotted against the velocity difference absorption minus emission instead of plotting it against the absorption velocity only (cf. Figure 7).

If we now combine all these results into one general picture of the atmosphere of P Cygni we find the following: Material from the stellar surface is driven away from the star. While moving outward it is accelerated unto a maximum velocity of about 240 km/sec. Beyond that point the velocity stays constant or may even decrease a little. The matter in the extended

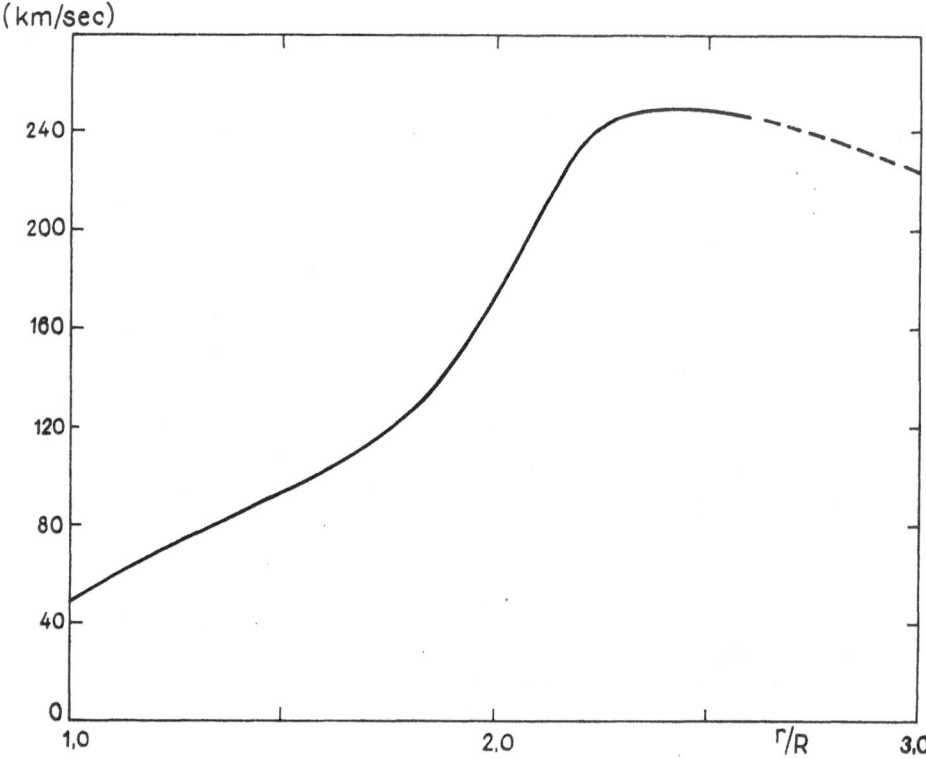

Fig. 8. Tentative picture of the outward velocity in the extended atmosphere of P Cygni against the distance to the stellar surface.

atmosphere is concentrated into three spherical shells of gas each giving rise to one of three absorption components. These shells are stationary; the particles move outward through the shells with high velocity. In the velocity range —80 to —140 km/sec the velocity increases not very much with the distance to the star but at higher levels (where the velocity is between —180 and —240 km/sec) the velocity changes more rapidly (see Figure 8). If now a varying opacity according to our previous second assumption permits one to see deeper into the atmosphere the result is that at high velocities one really sees into a layer with smaller velocity, while in the deeper layers one sees about the same velocity This explains why the velocity of the third component is varying while the first and second components only scatter about their mean value.

The next step is to find out if and how it is possible to fulfil the equation of continuity in this case. Furthermore, it is possible from spectroscopic criteria about the relation between emission intensity and dilution factor to give a more accurate height scale to Figure 8. From the study of absorption equivalent widths it is then possible to deduce values for the densities of the different shells which will complete the present provisional picture. This work is hoped to be completed in the next few months.

I am indebted to the Mount Wilson and Palomar Observatories, to the Dominion Astrophysical Observatory, to the Lick Observatory and to the Haute Provence Observatory for the spectrograms which form the underlying material for this investigation.

The stimulating remarks and comments of Prof. Anne B. Underhill I gratefully acknowledge.

REFERENCES

Alexander, Th. and Wallerstein, G., 1967, Pub. astron. Soc. Pacific **79**, 500.
Kharadze, E. K., 1936, Z. Astrophys. **11**, 304.
Magalashvili, N. L. and Kharadze, E. K., 1967a, Inf. Bull. Var. Stars No. 210.
Magalashvili, N. L. and Kharadze, E. K., 1967b, Observatory, **87**, 295.
Nikonov, V. B., 1936, Abastumansk. astrofiz. Obs. Gora Kanobili Bull. **1**, 35.
Nikonov, V. B., 1937, Abastumansk. astrofiz. Obs. Gora Kanobili Bull. **2**, 23.
Struve, O., 1935, Astrophys. J. **81**, 66.

DISCUSSION

Fernie: I think one can dismiss the W UMa hypothesis for this star more simply. From its position on the H$-$R diagram, P Cyg must have a radius of about 100 $R_\odot$. For a binary companion to have a period of 0.5 day, the primary would have to have a mass of the order of 10^5 $M_\odot$, and the companion an orbital velocity of $\sim 10^4$ km/sec.

An alternative explanation might be that P Cyg is a β CMa star. However, this too would require large and rapid variations in radial velocity, which your observations do not show. Also, current photometry at Toronto is in agreement with Wallerstein's finding that there are no significant short-period light variations in P Cyg at present.

De Groot: From the observations made by the Russian observers it seems to me that there certainly are brightness fluctuations on P Cygni. As you said, another possibility could be that P Cygni is a β CMa star. It might well be that an even more detailed spectrophotometric study would reveal the radial-velocity variations you mentioned. Although P Cygni lies somewhat out of the general region of occurrence of the β CMa stars we should have an open mind for new findings. And why not find a hot, overluminous β CMa star some day?

Fernie: That is true, but the β CMa variables are confined to the low-luminosity classes near the main-sequence. P Cygni is too bright to be a β CMa variable. However, I agree you that one should always have an open mind for new findigs.

Hutchings: The preliminary velocity field presented by Dr. De Groot is in general agreement with similar results I have obtained for this star and three others of similar type.

Note added on May 20, 1969: For a more thorough discussion of the light variations of P Cygni see Luud's contribution at this colloquium and M. de Groot, 1969, Bull. Astr. Inst. Netherlands *20*, 225—273.

VARIATIONS IN THE CONTINUOUS SPECTRUM OF γ CAS

N. L. IVANOVA, I. D. KUPO and A. CH. MAMATKAZINA

Alma Ata Observatory, USSR

After an instability period late in the 30ths some more or less long runs of observations of γ Cas, giving an idea of the behaviour of the continuous and line spectrum of the star at different years (Vandekerkhove 1967, 1961; Barber 1959) were accomplished. Unfortunately, when investigating the continuous spectrum, little attention was payed to the ultraviolet spectrum, and as a result all observations referred only to the photographic and visual regions. The resulting curve for the ultraviolet spectrophotometric gradient in the years 1948—1964 contains only two points (Chalonge, Divan 1952; Dibai 1956). In about 1963 a new increase of the brightness of γ Cas was marked. About 1966 the continuous brightening stopped (Kalish 1966) but the instability is still expressed in short-periodic oscillations in the star's luminosity. It seemed interesting to compare the available observational data and to investigate the question if the behaviour of the star in the contemporary stage fits the conception of a variable shell, in agreement with data on the previous flare of γ Cas (Grobatzkij, 1949).

We obtained photographic spectra of γ Cas, beginning 1958 in Alma-Ata (slit spectrograms with a dispersion of 140 Å/mm at H_γ) and during the years 1964—1967 in Byurakan (slitless spectrograms 175 Å/mm at H_γ). In 1965 we began to observe γ Cas in Alma-Ata with a diffraction spectrograph (30 Å/mm) attached to the 28″ reflector AZT-8, and in autumn 1966 we started to scan the star's spectrum on a 20″ reflector. The observations were conducted as far as possible parallel to perform simultaneous analysis of the continuous and line spectrum. δ Cas was chosen as a comparison star for the investigation of the continuous spectrum. We tried to conduct our observations on γ Cas for as long as possible, to find eventual short-periodic variations.

69 spectroelectric records, 22 slit spectrograms (1958—1964) and 23 slitless spectra in the range 3200—6600 Å were treated. Relative spectrophotometric gradients were determined in the usual way for three spectral regions: Φ_1 ($\lambda\lambda$ 4000—4800 A), Φ_2 ($\lambda\lambda < 3700$ A) and Φ_3 ($\lambda\lambda > 4800$ A). To get absolute gradients for γ Cas, average gradient values for the sepctral class of the comparison star were used (Aller 1955).

Average absolute gradients of γ Cas for each night are given in Table I. We notice essential differences in the averaged gradient values from to date. Even more astonishing are the considerable differences, occurring during several hours, for example, on November 15 1966, which cannot be explained by observational errors.

In Table I still another important fact should be noticed. For the whole period covered by our observations the relation $\Phi_2 < \Phi_1 < \Phi_3$ was satisfied

Table I

Date		D	Φ_1	Φ_2	Φ_3	n
1958						
October	21/22	—	0.70	—	—	4
November	6/7	—	0.59	—	—	4
	24/25	—	0.70	—	—	7
1964						
September	2/3	—	0.63	—	—	4
October	12/13	—0.06	0.75	0.74	—	2
December	15/16	—	0.74	—	—	3
1966						
October	28/29	—0.06	0.87	0.81	1.36	5
	29/30	—0.06	1.07	0.80	1.77	4
November	13/14	—0.06	0.92	0.92	1.38	3
	15/16	—0.07	0.83	0.98	0.84	5
	15/16	—	0.88	0.34	1.26	5
	23/24	—0,07	0.38	0.28	0.84	6
December	3/4	—0.07	1.28	0.75	0.72	2
1967						
March	13/14	—0.10	0.76	0.74	—	6
	14/15	—0.11	0.75	0.63	—	5
July	20/21	—0.05	0.98	0.77	1.12	2
	21/22	—0.05	0.88	0.92	1.31	3
	31/1	—0.06	1.07	0.64	1.25	3
September	12/13	—	—	0.56	—	3
December	10/11	—0.08	0.91	0.83	1.43	3
	10/11	—0.05	0.86	0.61	1.22	3
	11/12	—0.06	0.96	0.57	1.34	12
	12/13	—0.06	0.90	0.50	1.21	6
	14/15	—0.06	1.17	0.48	1.44	3
1968						
January	5/6	—0,07	0.57	0.32	1.65	2

on the average. The UV-gradient Φ_2 was smaller than the blue one. Formally it corresponds to an ultraviolet spectrophotometric temperature about 5000° higher than the temperature determined from the blue part of the spectrum. The relation $\Phi_2 < \Phi_1$ contradicts to our recent knowledge on the structure of stellar photospheres, requiring opposite correlation (Mustel 1941; 1944). At the same time our results, based on a sufficient number of observations, and obtained with different instruments and methods, seem to be reliable.

From Gorbatzkij's (1949) research it follows, that at the late 30ths the gradient ratio was in agreement with theory. It was of interest to confront all available observations executed during a longer period, for example during the last 20 years, to reveal the character of changes, which the gradients have undergone, to see at which time the UV and blue gradients obtained their recent abnormal ratio and to test the reality of our results. In Fig. 1 averaged half-year values of the spectrophotometric gradients, based on available data are represented. The diameters of the marks accord with the num-

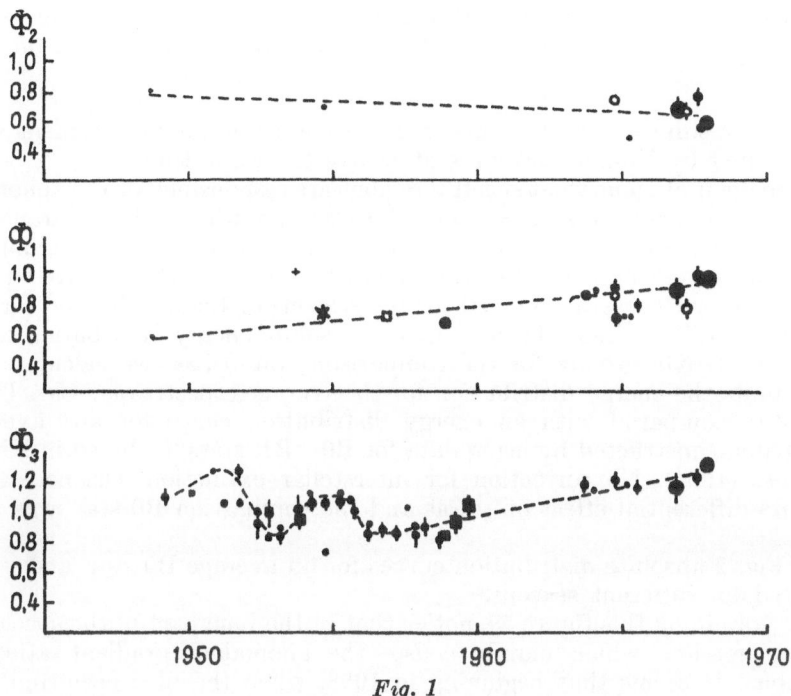

Fig. 1

ber of observations used, the scattering, as usual, is marked by a vertical line.
(Vandekerkhove 1957, Barber 1959, Zilevicùte 1965, Taffara 1957, Chalonge
and Divan 1952, Dibai 1956, Boyarchuk 1958.)

From Fig. 1 the following conclusions can be drawn:

1. During the last 20 years the spectrophotometric gradients of γ Cas
showed a smooth change at all investigated parts of the spectrum.

a) The ultraviolet gradient showed a distinct decrease, in accordance
with the increase of the spectrophotometric temperature of γ Cas by 8000°.

b) On the other hand, the blue gradient increased from ~ 0.6 in 1948
up to an average 0.9 in 1967. That corresponds formally to a cooling by more
than 15000°.

c) In the first half of the 50ths the red gradient showed a distinct wave
with an amplitude larger than 0.3. Since 1956 a stable increase-tendency was
marked, corresponding to a "cooling" by more then 20000°.

We want to draw special attention to the fact, that the gradient-changes
have a systematic character. It seems that a smooth process is proceeding in
the inner parts of the star and nobody can yet say to what results it will lead
in the near future.

2. Comparing the run of the gradients Φ_1 and Φ_2, we see, that till the
middle of the 50ths a normal ratio $\Phi_2 > \Phi_1$ was observed. Apparently in the
time interval 1956 to 1958 the numerical values of the gradients became
equal and since then the anomaly $\Phi_2 < \Phi_1$, in a continuously increasing degree
(due to the steady decrease of Φ_2 and especially to the growth of Φ_1) has set

in. It was already marked that at the same time, about 1956, the wavy
changes in the visual spectrum described by the gradient Φ_3, passed on to a
continuous growth.

3. Our results for the blue and red spectra are in good agreement with
those obtained by Vilnius observers at nearly the same time.

To make it obvious what spectral regions are responsible for the abnormal
structure of the continuous spectrum of γ Cas and what the appearance of
the anomaly, is, we have constructed curves of absolute energy distribution
for different years, using our observational data. Curves of the relative energy
distribution of γ Cas were averaged for the seasons of 1958, 1966, and for the
summer and winter of 1967. Then, using the absolute energy distribution tabulated by A. V. Kharitonov for the comparison star δ Cas, we calculated in
absolute units the energy distribution for the averaged spectra of γ Cas. These
results were compared with an energy distribution curve for an average
B0 spectrum, constructed by using data for B0—B1 stars, included in Kharitonov's list (1964). No correction for interstellar extinction was necessary,
because its differential effect on γ Cas and the comparison B0 star was negligible.

In Fig. 2 absolute distribution curves for an average B0 star and γ Cas
are plotted for different seasons.

By examining this figure we notice that in the blue part of the spectrum
a deep depression, which mainly causes the anomalous gradient-ratios, is
conspicuous. It seems that beginning in 1958, when the blue spectrum was
practically normal, spectral changes developed, deepening the depression,
more pronounced in the years 1966—1967. The energy distribution in the
UV spectrum seemed practically normal, in 1966 it completely corresponded
to the distribution of a B0-star, in 1967 it became somewhat steeper, "hotter".
The intensity in the red spectrum is larger than in the case of a normal star
and the discrepancy is growing with the wavelength. It was desirable to get
at least a rough idea of the infrared spectral part, inaccessible for our observations. For this purpose the results of multicolour photometry were used.
In Fig. 3 energy distribution curves based on 8-colour photometric data (Johnson 1964, Johnson et al. 1966) for γ Cas in 1963 and 1966 and for the average
B0-star, obtained by averaging the colour-indices of the listed B0 and B1
stars (Johnson 1966) of the luminosity classes IV—V, are plotted. Examining
Figure 3 we draw the following conclusions.

1) The intensity increase with increasing wavelength, noticed in the
visual range, is continuing in the infrared part of the spectrum as far as
observations exist. Reddening of γ Cas in comparison with other early B-stars
has already been marked (Lunel, 1954).

2) The reddening of γ Cas is intrinsic. This becomes obvious when analys-
ing the experimental law of interstellar extinction and especially from the
variability of the energy excess in the red spectral region, clearly seen in
Fig. 3.

3) The reddening increased considerably in 1966 as compared with 1963.
This agrees with the detected tendency of intensity increase in the red spectrum.

Hβ emission intensity variations as well as changes of gradients and
Balmer discontinuities, occurring during one night, were observed several
times. V. Liutij found brightness oscillations of γ Cas, which reached 0.1 mag

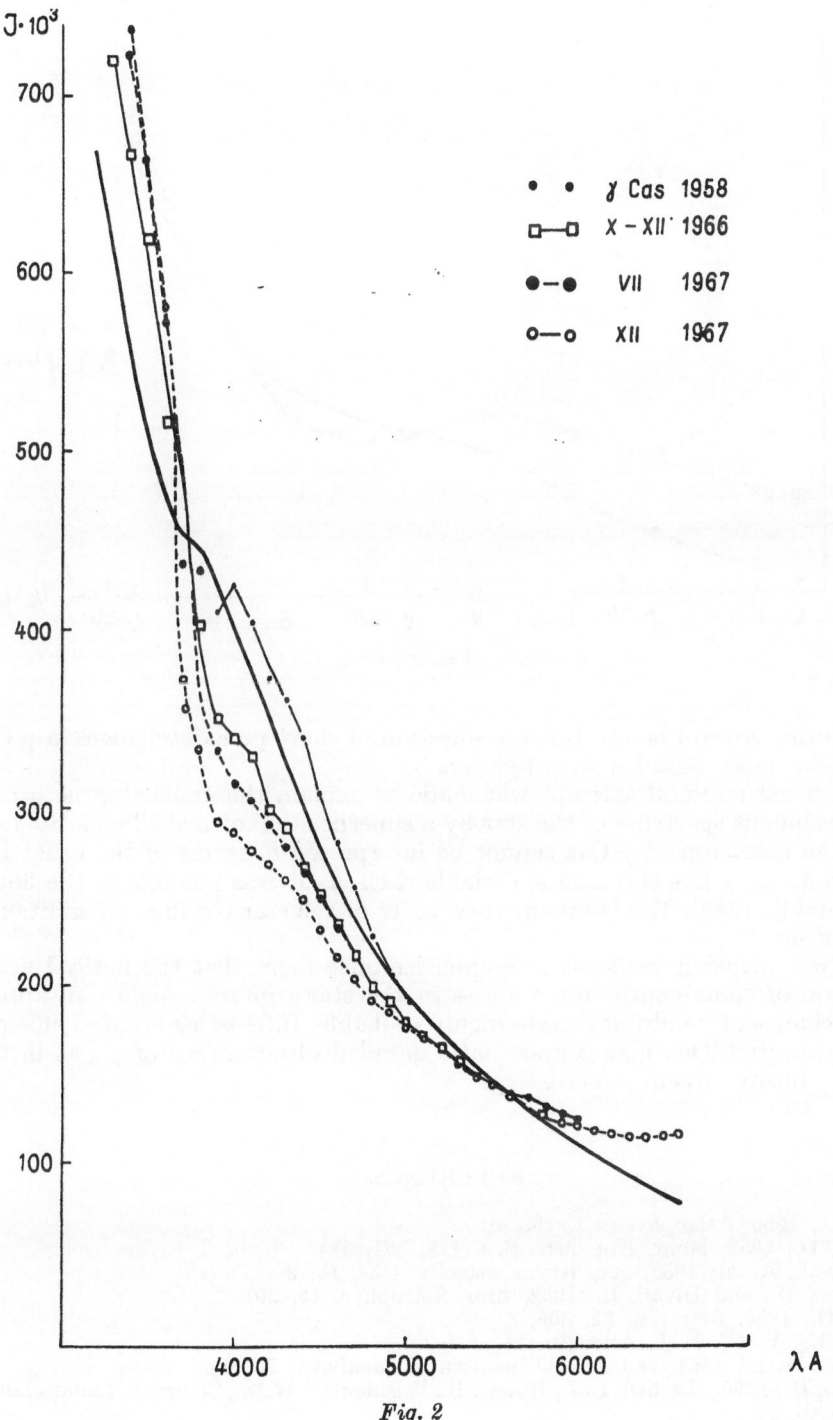

Fig. 2

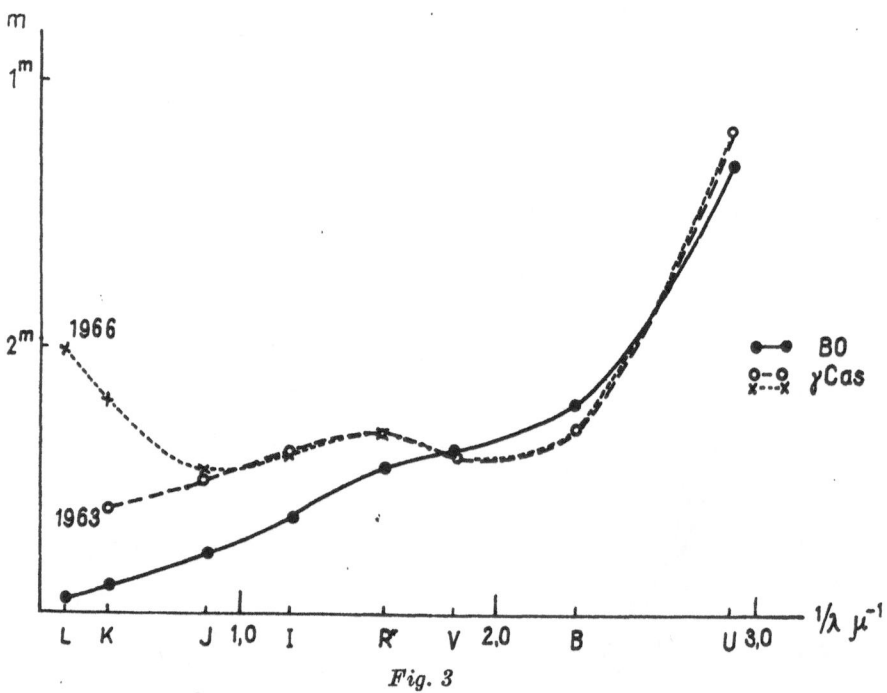

Fig. 3

in U during several hours. But the question of short-periodic changes in γ Cas requires a more detailed investigation.

An unsuccessful attempt was made to explain the unusual structure of the continuous spectrum of the star by a superposition of a shell on a B0-star. Thus the spectrum of γ Cas cannot be interpreted in terms of the usual Be-star model of a hot star and a variable shell, as .it was possible in the 30ths (Gorbatzkij, 1949). The contemporary state of the star requires an additional mechanism.

In conclusion we want to emphasize once more that the noticed accumulation of some continuous process in the star's interior makes an appreciable change of its physical state highly probable. Interesting spectral changes can be awaited. This makes systematic detailed observations of γ Cas in the nearest future urgently necessary.

REFERENCES

Aller, L., 1955, Astrophysics **1**, 183.
Barber, D., 1959, Mont. Not. astr. Soc. **119**, 381, 534.
Boyarchuk, A. A., 1958, Izv. Krym. astrofiz. Obs. **18**, 3.
Chalonge, D., and Divan, L., 1952, Ann. Astrophys. **15**, 201.
Dibai, O., 1956, Astr. Zu. **33**, 506.
Gorbatzkij, V. G., 1949, Astr. Zu. **26**, 307.
Johnson, H. L., 1964, Bol. Obs. Tonantzintla Tacubaya, **3**, 305.
Johnson, H., 1966. Mitchell, L. I., Iriarte, B., Wisniewski, W. Z., Commun. Lunar planet. Lab. 99.
Kharitonov, A. V., 1964, Izv. AN Kaz. SSR, Astrofysika, **17**, 28.

Kharitonov, A. V., 1964, Kariagina, Z. V., 1964, Astrofysika **17**, 10.
Kopilov, I. M., 1954, Izv. Krym. astrofiz. Obs. **12**, 162.
Kopilov, I. M., Boyarchuk, A. A., 1955, Izv. Krym. astrofiz. Obs. **15**, 190.
Kalish, L., 1966, Sky Tel. **32**, 265.
Lunel, M., 1954, Ann. Astrophys. **17**, 153.
Mamatkazina, A. Ch., 1967, Astr. Cirk. Izdav. bjuro astr. Soobsc. Kazah, No. 399, 2.
Mustel, E. R., 1944, Astr. Zu. **18**, 297, 1941. **21**, 133.
Taffara, S., 1957, Contr. Oss. astrofis. Univ. Padova No. 86, 3.
Vandekerkhove, E., 1957, Belgique Com. No. 126.
Vandekerkhove, E., 1961, Observatory, **81**, 144.
Zilevicûte, Z., and Straizis, V., 1965, Bjul. Vilnius Obs. No. **15**, 49.

VARIATIONS RAPIDE, PROBABLEMENT NON PÉRIODIQUES D'ENVELOPPES D'ÉTOILES Be

A. M. DEPLACE, R. HERMAN, A. PETON

Observatoire de Paris

D'après la littérature, il semble souvent que les variations observées dans les étoiles Be soient eratiques. Il est très difficile de savoir si cela est exact ou non, en raison de la lenteur de ces variations. Depuis 18 ans que nous avons observé un nombre relativement petit ($\sim$ 250) d'étoiles Be en les photographiant régulièrement, il nous semble que les variations de E/C et de V/R soient au moins pseudo-périodiques (Lacoarret, 1965). Mais, a côté de ces variations pseudo-périodiques, il semble bien qu'il y ait également des variations rapides. Nous allons en donner quelques exemples typiques.

ζ Tauri (HD 37202) a toujours été donnée comme binaire avec la période bien connue (Adams 1905, Losh 1932, Underhill 1952) de 133 jours. Une étude prolongée des vitesses radiales des raies d'enveloppe de cette étoile a permis de montrer, que, actuellement, l'axe de la période se déplace régulièrement pendant 7 ans (fig. 1). Le spectre pris en 1967 ressemble étonnament à celui de 1960. Il n'a pas été possible d'affirmer que cette période de 7 ans ait existé dans le passé, comme A. M. Deplace l'expliquera ailleurs. Néanmoins, la plupart des auteurs qui ont pu mesurer plusieurs cycles de 133 j. ont observé des variations de l'axe "γ" (Hynek et Struve 1952).

Ces auteurs ont également noté l'existence d'une perturbation dans la courbe de vitesse (133 j.) vers la phase 0.60. Dans la période de 7 ans signalée plus haut, nous avons observé de nombreux cycles de 133 j. Malheureusement, il est très difficile d'observer régulièrement une étoile en raison du peu d'heures d'observation disponibles. Néanmoins dans la période de 7 ans, 3 cycles présentent plusieurs mesures autout de la phase 0.60, dans le cycle 168, 7 mesures de vitesse radiale se trouvent entre les phases 0.40 et 0.60, dans le cycle 169, 5 mesures entre les phases comprises entre 0.5 et 0.7, dans le cycle 1964, 4 mesures entre les phases 0.5 et 0.6. Sur la figure 2, on voit que dans le cycle 164, il n'y a pas d'écarts à la courbe de vitesse radiale, alors que dans le cycle 168 on observe des variations rapides et importantes de vitesse radiale et, dans le cycle 169, la variation subsiste, plus faible et un peu décalée dans le temps.

Ces variations ne sont pas dues à la période de 133 j. comme la plupart des auteurs précédents le pensaient mais, plus probablement à la grande période.

Il est intéressant de noter que ce phénomène a lieu dans les cycles 168 et 169, c'est à dire à une époque où la vitesse moyenne diminue (voir figure 1). A cette époque, l'étoile est comprimée depuis assez longtemps, en raison de la récession de l'enveloppe et il apparaît une éjection de matière montrée par le mouvement d'expansion de l'enveloppe. Des couches à vitesses hété-

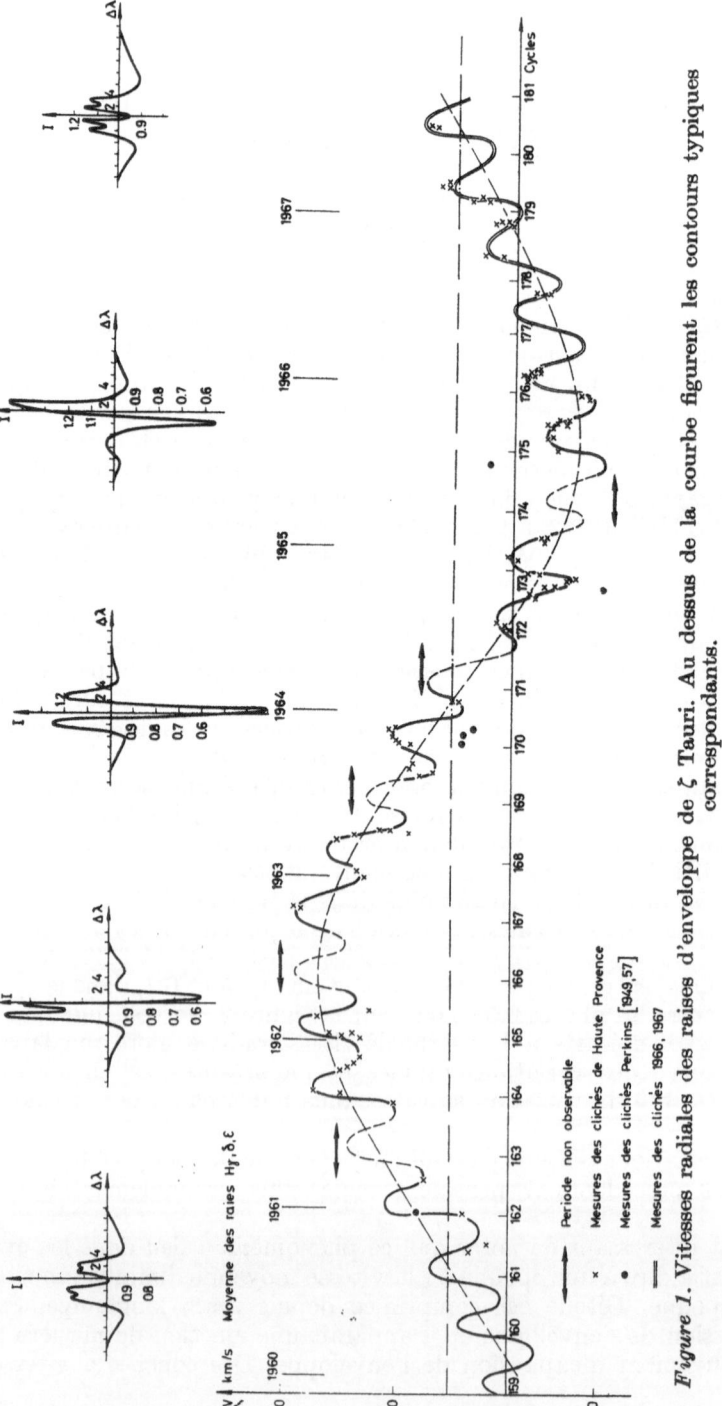

Figure 1. Vitesses radiales des raies d'enveloppe de ζ Tauri. Au dessus de la courbe figurent les contours typiques correspondants.

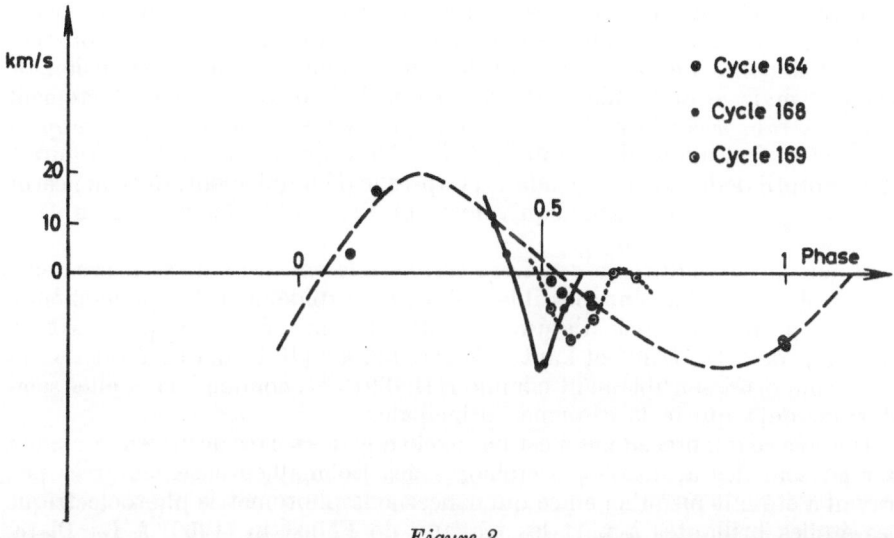

Figure 2

rogènes sont indiquées par le dédoublement des raies d'absorption d'enveloppe. L'un de nous avait déjà souligné ces faits au Colloque de St Michel (Herman 1964) mais nous n'avions pas pu expliquer ce phénomène. Nous l'avions simplement comparé à des dédoublements semblables qui se présentaient dans les spectres de P Cyg.

En ce qui concerne d'autres étoiles Be, nous allons tout d'abord parler de HD 45910 (AX Mon). De nombreux auteurs ont remarqué ce dédoublement des raies d'enveloppe (Struve 1943, Merrill 1948). Un travail assez complet

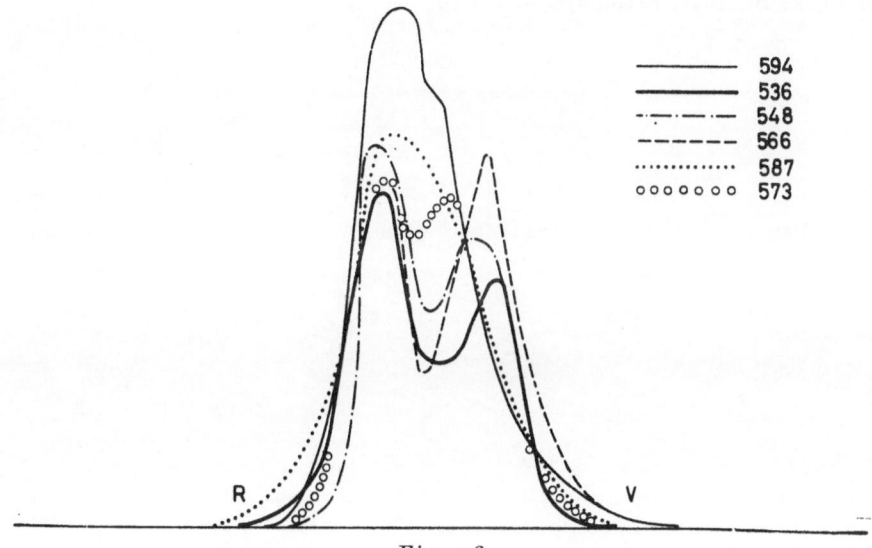

Figure 3

concernant cette étoile a été exposé en 1964 à Michigan par P. Cowley (1964). Néanmoins, de nombreux points restent obscurs. Comme pour ζ Tau, on observe des dédoublements de raies qui ont lieu dans une période d'expansion de l'enveloppe mais le dédoublement est tel que les deux raies sont fortement séparées, la raie secondaire correspond à une vitesse relative, par rapport à la raie primaire, allant de —150 km/s à —50 km/s en 13 jours. La figure 3 montre le profil de la raie H_{14} pendant et après le dédoublement. Ceci concerne bien l'enveloppe car la largeur équivalente de l'ensemble des 2 raies ne varie pas pendant ce phénomène.

Il nous est très difficile de suivre cette étoile en raison de sa basse déclinaison et il ne semble pas possible d'élucider rapidement les phénomènes. Nous nous proposons donc d'étudier en détail deux séries de spectres à 10 A/mm, l'une de HD 218393 et l'autre de HD 142983 (48 Lib). Ces étoiles semblent se comporter sensiblement comme HD 45910 et, comme ζ Tau, elles semblent avoir déjà quitté la séquence principale.

D'après ce qui précède, il n'est pas exclu que des sortes de "flash" puissent exister au sein des atmosphères stellaires des Be, malheureusement très peu de travail a été fait jusqu'ici en ce qui concerne la photométrie photoélectrique de ces étoiles brillantes à part les travaux de Feinstein (1968) à La Plata.

BIBLIOGRAPHIE

Adams, W. S., 1905, Astrophys. J. **22,** 115.
Cowley, P., 1964, Astrophys. J. **139,** 917.
Feinstein, A., 1968, Z. Astrophys. **68,** 29.
Herman, R., 1964, Ann. Astrophys. **27,** 507.
Hynek, J. A. et Struve, O., 1942, Astrophys. J. **96,** 1562.
Lacoarret, M., 1965, Thèse, Paris.
Losh, H. M., 1932, Publ. Obs. Univ. Michigan, **4,** no. 14, 1.
Merrill, J. E., 1948, Astrophys. J. **108,** 481.
Struve, O., 1943, Astrophys. J. **43,** 212.
Underhill, A. B., 1952, Astrophys. J. **57,** 168.

THE PROBLEM OF IRREGULAR VARIATIONS IN MAGNETIC STARS

Introductory Report by

T. JARZĘBOWSKI

Astronomical Institute, Wrocław, Poland

At the prelude of this short introductory report I would like to pay attention to the fact that the magnetic stars do not seem to be the proper subject for our Colloquium devoted to "non-periodic phenomena in variable stars". As a matter of fact, the investigations of the latest few years have revealed periodicity in many magnetic stars, which were hitherto treated to be irregular; thus the number of non-cyclic phenomena in magnetic stars has considerably decreased.

We shall present in this paper the most important observational data on the variability of the magnetic stars, emphasizing the facts that indicate non-periodicity of variations*.

OBSERVATIONAL DATA

The presence of a magnetic field has been firmly established in the case of about 100 stars (Babcock 1958a, b, Gollnow 1962, Preston and Pyper 1965). It is generally believed that all peculiar stars possess magnetic fields (although the magnetic field can only be measured in the case of sharp-line peculiar stars); the number of stars with peculiar spectra exceeds 800 (Bertaud 1959, 1960, 1965). In this paper we shall include under discussion all investigated peculiar stars.

The variations of the following parameters have been observed in magnetic (peculiar) stars:
1. the magnetic field
2. the light
3. the color index
4. the equivalent width
5. the line profile
6. the cross-over effect
7. the radial velocity
8. the polarization of the light (probably)

So far, 47 stars are known, in the case of which some of these parameters are varying periodically. The stars and their periods are listed in Table 1.

Variations of the magnetic field. Periodic variations of the magnetic field have been stated in the case of 12 stars; in the case of 9 others the

---

* All problems on magnetic stars are presented in detail in the following reviews: Babcock 1958a, 1960, Gollnow 1965, Jarzębowski 1965, Mestel 1965, Ledoux and Renson 1966, Jarzębowski 1969.

Table 1
Periodic A_p Stars

Star	Period	Variable param.	Star	Period	Variable param.	Star	Period	Variable param.
HD 124224	$0^d.521$	l, s	21 Per	$2^d.883^*$	m*, l*	HD 125248	$9^d.296$	m, l, s, v
κ Psc	0.580	l	HD 224801	3.740	l	HD 215441	9.488	l
HD 219749	0.723*	l*	ADS 16252	3.770	s	10 Aql	9.78*	l*
56 Ari	0.728	l, s, v*	HD 25354	3.900	l	HD 173650	10.1	m*, l, s*, v*
52 Her	0.96*	m*, l*	15 Cnc	4.116*	l*	41 Tau	11.94	l
HD 133029	1.054*	m*, l*	HD 10783	4.133	m, l, v*	78 Vir	12.26*	m*, l*
HD 140728	1.305*	l*	ε UMa	5.089	l, s, v	HD 192678	18*	l*
χ Ser	1.596	l, s	17 Com	5.09*	m*, l*	β Cr B	18.50	m
21 Aql	1.71*	l*	49 Cnc	5.43*	l*	73 Dra	20.275	m, l, s, v*
ι Cas	1.741	l, s	α² CVn	5.469	m, l, s, v	HD 4174	40.0	m*, l*
HD 215038	2.036	l	HD 98088	5.905	m, s, v (b)	HD 8441	106.27	m*, v (b)
HD 4778	2.156	s	HD 153882	6.009	m, l	HD 221568	160	l, s, v
21 Com	2.2*	l*, s*	HD 32633	6.431	m, l	HD 188041	226	m, s
π Boo	2.444*	s*	HD 71866	6.800	m, l, s	γ Equ	314*	m*
HD 34452	2.466	l, s*	κ Cnc	6.91*	l*	HD 187474	2500	m
γ Ari	2.607	l, s	53 Cam	8.028	m, l, s			

The symbols denote: m = periodic magnetic field variations, l = periodic light variations, s = periodic spectral variations, v = periodic radial velocity variations (b = a binary star with the same period).
The asterisk indicates an uncertain value.
The new data on the periods have been taken from the following papers: Jarzębowski 1960a, b, 1961, 1964, Wehlau 1962, Rakos 1962b, 1963, Steinitz 1964, Renson 1965, 1966, 1967, Osawa et al. 1965, Stępień 1968b.

periodic field variations are probable but not firmly established. In nearly all remaining magnetic stars (about 80) the observations indicate variations of the magnetic field. It has not been shown, as yet, that there is a star with constant magnetic field. On the other hand — it has not been proved that there exist magnetic stars with irregular field variations*.

We have, however, to call our attention to some facts:

1. In the case of several magnetic stars — e.g. HD 173650 ($P = 10^d1$), or HD 4174 ($P = 40^d$) — there are difficulties in fitting the magnetic observations to the period derived from luminosity variations. This seems to indicate the possibility of some irregularities.

2. In the magnetic star 73 Dra the period of variations (20^d3) is known for a long time. The magnetic measurements by Preston (1967a) follow this period, but the measurements by Babcock (made over ten years earlier) do not form a smooth curve and indicate only the negative polarity (Berg 1967). This is illustrated in Figure 1. We have rather to eliminate the supposition that Babcock's measurements are of lower accuracy, and thus, we have to accept that some secular magnetic field variations are taking place.

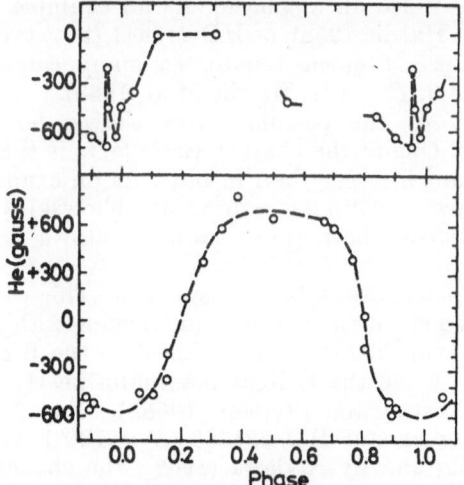

Fig. 1. The variations of the magnetic field of 73 Dra. Bottom: the measurements by Preston plotted with the period of 20^d2754. Top: Babcock's magnetic field observations plotted with the same period.

3. Babcock's magnetic measurements of the star 78 Vir seem to follow the probable 12-day period of luminosity variations, but the dispersion of individual points is very large, bringing into question the periodicity (Stępień 1968b).

4. Recent observations by Preston (1967a) indicate that the amplitude of magnetic variation of β Cr B ($P = 18^d5$) is varying; there is evidence of some secular changes of the amplitude in a period of the order of 10 years.

---

* Babcock has subdivided the magnetic stars into three classes denoted as α, β and γ (α—periodic magnetic variables, β—irregular variables with the reversing of polarity, γ—irregular with constant polarity). The sense of this division is very doubtful now.

16

In this place we have to mention that also Adam (1965) reported secular decrease of the amplitude of the magnetic field of HD 125248 ($P = 9^d3$). Adam's result seems to be, however, inconclusive, because the successive measurements were made by using different apparatus — and in such a case some divergence of the results can be expected (cp. e.g. Preston and Pyper 1965).

Variations in light and color. Periodic luminosity variations have been stated in the case of about 40 peculiar stars. The amplitudes are very small —- most often of the order of one, two or three hundredths of a magnitude, and do not exceed 0^m2. The small value of the amplitude does not permit, very often, to distinguish the large scatter and the true deviations from periodicity (e.g. HD 153882, Chugainov 1961) — although the observational data indicate in some cases such deviations. On the other hand, we cannot say — by the same reason — that the light of any peculiar star is constant.

We shall now summarize the most interesting observational data.

1. In several magnetic (peculiar) stars the variations are very regular and the amplitude is increasing towards the short wave length (the curves U, B, V, U—B, and B—V are all in phase). As an example we can here name HD 124224 ($P = 0^d5$, Hardie 1958), or HD 215441 (the star with the strongest magnetic field, $P = 9^d5$, Stępień 1968b). In many cases a double wave is conspicuous (e.g. 56 Ari, $P = 0^d7$, Hardie et al. 1963).

2. The majority of the peculiar stars shows, however, non-typical luminosity variations. One of the characteristic facts is that the amplitude in yellow is sometimes much larger than in blue. As an example let us mention the two peculiar stars: α^2CVn ($P = 5^d5$, Jarzębowski 1969), and 73 Dra ($P = 20^d3$, Stępień 1968a); both the stars show also a double wave on the V curve.

3. In a few cases a phase shift is conspicuous. This was firstly stated for 53 Cam (the star with a strong magnetic field, varying with a period of 8 days); the V curve lags here considerably with respect to the B curve (Jarzębowski 1960b, Rakos 1962b), while the U light maximum nearly coincides with the V light minimum (Preston and Stępień 1968b), Fig. 2. A similar case is presented by the peculiar star HD 221568 ($P = 160^d$) recently investigated by Osawa et al. (1965) and by Kodaira (1967); the changes in V are almost exactly reflected in the changes in B in an opposite sense — as we see in Figure 3. The authors also report the possibility of some irregularities in the luminosity variations of this star.

4. An interesting case is presented by the M-type magnetic star HD 4174, in the case of which the deviations from strict periodicity are undoubtful (Jarzębowski 1964, 1969). Three sets of photometric observations of this star are presented in Figure 4. We see that the observational data of 1962/63 and of 1965 can be described with the slightly variable period of 40 days, but the data of 1964 do not fit this period and are rather irregular. It may be noted that the magnetic variations of this star probably also follow the semi-regular 40-day luminosity period.

5. Besides the luminosity variations discussed on here (with periods given in Table 1) some of the peculiar stars show additional brightness fluctuations over very short intervals. It was stated in the case of four stars that these variations can be described by a period (Bahner et al. 1957, Rakos 1962a,

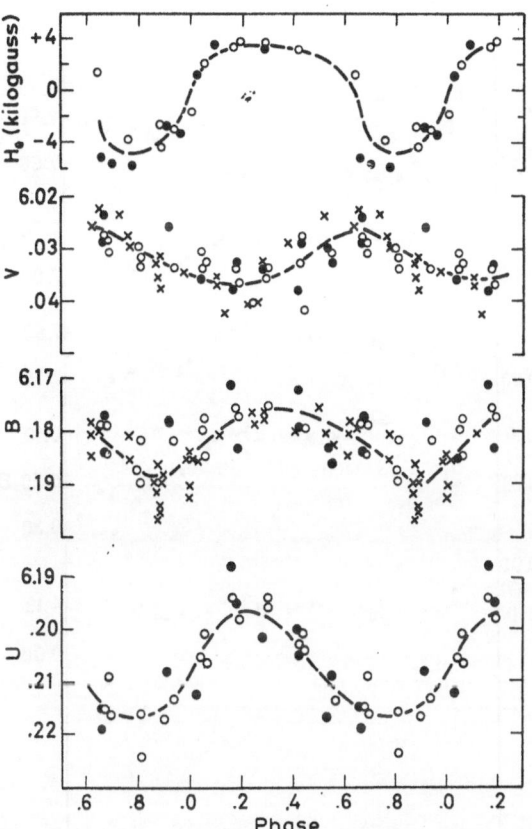

Fig. 2. The magnetic and light variations of 53 Cam, based on the data of Babcock,
Preston, Stępień, Jarzębowski, and Rakos (Preston and Stępień 1968b).

1963). The following values of the short periods were found: $0^h 32^m$ (21 Com),
$1^h 34^m$ (HD 71866), $1^h 46^m$ (HD 32633), and $2^h 04^m$ (HD 224801). These short-
period fluctuations in some phases disappear and display conspicuous irregularities (Fig. 5).

Spectral variations. Apart from the fact that the lines of elements such
as Si, Cr, Mn, Sr, and the rare earths are abnormally intense — spectral variations have been well established in many peculiar stars.

The following variations have been observed in the spectra of peculiar
stars:

1. The line intensity variations. This phenomenon is generally most
conspicuous in the case of those elements which are abnormally strong (i.e.
Si, Cr, Mn, Sr, rare earths). In about 20 cases periodicity of these variations
has been established (see Table 1). The phase relationship between the variations
of different elements may change from star to star and, further, some elements
may show a double wave. There is also no uniform relationship between the
line intensity and the magnetic variation.

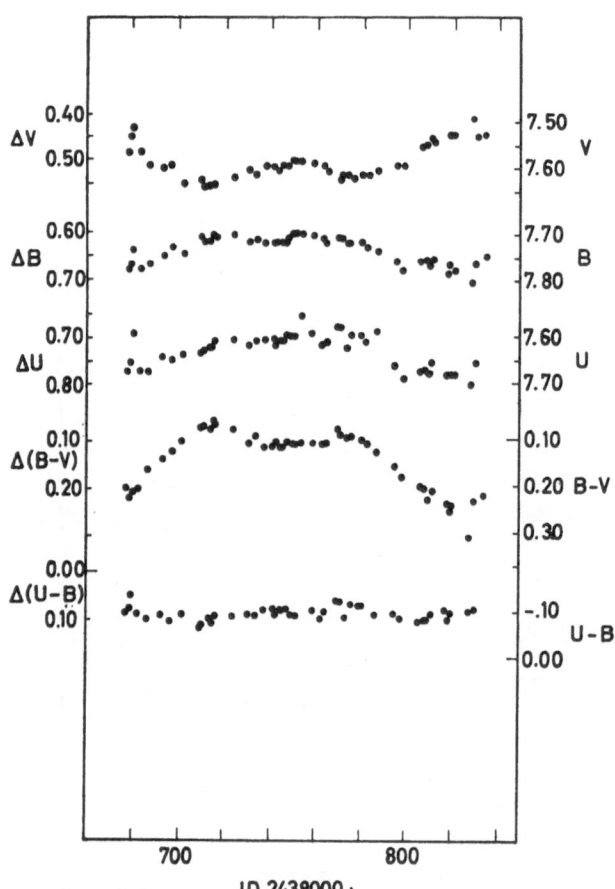

Fig. 3. Light variations of HD 221568 from October 1964 to March 1965 (Osawa et al. 1965).

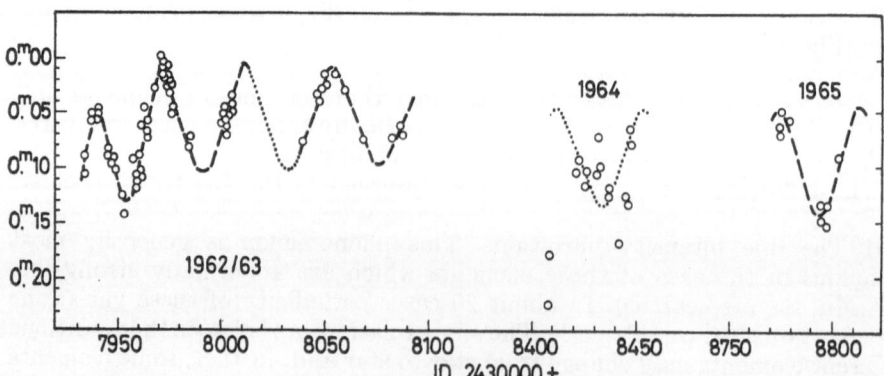

Fig. 4. Luminosity variations of the M-type magnetic star HD 4174 during three sets of observations (Jarzębowski 1969).

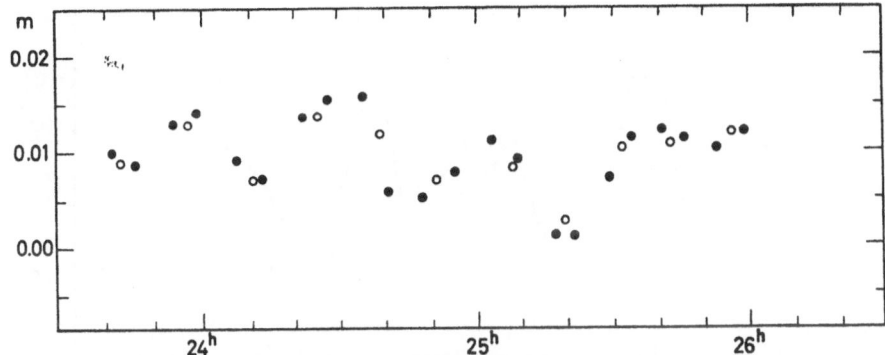

Fig. 5. Short interval brightness fluctuations of 21 Com observed on 28 III 1956 (Bahner et al. 1957).

2. The profiles of some magnetic stars show remarkable variations in width and depth. The profile variations are generally correlated with the intensity variations. In several cases it was possible to detect periodicity; for instance in 53 Cam ($P = 8\overset{d}{.}0$) most of the lines being much wider when the field is of negative polarity.

3. The cross-over effect. This effect appears as a difference in sharpness of the profiles when the right-handed polarized spectrum is compared with the left-handed (in the first spectrum the lines tend to be broad and shallow, in the other — sharp and deep). The cross-over effect is mainly observed in periodic magnetic variables; the sharpness of the profiles is periodically varying with the sign of the magnetic field. Let us note, however, that in the case of 78 Vir it was not possible to find the correlation between this effect and the presumable 12-day period of variations (Stępień 1968b).

Are there irregularities in the spectral variations?

This problem is rather undecided (here the same questions arise as in the case of luminosity and magnetic variations). The present observational data do not contradict the possibility that there are magnetic (peculiar) stars showing irregular spectrum variations, or variations that are not simply correlated with the magnetic changes (Babcock 1960). It may also be noted in this place that in several magnetic stars the possibility of some secular spectral changes was reported (for instance the lines of some elements are intense according to the HD Catalogue and weak according to Babcock etc.).

Finally, we have to announce a very interesting fact recently found by Wood (1965, 1967, 1968): the rapid Balmer-line variations. Using the photo-electric narrow-band photometer Wood has discovered the variations of the equivalent widths of Hβ, Hγ, and Hδ in several peculiar stars (HD 215441, HD 224801, 73 Dra, ε UMa). The variations seem to be irregular, but the existence of quasi-periods of the order of half an hour is not excluded. In the case of HD 224801, for instance, time variations of about 10 per cent with a quasi-period of about 35 min have been observed (the oscillations may become unrecognizable after several cycles), Fig. 6.

Radial velocity variations. Over ten magnetic stars belong to spectroscopic binaries. In HD 98088 the period of revolution equals the period of magnetic variation; a similar case presents, probably, HD 8441 (Renson 1966). In the

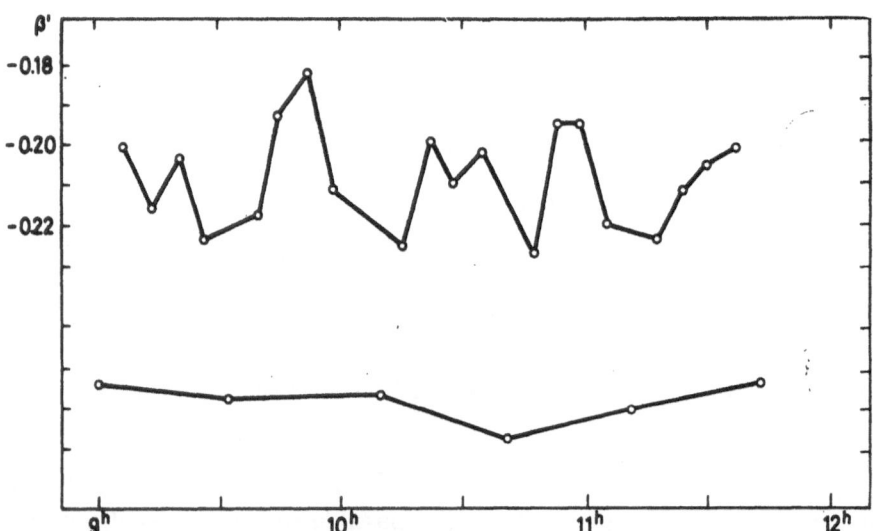

Fig. 6. Fluctuations of the equivalent width of Hβ in the spectrum of magnetic star HD 224801 on 29 VIII 1963 (bottom: the comparison star). From the paper by Wood, 1968.

other spectroscopic binaries the period of velocity variations differs from the period of magnetic, light or spectrum variations. For example, in β Cr B the period of magnetic variation is 18.5 days while the period of orbital revolution — 3833 days, in HD 125248 the corresponding values are 9.3 days and 1618 days; in ϰ Cnc this difference is much smaller: the period of orbital revolution being 6.39 days while the period of light variation seems to amount to 6.91 days (Stępień 1968b). The problem of the possible binary nature of magnetic stars matters much for the "binary-star hypothesis" which was put forward by Renson (1963). The binary nature of peculiar stars has also been set up by van den Heuvel (1967) and by Guthrie (1968).

Apart from the spectroscopic binaries, most of the peculiar stars show small fluctuations in differential radial velocity of the order of a few km/s. These radial displacements depend on the chemical element considered and are often correlated with the variations in line intensity. In several cases the variations follow the period of magnetic or spectrum variations (see Table 1). This periodicity is especially conspicuous in α²CVn, but we have to note that: 1. it is evident only for a group of elements, 2. for some elements secondary maxima occur, 3. there are elements showing nearly no velocity variations (cp. Jarzębowski 1965). In other stars the probable period of velocity variation differs from the main period (e.g. in HD 224801, Rakos 1963), while in the majority of cases no regularities in the velocity fluctuations could be detected.

The polarization of the light. Thiessen, in 1961, reported the variations in polarization for the star HD 71866 of the same period as the variations in brightness and magnetic field. Polosuhina and Lebedeva (1966) have investigated the 34 kG magnetic star HD 215441; they reported the existence of a possible correlation between the percentage of polarization and the magnitude in the $9^d.5$ photometric period.

On the other hand, the investigations by Serkowski (1965), Elvius and Engberg (1967), Hiltner and Mook (1967) did not reveal definite indication of variable plane polarization in magnetic stars.

With reference to the problem of the variations of polarization we have to keep in mind that the degree of polarization we are observing for the magnetic variables remains very small and seldom exceeds the mean error. Thus we have rather to conclude that within the accuracy of the present technics of observations there is probably no variable polarization in stars with variable magnetic fields.

CONCLUSIONS

Recapitulating the here discussed observational data we conclude that there are still many magnetic stars in the case of which the variations of magnetic field, light, radial velocity, or the spectral variations are non-periodic. It may well be, however, that thorough observations can reveal in the future the periodicity of variations.

The results of investigations of the latest few years seem to lead to the following statement: if a star is carefully studied, there is the probability of finding periodic or semiperiodic variations.

When considering this question we have, however, to take note of the fact that, generally speaking, it is much easier to find a period of variations than to prove that the variations are irregular.

On the other hand, it is sometimes very difficult to distinguish between true irregularity and the errors of observations (though, in some cases the irregularities seem to be undoubtful — as for instance in the case of HD 4174, Fig. 4).

At the end we shall add one general remark. The conclusion on the periodicity of variations in magnetic stars may be regarded as a new support for the oblique rotator theory. This theory requires, of course, constancy of the period of variations. Such constancy of the period has been stated in the case of stars that are investigated for a long time (e.g. α^2 CVn, 73 Dra, HD 125248). Rakos (1968) suggested variations of the period in a few stars, but the deviations seem to lie within the limits of errors. On the basis of the present observational data we have rather to conclude that the mean value of the period is constant, but overlapping of some irregularities is possible.

REFERENCES

Adam, M. G., 1965, Observatory, **85**, 204.
Babcock, H. W., 1958a, Astrophys. J. **128**, 228.
Babcock, H. W., 1958b, Astrophys. J. Suppl. Ser. **3**, 141.
Babcock, H. W., 1960, Stars and Stellar Systems (The University of Chicago Press), vol. 6, chapter 7.
Bahner, K., Mavridis, L., 1957, Z. Astrophys. **41**, 254.
Berg, R. A., 1967, Publ. Leander McCormick Obs. **15**, part 4.
Bertaud, Ch., 1959. J. Observateurs, Marseille, **42**, 45; 1960, ibid, **43**, 129; 1965, ibid, **48**, 211.
Chugainov, P. F., 1961. Perem. Zvezdy, **13**, 255.

Elvius, A., Engberg, M., 1967, Uppsala astr. Obs. Medd. No. 161.
Gollnow, H., 1962, Publ. astr. Soc. Pacific **74**, 163.
Gollnow, H., 1965, Stellar and Solar Magnetic Fields (Amsterdam: North-Holland Publ, Company), p. 1, and p. 23.
Guthrie, B. N. G., 1968, Publ. R. Obs. Edinburgh, **6**, No. 6.
Hardie, R. H., 1958, Astrophys. J. **127**, 620.
Hardie, R. H., Schroeder, N. H., 1963. Astrophys. J. **138**, 350.
Hiltner, W. A., Mook, D. E., 1967, The Magnetic and Related Stars (Baltimore: Mono Book Corporation), p. 123.
Jarzębowski, T., 1960a, Acta astr. **10**, 119.
Jarzębowski, T., 1960b, ibid, **10**, 237.
Jarzębowski, T., 1961, ibid, **11**, 191.
Jarzębowski, T., 1964, ibid, **14**, 77.
Jarzębowski, T., 1965, Stellar and Solar Magnetic Fields (Amsterdam: North-Holland Publ. Company), p. 64, and p. 126.
Jarzębowski, T., 1969, Acta astr. (to be published).
Kodaira, K., 1967, Ann. Tokyo astr. Obs., **10**, No. 4.
Ledoux, P., Renson, P., 1966, A. Rev. Astr. Astrophys., **4**, p. 293.
Mestel, L., 1965, Stellar and Solar Magnetic Fields (Amsterdam: North-Holland Publ. Company), p. 87.
Osawa, K., Nishimura, S., Ichimura, K., 1965. Tokyo astr. Obs. Reprint, No. 275.
Polosuhina, N. S., Lebedeva, L., 1966, Astr. Zu. **43**, 513.
Preston, G. W., Pyper, D. M., 1965, Astrophys. J. **142**, 983.
Preston, G. W., 1967a, Astrophys. J. **150**, 871.
Preston, G. W., 1967b, The Magnetic and Related Stars (Baltimore: Mono Book Corporation), p. 111.
Preston, G. W., Stępień, K., 1968a, Astrophys. J. **151**, 577.
Preston, G. W., Stępień, K., 1968b, Astrophys. J. **151**, 583.
Rakos, K. D., 1962a, Z. Astrophys. **56**, 153.
Rakos, K. D., 1962b, Lowell Obs. Bull., No. 117.
Rakos, K. D., 1963, ibid, No. 121.
Rakos, K. D., 1968, Astr. J. **73**, S 114.
Renson, P. 1963 Inst. Astrophys. Liège, Coll. in 8°, No. 460, and 463.
Renson, P., 1965, ibid, No. 491 and 496.
Renson, P., 1966, ibid, No. 517.
Renson, P., Trans. IAU, vol. 13 A, p. 524 (in Herbig's report).
Serkowski, K., 1965, Astrophys. J. **142**, 793.
Steinitz, R., 1964, Bull. astr. Inst. Netherl. **17**, 504.
Stępień, K., 1968a, Astrophys. J. **153**, 165.
Stępień, K., 1968b, Astrophys. J. (in press).
Thiessen, G., 1961, Astr. Abh. Hamburger Sternwarte, **5**, No. 9.
Van den Heuvel, E. P. J., 1968, Highlights of Astronomy, p. 420, and Utrechtse Sterr. Overdr., No. 81.
Wehlau, W., 1962, Publ. astr. Soc. Pacific, **74**, 137.
Wood, H. J., 1965, Publ. Leander McCormick Obs., **15**, part 2.
Wood, H. J., 1967, The Magnetic and Related Stars (Baltimore: Mono Book Corporation), p. 485.
Wood, H. J., 1968, Astrophys. J. 152, 117.

DISCUSSION

Detre: If the magnetic period is not constant, this fact alone must not be regarded as contrary to the oblique rotator theory because the magnetic fields can show some displacements relative to the photosphere. Theoretically, the oblique rotator theory is not quite satisfactory because an oblique rotator represents an unstable configuration. But as recent results by Preston and Böhm Vitense have given values near to $\pi/2$ for the angle of obliquity we may unite the oblique rotator model with the solar model: the star has a relatively weak polar dipole field and a strong equatoreal dipole field. According to Mestel, such a configuration might be stable.

Jarzębowski: Yes, if we accept the oblique rotator model, we should have to make the assumptions mentioned by you.

The sun, as we know, does not however fit to this model. Bumba, Howard and Smith have recently investigated how the sun would appear as a magnetic star viewed near the plane of its equator. They found that the 27-day rotation period is impossible to detect.

PHOTOMETRIC SEARCH FOR PERIODICITY AMONG MAGNETIC STARS

KAZIMIERZ STĘPIEŃ

Warsaw University Observatory, Warsaw, Poland

ABSTRACT

The results of the search for periodicity indicate that in almost every case when a careful photometric investigation is carried out periodicity can be found. There is no evidence of period variations of the magnetic stars. The in-phase as well as antiphase relationship between the light and magnetic curves was observed for different stars.

After extensive observations of a number of magnetic stars Babcock divided them into three groups. The stars from two of them were supposed to vary irregularly. Since that many members of these two groups were found to vary periodically. The problem of periodicity of other magnetic stars became open again. On the other hand the theory of oblique rotator requires all the magnetic stars to be periodic and it was often used as an argument against this theory that only a few magnetic stars were periodic while the rest seemed to vary irregularly. In addition photometric observations showed a periodicity of variations of many magnetic stars while the magnetic observations did not follow these periods. Hence, it seemed important for better understanding of phenomena taking place in these stars to reconsider the problem of periodicity among the magnetic stars both from magnetic and photometric point of view. Such an attempt was undertaken at the Lick Observatory. Here, the results of the photometric part of program are shortly presented.

Originally 23 stars were chosen mainly from the list of stars observed spectroscopically by G. W. Preston of the Lick Observatory. Later one of the comparison stars turned out to be a variable itself and was also included into the program. Of these 24 stars 10 had previously known photometric periods. The observations were carried out with the 24 inch reflector of the Lick Observatory and the standart UBV equipment. Two comparison stars were usually chosen for each variable and the constancy of light of these stars was checked against each other. On average 75 observational points in three colors were obtained for each magnetic star. The mean error of one point was about 0.005 of a magnitude. The variability and periods were confirmed for all 10 stars which were observed in the past. Only in case of 21 Per the variations fit to the period of around 3 days instead of 1.7 days as previously found. The presence of secondary maxima on V — curves of 73 Dra and HD 224801 was confirmed and such a maximum was found for HD 71866. Of the remaining 14 stars two, HD 2453 and HD 9996, were found to be constant within the error of measurement during the period of observations. For one star, μ Lib, the number of observations was insufficient to try to find a period. The star did not show large variations and the observations were difficult because of rather large negative declination of the star. In all the other cases it was possible to find periods of light variations although

for a few stars the results are uncertain due to the small amplitude of
variations and a low number of observations. List of observed stars and their
periods are given in Table 1.

Table 1
List of observed stars and their periods

Name or HD	P (days)	Name or HD	P (days)	Name or HD	P (days)
2453	—	71866	6.80	153882	6.01
9996	—	49 Cnc	5.43	10 Aql	9.78
10783	4.13	ϰ Cnc	6.91	21 Aql	1.7
21 Per	2.88	17 Com A	5.09	192678	18.
19216	7.7	111133	11.	73 Dra	20.3
32633	6.43	78 Vir	12.3	215038	2.04
53 Cam	8.03	μ Lib	—	215441	9.49
15 Cnc	4.12	52 Her	0.96	224801	3.74

A few stars (like HD 224801, 73 Dra, HD 71866) were observed often
enough in the past so the discussion of a constancy of their periods seemed
to be justified. It was always possible to phase all the existing photometric
data into one curve without necessity of introducing any period variations.
Hence, one can draw the conclusion that at the present the magnetic stars
do not present evidences of variations of period. For some stars concurrent
or close in time series of magnetic observations were obtained by G. W.
Preston and the magnetic curves were formed. In these cases it was possible
to discuss the phase relationship between light and magnetic curves. It turned
out that the minimum of the magnetic curve can coincide with the minimum
of the light curve as well as with the maximum. Sometimes it depends on the
color discussed. Hence, the suggestion that the light curve and the magnetic
curve are always in antiphase does not hold any more.

Discussing the problem of irregularity among the magnetic stars one
must keep in mind a few important facts. This investigation showed that
chances of detecting light variables with large amplitudes are rather small.
In no case the amplitude of a new discovered variable exceeded in any color
0.04 of a magnitude and typically it was of the order of 0.01—0.02 of a magnitude. It seems that most of variables to be discovered have such amplitudes
or even smaller. There is no doubt that a periodic variable with the amplitude
of 0.01 of a magnitude and a period of several weeks or more will be very
hard to detect. New techniques should be employed permitting diminishing
the error of measurement. Periodic variations can be detected only if the
amplitude of variations is at least a few times larger than the probable error
of one observational point (otherwise a very large number of observations is
required to apply the statistical analysis). The same considerations hold for
magnetic measurements except that here, contrarily to the photometry, the
probable error of one observational point varies strongly from one star to
another ranging from 30 gauss to about 1000 gauss. Because many stars
show magnetic variations of the order of a few hundred gauss it is clear that

the detecting of periodicity is often difficult if not impossible. Again new techniques should be looked for to permit to measure the magnetic fields of stars with broad spectral lines. This investigation strongly suggests that the percentage of periodic magnetic stars, probably with constant periods, is high.

More detailed discussion of the photometric behaviour of magnetic stars will appear in the Astrophysical Journal, vol. 154.

This is the pleasure of the author to thank the staff of the Lick Observatory and particularly Dr. George W. Preston for hospitality and help during his stay at the Lick Observatory where this investigation was carried out.

DISCUSSION

Herbig: Does there remain any one well established case of irregular variability among the magnetic variables?

Stępień: I think we cannot say for the present unambiguously that a given magnetic star is irregular. Even if long series of observations exist and the period cannot be found it does not yet prove that the star is really irregular. An example of such a star is 78 Vir. Babcock obtained a large amount of measurements for this star and the periodicity could not be found. Now, it seems that the period of around 12 days may satisfy the data although a large scatter is present. However, because the amplitude is not large, such a scatter is expected. Besides the scatter, some irregular fluctuations may exist in many stars obscuring existing periodicity. There must be long series of observations, variations must be much larger than the error of one measurement and such variations must be confirmed by another investigator (preferably with another equipment) before the conclusion about irregularity is reached. Taking all of this into account I can say I do not know any star which was proved beyond doubt to be irregular.

Plagemann: Your observations that the maximum of light do not always coincide with a minimum of magnetic field — which of the theories discussed by Dr. Jarzebowski would this tend to support or reject?

Stępień: It does not influence any particular theory of magnetic stars more than the other. If the light and magnetic curves were always in antiphase it would be very difficult to explain this fact on the ground of the basic theory of magnetohydrodynamics because the equations governing the motion and the behaviour of matter do not depend on the sign of the magnetic field. These equations must be used in any attempt of theoretical explanation of light variations. As a result the preference of the light and magnetic curves to being in antiphase would cause equal troubles for every theory of magnetic stars and a lack of such a preference removes these troubles from all of them.

PHOTOMETRIC RESEARCH ON MAGNETIC STARS AT THE CATANIA ASTROPHYSICAL OBSERVATORY

C. BLANCO, F. CATALANO, and G. GODOLI

Catania Astrophysical Observatory, Italy

SUMMARY

The results recently obtained at Catania from the observations of the magnetic stars SX Ari, 41 Tau, CU Vir, HD 173 650, HD 184 905, HD 219 749, 8k Psc, HD 224 801 are summarized.

INTRODUCTION

In 1967 systematic photoelectric observations of magnetic variables were started at the Catania Astrophysical Observatory. These observations have been undertaken in the course of the Catania research programme on stellar activity of solar type (Godoli, 1967, 1968). Some results have already been published (Blanco and F. Catalano 1968; Godoli 1968). Here the observations made and the results obtained until August 1968 will be summarized.

OBSERVATIONS

The instruments used for the observations are a 91 cm and two 30 cm (distinguished by N and S) conventional Cassegrain reflectors (Fracastoro 1967).

The characteristics of the stars observed (August 1968) according to the Ledoux and Renson Catalogue (1966) and of the instruments used are reported in Table 1. The stars observed have periods of 0.5 to 12 days.

The photoelectric observations have been carried out in the three colours U, B, V. For each colour, sequences of measurements were made in the order $S\,C_1\,S\,V\,S\,V\,S\,C_2\,S\,C_1\,S$ (S indicates the sky, C_1 and C_2 the comparison stars, V the variable under study), every measurement lasting one minute. Such a sequence was used since we were interested in the small (short period or secular) variations of these stars.

The comparison stars are listed in Table 2.

The observations are corrected for atmospheric extinction using a statistically determined mean absorption coefficient. The data are elaborated by the IBM computer of the Faculty of Sciences of Catania University. In Table 3 the sequences obtained in the U, B, V bands until August 31 are reported for every star.

METHOD OF REDUCTION

First of all, the diagrams $m_{c_1} - m_{c_2}$ versus JD have been plotted in order to control the constancy of the comparison stars C_1 and C_2. All the observations for each star are plotted in a diagram $m_V - m_{c_1}$ versus JD for each colour.

Table 1

Characteristics of the observed magnetic stars (August, 1968), according to the Ledoux and Renson Catalogue (1966) and the instruments used

Magnetic star	Coordinates (1950)		m_v	Δm		Sp.	Period days	Instrument
	R. A	D.		B	V			
HD 19832 (SX Ari)	3^h09^m3	$+27°04'$	5.7	0.06	0.04	B8pV	0.7279	30 N
HD 25823 (41 Tau)	4 03 .5	$+27$ 28	5.2	0.03	0.035	A0pV	11.9	30 N
HD 124224 A (CU Vir)	14 09 .7	$+$ 2 39	4.9	0.08	0.07	B9pV	0.52067	30 N
HD 173650 (BD+21°3550)	18 43 .5	$+21$ 56	6.4	—	—	A0pV	10.1	30 S
HD 184905 (BD +43°3290)	19 33 .2	$+43$ 50	6.6	—	—	A0pV	1 — 2	91
HD 219749 (BD +44°4373)	23 15 .6	$+45$ 13	6.3	0.03	0.02	B9pV	0.723	30 S
HD 220825 (8k Psc)	23 24 .4	$+$ 0 59	4.9	0.020	0.015	A2p	0.58	30 N
HD 224801 (BD +44°45)	23 58 .2	$+44$ 58	6.3	0.03	0.05	A0pV	3.74	91

Table 2

Comparison stars

Magnetic star	C_1			C_2		
	comparison star	m_v	Sp.	comparison star	m_v	Sp.
HD 19832 (SX Ari)	HD 19548	5^m62	B7V	HD 19600	6^m36	A0V
HD 25823 (41 Tau)	HD 25867	5^m32	dF1	HD 26322* HD 25626**	5^m40 7^m9	dF3 A2
HD 124224 A (CU Vir)	HD 121607	5^m86	A3	HD 125489	6^m14	A3
HD 173650	HD 173494	6^m10	dF5	HD 174160	5^m68	F5
HD 184905	BD +42°3386	5^m8	A	BD +41°3398	7^m0	A
HD 219749	HD 219891	6^m40	A2	HD 219668* BD +44°4421**	6^m40 7^m0	sgK0 A
HD 220825 (8k Psc)	HD 222603	4^m52	A7V	HD 221950	5^m60	dF0
HD 224801 (CG And)	HD 224559	6^m52	B3IV	BD +45°4363* BD +46°4231**	6^m7 7^m0	B A

    * For the observations made during 1967.
  ** For the observations made during 1968.

The curves m_v—m_{c_1} versus phase φ have been deduced from these diagrams. For this deduction previous values of the period, when available, have been taken into account.

From m_v—m_{c_1} versus phase (φ) curves the normal points have been calculated in order to determine the U — B and B — V colour index curves. When possible we have plotted the diagram m_v—m_{c_1} versus φ doubling the period.

Table 3

Number of observations in the three colours in the order U, B, V

Magnetic star		Number of observations for different nights
SX Ari	1967/XI	2(11; 10; 9), 9(4;4;3), 10(4;3;3), 14(20;18;18), 22(3;3;2), 24(10;9;9).
	1967/XII	7(7;6;6), 16(3;3;0), 17(3;2;2), 29(2;1;1).
41 Tau	1967/X	28(2;2;2).
	1967/XI	9(1;2;2), 14(1;1;1), 24(2;2;2).
	1967/XII	3(1;1;1), 7(3;3;3).
	1968/I	29(1;1;1), 30(4;5;4).
	1968/III	1(4;4;0), 19(1;1;0).
CU Vir	1968/II	26(3;3;3).
	1968/III	1(3;2;2), 18(6;5;4), 21(3;3;2), 23(6;6;7), 25(8;8;7), 28(7;6;6).
	1968/IV	5(9;7;7), 6(1;1;1), 8(6;5;4), 18(7;7;6), 22(3;3;3), 23(5;5;4), 24(1;2;1), 30(2;2;1).
	1968/V	4(2;3;1), 16(4;4;4), 17(6;5;6), 18(8;8;8), 21(3;2;2), 27(4;3;3).
	1968/VI	3(2;2;1), 8(1;0;0), 13(3;2;2), 15(1;1;0).
HD 173650	1967/VII	11(1;1;0), 15(2;2;2), 18(1;1;1), 20(1;1;1), 22(3;3;3), 24(2;2;2), 25(2;2;2), 26(2;2;2), 27(2;2;2), 28(2;2;2).
HD 173650	1967/VIII	1(1;1;1), 2(2;1;2), 3(1;1;0), 6(2;2;1), 7(2;2;2), 8(2;2;2), 9(1;0;0), 10(2;1;2), 12(2;2;2), 13(1;1;0), 14(2;2;2), 18(0;2;2), 19(4;0;4), 20(1;2;2), 24(2;2;2), 25(2;2;2), 26(2;2;2), 27(2;2;2), 28(2;2;2), 29(2;2;2), 30(1;1;1), 31(2;2;2).
	1967/IX	1(2;2;1), 5(1;1;0), 8(1;2;1), 10(2;2;2), 13(1;2;0), 16(2;1;2), 17(2;1;2), 19(2;2;2), 20(2;2;2), 21(2;2;2), 22(2;2;2), 26(2;2;2), 27(2;2;2), 28(2;1;2), 29(2;2;2), 30(2;2;2).
	1967/X	1(2;2;0), 2(2;2;2), 7(2;2;2), 11(2;2;1), 12(2;2;2), 16(2;2;2), 21(2;2;2) 22(2;2;0), 23(1;1;0), 29(1;0;0).
	1967/XI	3(2;1;0).
HD 184905	1967/VIII	18(4;4;4), 19(5;5;5), 20(5;4;5), 21(2;2;2), 24(5;5;5), 25(4;4;3), 30(4;4;4).
	1967/IX	1(7;7;6).
	1968/VII	9(4;4;5), 10(7;6;6), 12(8;8;8), 19(8;7;7), 20(7;9;8), 21(5;5;5), 27(2;2;2), 30(3;3;3), 31(9;9;9).
	1968/VIII	2(4;4;4), 10(1;0;0), 14(1;1;1), 17(6;6;6), 18(4;4;4), 19(8;8;8).
HD 219749	1967/IX	17(10;11;8), 19(6;5;5), 20(7;7;7), 21(8;8;8), 22(2;1;2), 23(7;6;6).
	1967/X	13(5;5;3), 16(2;1;1), 21(7;7;7), 22(11;11;0), 28(3;3;3), 9(9;9;9), 14(8;7;7).
HD 219749	1968/VII	15(5;5;5), 18(4;4;4).
8k Psc	1967/IX	7(5;5;4), 20(2;2;2), 21(2;0;1), 23(9;0;9), 26(6;4;6), 27(9;9;9), 28(6;6;6), 29(8;8;8).
	1967/X	1(2;2;2), 2(2;2;2), 11(2;2;2), 13(5;5;5), 16(6;0;6).
HD 224801	1967/IX	7(2;1;2), 15(1;1;1), 17(7;6;7), 19(3;2;2), 20(4;3;3), 21(6;5;5), 22(1;0;1), 23(9;10;9), 26(6;6;5), 27(4;4;4), 28(1;2;1).
	1967/X	11(4;4;4).
	1967/XI	2(4;3;3), 3(1;1;1), 4(6;5;6), 5(3;3;3), 9(6;5;6), 14(4;2;7), 15(5;5;5), 18(1;0;0), 22(2;2;2), 24(5;5;5).
	1967/XII	7(2;2;2).

RESULTS

1. *Comparison stars*

For the magnetic stars 41 Tau, HD 219 749 and HD 224 801 the comparison stars show variations larger than the usual dispersion. For this reason a third comparison star, C_3 was chosen for the 1968 observations.

2. *Periods*

SX Ari: Our observations are in agreement with the epoch and period given by Provin (1953). According to our observations:

$$\text{Min. light} = \text{JD } 243\ 9792.49 + 0^{d}7279 \text{ E.}$$

CU Vir: Our observations are in agreement with the epoch and period given by Hardie (1958). According to our observations:

$$\text{Min. light} = \text{JD } 243\ 9887.56 + 0^{d}52068 \text{ E.}$$

HD 173 650: Starting with the epoch given by Wehlau (1962) an epoch for our observations and a new period have been obtained:

$$\text{Min. light} = \text{JD } 243\ 9685.82 + 10^{d}1353 \text{ E.}$$

HD 219 749: Our observations are not complete at present, but the best agreement is found for the period of $2^{d}604$ given by Rakos (1962).

Light curves of all the other stars are not completely studied at present. The period of 8k Psc is difficult to determine, because of the very small light variations.

3. *Light curves*

The light curves and the curves of the colour indices for the completely observed stars are reported in Figs 1—4.

For all the stars observed, the maximum variation of luminosity occurs in the U light. We did not notice any displacement of the phase of the maxima in different colours. For all these stars the U — B curves are in phase with the light curve, except for CU Vir, where the minimum of the U — B curve shows a displacement of 0.1 P.

We noticed for none of the stars short period or secular variations. No peculiarity for any star has been found for odd and even cycles.

4. *Comparison between light curves and magnetic curves*

We have only for HD 173 650 the magnetic curve. For this star the light curve is in phase with the curve of the magnetic field given by Wehlau (1962).

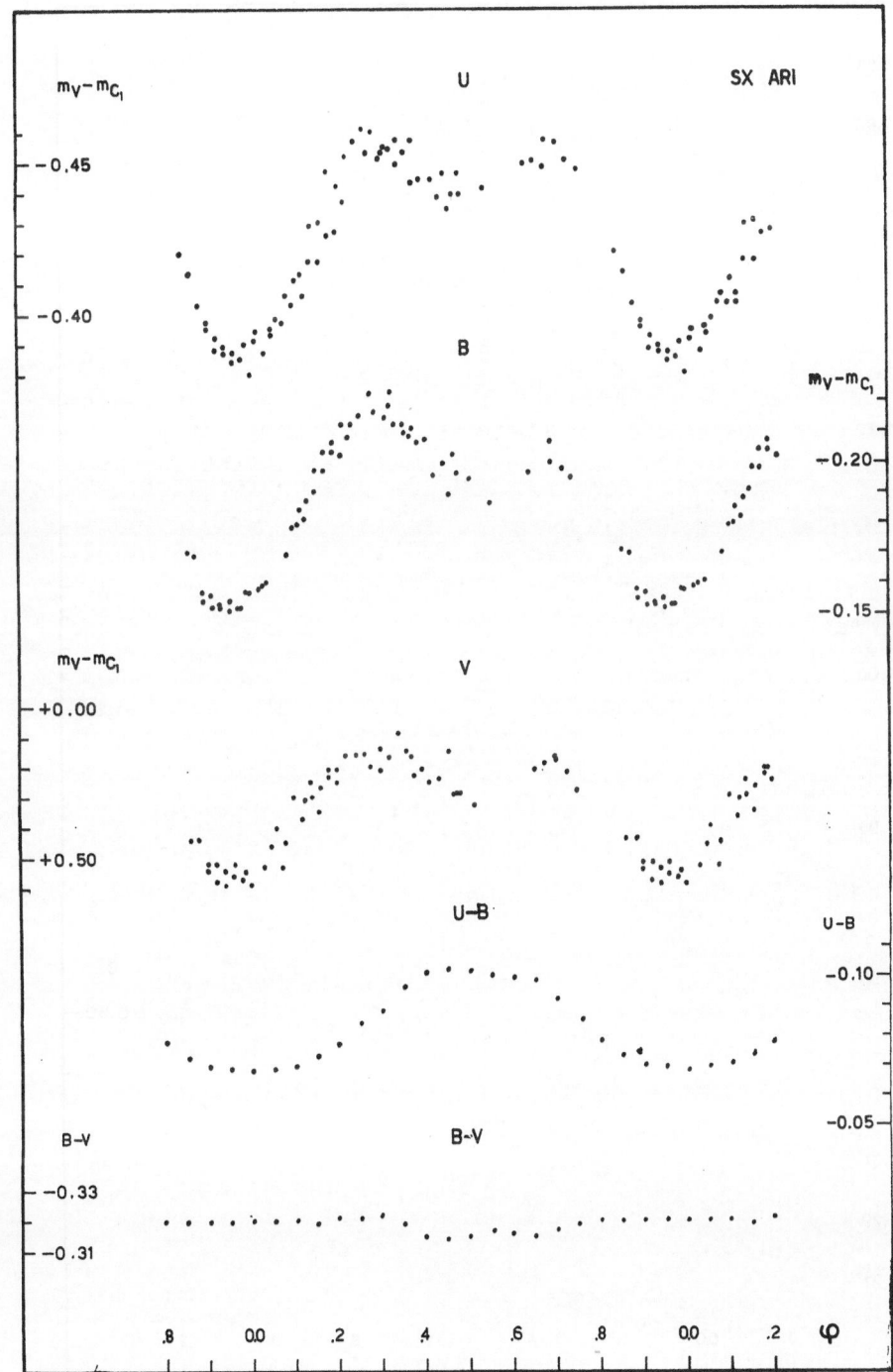

Fig. 1. Light curves and curves of the colour index for SX Ari

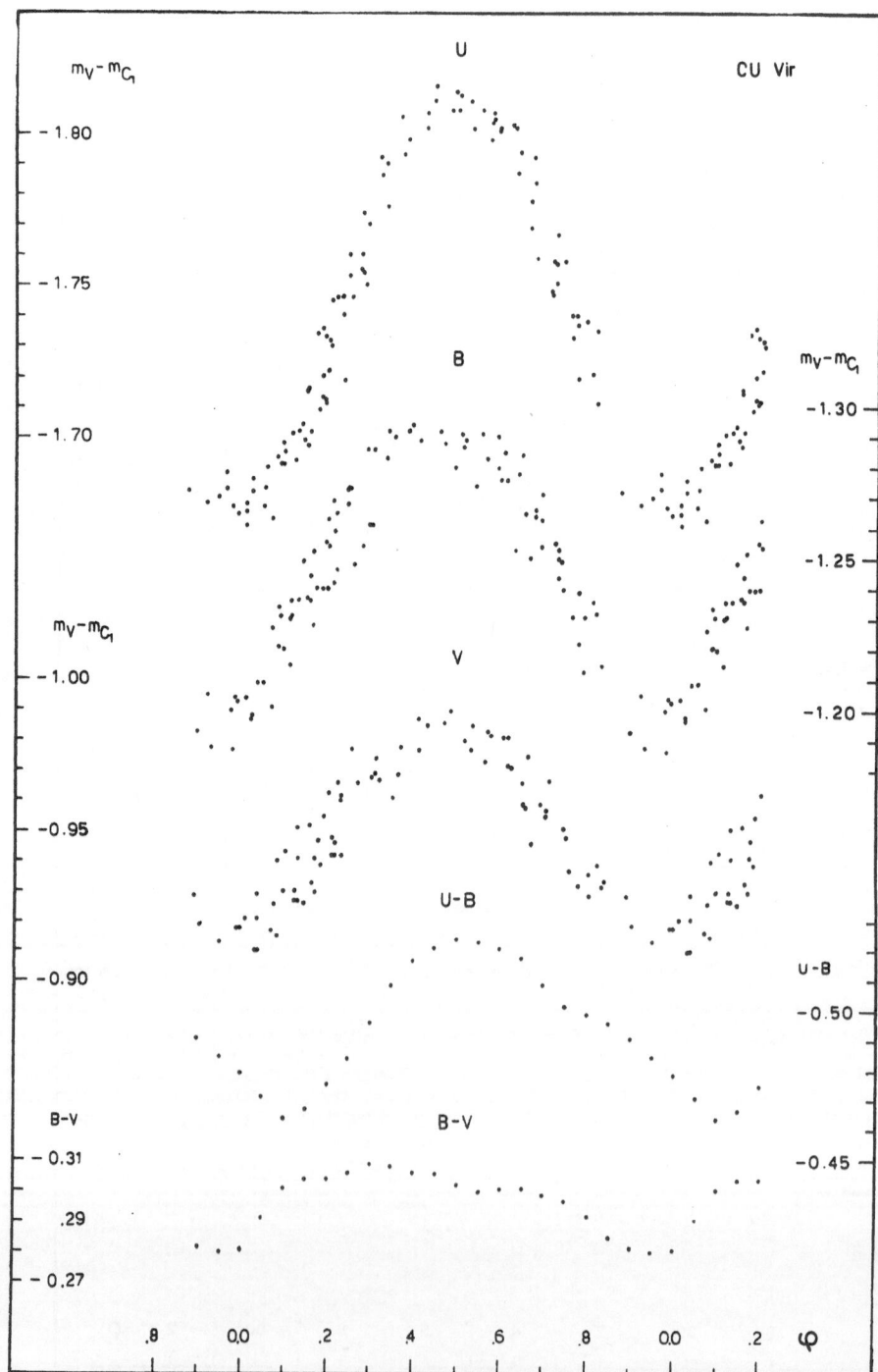

Fig. 2. Light curves and curves of the colour index for CU Vir

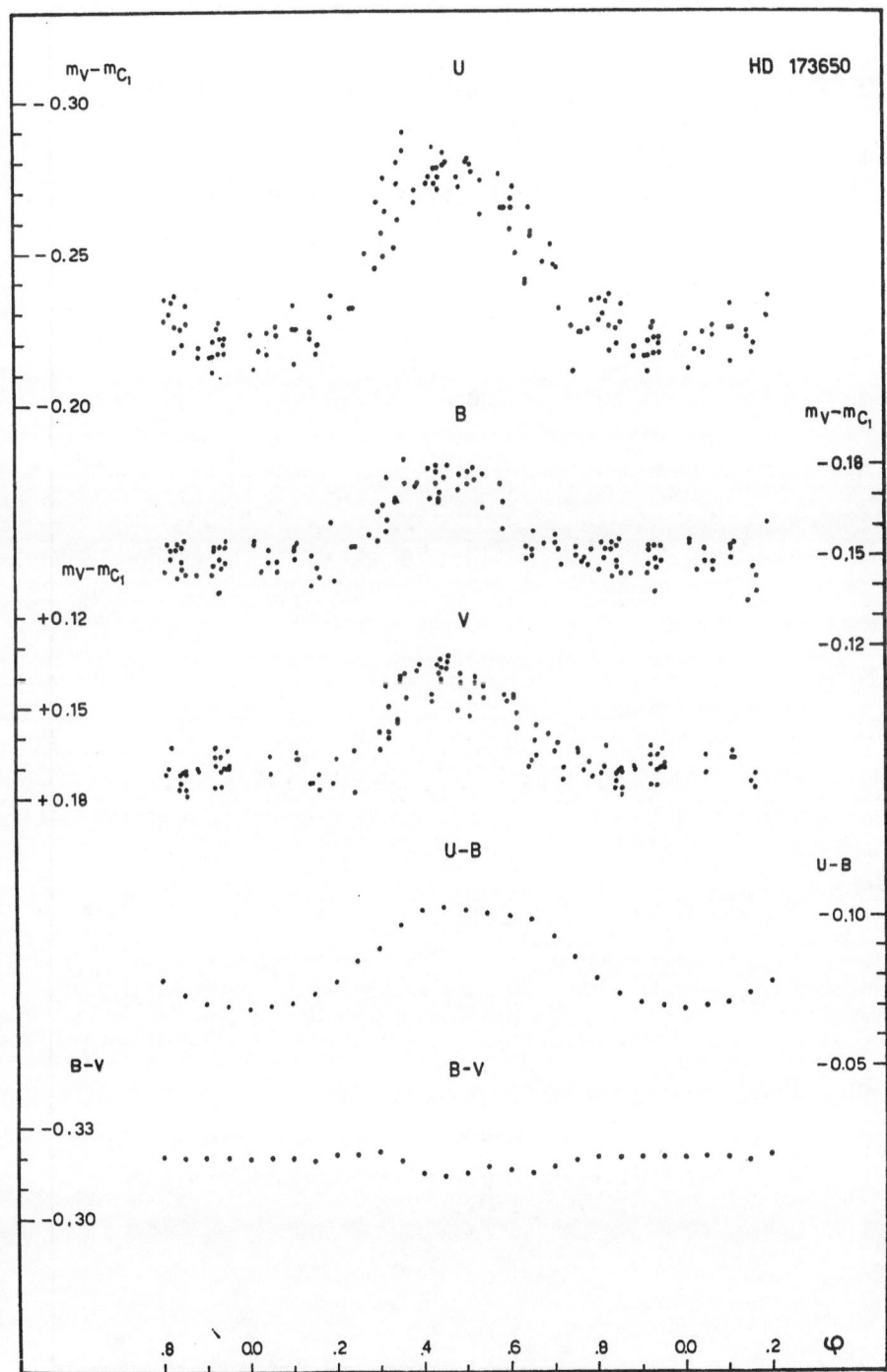

Fig. 3. Light curves and curves of the colour index for HD 173650

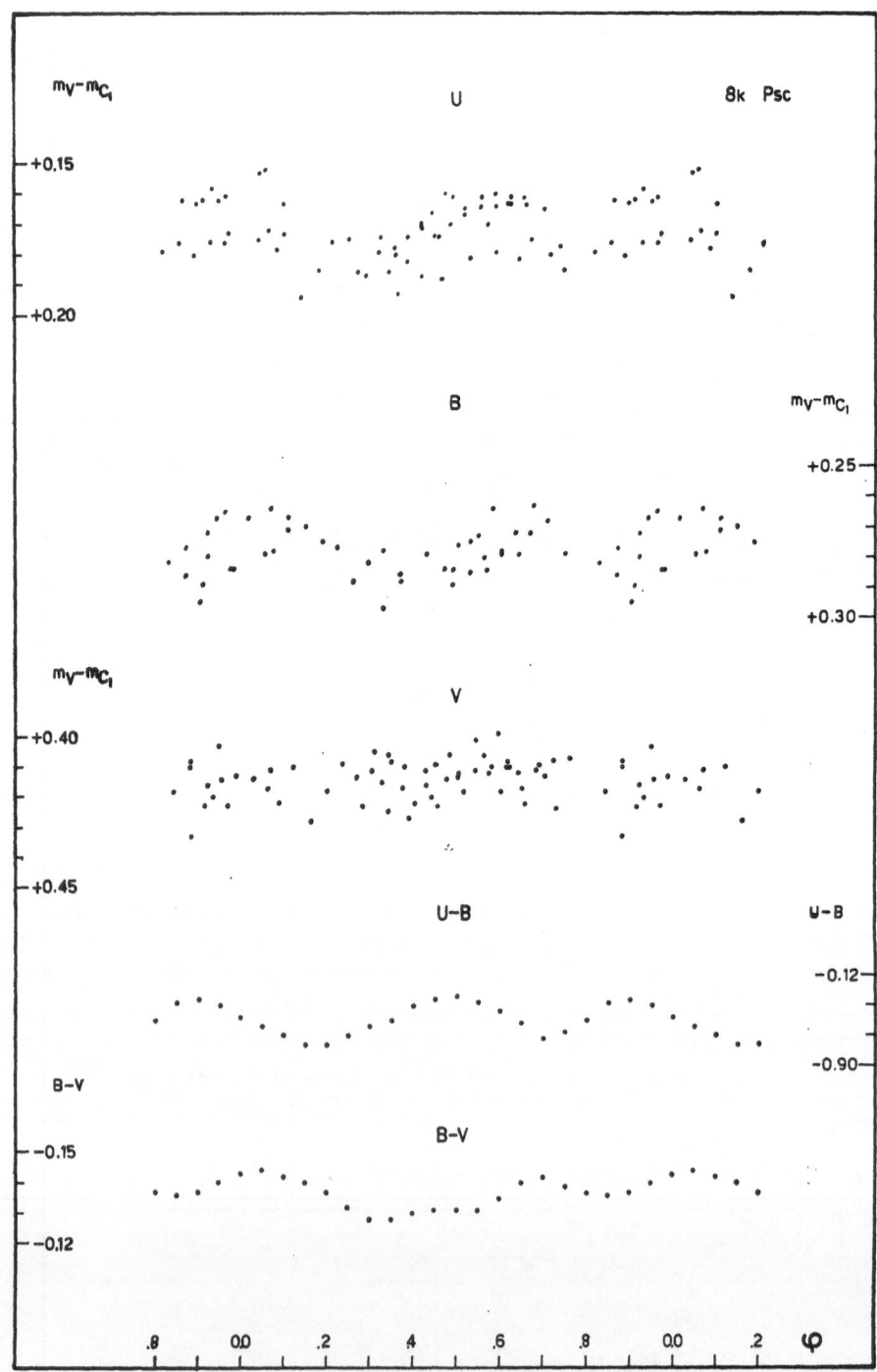

Fig. 4. Light curves and curves of the colour index for 8k Psc.

5. Summary

The periods and Δm values deduced from our observations are compared with those from the Ledoux and Renson Catalogue in Table 4.

Table 4

Periods and Δm according to the Ledoux and Renson Catalogue and according to our observations

Star	P		Δm					
			U	B			V	
	L & R	our obs.	our obs.	L & R	our obs.	L & R	our obs.	
SX Ari	0.7279	0.7279	0.08	0.06	0.06	0.04	0.05	
41 Tau	11.9	—	—	0.03	—	0.035	—	
CU Vir	0.52067	0.52068	0.14	0.08	0.09	0.07	0.07	
HD 173650	10.1	10.1353	0.06	0.039	0.03	0.045	0.04	
HD 184905	102 ?	—	—	—	—	—	—	
HD 219749	0.723	2.604	0.07	0.03	0.035	0.02	0.02	
8k Psc	0.58	0.58	0.025	0.007	0.02	0.012	0.015	
HD 224801	3.74	—	0.08	0.03	0.05	0.05	0.07	

REFERENCES

Blanco, C., Catalano, F., 1968, Mem. Soc. astr. ital., **39.**
Fracastoro, M. G., 1967, Oss. astrofis. Catania Pubbl. No. 107.
Godoli, G., 1967, Oss. astrofis. Catania Pubbl. No. 115.
Godoli, G., 1968, In "Mass motion in Solar flares and related phenomena" IX Nobel Symposium, Anacapri.
Hardie, R., 1958, Astrophys. J., **127,** 620.
Ledoux, P., Renson, P., 1966, A. Rev. Astr. Astrophys., **4,** 293.
Provin, S. S., 1953, Astrophys. J., **118,** 281.
Rakos, R., 1962, Lowell Obs. Bull., **5,** 227.
Wehlau, W., 1962, Publ. astr. Soc. Pacific, **74,** 137.

RY SGR DURING THE 1967-8 MINIMUM: A BLUE CONTINUUM AND OTHER SPECTROSCOPIC AND PHOTOMETRIC OBSERVATIONS*

M. W. FEAST

Radcliffe Observatory, Pretoria, S. Africa

Apart from visual estimates of magnitudes, the amount of information available on the R CrB type variables is surprisingly small. This is particularly true of the changes that occur during light minima. The principal published data are a very valuable series of spectra taken by Herbig (1949) during the decline of R CrB itself to minimum in 1949, and an analysis by Mrs. Payne-Gaposchkin (1963) of several coudé plates taken by Greenstein during the 1960 minimum of the same star. There is evidently a considerable need for rather detailed spectroscopic and photometric studies of individual minima of the variables, especially as these may vary from star to star and from one minimum to another. This is a preliminary report of observations of RY Sgr during the present minimum.

At maximum light RY Sgr is very similar to R CrB itself, as is shown by Danziger's (1965) high dispersion study. The star has been under visual observation by de Kock for many years and he observed a decline in brightness during July 1967. Since that time medium and low dispersion spectroscopy and UBV photometry has been carried out with the Radcliffe 74-inch reflector and the Cape 40-inch reflector as often as other work permitted. The present minimum is the deepest recorded for 20 years and we are fortunate in having secured a fairly good coverage of it. The star has not yet returned to maximum brightness and a detailed analysis of the observations may take some while. However a number of facts have emerged which seem to be of particular interest.

The initial drop of the star was very rapid, about $3^{m}5$ in 15 days. The star was first observed spectroscopically at about $9^{m}5$ (V) during this decline (that is about 3 magnitudes below maximum). It then showed a rich emission spectrum which strengthened relative to the continuum as the star faded further. At the first minimum ($\sim 10^{m}3$) the spectrum (at 48 A/mm) showed some 350 emission lines on a fairly weak continuum. Most of the emission lines correspond to lines seen in absorption at light maximum and are of Fe II, Ti II, Sc II etc. There seems to be a qualitative correspondence to the emission spectrum reported by Mrs. Payne-Gaposchkin in the initial phase of the decline of R CrB itself in 1960 though a detailed comparison has not yet been

---

* This discussion is based on observations in Pretoria and the Cape chiefly by J. B. Alexander, P. J. Andrews, R. M. Catchpole, P. Corben, D. H. P. Jones, R. P. de Kock, T. Lloyd Evans, J. W. Menzies, E. N. Walker and the writer. Several of these workers are participating in a study of the observations and it is hoped to publish a detailed joint paper at some later date. I am very grateful to my colleagues for allowing me to use the data for this preliminary discussion and for valuable conversations.

made. The continuum at this phase shows an interesting phenomenon which
has not apparently been reported before. From the red limit of ordinary
photographic plates to about 4000A, the continuum is weak and provides
a relatively low background to the emission lines. This could well be the residual stellar continuum. By 4030A this continuum is quite weak and one would
not expect to see any continuum to shorter wavelengths even for an early
type star. However at about 4000A the continuum intensity increases considerably and absorption lines are seen. The principal absorptions are H and K
of Ca II. These are not the normal broad stellar H and K lines seen at maximum light, but much narrower lines showing structure suggestive of blending
of several discrete components. Narrow emission complicates the appearance
but the absorption components have predominantly negative velocity shifts
(up to $\sim$ 200 km/sec).

Apparently these absorptions are formed in shells or streams moving
away from the stellar surface.

The increase in the continuum intensity shortward of 4000A together
with the absence of the normal wide stellar H and K absorptions makes it
rather difficult to explain the continuum in this region as normal (thermal)
radiation from the star. Rather it appears necessary to look for some interpretation involving an atomic or a molecular emission continuum. In addition
to this continuum there is an indication of banded structure in the 3880A
region. These are almost certainly emission bands of CN. CN was first seen in
emission by Herbig in R CrB during the 1949 minimum.

The most likely kind of process for the continuum is perhaps free-bound
transitions (of electrons), and if we consider only transitions involving ground
states then the energy involved (i.e. 3.1 eV for a wavelength cut off of about
4000A) drastically limits the possible emitters. The predicted continuum which
appears most plausible as an identification is the electron attachment spectrum
of CN. The electron affinity of CN is 3.1 eV and the inverse process of the
one considered here (i.e. electron detachment of CN^-) has been considered
by Branscomb and Pagel (1958) as a possible source of opacity in cool carbon
stars. Quantitative work on the shape of the continuum for a comparison
with theory might enable this suggested identification to be tested. A possible
contribution at shorter wavelengths ($<$ 3900A) from the electron attachment
spectrum of C_2 should also be borne in mind.

After the steep fall to 10^m3 the light curve goes through a substantial
hump. The chief spectroscopic characteristic of this hump is the increase of
the normal continuum with respect to the emission lines. If we follow Mrs.
Gaposchkin's interpretation of R CrB then this would correspond to a brightening of the underlying star but it should be noted that the H and K absorptions
still have the abnormal appearance discussed above and the continuum in
this region may still be chiefly non-thermal (CN^- ?). Quantitative work should
decide whether Mrs. Gaposchkin's explanation in terms of the intreplay of
independent absorption and emission spectra holds in the present case. Certainly the appearance of the spectrum is most peculiar at certain phases. For
instance near the top of the hump ($\sim 9^m$) there is a practically continuous
spectrum, the only very strong absorptions being Mg II and C I (besides
shell, Ca II). Other emissions and absorptions are very weak. As the star
fades again the continuum drops relative to the emission. The non-thermal
continuum now seems to have gone and the continuum observed is probably

a residual stellar continuum. The sharp emission spectrum is now quite different from that observed earlier, being qualitatively similar to that observed by Herbig during the faint phase of R CrB in 1949. The main feature of this spectrum is the weakness of Fe II and the strength of Ti II and Sc II emissions. Actually there is no abrupt change in the sharp emission spectrum but a gradual smooth change from the earliest observations onwards. The overall strength of this emission spectrum seems to fall off with time rather independent of the magnitude of the star (e.g. independent of the occurrence of the hump). This is similar to the fall off in the strength of the emission spectrum of R CrB in 1960 (as reported by Mrs. Gaposchkin). In addition in RY Sgr there is at the same time a gradual change in the relative intensities of the lines. The changes effect adjacent lines and cannot therefore be attributed to the interplay of a fading emission spectrum of fixed relative intensities with a fixed absorption spectrum showing changing reddening. The general behaviour is a reduction of the relative intensities of lines of high excitation potential but the detailed behaviour is rather complex.

These observations may well be consistent with Mrs. Gaposchkin's suggestion that the emission lines originate in the outer parts of the star (a chromosphere) from which the source of excitation has been cut off (by obscuration of the central star or otherwise). Two points are worth noticing in this connection. Firstly, because the emission spectrum is found in RY Sgr to be varying quite rapidly with time, it is doubtful whether the physical conditions of the chromosphere, even in the earliest phases, correspond to chromospheric conditions at light maximum. This probably also applies to the chromosphere parameters deduced by Mrs. Gaposchkin for R CrB. Secondly, the decay of an emission region suddenly cut off from its source of excitation would appear an ideal place for the production of an electron attachment spectrum, such as that of CN was postulated earlier.

As RY Sgr fades towards minimum broad emission lines of H and K (Ca II), 3888 (He I) and the D lines were seen. The star faded to 13ᵐ6 (V) before going into conjunction with the sun and was picked up at nearly the same magnitude a few months later.

The recovery in brightness is characterized spectroscopically by the gradual disappearance of the last vestiges of the emission line spectrum and a return to essentially the normal maximum absorption spectrum (by $\sim$ 9ᵐ5 (V)). Photoelectrically the rise is seen to progress in a number of smooth waves with a period of about 30 to 40 days. Similar waves in RY Sgr were suggested by Jacchia (1933) from visual observations. These waves are seen in V, B—V and U—B. No extensive photometry at maximum light has apparently been made but visual observations indicate variations of amplitude $\sim$ 0ᵐ5 and period $\sim$ 39 days, quite similar to those observed during the rise. It has been suggested that these are cepheid like variations (i.e. pulsations). If this is so then the star appears to be pulsating in a normal (maximum) fashion when at least 5 magnitudes below maximum. This (together with the recovery of the normal spectrum well below maximum) would suggest that, at least at this stage, the diminution of light cannot be due to some phenomenon in the star itself but must be caused by some form of obscuration above the stellar surface (e.g. O'Keefe's (1939) hypothesis of graphite absorption).

The colour changes in the UBV system are remarkable. At the bottom of the first rapid decline the star is quite blue (B—V $\sim$ + 0.24, U—B $\sim$ —0.60).

This is at least partly due to the effects of the (non-thermal) blue continuum. The complex variations in the colours on the descending branch of the light curve are probably largely caused by emission line effect. During the later stages, as the star rises again and when the emission lines are probably not affecting the colours to any great extent, the star is very red (reaching B—V $\sim +1.48$ U—B $\sim +1.32$) and is also performing loops both in the two colour diagram and the V, B—V diagram. These results are best interpreted as due to some form of intrinsic stellar variability together with heavy overlying reddening and obscuration.

While RY Sgr was faint it was noticed to have a faint companion about 12″ distant V = 15.69 B—V = +0.79 U—B = +0.15. After allowing for a small amount of reddening (cosecant law) this star lies close to the unreddened main sequence relation in the two colour plot and fitting to the main sequence indicates an absolute magnitude of about $+5^m\!.1$ (V). If the star is a physical companion then the absolute magnitude of RY Sgr at maximum is about -4^m (V). This would be quite an acceptable absolute magnitude and it would appear worthwhile making a determined effort to decide if the companion is really physical. Some direct photographs have been taken by Dr. P. J. Andrews to serve as first epoch plates for the purpose.

REFERENCES

Branscomb, L. M. and Pagel, B. E. J., 1958, Mon. Not. R. astr. Soc., **118**, 258.
Danziger, I. J., 1965, Mon. Not. R. astr. Soc., **130**, 199.
Herbig, G. H., 1949, Astrophys. J., **110**, 143.
Jacchia, L., 1933, Publ. Oss. astr. Univ. Bologna, **2**, 173.
O'Keefe, J. A., 1939, Astrophys. J., **90**, 294.
Payne-Gaposchkin, C., 1963, Astrophys. J., **138**, 320.

DISCUSSION

Fernie: Have you made observations of any other southern R CrB stars?

Feast: A few spectroscopic observations were made some years ago mainly of S Aps and W Men (a member of the Large Magellanic Cloud). It is intended to carry out further spectroscopic, and possibly also photoelectric work.

Non-Periodic Phenomena in Variable Stars
IAU Colloquium, Budapest, 1968

SPECTROSCOPIC OBSERVATIONS OF THE RECURRENT NOVA RS OPHIUCHI FROM 1959 TO 1968

R. BARBON, A. MAMMANO, L. ROSINO

Astrophysical Observatory Asiago, University of Padova

ABSTRACT

Spectra of RS Oph have been taken at Asiago with the 122 cm telescope during the 1959—67 minimum and on the occasion of the 1967—68 outburst. At minimum the variable has a composite spectrum, one component being of spectral type around M2, while the other component gives a blue continuum with emission lines of medium or high excitation. During the outburst the star has shown the same spectral evolution as in the 1933 and 1958, with the development of bright nebular and coronal lines, indicating an extremely high degree of ionization. Further details will be given in a forthcoming paper.

A preliminary account is given here of the spectroscopic observations of the reccurrent nova RS Oph carried out at Asiago during its minimum from 1959 to 1967 and on the occasion of its recent outburst in 1967—1968. It is not necessary perhaps to recall that RS Oph was caught at maximum four times, in 1898, 1933, 1958 and more recently in 1967, rather surprisingly in view of the short time-interval passed from the precedent maximum.

The nova was extensively observed at Asiago from July to October during its 1958 maximum. A detailed report on its spectral evolution is published in the Contributions of Asiago, No. 113 (1960). At the end of October 1958, 104 days past maximum, RS Oph had reached its highest stage of ionization, with an emission spectrum characterized by the presence of strong coronal lines of [FeXIV], [FeX] and [AX], forbidden lines of [OIII] and [NII] and permitted lines of NIII, HeI and HeII, besides of course the Balmer series of hydrogen. H_α, H_β, [FeX] 6374, HeI 5875 and HeII 4686 were by that time prominent in the spectrum of the nova.

THE MINIMUM 1959—1967

When the nova was again observed in 1959, after an intermission of several months due to the seasonal period of invisibility, the situation was radically changed. RS Oph had reached its minimum magnitude at about 12.5 and all of the lines due to high ionized atoms had faded out or disappeared. The object was since then included in a regular program of observations of old novae, carried out at first with the prismatic spectrograph attached to the newtonian focus of the 122 cm telescope (dispersion: 130 A/mm at H_γ) and later with an RCA Carnegie intensifier applied to the cassegrain spectrograph giving a dispersion of 60 A/mm at H_γ. Some cassegrain spectra were also taken at 75 and 180 A/mm.

From 1959 to 1962 RS Oph, as shown by the AAVSO observations, was constantly at a mean magnitude of about 11.5 visual, with slow irregular fluctuations of luminosity from 10.5 to 12.5. The spectra obtained during this period, independent of the intrinsic brightness of the star, show a gradual decline in the degree of excitation, a progressive fading of the blue-visual continuum and consequently a decisive strengthening in the red. In the 1959—1960 spectra, the following emission lines have been recorded: the Balmer series from H_α to H_δ, with very steep decrement; HeI 6678 and 5876; [OIII] 5006 very faint, partially in blend with HeI 5015; [NII] 5755. Some FeII permitted emission lines of multiplets 26, 27, 37, 38, 48, 49 and 74 are also represented. HeII 4686 was not recorded, even as a trace, and also NIII 4640 and the coronal lines had completely disappeared.

In 1961—62 a further general weakening of the emission lines was observed: the FeII emissions faded out and only H_α, H_β and λ 5876 HeI, the last two very faint, were still visible in our spectra. By this time wide absorption bands of moderate strength (faintly recorded also in 1959—60) emerged clearly at $\lambda\lambda$ 6350—6158, 6030—5890, 5780—5635, 5460, 5167 etc. The NaI doublet 5890—96 was prominent. Although some of the dark lines might be tentatively attributed to interstellar absorption, the comparison with standard spectra of advanced type leaves no doubt that the predominant spectrum of RS Oph in 1962 is an early M type, very likely M2—III. The symbiotic nature of the nova is therefore confirmed by the observations at minimum. In 1962 the red companion was dominant and masked the spectrum of the blue component, except for the H emission (Fig. 1).

A new phase of activity was displayed by RS Oph in 1965. Mrs. Mayall announced on March 26 that the star was of magnitude 9.7. On spectra taken at Asiago on April 2 a neat increase of the degree of excitation was clearly indicated by the strengthening of the continuum at wave lengths shorter than 5500 and by the emergence of relatively strong emission lines of H, HeI and FeII. During the following weeks, while the nova maintained a visual magnitude of about 10.2, the degree of excitation further increased. Noteworthy was the reappearance of HeII 4686, about as strong as H_β, which had never been observed since 1958. By May 14, however, 4686 faded out and became just visible as a trace, while the Balmer lines, HeI and FeII still appeared rather strong. Faint forbidden lines of FeII, SII and the blend NIII 4640 were also weakly recorded. The degree of excitation, although less than in April and May, was higher than at minimum, when the observations were interrupted in July 1965.

THE 1967 MAXIMUM

The announcement of the new explosion of RS Oph was given by Dr. Beyer of Hamburg-Bergedorf and by Fernald on October 26, 1967 (IAU Circ. 2040), who estimated the nova of magnitude about 5.9. Three days before, the star was still at its minimum, magnitude 10.7. I shall not speak here of the light curve, which was found to be almost exactly the duplicate of the light curves of 1933 and 1958.

Four months before maximum, on June 27—29, some spectra of RS Oph had been taken at Asiago with the new Carnegie intensifier. They show an intense continuum, also in the blue, with emission lines of H, HeI (6678, 5876, 4922, 4471, 4384) and FeII, moderately strong. The spectra do not look different from those taken during the secondary outburst of 1965, except

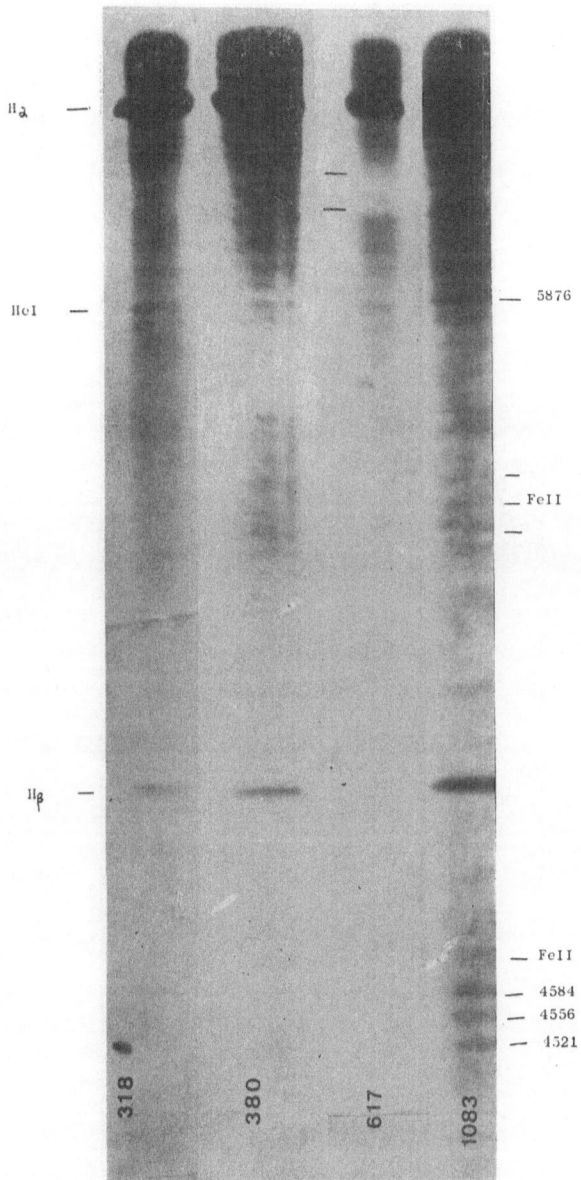

Fig. 1. Spectra of RS Oph at minimum, from 1959 to 1965.

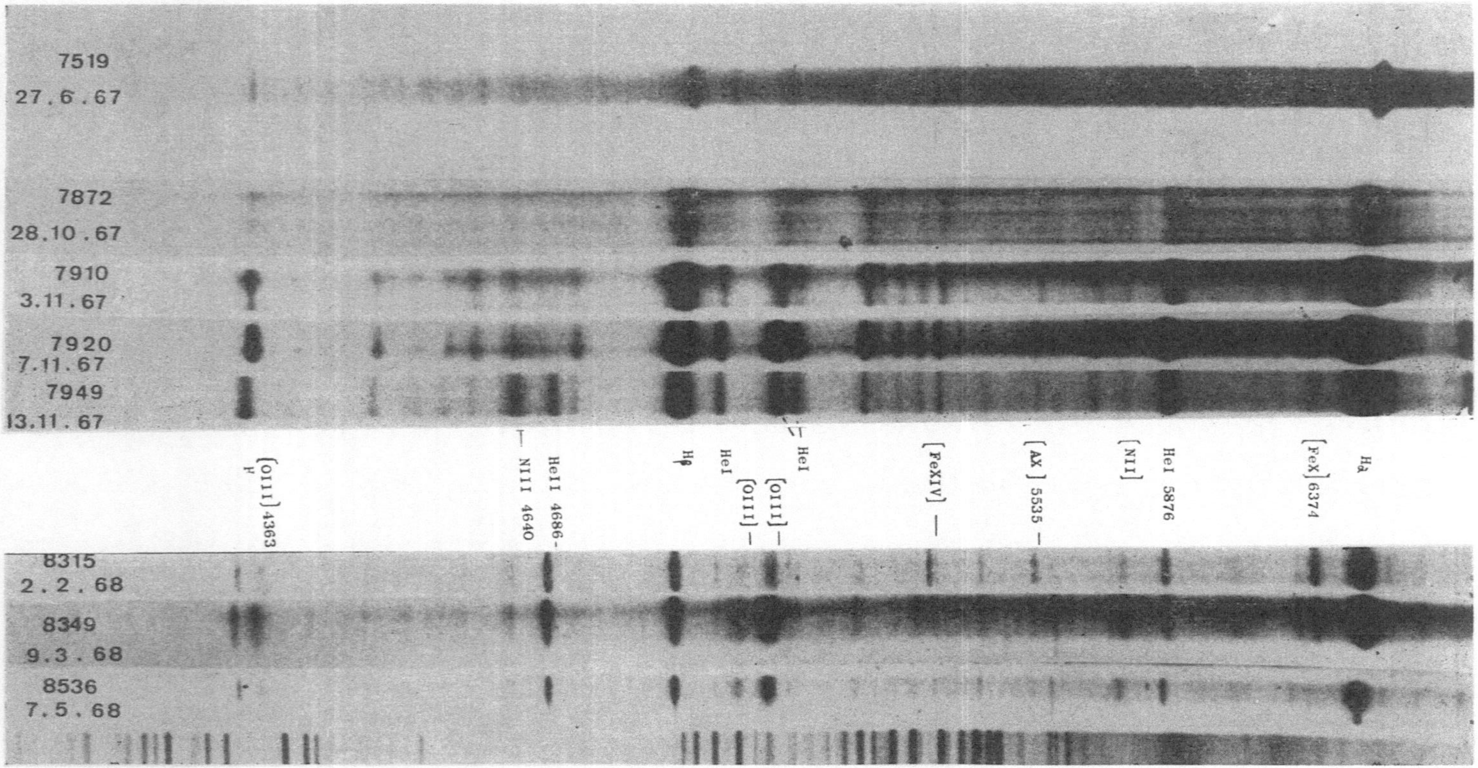
7519
27.6.67
7872
28.10.67
7910
3.11.67
7920
7.11.67
7949
13.11.67
[OIII] 4363
HeII 4686
NIII 4640
Hβ
HeI
[OIII]
[OIII]
HeI
[FeXIV]
[AX] 5535
[NII]
HeI 5876
[FeX] 6374
Hα
8315
2.2.68
8349
9.3.68
8536
7.5.68

Fig. 2. Spectral evolution of RS Oph during the 1967—68 outburst. On the top a premaximum spectrum taken on 27 th June 1967.

for the absence of HeII 4686, and indicate an unusual state of activity of the nova.

After the explosion, the nova was followed from October 27 to Dec 2 and again, from January 30 to July. It is presently under observation. Its spectral evolution can be shortly described as follows. On October 27, one day after maximum, the spectrum is characterized by wide emission bands (halfwidth 30—60 A) bordered on the blue side by two systems of broad absorptions, to which correspond mean radial velocities of —2700 and —3900 km/s. The continuum is strong. Near the center of the emission bands a narrow absorption line is visible, bordered on the red by a sharp emission. On the highest member of the Balmer series only the narrow absorption is recorded. The radial velocity of these sharp components is nearly —40 km/s, coincident with the radial velocity of the star, as given by Sanford. There seems to be no doubt that the sharp emission-absorption components, which were observed also in 1958, are originated in a stationary envelope which surrounds the nova.

On the following days, the absorption systems fade out and finally disappear. At this point the flaring of the HeI lines takes place. In the space of a few days, from October 27 to November 7, the HeI emission bands become outstanding in strength surpassing all other emissions except H_α and N_β. While the excitation further increases, the HeI lines start to weaken and by November 8—9 some lines of highly ionized atoms, as HeII 4686 and NIII 4640 make their appearance, and rapidly increase in intensity. Forbidden lines of [NII] 5755, [FeII], [OI] weakly appear from November 3 to 13. The [FeX] 6374 coronal line was already present from November 3, in blend with [OI] 6364. When the observations were interrupted, on December 3, thirty days after maximum, [FeX] was prominent, stronger than H_β.

The new series of spectra, taken from February 3, 1968, one hundred days after maximum, at about the same phase at which the observations were interrupted in 1958, show the latest stages of evolution of the nova. The early spectra strongly resemble those taken in October 1958. The HeI lines (with the sole exception of 5876) and the FeII lines have faded out and the spectrum discloses the extremely high excitation of the nebular envelopes ejected by the star. Besides the Balmer lines and HeII 4866 the most prominent features in the spectrum are: 6827, [FeX] 6374 still stronger than H_β [AX] 5535, [FeXIV] 5303, [OIII] 5006—4959, [OIII] 4363, [NiXIII] 5116, [NiXII] 4231.

By March however, with the star well at its minimum, the excitation begins to decrease. The coronal lines are rapidly weakening: [FeX] 6374 is still conspicuous, in blend with 6364, but [AX] 5535 and [FeXIV] 5303 have faded. It is the moment of the nebulium flaring. [OIII] 5006—4959 with 4363 become outstanding. As in 1958 they are rather fuzzy, as well as NII 5755, now stronger than 5876 HeI. Also [FeVII] 5160, which was barely visible in February has become conspicuous. The last spectrum represented in Fig. 2 was taken on May 7 when the nova, as announced by Locker and confirmed by Mrs. Mayall, was fainter than its normal brightness, about 13.3. The coronal lines, FeX included, have completely disappeared, while [OIII] 5006 is still conspicuous, stronger than H_γ, and so are HeII 4686, HeI 5876, [NII] 5755 and, in the ultraviolet, the two [NeIII] lines 3967, 3689. They are still present, although very weak, on the last spectra of the nova, taken in August.

A more detailed description of the spectra obtained at Asiago from 1959 to 1968 and the complete discussion of the material will be given in a forthcoming paper.

DISCUSSION

Bakos: Is the spectroscopic orbit of the system known?

Rosino: No, the N type secondary was never distinctly observed before 1962.

Herbig: Why should only RS Oph among all recurrent novae have these sharp stationary emission lines near maximum?

Rosino: The presence of a stationary evelope has been supposed by Wallerstein and by the writer. However, it is not clear why other recurrent novae do not display this peculiarity. The astonishing fact in the recurrent novae is the presence, in advanced phase of evolution past maximum, of coronal lines, which may eventually derive by collision with a stationary envelope around these novae.

Mrs. Mayall: Do you have any information on the magnitude of RS Oph at the time of your last spectrum (August 3)?

Rosino: No, we did not make photometric observations. The star was very faint, probably 12—13.

Mrs. Mayall: The last observations I received before I left Cambridge were made in July, and it was still below 12.5, — unusually faint.

SPECTROSCOPIC OBSERVATIONS OF THE RECURRENT NOVA T PYXIDIS DURING THE 1967 MAXIMUM

G. CHINCARINI, L. ROSINO

Astrophysical Observatory Asiago, University of Padova

ABSTRACT

Spectroscopic observations of the recurrent nova T Pyx have been made at Asiago in the first months of 1967, during the slow decline of the star from maximum. The early spectra show wide emission lines of H, HeI, CII, NII, NIII, OIII, FeII, etc. with violet-shifted absorption components. The mean expansion velocity derived from the dark lines is about 1800 km/s. Spectra taken in March and April indicate an increasing degree of ionization, as shown by the strengthening of the emission bands of HeII, NII, NIII, OIII. The absorption lines weaken or disappear. Although the forbidden lines of [OIII] and [FeX] 6374 are already present, the star has not yet reached its highest degree of ionization, as observed by Joy in 1944, when the observations were interrupted.

INTRODUCTION

A fifth outburst of the recurrent nova T Pyxidis ($\alpha = 9^h2^m36^s$; $\delta = -32°9'.5$; 1950.0) was announced on Dec. 7, 1966 by A. Jones, New Zealand, who observed that the star had brightened to magnitude 12.9. The variable reached magnitude 9.0 on Dec. 10 and then slowly increased in luminosity reaching the maximum on January 9. The light-curve (Fig. 1) derived by M. Mayall (1967) plotting AAVSO observations, is remarkably similar in shape

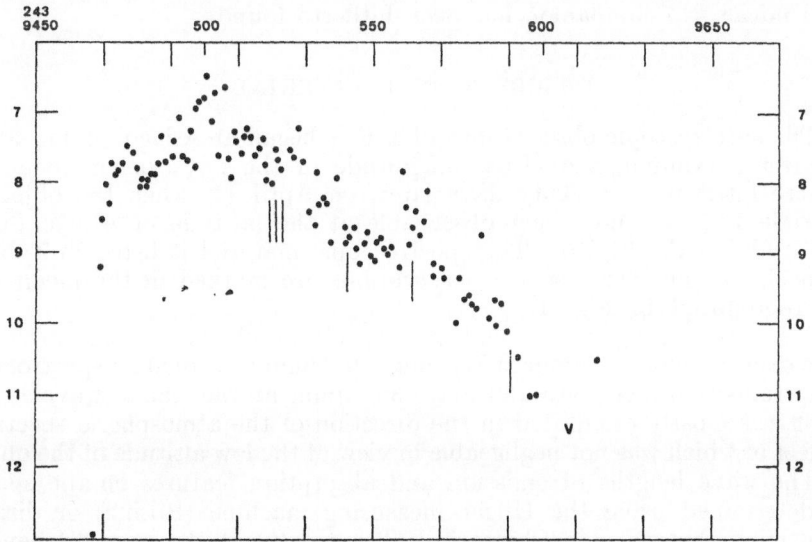

Fig. 1. Light curve of T Pyx in 1966—67 (Mavall, JRAS Canada *61*, 350). Vertical lines sign the epochs of the spectroscopic observations.

to the light-curves obtained combining the observations made during the four previous maxima in 1890, 1902, 1920 and 1944. The similarity of the light curves in successive outbursts was also noticed for the recurrent novae RS Oph, T CrB and WZ Sge. At maximum, in 1967, T Pyx reached visual magnitude 6.9. At minimum the star has magnitude 14.0, so the amplitude is about 7^m.

While other recurrent novae (T CrB, RS Oph) have an extremely fast variation during the outburst, T Pyx can be classified as a relatively slow nova, with a flat maximum and slow decline of brightness. In 1966—67 it took the nova about 30^d from the epoch of maximum to decline by one magnitude, 50^d by two magnitudes and 80^d by three magnitudes. The corresponding figures for RS Oph are: 4^d for a decline of two magnitudes; 9^d for three magnitudes. RS Oph is therefore about ten times faster than T Pyxidis.

Although the relatively slow decline of T Pyxidis offers the possibility of securing good spectra for months after its outburst, the information on the spectral evolution of the nova is rather fragmentary. A few spectra of the nova were obtained by Adams and Joy (1920) during the maximum of 1920. They show an emission spectrum with wide lines of H, OII, NII, FeII, etc. accompanied on the violet side by absorption components indicating a radial velocity of —1700 km/s. Joy in 1945 observed the nova about four and a half months after maximum and found a high excitation emission spectrum with bright lines of H, HeI and HeII, forbidden lines of NII, OI, OII, OIII, NeIII, NIV, SII, FeV, FeVI and coronal lines of FeX and FeXIV. The expansion velocity, estimated by the halfwidth of the emission bands was near 1700 km/s as in 1920. At minimum, the spectrum was studied by Humason (1938), Babcock and Elvey (1943) who found, over a continuum of variable intensity, emission lines of moderate strength due to H, HeII 4686 and OIII, as in other normal novae. No companion has been hitherto found.

OBSERVATIONAL MATERIAL

The spectroscopic observations of T Pyx began at Asiago on Jan 31, 22 days after maximum, the visual magnitude of the nova being about 7.5, and were interrupted seventy days later, on April 11, when the object at magnitude 10.2 was no longer observable at the latitude of Asiago for its low altitude on the horizon. The spectroscopic material is listed in Table I. The epochs of the spectroscopic observations are marked in the mean light curve reproduced in Fig. 1.

Two spectrographs were emloyed: spectrograph AI (cassegrain) which gives a dispersion of 75 A/mm at Hγ and 330 A/mm at λ 7000 A; spectrograph BI (newtonian) which gives 130 and 630 A/mm at the same wavelengths. The slit was mostly orientated in the direction of the atmospheric spectrum, the effect of which was not negligeable in view of the low altitude of the object.

The wave lengths of emission and absorption features on the spectra were determined using the Hilger measuring machine (Rosino) or directly on the microphotograms (Chincarini). The results of the two independent determinations were always found in good agreement.

The final results of measurements and identifications are reported in Table II, which gives, in the successive columns: I, the mean observed wave length; II, the weight, determined by the number of the spectra used in the

Table I

Observational material

Plate No.	Date		UT		Phase	Exp.	Emul.	Camera
7357	1967 Jan.	31	1^h 45^m		22^d	90^m	103a-F	AI
7361	1967 Feb.	2	22	35	24	30	103a-F	AI
7362	1967 Feb.	2	23	40	24	80	IIa-O bkd	AI
7363	1967 Feb.	3	0	40	24	20	IN sens.	AI
7364	1967 Feb.	3	1	30	24	60	IN sens.	AI
1198	1967 Feb.	6	22	59	28	25	103a-F	BI
1199	1967 Feb.	6	23	48	28	40	IN sens	BI
7381	1967 Feb.	24	23	10	46	70	103a-F '	AI
1222	1967 Mar.	15	21	30	65	40	103a-F	BI
1223	1967 Mar.	15	22	00	65	15	103a-F	BI
1224	1967 Mar.	16	22	00	66	45	IN sens.	BI
1227	1967 Apr.	11	19	50	92	40	103a-O	BI

Note: Objective prism spectra (dispersion 450 and 630 A/mm at H$_\gamma$) have also been obtained with the two Schmidts of Asiago on Feb. 2, 3, 6, 8, 9 and March 15. The quality is poor because of bad seeing, due to low altitude. They have been used only for the continuum.

mean wave length (Cam. AI, weight 2; Cam. BI, weight 1); III, mean visually estimated intensity (H$_\beta$ = 30; trace = 0, on 103a—F for the blue-visual, IN for the infrared). IV, V, VI, wave length, ion and multiplet of the suggested identification; VII, difference O—C between the observed and laboratory wave length. Further comments are given in the foot notes.

DISCUSSION

The spectra of T Pyx obtained at Asiago from Jan. 31 to Feb. 6 are typical of novae in the early decline. The continuum, as shown by objective prism spectra, is relatively strong and, in spite of the intense atmospheric extinction, is well extended in the ultraviolet, like that of a B—A type star. Wide emission lines due to H (from H$_\alpha$ to H$_{11}$), HeI (4471, 5015, 5084, 5876, 6678), HeII (4686), CII, NII, NIII, OI, SiII, FeII appear over the continuum and are accompanied on the blue side by a system of displaced absorption components. The measured half-width of the emission features is approximately 12—15 A. Some of the emissions, particularly on the spectral region between H$_\beta$ and H$_\gamma$ are blended together.

The absorption system associated with the Balmer emissions is particularly sharp and strong and can be recorded up to H$_{12}$. Lines of others atoms have absorption components much weaker and rather diffuse.

Measurements of the central wave lengths of the hydrogen absorption components yield expansion velocities of 1535 $\pm$ 85 km/s on Jan. 31, 1760 $\pm$ $\pm$ 30 km/s on Feb. 2 and 1820 $\pm$ 30 km/s on Feb. 6. The other absorption lines give values slightly higher. The mean radial velocity over the period Jan. 31—Feb. 6 determined from all the measurable absorptions lines, is: —1810 $\pm$ 40 km/s in good agreement with the value given by Adams and Joy in 1920, near the maximum (—1700 km/s).

18*

Table II

Identification of emission lines

λ obs	Weight	Int.	Ident.	Atom	Mult.	Notes
3771	2	2	3770.6	H_{11}	2	
3798	2	2	3797.9	H_{10}	2	
3887	3	2—3	3889.1	H_8	2	
3969	4	4	3970.1	H_ε	1	
3995	1	3	3995.0	NiI	13	
4057	1	3—4	4068.6	[SII]	1F	1
4102	9	8—11	4097.3	NIII	1	2
			4101.7	H_δ	1	
			4103.4	NiII	1	
4191	1	1—2	4195.7	NIII	6	1
			4200.4			
4240	1	0—1	4237.0	NII	47, 48	
			4241.8			
4271	1	0	4267.0	CII	6	3
			4267.3			
4318—25	1	2—3	4317.19	OII	2	4
			4325.8	CII	28	
4341	11	14	4340.5	H_γ	1	5
4369	3	3—20	4363.2	[OIII]	2F	
4415	3	3	4411.4	CII	39	
			4415.17	OII	5	
4446	1	4	4447.0	NII	15	6
4470	3	4—5	4471.5	HeI	14	
4514—24	4	2—20	4520	NIII	3	7
4537—51	2	2	4541.6	HeII	2	8
			4535.47	NIII	3	
			4549.56	FeII	38.37	
4595	7	2—3	4583.9	FeII	38	8
			4591.0	OII	15	
			4596.2			
4647	10	3—45	4634.41	NIII	2	9
			4647.51	CIII	1	
4686	5	3—10	4685.7	HeII	1	10
4702	7	5	4699	OII	40	
			4703	HeI	12	
			4713			
4798	5	2—3	4788	NII	20	11
			4803			
4861	11	30	4861.3	H	1	
4880	4	1—2	4881.84	NIII	9	
			4891	OII	28	
4925	8	1—2	4921.9	HeI	48	
			4924.6	OII	28	
4959	1	0—1	4958.9	[OIII]	1F	12
5007	8	2—25	5006.8	[OIII]	1F	13
5015	6	3—4	5015.7	HeI	4	
5048	4	1—2	5045.1	NII	4	
			5047.7	HeI	47	
5175	8	3—4	5171—79	NII	70.66	14
5312	2	1—2	5314.5	NIII	15	
5584	2	2	5577.4	[OI]	3F	15
5682	8	5—17	5679.6	NII	3	16
			5677.0	[FeVI]?	1F	
5756	4	4—25	5754.6	[NII]	3F	17

Table II (Cont.)

λ obs	Weight	Int.	Ident.	Atom	Mult.	Notes
5816	2	2—3	{ 5801.5 { 5812.1	CIV	1	18
5878	9	5—9	5875.6	HeI	11	19
5942	6	4—5	5941.7	NII	28	20
6002	1	2—3	5991.4	FeII	46	21
6022	1	2	6021.2	FeII?	24	21
6083	1	1—2	{ 6074.1 { 6084.1	HeII FeII	8 46	
6160	3	3—4	6156—58 6167.8	OI NII	10 36.60	
6207	1	3	—	—	—	22
6310	1	1—2	6300 6287—6312	[OI] SII	1F 26	23
6349	2	3—4	{ 6340.7 { 6347.1 { 6357.0	NII SiII NII	46 2 46	21
6375	4	3—4	6372.9	[FeX]	1F	24
6481	6	3—10	{ 6467 { 6482	NIII NII	14 8	25
6563	6	50	6562.8	Hα	1	
6675	4	—	6678.1	HeI	46	
6720	1	—	6716—30	[SII]	2F	
7077	1	—	7065.2	HeI	10	
7223	1	2	7231—36	CII	3	
7324	2	2—3	7319—30	[OII]	2F	26
7494	2	10	7476—80	OI	55	27
7772	4	20	7772—75	OI	1	
8034	1	2	—	—	—	28
8239	4	20	{ 8216 { 8232—35 { 8237	NI OI HeII	2 34 6	29
8445	4	30?	8446	OI	4	30
8657	4	30?	{ 8629 { 8680	NI	8.1	31

Notes:

1. The identification is doubtful.
2. Broad diffuse band.
3. Probably in blend with multiplets 67, 68 of OII
4. The identification is somewhat doubtful. Many other components of OII and CII are probably in the blend.
5. These two lines are partially in blend. In the first spectra H_γ is prominent, but in April λ 4363 is definitely stronger than H_γ. The estimated intensities are very rough.
6. Partially in blend.
7. Diffuse blend of mult. 3 of NIII. Its strength increases in March and April.
8. Diffuse blend. The contribution of FeII is doubtful.
9. Wide band of complex structure partially in blend with 4686 HeII. The chief contributor is mult. 2 of NIII. The displaced wave length of the band suggests CIII as a possible contributor. The intensity of the band increases enormously in March and April.
10. The line strengthens in March and April. Partially in blend with 4647 and 4702.
11. Blend of NII mult. 20.
12. Weakly represented in March.

13. In blend with 5015 HeI and probably also with 5003—11 NII. By March its intensity
    considerably increases and in April it becomes the chief contributor of the blend.
14. Broad blend of NII lines. [FeVI] 5176 may be a possible contributor in the late stages.
15. Partly atmospheric.
16. Blend of the complete mult. 3 of NII. The emission band which represents the blend
    increases considerably in intensity by March. The presence of [Fe VI] 5677 may
    explain the strengthening of the band.
17. The auroral line [NII] 5755 is present from the beginning and gradually increases
    in strength. By March the line is nearly comparable in intensity to H_β.
18. The representation with CIV is somewhat doubtful. No better identifications have
    been found.
19. Gradually increasing in strength.
20. Blend of the mult. 28.
21. The identification is somewhat doubtful.
22. Unidentified.
23. The presence of SII is somewhat doubtful.
24. Comparisons with spectra of RS Oph show that the identification is correct. The line
    is present from February, slowly increasing in strength. Possibly in blend with
    6364 [OI].
25. In blend; however, in some spectra the maxima appear separated. The intensity
    of NIII increases in March.
26. The line emerges on the spectra taken in March. Increasing intensity.
27. The identification is somewhat doubtful.
28. The wave length is uncertain. Unidentified.
29. The main contributors of this strong band are probably NI and OI.
30. The estimate of the intensity is very uncertain.
31. Broad blend of multiplets 1, 8 of NI.

The infrared spectrum has been obtained on hypersensitized IN emulsion.
It consists of fairly strong emission lines of HeI, OI and NI, which were also
observed by Joy in 1944 and are common in normal novae.

In conclusion, the spectra obtained during the period Jan. 31—Feb. 6
indicate a state of moderate excitation in the expanding envelope around the
nova. Forbidden lines are represented by [NII] 5755 and possibly by [OIII]
5006. NIII 4640 and a faint HeII 4686, are blended in a wide band of low
intensity.

The situation radically changes on March and April 1967. In the spectra
taken on March 15—16 and April 11, the continuum has become weak, the
absorption lines have disappeared, or faded to a diffuse trace which cannot be
measured. A general increase in the degree of ionization is indicated by the
appearance of new lines and by the strengthening of permitted and forbidden
lines of ions with a higher ionization potential. The star has entered the nebular
stage. The NII and NIII permitted emission lines have had a flash-like increase
of intensity. On April 11, the NIII blend at 4640 represents the strongest
emission feature in the blue-violet spectrum of the nova. [NII] 5755 is also
an outstanding feature, while [OIII] is represented by λ 5006 rapidly growing
in strength (with 4959) and by the auroral line 4363, which in April appears
much stronger than $H\gamma$, although in blend with it. HeII 4686, partially in
blend with 4640, has also strengthened as have the other HeII lines. The
infrared spectrum, on the contrary, has not substantially changed, except
for the appearance of OII 7324 in emission.

An interesting point is the presence of coronal line 6374 of [FeX] which
at first appeared somewhat doubtful. However, its identification with a sharp
emission line at about 6374, well visible in the spectra of March and also,

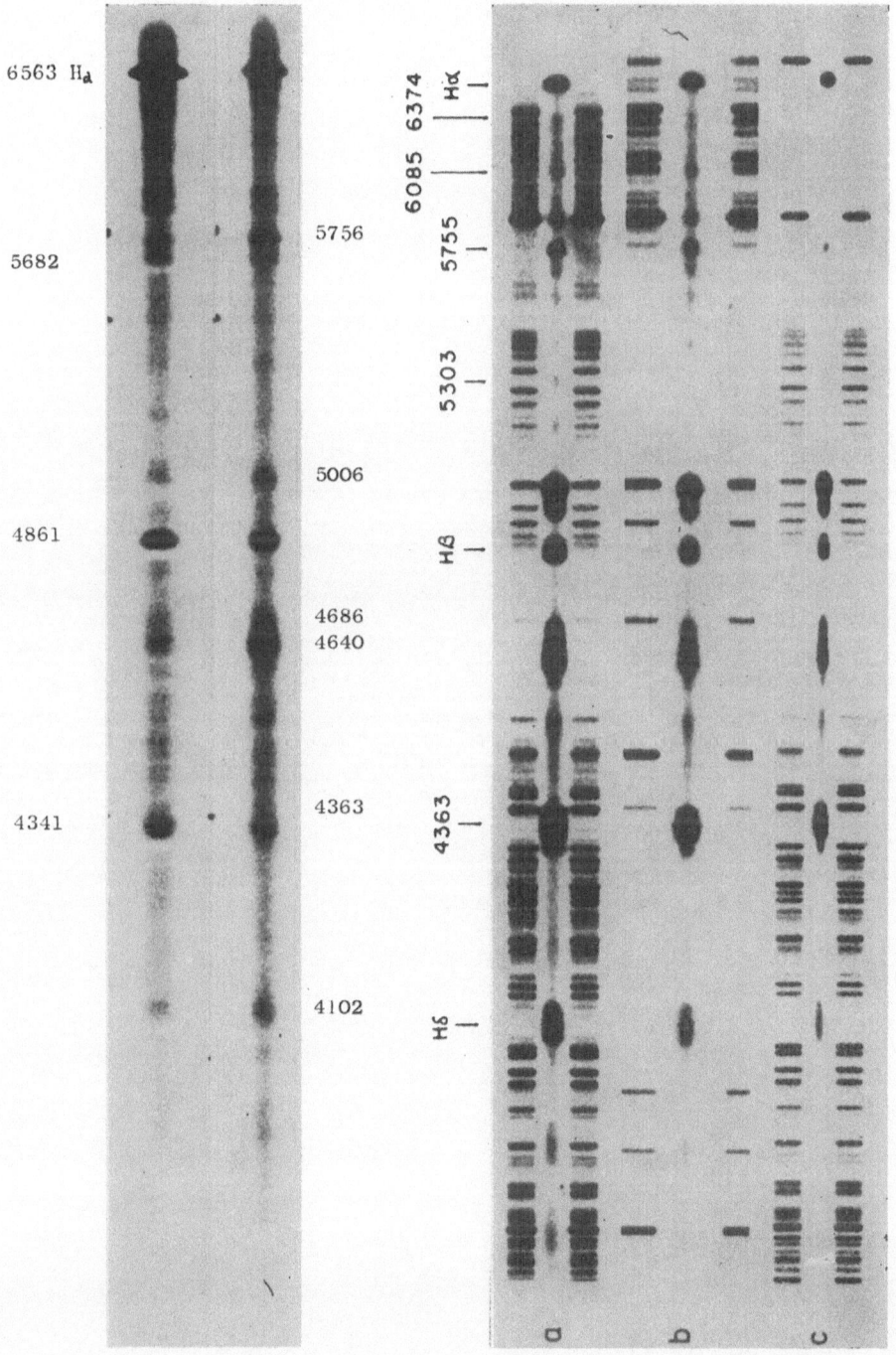

Fig. 2. On the left: T Pyx: 1967 Feb. 6; 1967 March 15 (122 cm telescope, Cam. BI). On the right: T Pyx: 1945 Apr. 4; Apr. 12; Apr. 17 (from Joy, PASP 57, Plate IX).

although fainter, in the spectra of February, seems to be well authenticated by the measures and by comparisons with spectra of RS Oph taken with the same instrument. It should be remembered that in RS Oph [FeX] 6374 made an early appearance and became later one of the most prominent lines of its spectrum.

Table III gives the variations of the relative intensities for some of the most representative lines, from Jan. 31 to Apr. 11. A comparison of the spectra of February and March is shown in Figs. 2 and 3 which reproduce some spectra and microphotometer tracings.

Table III

Variations in the relative intensity of emission lines in T Pyxidis (1967)

	Atoms	Jan. 31— Feb. 2	Feb. 24	Mar. 15	Apr. 11
4102	H δ	9	9	10	11
4341—4363	H γ + [OIII]	14	13	14	25
4514—4524	NIII	—	2	7	20
4640—4686	NIII + HeII	8	16	28	45
4861	H β	30	30	30	30
5006—5015	[OIII] + HeI	8	16	17	25
5679	NII	5	12	17	—
5755	[NII]	4	7	20	—
5876	HeI	5	7	9	—

Unfortunately, the southern declination of the nova prevented us from continuing the observations just in the most interesting phase, when the nebular spectrum was attaining its maximum strength, and forbidden lines of highly ionized atoms were on the point of appearing. This phase, however, was covered by Joy (1945) in 1944. In Fig. 2 we have reproduced together with our spectra and in the same scale, three of Joy's spectra taken one hundred and thirty days past maximum, which illustrate the successive development of the spectrum of T Pyxidis. So, all of the stages in the spectral evolution of this nova, except the latest, are now covered by observations. Continuous spectroscopic control of T Pyx at minimum from a southern Observatory should be highly desirable for a better understanding of the phenomena involved in the periodic outburst of this and other recurrent novae.

In conclusion, although different from RS Oph and T CrB for the slower photometric and spectroscopic evolution and the apparent absence of a companion, T Pyx displays some of the same phenomena which have been observed in normal and recurrent novae, such as the development near maximum of absorption systems of large radial velocities and, in a more advanced phase of evolution, of broad emission lines of high excitation. The most astonishing fact is the appearance of coronal lines (FeX, FeXIV) during the decline of brightness, which seems to be a common characteristic of all recurrent novae and was observed in T Pyx as well as in T CrB and RS Oph. The source of such a high degree of ionization is still unknown, and can be tentatively attributed to collision with material surrounding the star, or to photoionization, or to ejection of material from the deep interior of the star at the moment of the outburst.

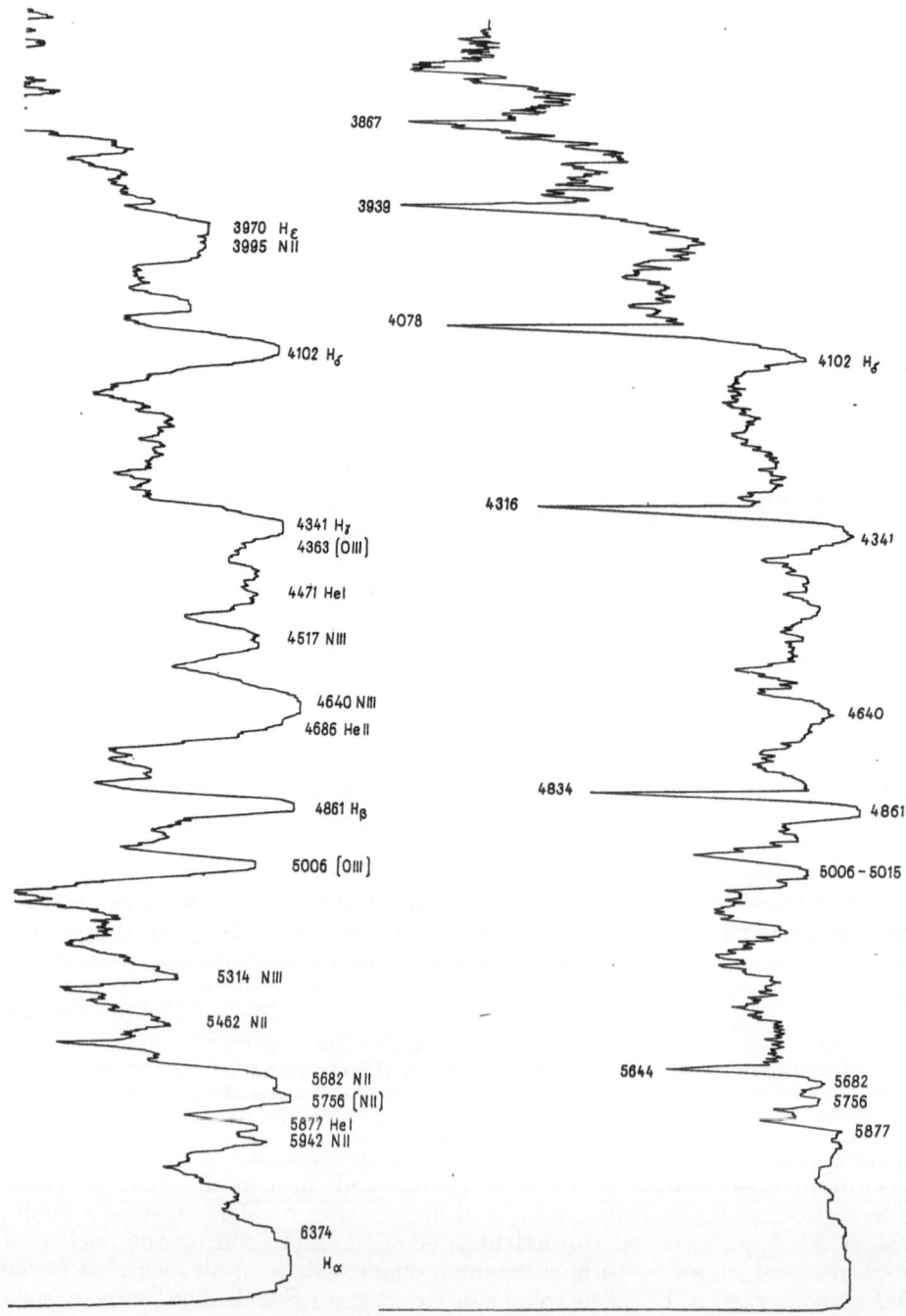

Fig. 3. Comparison of microphotometer tracings of spectra No. 1198 and 1222.

REFERENCES

Adams, W. S. and Joy, A. H., 1920, Pop. Astr. **28,** 514.
Babcock, H. W., Elvey, C. T., 1943, Astrophys. J. **97,** 412.
Humason, M. L., 1938, Astrophys. J. **88,** 228.
Jones, A., 1966, IAU Circ. 1983.
Joy, A. H., 1945, Publ. astr. Soc. Pacific **57,** 171.
Mayall, M. W., 1967, J. R. astr. Soc. Can. **61,** 350.

COMMENT

Feast: A series of spectra of T Pyx was obtained at the Radcliffe Observatory and these have been discussed by Mr. R. Catchpole whose paper will appear shortly. Evidence for coronal lines was obtained although they never became very strong.

PRELIMINARY REPORT ON THE SPECTRUM OF NOVA VUL 1968

A. MAMMANO, R. MARGONI and L. ROSINO

Astrophysical Observatory of Asiago

ABSTRACT

Spectroscopic observations of Nova Vul 1968 made at Asiago from April to August are reported in this paper. The nova belongs to the fast type. Absorption systems with velocities of -680, -800, -1380 and -2500 km/s have been observed. Some peculiarities of the emission components of CaII $\lambda\lambda$ 8498, 8542 and 8662 (mult. 2) are pointed out. The evolution of the nova from the premaximum to the nebular stage is shortly described.

1. Nova Vulpeculae 1968 has been discovered by Alcock on April 15 1968 during its rise to the maximum, reached on April 17 (Candy, 1968). A comparison of blue photographs obtained at Asiago with the Palomar Sky Atlas indicates that the prenova was fainter than 16^m, in agreement with Herbig's (1968) estimate of 16^m5. Herbig pointed out that the prenova is one component of a double star and that it is *not* the remnant of Nova Vul 1670.

The spectra discussed here have been obtained with the Cassegrain spectrograph attached to the 122 cm reflector of Asiago. The following combinations have been mostly used: Camera III, dispersion 40 A/mm at H_γ; Cameras VII and VI, associated with the Carnegie Intensifier, dispersion 45 and 60 A/mm at H_γ. Kodak blue, red and infrared sensitive material was employed.

The photoelectric light curve of the nova, kindly supplied by Prof. P. Tempesti, Teramo Observatory (1968) is reproduced in Fig. 1; short vertical lines indicate the epochs at which the most significant spectra have been taken.

2. At the moment of its maximum brightness, on April 17, the nova is still characterized by a premaximum spectrum, with many absorption lines, shortwards displaced of about -670 km/s. The lines are rather broad and diffuse and the spectrum can be classified as a peculiar F5—F8. Noteworthy is the presence of the infrared triplet 8498, 8542, 8662 of CaII (mult.2) relatively weak, while H and K are very strong and wide. The two absorption bands of OI at 7774 and 8446 have about the same intensity. Emission features are represented by H_α with moderate intensity and by a weak component of CaII 8542 and OI 7774 (fig. 2).

On April 18, one day after maximum, the nova develops the principal spectrum (-1380 km/s), together with emission components of H, FeII, OI (7774, 8446), etc. The number and strength of the emission bands increase on April 19, when emission components of CaII 8498 and 8662 also appear (Fig. 2). The infrared triplet of CaII is therefore present with emission and absorption components; the emission became later stronger than the absorptions. On the following day wide emission bands appear also for H and K, the absorptions remaining, however, much stronger than the emissions.

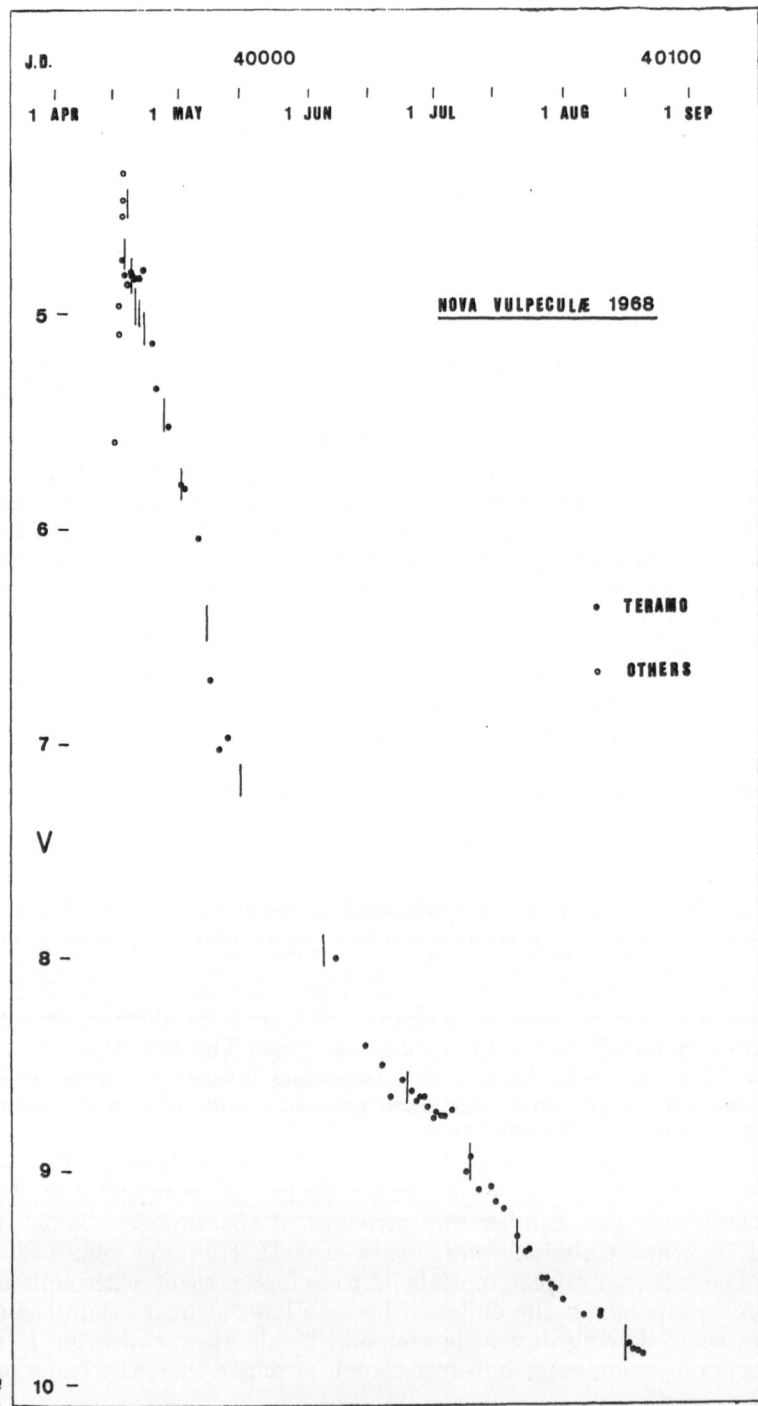

Fig. 1. Light curve of N Vul 1968 obtained at Teramo Observatory photoelectrically (courtesy of Prof. P. Tempesti). Short vertical lines indicate the epochs of the most important spectroscopic observations at Asiago.

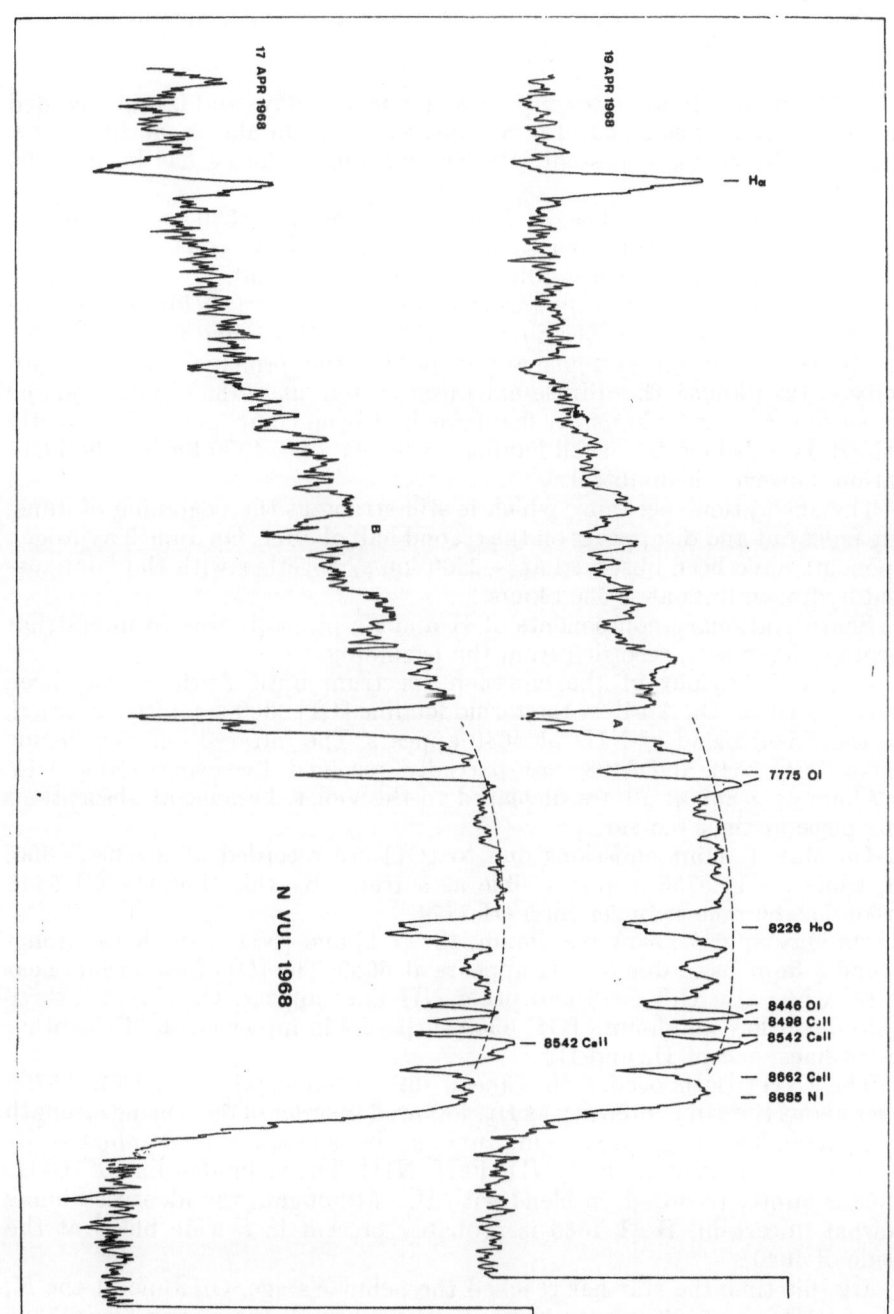

Fig. 2. Microphotometric tracings of two near-infrared spectra of N Vul 1968, obtained with camera III. Original dispersion 400 A/mm at λ 8500 A.

From now on, the evolution of the absorption and emission lines will be separately described.

3. The premaximum system persists for several days and is still recorded when the nova has weakened of $0^{m}8$ (Apr. 24—25). On May 1, at $1^{m}3$ below maximum, it is no more present. Its last recorded velocity has been —800 km/s.

The principal system has, at first, a velocity of —1380 km/s. Later it splits into two components, whose velocities arc —1000 and —2000 km/s, the second being the strongest (diffuse enhanced system).

The Orion system, faintly recorded on May 6 ($2^{m}4$ below maximum) strengthens on May 15, when the nova has fallen by three magnitudes. At this phase three absorption systems are recorded: the principal system, with velocity —1270 km/s; the diffuse enhanced system at —1850 km/s and the Orion system at —2330 km/s. A few faint hydrogen components, if correctly identified, may belong to a shell having a velocity of —4000 km/s. The identification, however, is doubtful.

The absorption spectrum, which is still strong at the beginning of June, slowly fades out and disappears on the second half of June. On June 3 hydrogen components have been observed at —2500 km/s, together with the faint suspected hydrogen lines at —4300 km/s.

Sharp stationary components of H and K, probably due to interstellar absorption, have been recorded from the beginning.

4. The behaviour of the emission spectrum until April 20 has been already described. On April 24 the forbidden line [OI] 6300 is faintly recorded, while the broad blend of NIII at 4640 appears. The infrared emission bands OI 8446, CaII 8542 and 8498 are partially resolved by using Camera IV (140 A/mm at λ 8600): all are displaced to the violet, because of absorptions taking place at their red side.

On May 1, faint emissions due to [OI] are recorded at λ 6364, 6300, 5578, while [NII] 5755 is just visible as a trace. By this time the OI 8446 emission has become stronger than OI 7774.

On May 8, $2^{m}4$ below maximum, the [OI] and [NII] 5755 lines strengthen and a faint band due to NII appears at 5680. The [OI] flash occurs near May 14, when also HeI 5876 and other NII lines appear. On May 14, three magnitudes below maximum, [OI] 6300 surpasses in intensity all of the other emission lines, except H_{α} and H_{β}.

The [NII] flash occurs on June 3 ($3^{m}7$ from maximum). [NII] 5755 reaches about the same intensity as H_{β}, followed in order of decreasing strength by [OI] 6300, 6364 and 5578. On June 9 the spectrum is dominated by wide emission bands of H, HeI, NII, FeII, NIII. The forbidden line of [OIII] at 4363 is faintly recorded, in blend with H_{γ}. Althought the identification is somewhat uncertain, HeII 4686 is probably present in a wide blend at the red side of 4640.

By this time the star has reached the nebular stage. On June 21 the N_{1} band of [OIII] at 5006 is prominent. In the infrared, the strong blend 7325 of [OII] is recorded; OI 8446 has become about ten times as strong as OI 7774 and a faint band due to HeI emerges at 7064.

A further strengthening of the nebular emissions is noticed on the spectra taken on June 27: H_{β}, however, is still brighter than [OIII] 5007, although

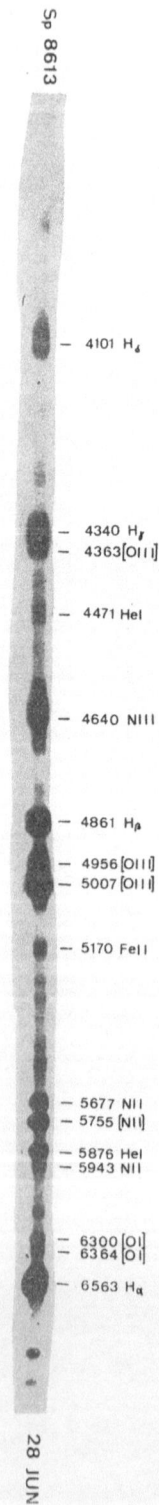

Sp 8613

4101 H$_\delta$

4340 H$_\gamma$
4363 [OIII]

4471 HeI

4640 NIII

4861 H$_\beta$

4956 [OIII]
5007 [OIII]

5170 FeII

5677 NII
5755 [NII]

5876 HeI
5943 NII

6300 [OI]
6364 [OI]
6563 H$_\alpha$

N   VUL 1968

28 JUN

Fig. 3. Spectrum of N Vul 1968 at nebular phase, obtained through the Carnegie-RCA intensifier. Dispersion 60 Å/mm at H$_\gamma$.

the band at 5007 appears stronger because of the contribution of HeI 5018. On July 6 the N_1 line becomes definitely stronger than H_β, while [OIII] 4363 has about the same intensity of H_γ. On July 21, the most conspicuous emissions are: H_α, [OIII] 5006, H_β, [OI] 6300 and 6364, [NII] 5755, [OIII] 4958, HeI 5874, [OIII] 4363 much stronger than H_γ, NIII 4640, NII 5680. Finally, in the last spectrum obtained on August 14, 1968 when the nova has fallen by $5.^m4$, 5007 [OIII] is about two times stronger than H_β. The emission bands are very wide, the expansion velocity derived by their halfwidth being of the order of 1200 km/s. It may be observed that the halfwidths of the emission bands have increased with time, passing from 700 km/s when the principal spectrum was first recorded, to 900 km/s at the appearance of the diffuse enhanced system and to 1100—1200 km/s at the nebular phase.

5. Although the nova is still far from minimum, we may attempt to sketch some of its main properties:

a) It is clear from the photometric and spectroscopic evolution that N Vul 1968 is a fast nova, without strong deviations from the normal type. From the velocity of decline (3 magnitudes in 30 days) a mean absolute magnitude at maximum of $-7.^m8$ can be derived; the amplitude is 12^m and the photographic magnitude at minimum about $+4$, as in other normal novae.

b) Infrared spectra taken near maximum have shown for the first time in a nova the absorption lines of the CaII infrared triplet. Since N Vul 1968 is a common nova it is likely that this infrared triplet may be visible also in the other normal novae, provided that infrared spectra are taken very close to maximum. Now, the fact that the infrared triplet of CaII appears in emission stronger than H and K is quite puzzling, because both multiplets have in common their upper term and the transition probabilities for H and K are much higher than for the infrared triplet lines. The same problem apply with a few long period variables and the accepted explanation is that of Herbig (1952), according to which, a self-absorption, due to a layer lying above the emitting region, reduces the emission intensities at H and K much more than at the infrared triplet of CaII. In the case of Nova Vulpeculae 1968 there is however no clear evidence of such an absorbing layer above the emitting region.

Worthy of mention is also the great variation of the intensity ratio of OI 8446 to OI 7774, which changed from 0.5 near the maximum of the nova, to about 20, during the last recorded nebular stage. It is likely that the theory of Pagel (1960) may account for the behaviour of this ratio during the first days close to maximum, while Bowen's fluorescence mechanism may become more important during the nebular stage.

A detailed study of the spectral evolution of this interesting nova will be published in a forthcoming paper.

REFERENCES

Candy, M. P., 1968, I. A. U. Circ. 2066.
Herbig, G. H., 1952, Astrophys. J. **116**, 369.
Herbig, G. H., 1968, I. A. U. Circ. 2072.
Pagel, B. E., 1960, Ann. Astrophys. **23**, 850.
Tempesti, P., 1968: Private communication.

ON NOVA DELPHINI 1967 AND SOME SLOW NOVA CHARACTERISTICS

WALTRAUT CAROLA SEITTER

Observatorium Hoher List, Daun (Eifel), Germany

A Schmidt telescope equipped with an objective prism is not the most suitable instrument for observing spectral changes of a single object; yet, lacking other spectroscopic equipment, we have used this combination to follow the development of Nova Delphini 1967 for more than a year. The major advantages and disadvantages of the instrumentation are summarized in Table I. It may be added that a low resolution applied to very complex objects occasionally facilitates the recognition of major trends which might be more difficult to extract from the wealth of detail seen at larger dispersions.

Table 2 gives the characteristics of our instrument.

155 spectrograms were obtained on Kodak I—N plates between July 16, 1967 and August 26, 1968. A complete atlas will be published elsewhere. Representative samples are shown in Fig. 1. The following history of the nova is assembled from brightness determinations published by various authors (IAU Circulars 1967—68, Sky Tel. *34*, 300, Rev. Pop. Astr. *62*, 23, 1968) and a visual inspection of our plates plus one set of radial velocity measurements. The spectral analysis is quite incomplete and subject to some changes, especially concerning the dates of first and last observations of lines, after quantitative measurements of our spectra become available. Only some of the more prominent features are described.

Two main periods of grossly different nova behaviour may be distinguished.

PERIOD I

During the first six months changes in both brightness and spectrum occurred in two successive stages, the second stage showing three subdivisions. The dates given refer to first or last appearance on the spectrograms. The actual changes may have occurred at earlier or later dates not covered by observations.

1. *July 16—August 20*

Mean brightness constant at about 6.5 magnitudes above pre-outburst brightness with oscillations of less than 0.2 magnitude.

Both emission and absorption spectra well developed.

Strongest absorption lines: Balmer lines, Na I 5890, 5896, Ca II 3934, 3968, Ti II 3759, 3761.

Strongest emission lines: H and Fe II

Table 1

Objective prism spectra with Schmidt-telescopes

Advantages	Disadvantages	Compensations
Large fields	Overlapping spectra	Different Position Angles of prism
All photographically accessible wavelengths covered simultaneously		
Optimum exposure times	Diffuse lines due to bad seeing For large instruments: Wavelength resolution limited by seeing disk in contrast to slit spectroscopes of same dispersions	
Comparison spectra for photometric calibration present		
	No Comparison spectra for radial velocity determinations	Zero point through atm. A-band scale from neighbouring stars Double exposures with direct and reversed instrument Superimposed scales
	Sky background (Moon, twilight)	

Table 2

Schmidt-telescope Hoher List

Aperture: 340 mm (Mirror: 500 mm)
Focal length: 1375 mm

Prism I

Flint glass
Angle of refraction: 7°31
Reciprocal linear dispersion in A/mm at

λ 3500	λ 3700	Hγ	Hα	λ 8000
85	120	240	850	1680

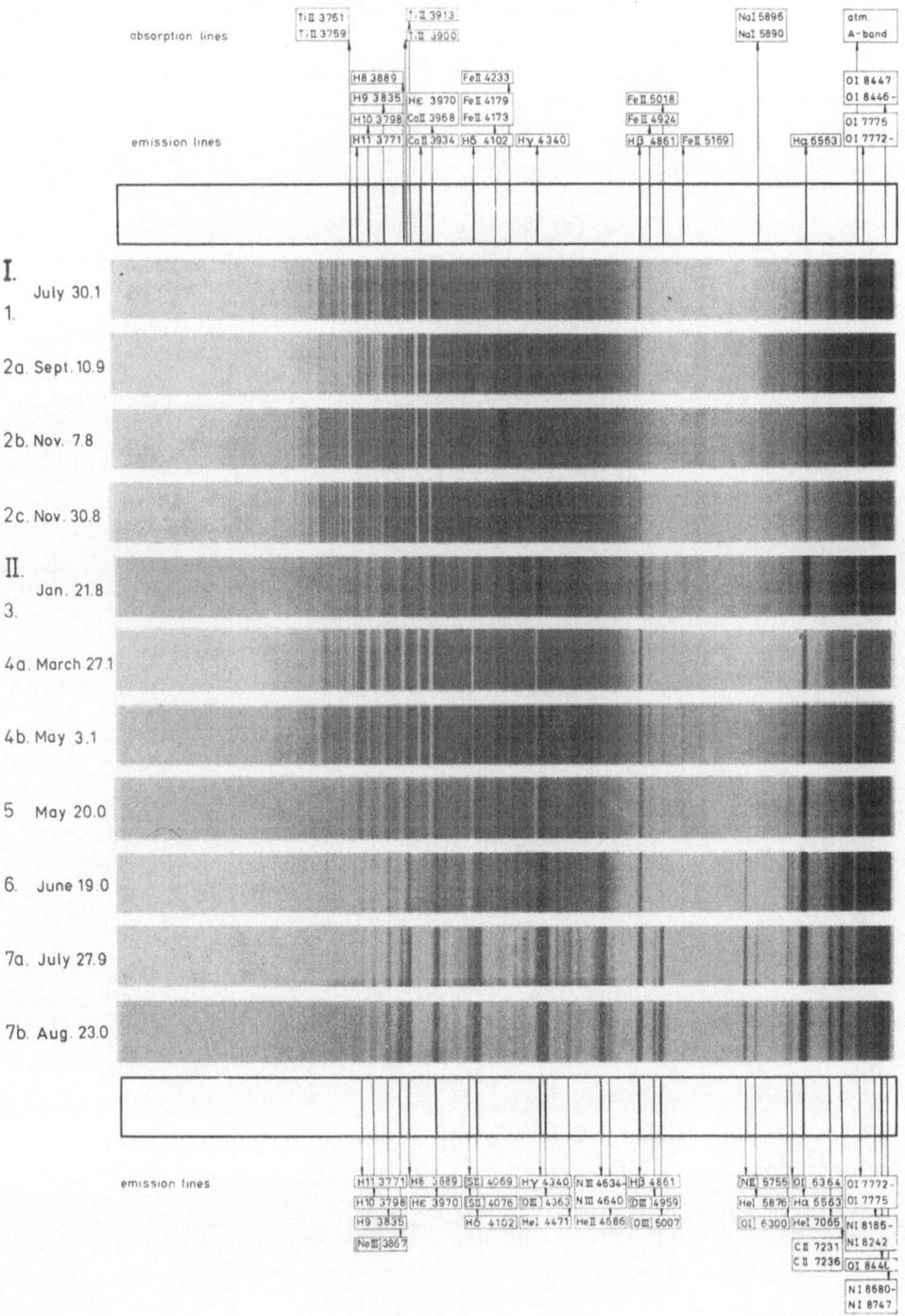

Fig. 1

Radial velocity measurements indicate —600 km/sec ($\pm$120 km/sec) for the H-lines. During the remainder of Period I no major velocity changes are suggested from the appearance of the spectra.

2a. *September 5—October 21*

September: Nova about 0.7 magnitude brighter than during stage 1 after an increase between August 18 and August 28.

October: Slow decline through 0.4 magnitude to broad minimum around October 16; small oscillations as in stage 1.

The absorption spectrum resembles that of Φ Cas F0Ia. The D-lines and numerous Ti II lines, however, are much stronger than in the supergiant and even more pronounced than in stage 1.

Emission spectrum weak.

2b. *November 8—November 12*

Nova has increased again to September brightness; small oscillations as in stage 1.

Certain absorption line intensities, e.g. between λ 3700 and λ 3760 have decreased as compared with stage 2a.

No apparent changes in the emission spectrum.

2c. *November 18—December 11*

Mean brightness slightly lower than during stage 2b and approximately 0.6 magnitude above that of stage 1. Oscillations as before. Rise towards secondary maximum starts December 4.

Both absorption and emission spectra practically identical with those of stage 2a.

PERIOD II

For Period II not enough brightness data are yet available to permit looking for a close correspondance between changes in magnitude and spectral changes.

Photometric Data

The secondary maximum reaching 2 magnitudes above the brightness of the first rise, occurred on December 13. It is followed by a symmetrical decline through about 2.3 magnitudes. During the following months the brightness oscillates about the new mean with amplitudes up to 1 magnitude. In early July the nova starts to decline steadily through about 1.5 magnitudes.

Spectroscopic Data

Five distinct stages and two subdivisions are found in the spectral development of the nova during Period II.

19*

3. *December 17—March 24*

Absorption spectrum:
Major changes following the December maximum are.
December 17: First doubling of H-lines observable at our small dispersion.
December 28: Ca II lines distinctly double.

After December 28 the more shortward displaced Balmer and Ca II components rapidly increase in strength and soon become dominant. By the middle of March the primary components seem to have disappeared and the secondary components stand separated from the unexpectedly narrow emission lines.

The behaviour of the secondary components is compatible with their being members of the principal absorption system.

The Sodium D-lines are weak.

One could suspect this to be caused by the bad definition of the plates taken at very low altitudes during the months preceeding and following the January 28 conjunction of the nova with the sun. The like behaviour of Ca II 3934, however, strengthens the assumption that the effect is real. Ca II 3934 lies in a spectral region of much better resolution and is not so much affected by bad seeing.

Ti II 3759, 3761 begin to weaken in early January and remain faint until late March.

Emission spectrum:
Permitted lines: The beginning of Period II is marked by the reappearance of a strong emission spectrum, especially of H and Fe II, similar to that seen after the July outburst. The behaviour of the hydrogen lines is unusual; they are much narrower than would be expected from the displacement of the absorption lines.

One of the most striking features is the strong O I 8446 emission first visible on December 18. The triplet had been absent or very weak before and comparable in strength to the O I triplet near λ 7774. Since the greater intensity of O I 8446 relative to the shorter wavelength triplet is attributed to a fluorescence effect due to Lyman β emission (Bowen, 1947), one must conclude that either Lyman β has considerably increased or that the underlying material has become much more transparent to UV radiation.

Forbidden lines: [O I] 6300 and possibly [O I] 6364 are first visible on our spectrograms on January 9. After some slight increase they stay more or less constant until April. 23.

[N II] 5755 is the next one of the forbidden lines to appear. It is first seen faintly on January 30.

The emission features, too, correspond to those seen in the principal spectrum of normal novae.

4a. *March 27—April 23*

Absorption spectrum:
The Balmer lines appear as in the latter part of stage 3. The D-lines and Ca II 3934 are strong again.

Ti II 3759, 3761 first show moderate strength on March 27. The lines display some variation, possibly synchronous with Ca II, approaching their former prominence on March 30 and April 17.

Since these lines are sensitive indicators of a tenuous atmosphere, conclusions relating to the density of the Ti II absorbing shell may be drawn.

A rich absorption spectrum is seen in the ultraviolet extending to the end of the recorded spectrum near λ 3400. It disappears after May 15.

Emission spectrum:

Permitted lines: The Balmer and Fe II lines appear as in stage 3. The strength of O I 8446 only slightly surpasses that of O I 7774.

Forbidden lines: As in stage 3.

4b. *May 3—May 15*

Absorption spectrum:

Na I 5890, 5896 and Ca II 3934 strong.

Ti II 3759, 3761 as prominent as during Period I.

Emission spectrum:

Permitted lines: as in stage 4a.

Forbidden lines: [O I] 6300, 6364 have disappeared; they are possibly masked by strong absorption lines.

[N II] 5755 has disappeared.

5. *May 20—June 2*

Absorption spectrum:

The Balmer lines are double again with a slightly more violet displaced component.

Sodium, Calcium and Titanium lines fade away rapidly.

Emission spectrum:

Permitted lines: O I 8446 is again sensibly stronger than the triplet near λ 7774.

Forbidden lines: [O I] 6300, 6364 are stronger than ever before [N II] 5755 reappears and becomes one of the most prominent emission lines in June and later.

6. *June 7—July 18*

Absorption spectrum:

The Balmer lines have disappeared; most of the other absorption lines weaken giving place to broad absorption bands.

Emission spectrum:

The N III 4640 emission first stands out clearly on June 7. At the same time a faint increase of Fe II 5018 signals the appearance of [O III] 5007.

The above and several other emission lines indicate that the nova has entered the 4640 stage.

7a. *July 20—July 28*

Absorption spectrum:

Broad bands.

Emission spectrum:

Permitted lines: H lines become much broader. Neutral and ionized Helium lines appear on July 18 and fully develop during stage 7a.

Forbidden lines: On July 20 [O III] 5007 suddenly begins to increase. [O III] 4363 is first indicated on July 20 and definitely present after July 25, when [O III] 4959 is also seen.

On July 28 the nebular spectrum has emerged including strong lines of [Ne III].

7b. *July 28 and later*

The nova shows a fully developed nebular spectrum.

CONCLUSIONS

From the above description it seems that Nova Delphini has some unusual features. For two reasons, however, I would refrain from calling it abnormal, as has been done by some authors.

1. Nova Delphini is a very slow nova and with less than ten adequately observed objects we have rather incomplete statistics concerning the spectral development of this class.

2. Among the very slow novae are objects such as η Carinae and FU Orionis which apparently adhere even less to the established rules of nova behaviour.

In spite of these limitations, however, it seems possible to extract from the meager data available some general pattern of slow nova development:

A comparison of the light curves of eight slow novae shows that their common characteristics is a flat topped maximum which is of longer duration the slower the nova.

For some of the novae secondary maxima are well established: η Car, RR Pic, V 849 Oph and N Del (Payne—Gaposchkin, 1957). (see Table 3).

Fig. 2 shows the first section of a generalized slow nova light curve.

Some of the novae have shown only or predominantly absorption spectra during the protracted maximum: FU Ori, DO Aql and N Del (Payne-Gaposchkin 1957). Because of this evidence and because the spectral development of Nova Delphini indicates the presence of the principal spectrum and

Table 3

Star	Pre-maximum			Maximum	
	time of rising	amplitude	duration	time of rising	amplitude
η Car 1848	> 8 a	8^m	150 a	10 a	3^m
FU Ori (1937)	120 d	6^{m}5	>30 a	—	—
RT Ser (1909)	—	>6^m	14.5 a	—	0^{m}5
RR Tel (1945)[x]	150 d	7^m	4.1 a	(< 260 d)	(1^m)
DO Aql (1925)[xx]	>30 d	>3^{m}5	200 d	40 d	0^{m}4
Nova Del 1967	25 d	7^m	150 d	12 d	2^m
RR Pic (1925)	< 45 d	4^m	50 d	5 d	2^m
V 849 Oph (1919)	—	—	>4 d	4 d	1^{m}2

[x] More likely explanation through superposition of periodic light change

[xx] Vorontsov—Veljaminov suggests earlier and brighter maximum from comparison with RR Pic

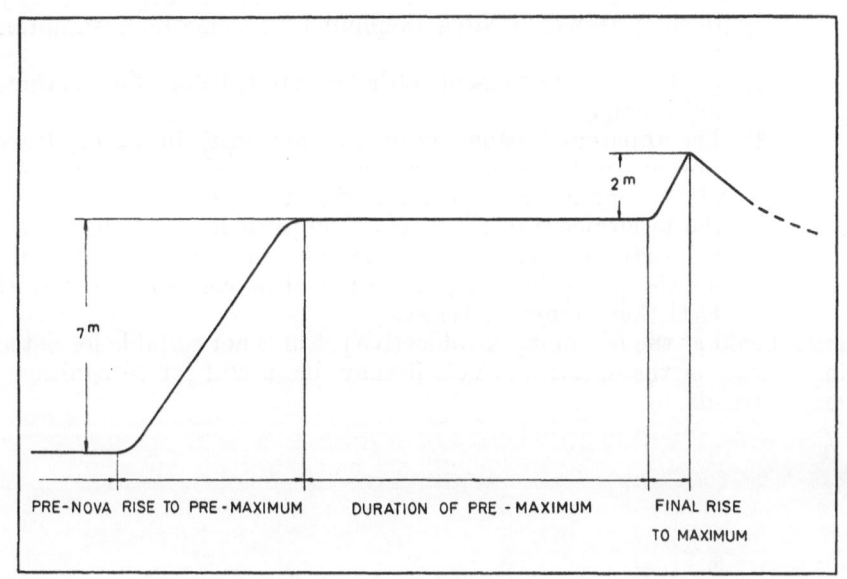

BRIGHTNESS

7 ᵐ

2 ᵐ

PRE-NOVA RISE TO PRE-MAXIMUM   DURATION OF PRE-MAXIMUM   FINAL RISE
                                                                    TO MAXIMUM

TIME

Fig. 2

the sequence of other typical postmaximum spectra only after the secondary outburst, the following hypothesis is adopted:

In slow novae the pre-maximum halt which is normally of short duration or absent (McLaughlin, 1960) appears to be protracted as a direct function of the slowness of the nova.

REFERENCES

IAU Circulars 1967—68, Sky and Telescope Vol. **34**, 300 (1967), Review of Popular Astronomy Vol. **62**, 23 (1968).
I. S. Bowen, 1947, "Excitation by line coincidence", Publ. astr. Soc. Pacific **59**, 196.
McLaughlin, D., 1960, "The Spectra of Novae", Stars and Stellar Systems Vol. VI, 585.
Payne—Gaposchkin, C., 1957, "The Galactic Novae" Amsterdam 1957.

DISCUSSION

Hutchings: An interpretation of the spectrum based on line profiles from a series of plates at 15 A/mm at Victoria indicates that the line spectrum during 1967 is formed in an extended envelope expanding together with the underlying photosphere. After the December outburst and until May the profiles correspond to shells ejected from a large and fairly stationary photosphere.

Mammano: 1) Recognition of the conventional absorption systems is rather difficult even at 40 A/mm, because many strong or faint systems appeared in Nova Delphini 1967 sometimes simultaneously.

2) [1] 6300 A was present, although faint, before the maximum in December.

3) The apparent broadening of emission lines, during the transition at your dispersion, is actually caused by the emergence of two more emission bands due to some condensations in the atmosphere of the nova. Each forbidden line has central band stronger than the two apart, while the contrary is true for the permitted lines. Details will be communicated at the IAU Colloquium in Triest.

Seitter: As I said at the beginning, an objective prism is not suitable for detecting details in the spectrum. Yet, it may be useful for recognition of major trends.

PART III

IRREGULAR ACTIVITY IN PERIODIC VARIABLES

A. β CMa, δ Sct VARIABLES

B. RR Lyr, δ Cep VARIABLES

C. LATE-TYPE GIANTS

TIDAL RESONANCES IN SOME PULSATION MODES OF THE BETA CANIS MAJORIS STAR 16 LACERTAE

W. S. FITCH

Steward Observatory, Tucson, Arizona, USA

Analyses by Fitch (1967) of the published observational data on the δ Scuti star CC Andromedae and the β Canis Majoris star σ Scorpii led to the suggestion that the often observed presence of multiple periodicities in these kinds of stars should be attributed to the influence on the normal pulsation mode of structural changes produced in the outer layers of the pulsating primary by tidal perturbations due to a faint companion. This suggestion is further supported by Preston's (1966) announcement at the 3rd Variable Star Colloquium that the δ Scuti star δ Delphini is a double-line spectroscopic binary.

In a continuation of work on this problem the β Canis Majoris star 16 Lacertae, whose intrinsic variability was discovered by Walker (1951), was selected because very extensive observational material had been published. This star was announced by Struve and Bobrovnikoff (1925) as a single-line spectroscopic binary with a period of 12.3106 days, but the observations here analysed are the radial velocity measures obtained on 26 nights in 1951 by Struve, McNamara, Kraft, Kung, and Williams (1952), together with the data contained in the radial velocity and blue light curves published by Walker (1951, 1952, 1954). Because the original lists of the photoelectric measures have been lost (Walker 1967), it was necessary to read the observations from photographic enlargements of the published light curves, and it is expected that these measures, which are consequently less accurate than the originals, will be made available to interested workers by inclusion in the IAU (27). RAS. file of the Royal Astronomical Society Library. All of Walker's blue magnitude measures (7 nights in 1950, 13 nights in 1951, and 10 nights in 1952) were employed, except that the two nights of 16 and 17 August 1951 were never included because of the very low signal-to-noise ratio evident in the light curves for these nights. A single velocity curve published by Walker (1954) was read in the same fashion and included in the velocity analysis, but a second velocity curve which Walker (1951) published was inadvertently skipped.

A preliminary periodogram analysis (Wehlau and Leung 1964; Fitch 1967) confirmed the primary period reported by previous workers, and a frequency of 5.9113 cycles/day (c/d) with a corresponding period of 0.169168 day was adopted as best fitting these observation. With this frequency the light and/or velocity measures on each night were individually fitted by least squares to obtain the amplitude, phase, and mean value best representing a sinusoidal variation on each night. The nightly velocity means were then fitted by least squares in our spectroscopic binary program to yield these orbital elements with their probable errors:

$$P \ = \ 12.079 \pm 0.005 \text{ days} \qquad e \qquad = \ 0.012 \pm 0.02$$
$$T_0 \ = \ \text{JD } 2433872 \pm 2.6 \text{ days} \qquad \omega \qquad = \ 121° \pm 79°$$
$$\gamma \ = \ -13.0 \pm 0.6 \text{ km/sec} \qquad f(M) \ = \ 0.0147 \text{ M}_\odot$$
$$K_1 \ = \ 22.7 \pm 0.4 \text{ km/sec} \qquad a_1 \sin i = 3.77 \times 10^6 \text{ km}$$

The corresponding velocity curve is shown in Figure 1.

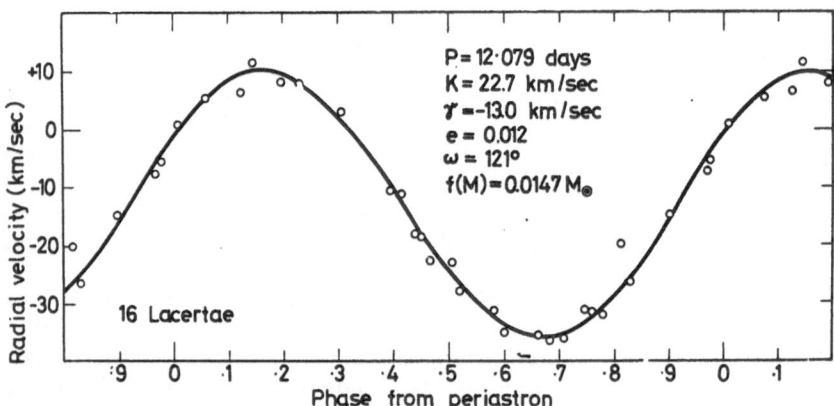

Fig. 1. The orbital radial velocity variation of the primary component of 16 Lacertae. The plotted points are mean velocity measures for individual nights and the full curve is the computed solution.

The nightly luminosity means vary strongly in an apparently erratic fashion which we were unable to correlate with the orbital motion, and since the comparison star 14 Lacertae has been reported to be variable (Walker 1952, 1953), it was necessary to attribute these variations in the mean light level to 14 Lacertae and to compensate for them by applying night corrections deduced from the mean values of the luminosity ratio. This arbitrary correction procedure automatically precludes the possibility of finding any light variation which depends solely on the orbital motion.

When the velocity and light measures had been freed of the principal slow variations by means of the known orbital motion (here represented as a sinusoidal velocity variation with frequency $n_0 = 0.08279$ c/d) and the night corrections, respectively, further periodogram analysis disclosed the existence of two more periodic variations (frequencies $n_2 = 5.8529$ and $n_3 = 5.4998$ c/d, in addition to the primary frequency $n_1 = 5.9113$ c/d) in both the light and velocity observations. The three photometric observing seasons were from J.D. 2433504 to 2433536, 2433870 to 2433926, and 2434231 to 2434242, while we have divided the principal velocity observing season in two halves running from 2433869 to 2433926 and 2433937 to 2433967. The principal data relating to these variations is summarized in Tables 1 and 2, in which it is seen that the velocity-to-light amplitude ratio is much larger for n_1 than for n_2 and n_3 (about 6.0, 3.0, and 1.5 km/sec percent mean light, respectively), and also that the amplitude of n_3 varies strongly and apparently

Table 1
Blue Light Variation of 16 Lacertae

| Year | Frequency | Blue Light Range (%) | | | Phase Zero Point (Periods) | | | M. E. of 1 obs. | No. of Obs. |
		5.9113	5.8529	5.4998	5.9113	5.8529	5.4998		
1950		4.8	2.3	3.1	0.590	0.281	0.565	1.6	242
1951		4.6	2.4	0.3	0.617	0.242	0.567	0.7	473
1952		4.5	2.5	2.7	0.618	0.225	0.550	0.8	452
1950, 1952		4.5	2.5	2.7	0.611	0.231	0.560	1.1	694
1950, 51, 52		4.7	2.4	1.9	0.612	0.235	0.560	1.0	1167

Table 2
Velocity Variation of 16 Lacertae

| Year | Frequency | Velocity Range (km/sec) | | | Phase Zero Point (Periods) | | | M. E. of 1 obs. | No. of Obs. |
		5.9113	5.8529	5.4998	5.9113	5.8529	5.4998		
1951—1st ½		27.5	6.8	0.8	0.857	0.503	0.895	3.5	234
1951—2nd ½		29.3	7.2	4.1	0.852	0.548	0.864	3.5	208
All 1951		28.3	6.9	2.2	0.854	0.516	0.876	3.6	442
1951, 1952		28.6	7.2	2.3	0.855	0.514	0.885	3.6	456

concordantly in both light and velocity measures. The former observation is illustrated by Figures 2 and 3, which show from the results of fitting to simultaneous light and velocity measures obtained on 11 nights that both amplitude and phase perturbations of the primary pulsation are much stronger in light than in velocity, while the latter conclusion suggests the possibility of long period interference effects arising from a close frequency doublet. Since the annual cycle count is uncertain due to the very short photometric observing seasons, the pulsation frequencies derived by periodogram analysis

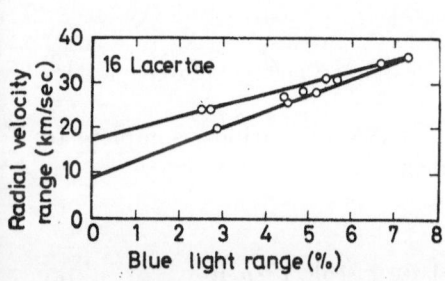

Fig. 2. A comparison of the variations in radial velocity and blue light ranges of the primary pulsation, as determined by simultaneous observations on 11 nights. Extrapolations of the straight lines bounding the observed region do not enclose the origin.

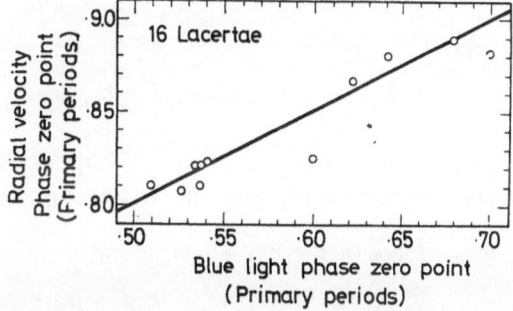

Fig. 3. A comparison of the variations in radial velocity and blue light of the primary pulsation's phase zero point, as determined by simultaneous observations on 11 nights. The straight line was drawn with slope $1/2$.

could be in error by one cycle per year $= 0.0027$ c/d, though they are otherwise accurate to ± 0.0002 c/d or better. A proper change of 0.0027 c/d in n_2
and in n_3 would in each case produce a nearly perfect tidal resonance, so exact
resonances were assumed and subsequent periodogram analysis disclosed the
existence in the light and velocity variations of the corresponding doublet
components of frequencies n_2 and n_3, as shown in Table 3.

Table 3.
Adopted Solution for 16 Lacertae

Frequency (c/u)	Description	Half Range		Phase Zero Point	
		(km/sec)	(% light)	R. V.	Light
5.9177	$n_3 + 5n_0$	0.3	0.26	0.11	0.41
5.9113	n_1	14.4	2.41	0.85	0.63
5.9048	$n_2 + 2/3n_0$	0.8	0.24	0.06	0.89
5.8561	$n_1 - 2/3n_0$	2.9	0.70	0.56	0.42
5.8496	n_2	2.2	0.76	0.57	0.20
5.5037	n_3	1.0	0.82	0.23	0.79
5.4973	$n_1 - 5n_0$	0.9	0.83	0.16	0.60
0.08279	n_0	22.9		0.88	
0.1656	$2n_0$	0.7		0.21	
0.2484	$3n_0$	0.5		0.39	
0.3312	$4n_0$	1.2		0.08	

To explain this result, we consider a system of N coupled linear oscillators,
in which the state of the kth oscillator (of natural frequency n_k) is specified
by the generalized coordinate q_k, and on which an external periodic force F
of frequency n_0 (where $n_0 \ll n_k$) acts. We assume we can expand each
component of F in a Fourier series harmonic in n_0, so that, neglecting both
the direct response of the system to the imposed force F and non-linear terms
in the coordinates q_k, we have

$$\frac{d^2 q_j}{dt^2} + n_j^2 q_j = \sum_{k=1}^{N} F_{jk} q_k, \qquad (j = 1, 2, \ldots N), \tag{1}$$

where

$$F_{jk} = \sum_{l=0}^{\infty} (a_{jkl} \sin ln_0 t + b_{jkl} \cos ln_0 t), \qquad (n_0 \ll n_1, n_2, \ldots n_N). \tag{2}$$

It is easily shown that the periodic solutions of these equations, complete to
the second order, may be written in the form

$$q_j = A_j \cos (n_j t + \alpha_j) +$$

$$+ \sum_{k=1}^{N} \sum_{l=0}^{\infty} B_{jkl} \left\{ \frac{\cos[(n_n - ln_0)t + \beta_{jkl}]}{n_j^2 - (n_k - ln_0)^2} + \frac{\cos [(n_k + ln_0)t + \gamma_{jkl}]}{n_j^2 - (n_k + ln_0)^2} \right\}. \tag{3}$$

Thus the solution will contain both the natural frequencies n_k and also the
combination terms $n_k \pm ln_0$, but of the latter the most important are likely
to be those (if any) for which a near resonance exists (i.e. $[n_k \pm ln_0]^2 \approx n_j^2$).
Of course, terms with the forcing frequencies ln_0, which were neglected in

equations (1), will also exist. Our foregoing argument for the resonance case requires, strictly speaking, that l be integer, but as is well known, resonance often occurs when l is rational, provided it is the quotient of two small integers (e.g. the Kirkwood gaps and the Hilda group in the case of the asteroids). In the present case, with P_0 as the orbital period in a nearly (or perhaps exactly) circular orbit, the tidal perturbations are nearly the same at intervals of $1/2\, P_0$, P_0 and $3/2\, P_0$, so that the strongest perturbing frequencies should be $2n_0$, n_0, and $2/3\, n_0$.

Since the two resonance pairs of difference frequencies $(n_2,\ n_1 - 2/3\, n_0)$ and $(n_3,\ n_1 - 5n_0)$ had been observed, in accordance with the preceding analysis the corresponding sum frequencies $n_2 + 2/3\, n_0$ and $n_3 + 5\, n_0$ were included in the least squares fitting for the adopted solution displayed in Table 3. That these sum terms are of very small amplitude (and perhaps not present) is not surprising, since they arise from tidal action on the very weak terms n_2 and n_3, whereas the difference terms here originate in the action of F on n_1. The ordinary, non-resonant combination terms $n_1 \pm ln_0$ were also included in a fitting with $l = 1$, 2, and 3, and are probably present but are of too small amplitude to be significant. We note that $n_3 - (n_1 - 5n_0) = +0.0064$ c/d and $n_2 - (n_1 - 2/3\, n_0) = -0.0065$ c/d, with errors probably not exceeding $\pm\ 0.0002$ c/d, so that the detuning effect on the primary frequency n_1 is almost exactly cancelled by the opposing actions of the two resonance terms.

In figures 4, 5 and 6 we illustrate the agreement between the adopted radial velocity solution and the observations, and in figures 7, 8 and 9 the same comparison is made for the light variation. For the velocities only on

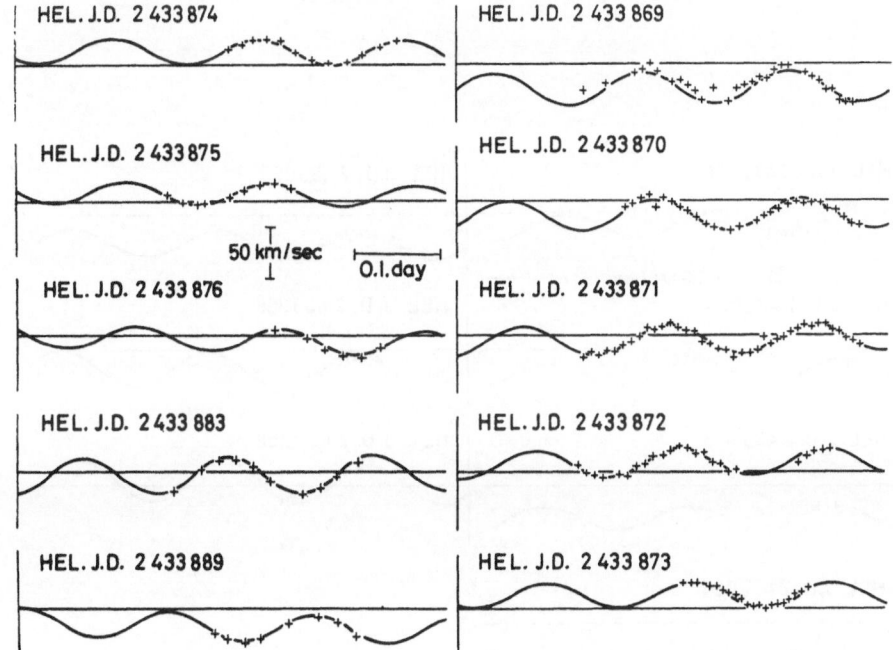

Fig. 4. A comparison of the observed (crosses) and computed (full curve) radial velocity variation of 16 Lacertae.

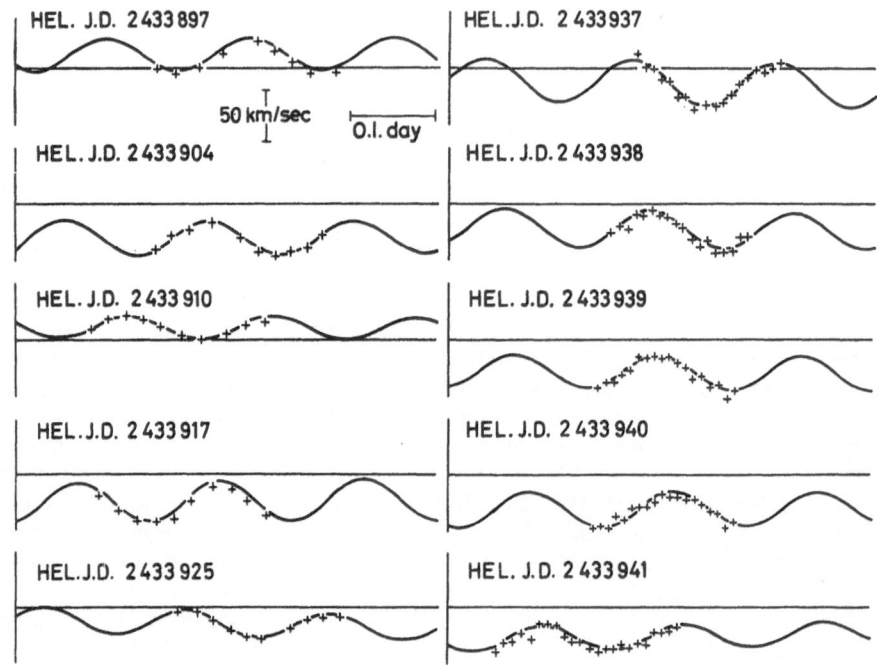

Fig. 5. Cont. of Fig. 4

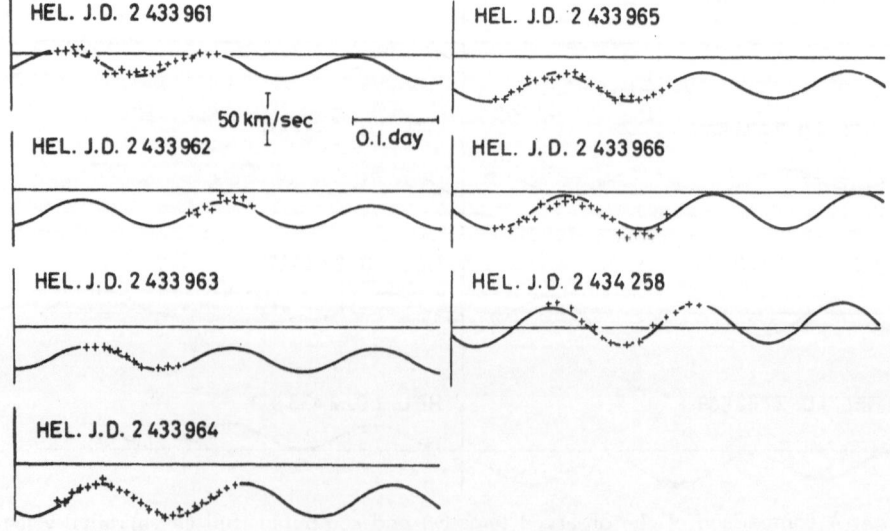

Fig. 6. Cont. of Figs. 4 and 5

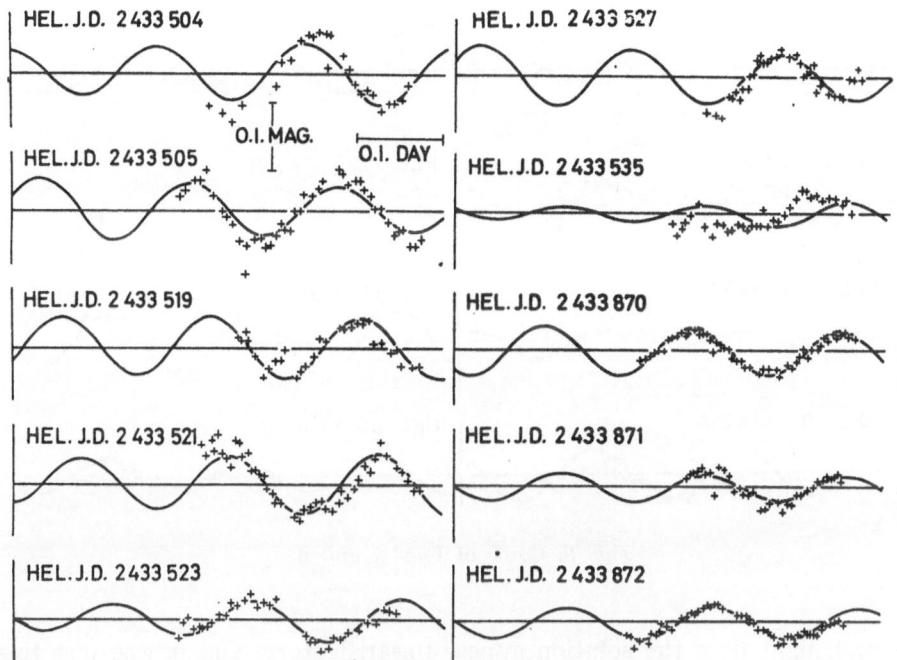

Fig. 7. A comparison of the observed (crosses) and computed (full curve) blue lightvaria-tion of 16 Lacertae.

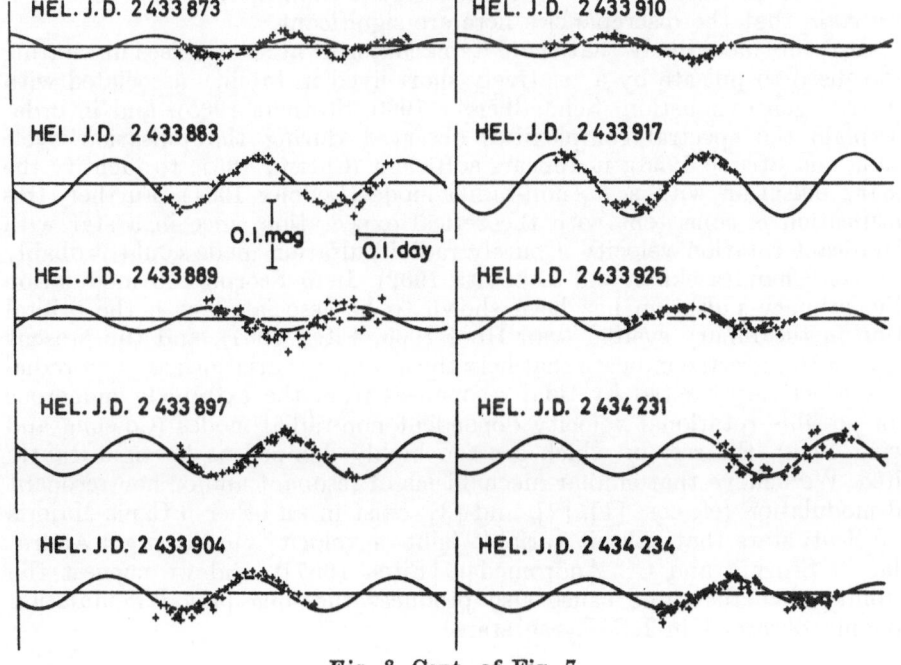

Fig. 8. Cont. of Fig. 7

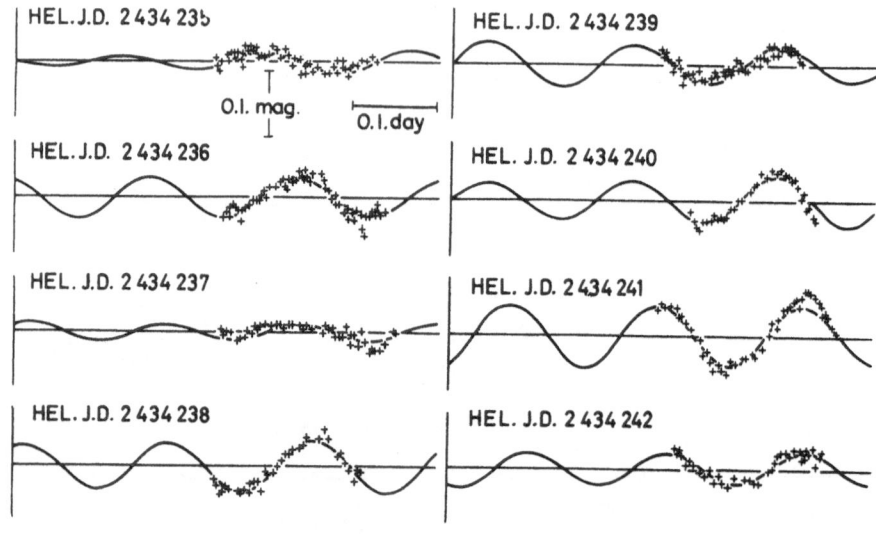

Fig. 9. Cont. of Figs. 7 and 8

the first night does the solution appear unsatisfactory, and it was just this night that showed the large residual on the radial velocity curve. The representation of the observed light variation appears generally satisfactory except for the last night in the first (1950) observing season. But since the signal-to-noise ratio for this first season is much lower than in 1951 and 1952, it is not certain that the discrepancies here are significant.

It seems most likely that the β Canis Majoris variation arises in rotating stars caused to pulsate by a relatively short-lived instability associated with core hydrogen exhaustion (Schmalberger 1960, Stothers 1965), and in order to explain the spectral complexities observed during the pulsation cycle (Huang and Struve 1955), it appears sufficient (Christy 1967) to identify the primary pulsation with a P_2 non-radial mode (Ledoux 1951). Further, this identification is consistent with theoretical expectation since in a star with a significant rotation velocity a purely radial pulsation mode could probably not exist (Chandrasekhar and Lebovitz 1962). In σ Scorpii the modulation of the primary pulsation has been shown to be associated with the orbital motion in the binary system (van Hoof 1966, Fitch 1967), and the present results on 16 Lacertae indicate that here the n_2 and n_3 variations are non-radial modes selectively excited by tidal resonances from the extremely numerous set of possible rotational velocity dependent non-radial modes (Cowling and Newing 1949), the rest of which are too highly damped to be significantly excited. We believe that similar mechanisms of resonant and/or non-resonant tidal modulation (cf. eqs. [1], [2], and [3]) exist in all other β Canis Majoris and δ Scuti stars that exhibit variable light or velocity curves (e.g. ν Andromedae, σ Scorpii, and CC Andromedae [Fitch 1967]), and we suggest the possibility that the same cause also produces the long period modulation commonly observed in RR Lyrae stars.

REFERENCES

Chandrasekhar, S., and Lebovitz, N. R., 1962, Astrophys. J., **136**, 1105.
Christy, R. F., 1967, Astr. J., **72**, 293.
Cowling, T. G. and Newing, R. A., 1949, Astrophys. J. **109**, 149.
Fitch, W. S., 1967, Astrophys. J., **148**, 481.
Hoof, A. van, 1966, Z. Astrophys., **64**, 165.
Huang, S. S. and Struve, O., 1955, Astrophys. J., **122**, 103.
Ledoux, P., 1951, Astrophys. J., **114**, 373.
Preston, G. W., 1966, Kleine Veröff. Remeis-Sternw. Bamberg, No. 40, p. 163.
Schmalberger, D. C., 1960, Astrophys. J., **132**, 591.
Stothers, R., 1965, Astrophys. J., **141**, 671.
Struve, O. and Bobrovnikoff, N. T., 1925, Astrophys. J., **62**, 139.
Struve, O., McNamara, D. H., Kraft, R. F., Kung, S. M. and Williams, A. D., 1952,
    Astrophys. J., **116**, 81.
Walker, M. F., 1951, Publ. astr. Soc. Pacific, **63**, 35.
Walker, M. F., 1952, Astrophys. J., **116**, 106.
Walker, M. F., 1953, Astrophys. J., **118**, 481.
Walker, M. F., 1954, Astrophys. J., **120**, 58.
Walker, M. F., 1967, (private communication).
Wehlau, W. and Leung, K. C., 1964, Astrophys. J., **139**, 843.

DISCUSSION

Detre: Is the pulsation excited, or only influenced by the companion?

Fitch: No, the primary pulsation is not excited but only influenced by the companion. It is caused by the evolution of the star carrying it into either the β Canis Majoris or Cepheid instability strip. However, in 16 Lacertae it appears that two very weak pulsation modes which would otherwise be changed out are excited by tidal resonances with the primary pulsation.

Detre: Have β CMa stars and δ Scuti stars erratic O−C diagrams, or smooth ones? According to my opinion there may be some difference in the behaviour of RR Lyrae variables and β CMa stars, the former ones having more erratic O−C diagrams. Further, RR Lyrae stars with secondary period have smaller amplitudes than RR Lyrae stars having stable light-curve. There is no such difference in the amplitudes of δ Sct stars and dwarf Cepheids.

Fitch: As you have shown, the O−C diagrams of the RR Lyrae stars are quite erratic, while so far as I am aware, the δ Scuti and β Canis Majoris stars have smooth O−C curves. However, this is not at all surprising, since the β Canis Majoris and δ Scuti stars are all very small amplitude pulsators with behaviour probably governed by the linear wave equation (they all have nearly sinusoidal, small amplitude variations), whereas the RR Lyrae star pulsation is, as Christy has shown, governed by an extremely non-linear set of equations. Therefore the amplitude, shape, and frequency of an RR Lyrae star's pulsation must be very sensitively determined by conditions in the outer layers of the star, and whenever these layers are very slightly disturbed by evolutionary readjustment (and perhaps also by the tides induced by a companion) the observed pulsation characteristics must change − hence the erratic O−C behaviour due to evolution and, perhaps, the long period modulation due to tides. With regard to the dwarf Cepheid or AI Velorum stars, I know

of none that show long period modulation, though many have two pulsation modes, simultaneously excited so that they show a very short period
beat. With regard to the small pulsation amplitude of RR Lyrae stars
having secondary periods, I think that probably when tides are present
the amplitude and phase of the normally radial pulsation vary in tidal
zones over the surface of the star, so that when the integrated light is
observed both light and velocity amplitudes will appear smaller than in
the normal case with no tides present.

Correction on March 7, 1969

I prepared for publication a paper which included an expanded version
of my Budapest lecture on 16 Lacertae, as well as an analysis of β Cephei, and
then learned I'd overlooked some velocity measures of 16 Lacertae published
by Mc Namara. The addition of the new data and a more careful examination
of the errors involved has shown that only one of the two weak pulsation
modes has its apparently excitation due to tidal resonance, and that the other
is apparently excited by rotational coupling. This does not distress me, for
the original data used were consistent with my (erroneous) interpretation, and
the changes will be duly published.

However, in the course of this work I've re-calculated the orbital elements with a new program we developed here, and found that there was an
error in the original program which we had copied from one then under development at Kitt Peak National Observatory. I assume they now have a correct version, but in the one we used a mistake had been made in programming
the calculation of phase from periastron for orbits of small eccentricity, with
the result that in my Figure 1 all phases are in error by 0.5 period.

VARIATIONS DANS LE SPECTRE DE γ PEGASI

JEAN-MICHEL LE CONTEL

Observatoire de Paris

Contrairement à d'autres étoiles de type β CMa, γ Pegasi ne présente qu'une seule période d'oscillation de la courbe de lumière et des vitesses radiales (3h 38 mm). Les amplitudes de variation sont faibles ($\Delta m = 0,015$ et $\Delta \vartheta_R = 7$ km/s). Le spectre a été très étudié, notamment par Aller (1949), Aller et Jugaku (1958) et Wright et al. (1964).

En 1966 et en 1967 nous avons entrepris des observations du spectre de cette étoile à très haute résolution spectrale et à grande résolution dans le temps, grâce à un nouveau spectrographe à très grande résolution installé au foyer coudé du télescope de 193 cm de l'Observatoire de Haute Provence. Le spectrographe permet d'utiliser une caméra électronique Lallemand comme récepteur (Baraune et al., 1967).

Le but de ces observations était de rechercher une variabilité des profils de raies dans le spectre de γ Pegasi.

La dispersion sur la photocathode du récepteur est de 2 A/mm; le champ du spectrographe de 25 A. Les temps de pose ont toujours été choisis inférieurs ou égaux a 12 mn. Pour chaque domaine spectral étudié, plusieurs séries de spectres ont été obtenues, chaque série s'étendant sur 4 heures environ. Pour une même série les temps de pose ont été constants; le lissage dû au temps d'intégration a ainsi été le même pour tous les profils.

Plusieurs phénomènes ont été observés.

1) Variations de la forme du profil des raies au long de la période:

Avant chaque extremum de vitesse radiale, la raie est presque symétrique et simple. Sur la partie descendante de la courbe de vitesse radiale, la raie présente une aile rouge importante.

Sur la branche ascendante, elle présente une large aile violette.

Ces variations sont liées à l'existence de composantes qui apparaissent au voisinage des extremums de vitesse radiale et sont parfaitement séparées ainsi que sensiblement equidistantes au voisinage de l'axe γ (figure 1). A ce moment, leur séparation est de 0.07 A, correspondant a une vitesse relative de 4,6 km/s. Il est absolument nécessaire de ne pas dépasser 12 à 15 mn de pose pour espérer détecter ce phénomène, quelle que soit la résolution spectrale dont on dispose. Cela est mis en évidence sur la figure (2) qui représente 2 enregistrements du doublet λ 4481.13, λ 4481.33 de MgII, l'un (a) obtenu en 11 mn en utilisant la caméra électronique et l'autre (b) obtenu en 3 heures avec le même spectrographe sur plaque IIaO chauffée. Le profil (b) est très

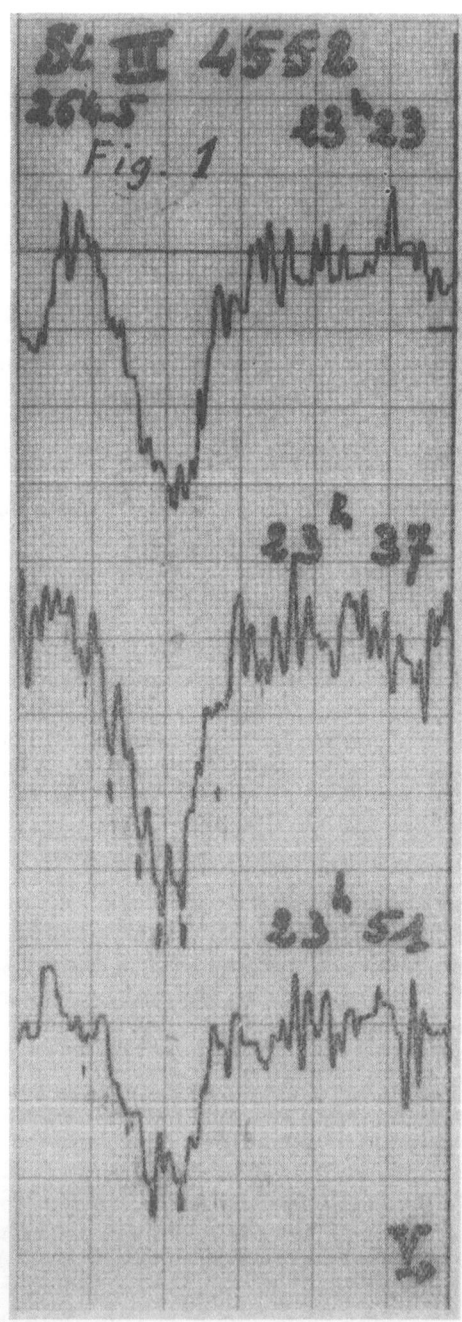

Fig. 1. Enregistrements de 3 spectres montrant les variations de la forme du profil de la raie Si III λ 4552 liées à la présence de composantes.

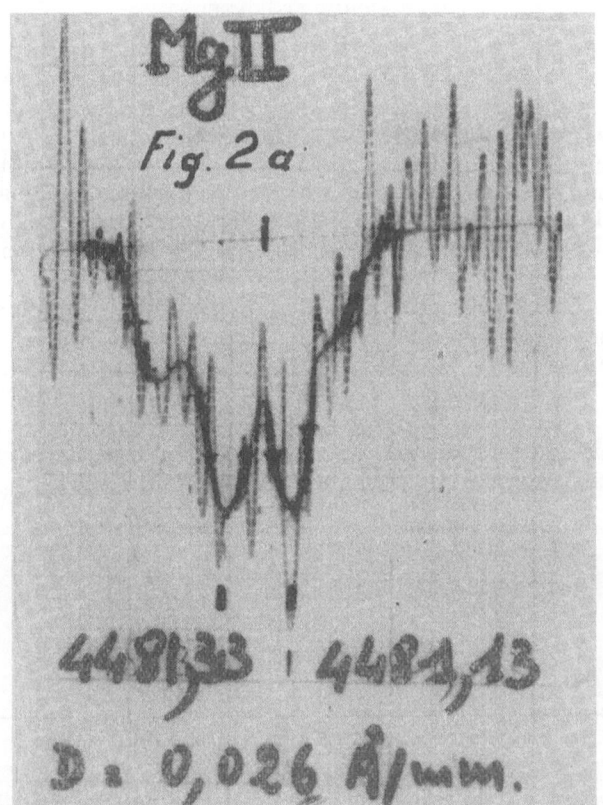

Fig. 2a. Enregistrement du doublet Mg II λ 4481, 13, λ 4481, 33 obtenu en Mmn avec la caméra électronique. Ce spectre est en opposition de phase avec le spectre à T. U. = = 23h37 m de la raie Si III λ 4552 (Fig. 1)

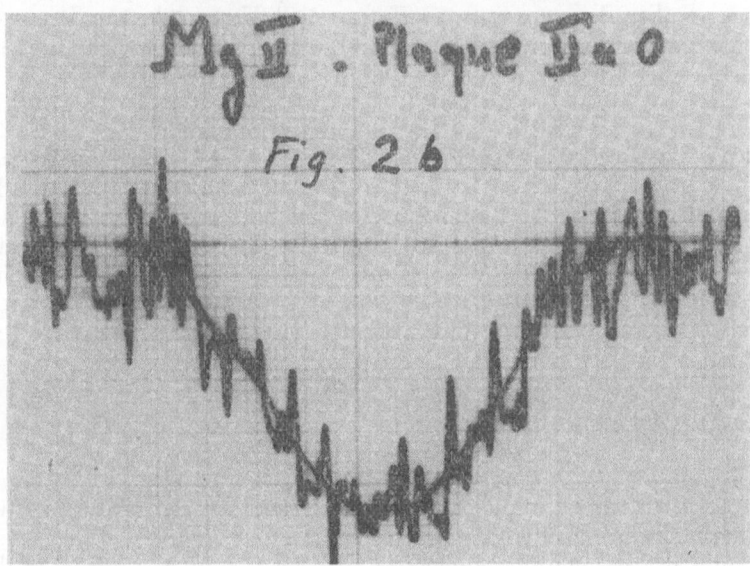

Fig. 2b. Enregistrement du même doublet de MgII obtenu au même foyer en 3 heures sur plaque Kodak IIaO chauffée.

semblable à ceux obtenus par Wright par exemple pour des temps de pose voisins de 20 mn et peut s'expliquer par un effet de lissage dû à la superposition des différents profils observés. En effet, le profil (a) correspond à un déphasage de π/2 par rapport au profil des SiIII obtenu à 23h 37 (fig. 1). A cette même heure, le doublet de MgII n'est plus séparé (ce même phénomène est observé sur le doublet de Al III λ 4479. 89, λ 4479. 97). Les observations montrent que l'interprétation du profil (b) par un effet de microturbulence donné par Miss Underhill (1966) n'est pas à retenir dans le cas de γ Pegasι et qu'on est plutôt en présence d'un champ de vitesse complexe associé à l'oscillation responsable des variations de lumière.

2) *Variations des largeurs equivalentes*

Sur la figure (3) sont portées les mesures effectuées sur les profils de la raie SiIII λ 4552 pour la nuit du 14 au 15 septembre 1967.

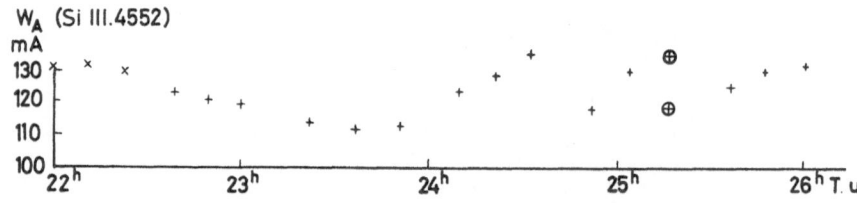

Fig. 3. Variations de la largeur équivalente de la raie Si III λ 4552 au cours de la période. Les explications concernant les points ⊕ sont données dans le texte.

Le minimum de largeur equivalente se produit lorsque l'on sépare le plus grand nombre de composantes (figure 1 T. U. = 23h 37). Il correspond approximativement au point de passage de l'axe γ sur la branche descendante de la courbe des vitesses radiales obtenue à partir des profils lissés.

Les discontinuités observées à T. U. = 24h 40 m et T. U. 25h 16 m le sont également pour la raie SiIII λ 4568 du même multiplet.

A T. U. = 25h 16 m, nous avons porté 2 valeurs de la largeur équivalente: $W_\lambda = 135$ et $W_\lambda = 118$. La différence entre les 2 valeurs est égale à la largeur equivalente $W_\lambda = 17$ m Å d'une composante nettement séparée dans l'aile violette de la raie. Une composante identique est également présente à la même distance de la raie λ 4568, ce qui permet de rejeter l'hypothèse de l'apparition d'une raie d'un autre élément. La distance de cette composante au point d'intensité maximum de la raie est de 0,3 Å dans les 2 cas (environ 20 km/s).

3) *Sur quelques spectres apparaissent de faibles raies de PII, NeII, FeIV, non présentes tout au long de la période*

Ces résultats, qui ont pu être obtenus grâce aux propriétés principales de la caméra électronique (linéarité de la réponse, gain en temps de pose, faible granularité des plaques nucléaires utilisées), sont à rapprocher de ceux obtenus par d'autres auteurs sur des étoiles du même groupe: ainsi Odgers and Kushwara (1960), a observé des composantes sur ces mêmes raies de SiIII dans le spectre de BW Vulpeculae de même que Struve et Su Shu Huang (1955) dans σ Scorpii. Stableford et Abhyankar (1959) ont mis en évidence des variations de profil et de largeur équivalente des raies (de 10% environ) dans le spectre de HD 21803. Il est d'ailleurs probable qu'une étude du spectre de cette dernière étoile, effectuée dans les mêmes conditions que pour γ Pégase, permettrait de mettre en évidence des composantes dans les raies dont les profils varient. En effet, l'amplitude de la variation des vitesses radiales est aussi faible pour HD 21803 (16 à 25 km/s), que pour γ Pegasi, alors qu'elle est supérieure à 150 km/s dans BW Vulpeculae. Mais la faible magnitude visuelle de HD 21803 la rend difficile à observer avec à la fois une grande résolution spectrale et une grande résolution dans le temps. L'importance des conditions d'observations dans l'étude des étoiles variables à courte période est confirmée par ces résultats. Les observations de Hill (1967) montrent que les β CMa ne sont pas toutes des rotateurs lents, nos résultats permettent de penser qu'elles montrent pout-être toutes des variations de profil.

D'autres spectres doivent encore être réduits, sur lesquels se trouvent notamment les raies de HeI, λ 4388, λ 4438, λ 4471. Ils devraient permettre de préciser la corrélation entre les variations observées et la période de γ Pegasi.

Par ailleurs, nous nous proposons de rechercher l'interprétation de ces phénomènes par un champ de vitesse en présence d'une oscillation non radiale. Il semble, en effet, difficile d'expliquer ces variations (de lumière, de vitesse radiale, de largeur equivalente) et l'existence des composantes dans les raies, si l'on conserve une symétrie sphérique à l'étoile tout au long de la période.

Dans le cas de γ Pegasi, pour laquelle on n'observe pas d'élargissement des raies par rotation, nous essayons de comparer des profils observés à des profils calculés dans l'hypothèse d'un modèle où l'étoile est vue «pole on», l'axe de l'oscillation étant perpendiculaire à l'axe de rotation.

REFERENCES

Aller, L. H., 1949, Astrophys. J. **109**, 204.
Aller, L. H., and Jugaku, J., 1958, Astrophys. J. **127**, 125.
Baranne, A., Bastie, J., Bijaoui, A., Duchesne, M., Le Contel, J. M., 1967, Publ. Obs. Hte-Provence **9**, no. 18.
Hill, G., 1967, Astrophys. J. Suppl. XIV, no. 130, 263.
Odgers, J. and Kushwaha, R. S., 1960, Publ. Dom. astrophys. Obs. Victoria XI, no. 6, 185.
Stableford, G. and Abhyankar, K. D., 1959, Astrophys. J. **130**, 811.
Su Shu Huang and Struve, O., 1955, Astrophys. J. **122**, 103.
Underhill, A. B., 1966, J. Quant. Spectrosc. Radiat. Transfer. **6**. p. 675—689.
Wright, K. D., Lee, E. K., Jacobson, T. V., Greenstein, J. L., 1964, Publ. Dom. astrophys. Obs., Victoria, **12**, no. 7, p. 173.

VARIATIONS IN δ SCUTI STARS

During the recent years, a few short period variables have been discovered on or near the main sequence. Very often, these objects show peculiar variations in light or radial velocity.

Many questions are raised concerning the structure of these stars, the nature of the pulsation and their relation to other groups of similar objects.

A group of French astronomers just began to study these stars from the theoretical and observational point of view.

Up to now, some observational work has been completed. The only reduced data concern γ Bootis and 14 Aurigae.

IS γ BOOTIS A SPECTRUM VARIABLE STAR?

A. BAGLIN
Institut d'Astrophysique de Paris

F. PRADERIE   M. N. PERRIN
Observatoire de Meudon   Observatoire de Lyon

I. γ Bootis has been recognised as a photometric variable for a long tim . It has been followed by several authors (Guthnick and Prager (1914), Guthnick and Schneller (1943) Guthnick and Fischer (1940), Magalashvili and Kumsishvili (1965)). But according to the Kukarkin and Parenago Catalogue, it does not belong to a well defined type of variable. Sometimes it experiences periodic variations and the period is very regular 0.2903137 days; sometimes it shows erratic variations or even constancy.

Due to its spectral type —A7— and luminosity class —III— it is tempting to classify the star as a δ Scuti or a dwarf cepheid.

It does not seem to be a spectroscopic binary but, as the rotational velocity is large —135 km/sec, measured by Slettebak (1965) — the binary motion, if any, would be difficult to detect.

As to the velocity field which could be connected with the light variation, no one has even reported (Miczaika 1952).

Simultaneous photometric and spectroscopic observations have been made on several nights in April 1968.

II. The photometric device has the same characteristics as in Chevalier et al. (1968), the same filter has been used (35 A centered on $H\beta$). Two comparison stars have been measured: HD 125 642 (A2, $m_v = 5.98$) and

HD 130817 (F0, m_v = 5.98). The large brightness of γ Boo does not allow to use comparison stars of the same magnitude as the star itself. Due to the large flux the photometer works in conditions not very far from saturation, and the precision of the measurements is less for this star than for the comparison star. The seeing conditions were good on only two nights: April 24th and 25th. During these two nights no variation of the brightness of γ Boo could be detected (Table I); the fluctuations are everywhere within the errors amplitude. The internal error per single measurement derived from the observations of the comparison stars is 0.005 mag.

Table I

Photometric results

Date	N_1	N_2	N_3	$\overline{m_1-m_2}$	$\overline{m_1-m_3}$
4—20—68	25	9	14	0.044	3.215
4—21—68	8	3	5	0.039	3.218
4—24—68	34	18	16	0.029	3.206
4—25—68	41	21	20	0.030	3.205

Subscript 1 refers to the comparison star   HD 125642
      2 ,, ,,   ,,   ,,   HD 130817
      3 ,, ,,   ,,   ,,   HD 127762 (γ Boo)

The bar indicates the mean value during the interval of observations on the corresponding night.

N is the number of measurements per star; columns 5 and 6 give magnitude differences in the chosen pass-band.

III. Spectroscopic observations have been made at the 120 cm telescope of the Observatoire de Haute Provence, with a prism spectrograph. The dispersion is 77 A/mm at $H\gamma$ —II a0 backed plates have been used.

The large rotational velocity does not permit a reliable study based on weak lines.

Kopylov (1959) has measured a few lines. Balmer lines seem to be normal, but the equivalent width of K line: W_K = 5.3 A is large for the spectral type.

Measurements of k index by Mc Namara and Augason (1962) and by Henry (1966) give the same value : k = 0.792, corresponding to an equivalent width of 4.53 A. This number is large as compared to the mean K line equivalent width for normal stars of the same b—y. As far as we know the K line has never been tested for variability in γ Boo.

One aim of this preliminary programm was to obtain spectra with short enough exposure times to avoid averaging effects over a large fraction of the period.

An unpredicted feature appeared : on two consecutive nights, the spectra show large variations of the width of the hydrogen and the K lines within a few hours (Fig. 1).

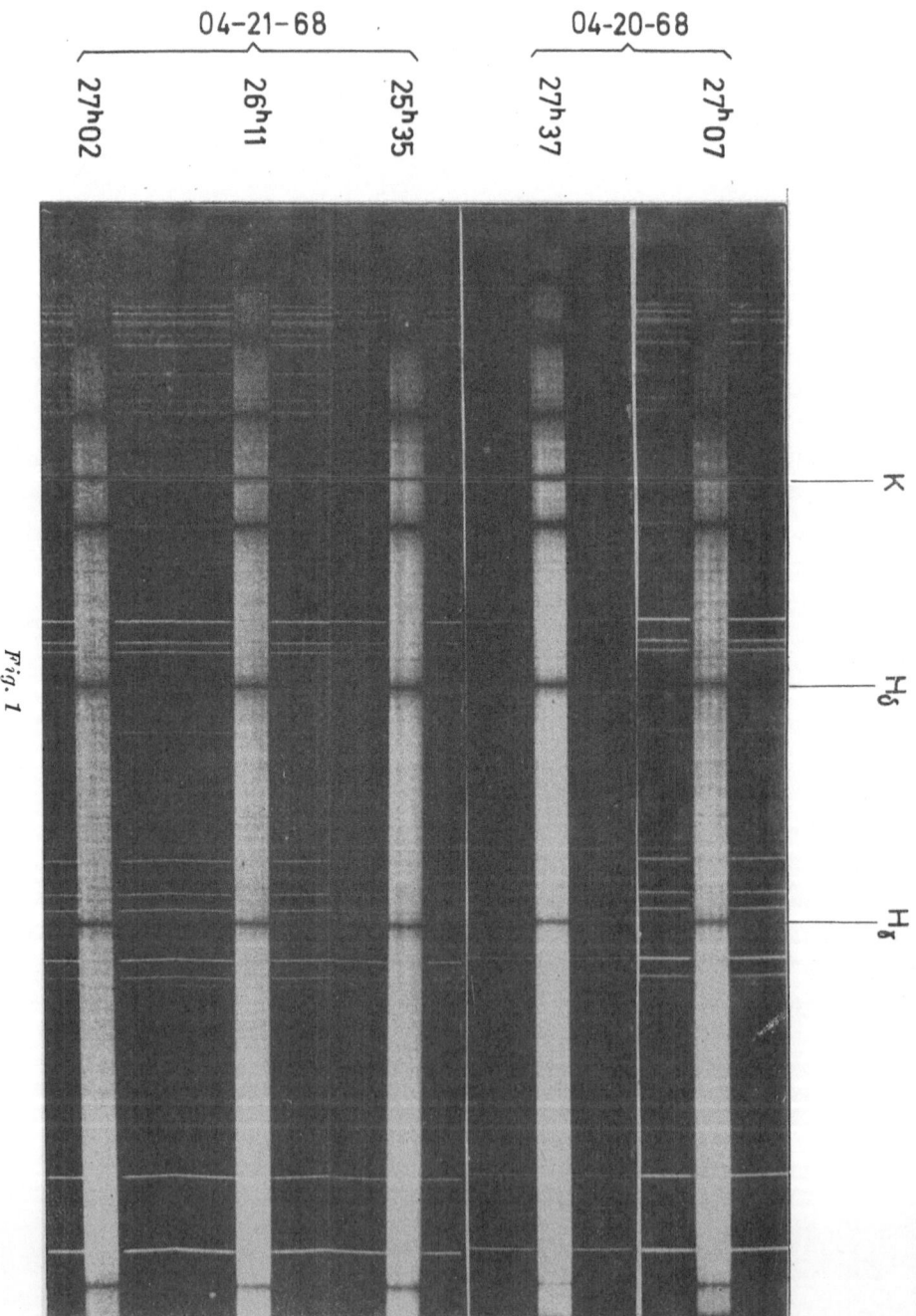

04-21-68

04-20-68

27ʰ02

26ʰ11

25ʰ35

27ʰ37

27ʰ07

K

H₅

H₆

Fig. 1

Equivalent widths of the lines H_δ, H_γ and K are given in Table II.

Table II

Variations of the equivalent width of the K, H_γ and H_δ lines in γ Boo.

Date	U. T.	W_K	$W_{H\delta}$	$W_{H\gamma}$
4/20/68	24h 30m	4.3:	13.6:	
4/20/68	27 07	3.5:	13.1:	11.8:
4/20/68	27 37	5.0	7.2	7.6
4/21/68	25 35	3.2	15.4	16.2
4/21/68	26 11	4.9	14.3	12.9
4/21/68	27 02	4.6	12.0	13.9
4/22/68	22 44	4.4	11.9	12.3
4/23/68	23 55	4.8	13.3	13.0
4/24/68	24 50	5.0	13.0	14.3
4/24/68	27 03	4.6	14.2	12.4
4/25/68	21 08	4.7	14.0	12.5
4/25/68	22 41	5.1	12.5	13.0
4/25/68	24 33	5.1	12.0	12.3
4/25/68	25 20	5.1	12.8	13.3
4/25/68	26 25	4.9	12.2	13.6

The symbol : indicates poor photometry (see Figure 1).

The main phenomenon is the decrease of the equivalent width of the K line by a factor 1.5. At the time of the K line minimum, the equivalent widths of H_γ and H_δ appear once to be reduced (4—20—68, 25H 35). On 4—20—68, 27H 37, the hydrogen lines are very weak.

During the following nights, no variations can be detected, taking into account the rather bad precision of the photometry. Particularly, it is seen on figure 1 that the spectrogram taken on 24—20—68, 27H 07 presents some features due to irregular sweeping during the exposure time. So that one cannot conclude from the first spectrograms of our series that there is an inverse behaviour of hydrogen and calcium II lines.

IV— Weakening of K line can be attributed to a very sudden increase of the surface temperature of the star or to a chromospheric activity. A sudden heating of the very outer layers of a region of the star can produce an emission core in the K line, which reduces the equivalent width, but cannot be seen as an emission at the resolution of our spectra.

For a few known spectrum variable stars, important variations in k index have been reported by Henry (1966) which are almost of the same order of magnitude as in γ Boo.

The question could be raised whether γ Boo belongs to the Ap stars; the Sr II lines should be studied in that perspective.

REFERENCES

Chevalier, C., Le Contel, J. M., Perrin, M. N., 1968, Astrophys. Letters vol. II.
Guthnick, P. et Prager R., 1914, Veröff. der Sternw. Berlin—Babelsberg, **1**, 45.
Guthnick, P. and Schneller, H., 1943 Astr. Nachr., **273**, 274.
Guthnick, P. and Fischer, H., 1940, Astr. Nachr., **271**, 80.
Henry, R. C., 1966, Princeton thesis (unpublished).
Kopylov, I. M., 1959, Izv. Krym. astr. Obs., **22**, 189.
Miczaika, G. R., 1952, Z. Astrophys., **30**, 134.
Magalashvili, N. L. et Koumsishvili, J. J., 1965, Abastumansk. astrofiz. Gora Kanobili-
    bull. **32**, 3.
McNamara, D. H. and Augason, G., 1962, Astrophys. J. **135**. 64.
Slettebak, A., 1955, Astrophys. J. **121**. 653

THE SHORT-PERIOD VARIABILITY OF 14 AUR (HR 1706)

C. CHEVALIER

Institut d'Astrophysique de Paris

14 Aur (HR 1706) has a magnitude $V = 5.05$, spectral type A 9 V. It is a close spectroscopic binary with an almost circular orbit, the period being 3.79 days (Harper 1916). Its short-period variability was first noticed by Danziger and Dickens (1967). According to their photoelectric U B V observations, the variation in the magnitude V was about 0.07 magn., with a period of 0.122 days ($\sim$ 3 hours).

We have obtained in 1968 on five nights photometric measurements and three nights radial velocity measurements at the Haute-Provence observatory. On two nights, on January 16 and 19, the observations were simultaneous.

PHOTOMETRIC OBSERVATIONS

They are described in the paper by Chevalier et al. (1968). The results are listed in Table I and plotted in Figs. 1 and 2.

RADIAL VELOCITY MEASUREMENTS

Condé spectra were taken with the 193 cm telescope at a dispersion of 10 A/mm. The plates were measured on a Ferranti machine at the Marseille Observatory. The measured lines are listed in table II, the results obtained are listed in table III and plotted in Figs. 1, 2 and 3.

COMMENTS

Both of the variations in magnitude and in radial velocity confirm the existence of a short-period pulsation of 14 Aurigae. In the two cases, the period obtained is about 2 hours 15 min. with an accuracy of about 15 min. The radial velocity curves seem to indicate the existence of secondary periods.

We notice opposite variations in the amplitudes of the light curve and the radial velocity curve.

Table I
Photometric Measurements

Date (1968)	U. T.	Δm (χ Aur-14 Aur)	Date (1968)	U. T.	Δm (χ Aur-14 Aur)
January 16	18 h 44 m	0.431	January 19	19 h 08 m	0.442*
JD: 2 439 872	19 39	0.407	JD: 2 439 875	40	0.388
	20 05	0.410		53	0.385
	29	0.423		20 04	0.380
	48	0.429		16	0.398
	21 07	0.439		25	0.398
	24	0.428		43	0.400
	42	0.413		53	0.402
	22 07	0.404		21 04	0.405
	27	0.415		15	0.400
	39	0.416		28	0.412
	43	0.420		38	0.402
	56	0.433		49	0.402
	23 26	0.447		57	0.402
				22 07	0.410
				15	0.410
				26	0.408
				39	0.410
				54	0.402
				23 07	0.395
January 18	19 h 30 m	0.390		18	0.400
JD: 2 439 874	20 14	0.390		29	0.400
	30	0.401	January 20	20 h 50 m	0.287
	54	0.424	JD: 2 439 876	21 00	0.386
	21 06	0.424		10	0.390
	18	0.422		20	0.396
	32	0.404		30	0.414
	44	0.394		40	0.427
	56	0.387		50	0.434

* Δm practically constant during this night.

Table II
Measured Lines

No.	Wavelengths (A)	Elements
1	4030.678	MnI (2) blend
2	4034.490	MnI (2)
3	4045.190	FeI (43) blend
4	4063.545	FeI (43), MnI (5)
5	4071.740	FeI (43)
6	4077.714	SrII (1)
7	4101.737	Hδ
8	4202.031	FeI (42)
9	4215.524	SrII (I)
10	4340.468	Hγ
11	4383.547	FeI (41)
12	4404.752	FeI (41)
13		{MgII (4) 0.129
	4481.228	{MgII (4) 0.327

Table III

Radial Velocity Measurements

Date (1968)	U. T.	Heliocentric radial velocity	Rms deviation	Number of lines	Exposure (min)	Plates
Jan. 16	18 h 33 m	−31.43 km/sec	0.19 km/sec	13	24	baked IIaO
	19 07	−28.54	0.29	13	14	baked IIaO
	28	−30.85	0.18	13	16	baked IIaO
JD 2 439 872	48	−32.07	0.20	13	15	baked IIaO
	20 10	−33.37	0.40	13	20	baked IIaO
	35	−30.40	0.40	13	20	baked IIaO
	21 02	−29.22	0.25	12	23	baked IIaO
	28	−29.46	0.41	12	26	baked IIaO
	22 21	−31.09	0.30	13	22	baked IIaO
	49	−30.50	0.49	13	27	baked IIaO
	23 16	−27.02	0.33	13	22	baked IIaO
	41	−27.28	0.40	13	22	baked IIaO
Jan. 19	19 h 40 m	−15.99 km/sec	0.52 km/sec	11	15	IaO
	20 02	−22.41	0.53	12	20	IaO
	27	−19.69	0.51	13	24	IaO
JD 2 439 875	51	−19.68	0.37	13	12	IaO
	21 10	−22.34	0.55	12	15	IaO
	30	−18.36	0.50	13	18	IaO
	50	−20.29	0.49	13	17	IaO
	22 11	−22.05	0.20	13	17	IaO
	28	−23.10	0.70	13	11	IaO
	48	−21.11	0.25	13	19	IaO
	23 08	−25.83	0.73	11	16	IaO
	25	−23.73	0.50	10	12	IaO
Jan. 21	19 h 20 m	+ 9.63 km/sec	0.37 km/sec	12	8	IaO
	31	5.45	0.71	11	9	IaO
JD 2 439 877	43	3.36	0.58	12	13	IaO
	59	1.80	0.45	10	16	IaO
	20 18	3.75	0.58	13	14	IaO
	35	3.19	0.33	13	11	IaO
	49	6.08	0.30	11	11	IaO
	21 03	9.61	0.53	12	15	IaO
	21	7.54	0.52	11	12	IaO
	36	9.27	0.47	12	16	IaO
	55	6.78	0.31	11	17	IaO
	22 15	4.33	0.40	11	19	IaO
	37	5.93	0.62	11	11	IaO
	51	7.61	0.52	11	11	IaO
	23 05	9.69	0.53	12	12	IaO
	20	12.42	0.42	11	11	IaO

On January 16, the light curve is perfectly regular with an amplitude of about 0.04 magn. The radial velocity curve is regular too, with an amplitude of 5 km/sec.

Three days later, on January 19, the light curve is practically flat and, on the contrary the amplitude of radial velocity curve is about 10 km/sec, that is, multiplied by the factor two.

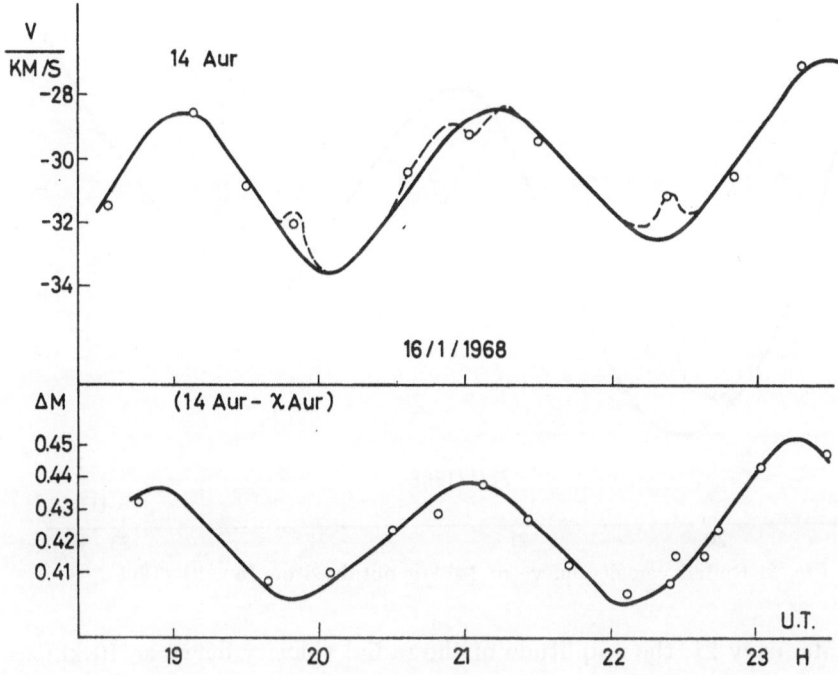

Fig. 1. Light curve and radial velocity curve of 14 Aur obtained simultaneously on Jan. 16 (1968).

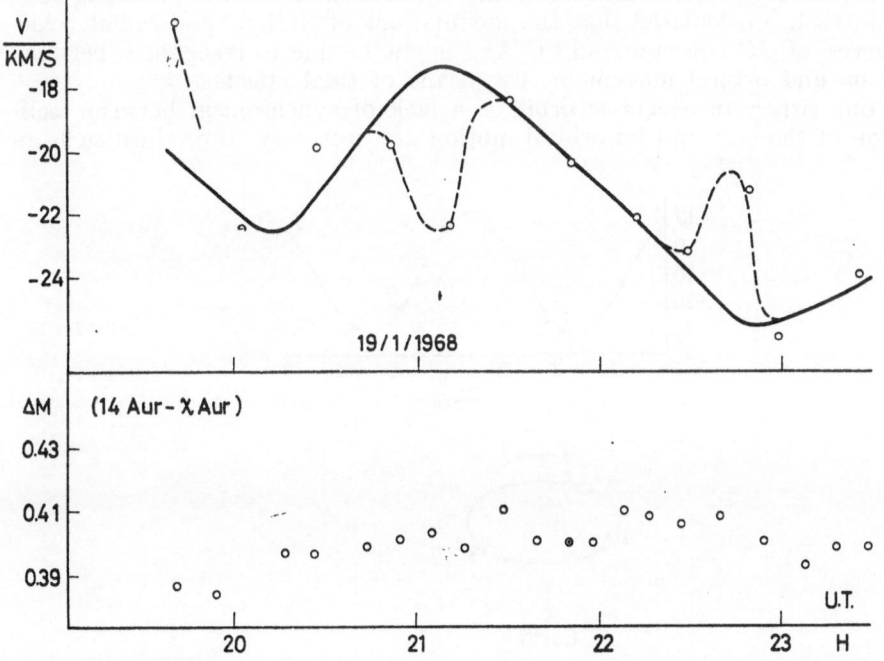

Fig. 2. Light curve and radial velocity curve of 14 Aur obtained simultaneously on Jan. 19 (1968).

21

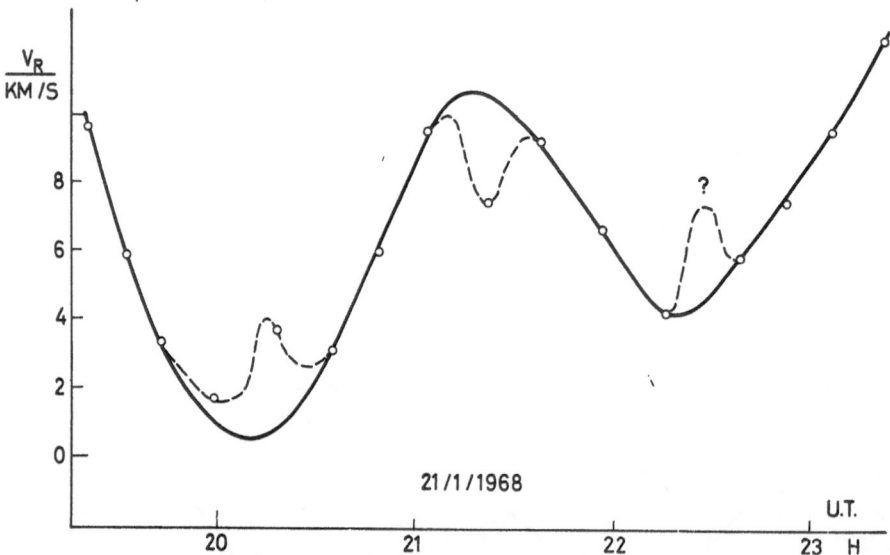

Fig. 3. Radial velocity curve of 14 Aur obtained on Jan. 21 (1968).

On January 21, the amplitude of the radial velocity curve is 10 km/sec again.

Our results have to be confirmed before attempting any interpretation, and particularly before calculating any detailed model of the pulsating star.

Fitch has calculated that the modulations of the light and radial velocity curves of 16 Lacertae and CC And might be due to resonances between pulsation and orbital movement, by means of tidal effects.

But either an excentric orbit or a lack of synchronism between self-rotation of the star and its orbital motion are necessary to produce such an

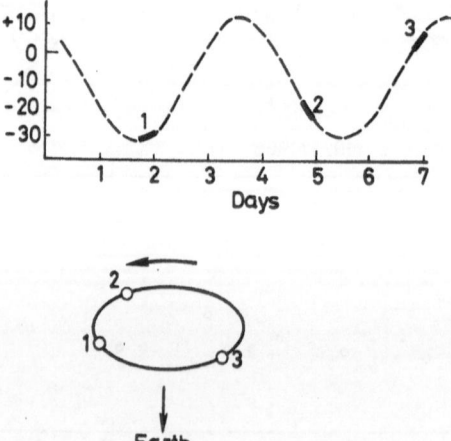

Fig. 4. Radial velocity curve of 14 Aur due to its orbital motion, (from Harper 1916), and positions corresponding to our observations.

effect. Crude calculations seem to indicate that the rotation of 14 Aur is probably synchronized with its orbital motion. Otherwise, its orbit is almost circular ($e = 0.033$, from Harper).

Another explanation of these opposite modulations of the light and radial velocity curves could be non radial oscillations of the star, the apparent amplitudes depending obviously on the position of 14 Aur in its orbit.

Anyway, it would be premature to make any conclusion.

REFERENCES

Chevalier, C., Le Contel, J. M., Perrin, M. N., 1968, Astrophys. Letters, vol. II.
Danziger, I. J. and Dickens, R., 1967, Astrophys. J., **149,** 55.
Fitch, 1968, Budapest Conference on Variable Stars.
Harper, W. E., 1916, J. R. astr. Soc. Canada **10,** 165.

DISCUSSION

Fitch: 1) Is 14 Aurigae a spectroscopic binary, and if so what is its period?

2) (In reply to the question from Baglin concerning possibility of tidal exploration of observed strong modulation in light and velocity of 14 Aurigae:)

Yes, I think it most likely that tidal effects can explain the variations you have observed in the light and velocity variations of 14 Aurigae.

3) With regard to your observation of large velocity variation simultaneous with nearly constant light on one night, a similar type of phenomenon occurs in 16 Lacertae (which is, of course, a very different kind of star). In that star one finds that the light-to-velocity amplitude ratios are very different in the various observed (probably non-radial) pulsation modes, so that sometimes one observes an appreciable velocity amplitude with almost no corresponding light variation.

Chevalier: Yes, 14 Aurigae is a close spctroscopic binary, its period is 3.79 days, its orbit is almost circular. Companion is unknown. It seems that the inclination of the orbit is not very high. Reference: Harper (1916)

Non-Periodic Phenomena in Variable Stars
IAU Colloquium, Budapest, 1968

ON THE NATURE OF SOME OF THE O—C DIAGRAMS OF THE RR LYRAE VARIABLES IN M5

C. COUTTS

Dunlap Observatory, Toronto, Canada

INTRODUCTION

The O—C diagrams determined from light curves of RR Lyrae stars in globular clusters have many different forms. Some are extremely irregular, some appear to be periodic in form, while others seem less complicated. The purpose of this investigation is to consider those diagrams which are reasonably regular (i.e. those to which it is reasonable to fit some simple curve).

When a star has a constant period, its O—C diagram (phase-shift plotted against time) is a straight line. If the period is not constant, the form of the diagram is different. If the period changes at a constant rate, the O—C diagram is a parabola. However, if the change of period is more complicated, then the diagram is more complicated. If we wish to learn something of the evolution of a star from the behaviour of its period, then it is convenient to assume as simple a form as possible for the diagram. The two simplest forms which come to mind are:

1. a parabola, implying that the period changes at a constant rate, or
2. two intersecting straight lines, implying that the period changes abruptly.

To learn about the evolution of the star, the former assumption seems the more advantageous because the rate of period change, β, can be deduced from the shape of the parabola. The determination of β has been discussed in more detail by Sawyer-Hogg and Coutts (1969). If we assume the latter (intersecting straight lines), we can deduce the net change of period, but not the rate. Thus it seems that the assumption of the parabola is the better one.

However, some recent investigations in M5 (Sawyer-Hogg and Coutts 1969) show that, of 66 RR Lyrae stars studied, 18 have had constant periods during a 70-year interval. If the usual period changes are abrupt, then it is possible that the stars that have had constant periods for 70 years would also exhibit changes if the range of observations were extended to 100 years, for example. Furthermore, if for all the stars which have exhibited period increases or decreases during the 70-year interval of observations, the observed period change represents the net period change for 100 years, then it is possible to calculate an average rate of period change from these abrupt period changes. Such calculations have been made for M5 (Sawyer-Hogg and Coutts 1969) and Figure 1 is a graph showing the relation between ΔP, the period change in days per million years, calculated in this manner and β, the rate of period change in days per million years calculated from the parabolas. A straight line of slope 1.1 has been drawn through the points. It seems that the rates of period change calculated from the two different assumptions are approximately

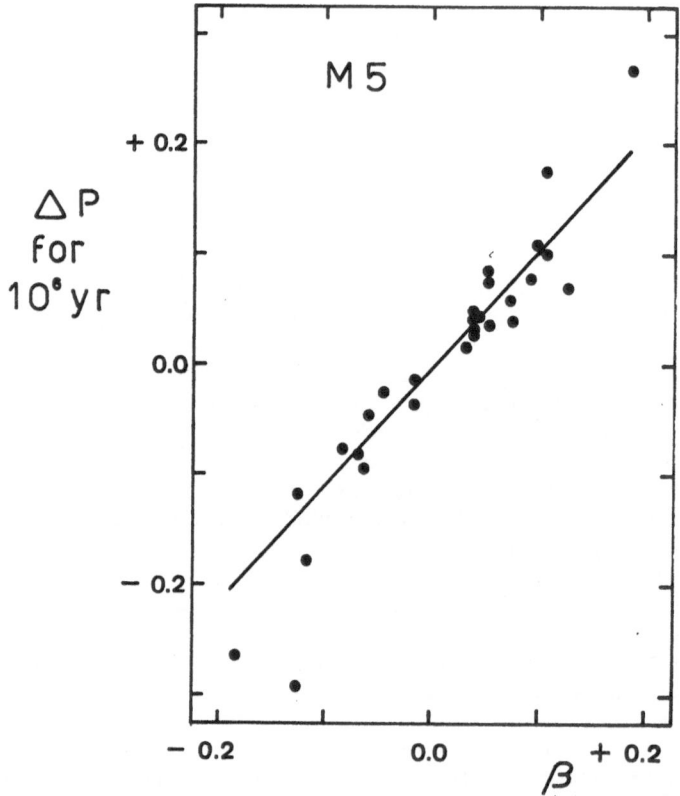

Figure 1. ΔP, period change in days per million years calculated assuming abrupt period changes, plotted against β, in days per million years, calculated assuming gradual period changes.

equivalent, and so the assumption of a parabola is not necessarily more convenient than that of intersecting straight lines.

To understand the significance of the constancy of period of so many of the stars, it is useful to know whether the changes occur in a uniform manner or abruptly. If the changes usually occur in a uniform manner, then the stars which have had constant periods over the 70-year interval might have, either periods which are truly constant, or, periods which are changing, but at a rate too slow to be detected with the present observations. On the other hand, if the changes are usually abrupt, this could imply that the stars with constant periods would exhibit changes as large as those observed in the other stars if the interval of observations were extended for a few years.

Clearly, if we are interested in understanding the changes of period, to learn of any possible evolutionary interpretation, we need to know which of these interpretations is the more probable. It must, however, be mentioned, that at present, it does not appear that observed period changes of RR Lyrae stars in globular clusters are caused by the evolution of the stars. Time scales for stars on the horizontal branch calculated by Faulkner and Iben (1966) do not agree with the observed period changes. This has been discussed for the variables of M5 by Sawyer-Hogg and Coutts (1969).

INVESTIGATION

The present investigation is only preliminary, based on 6 RR Lyrae variables in M5. The purpose is to determine whether the O—C diagrams of those stars which show increasing or decreasing periods, are best represented by parabolas or intersecting straight lines.

When we construct O—C diagrams, we must assume a certain period of light variation. However, if this period is varying, it is difficult to know what is the best period to assume. If the change of period occurs in a uniform manner, the O—C diagram is a parabola and the vertex of the parabola occurs at the time when the assumed period is actually the real period. Thus, as different periods are assumed for a particular star, the O—C diagram is a parabola, always with the same shape, but with the vertex shifted along the time-axis. On the other hand, if the period change is sudden, the O—C diagram is two straight lines which intersect at the time when the period change occurs. If the diagram is constructed assuming a range of periods, the time of intersection of the two lines is always the same. The slopes of the two lines vary according to the period assumed, but the difference of the two slopes always remains constant and corresponds to the magnitude of the period change.

It would therefore seem that one could see, from a series of O—C diagrams for one star, constructed using a range of different periods, which interpretation is better. Figures 2 show the O—C diagrams for six RR Lyrae variables in M5. For each star, O—C diagrams are shown for three different periods. In the upper three diagrams, for each star, parabolas have been fitted to the points. In the lower three, intersecting straight lines have been fitted. It is not readily observable from the diagrams which interpretation is better, except for the case of variable 19 for which two straight lines obviously fit the points better than the parabolas.

For each star studied here, O—C diagrams have been constructed for 10 different periods. In each case, parabolas corresponding to a certain value of β, and intersecting straight lines have been fitted to the points. The deviation of each point from the curve has been measured and the mean sum of these residuals has been calculated, first for the intersecting straight lines and then for the parabolas. These values are listed in Table 1, which contains in

Table I

Var.	Period	β	N	Error	Sum of Residuals 2 lines	parabolas
7	0.49	0.07	15	0.00140	0.0393	0.0413
12	0.47	—0.06	17	0.00078	0.0233	0.0148
15	0.34	0.03	16	0.00136	0.0243	0.0451
19	0.47	0.16	16	0.00183	0.0298	0.1756
29	0.45	—0.12	16	0.00067	0.1232	0.1595
41	0.49	—0.04	16	0.00074	0.0162	0.0273

VAR 7

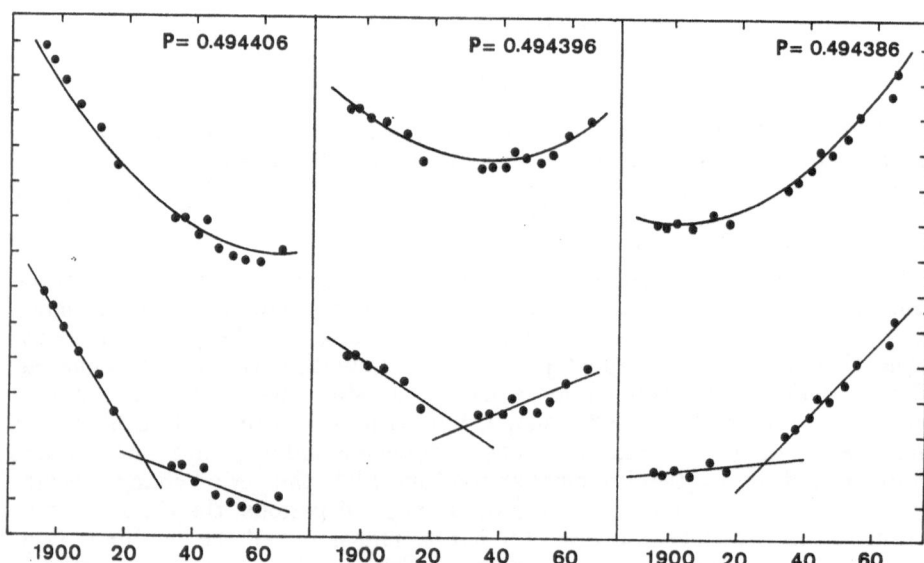

Figure 2a. O—C diagram for Var 7 in M5. The marks on the vertical axes are two-tenths of a cycle apart; those on the horizontal axes are twenty years apart.

VAR 12

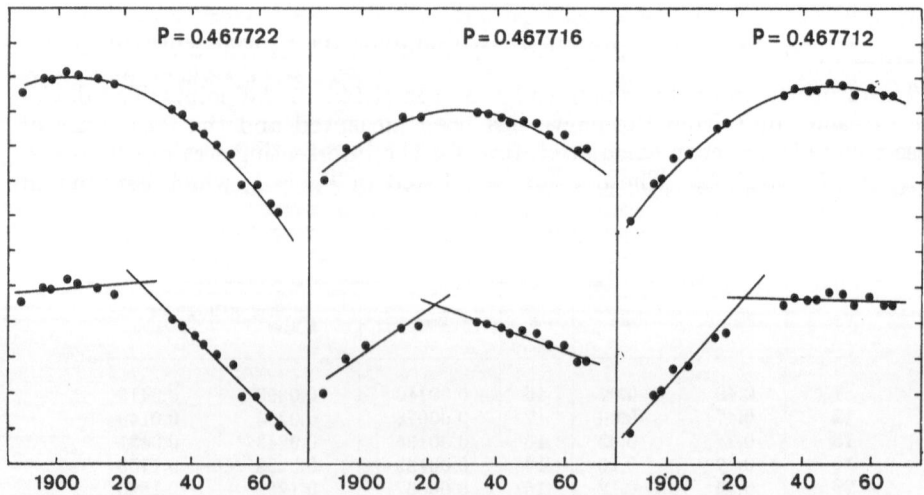

Figure 2b. O—C diagram for Var 12 in M5. The marks on the the vertical axes are two-tenths of a cycle apart; those on the horizontal axes are twenty years apart

VAR 15

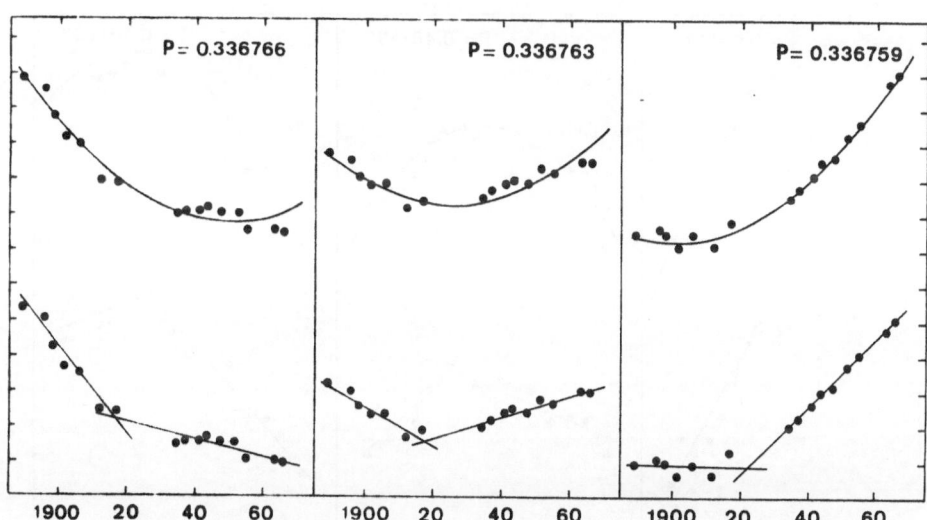

Figure 2c. O—C diagram for Var 15 in M5. The marks on the the vertical axes are two-tenths of a cycle apart; those on the horizontal axes are twenty years apart

VAR 19

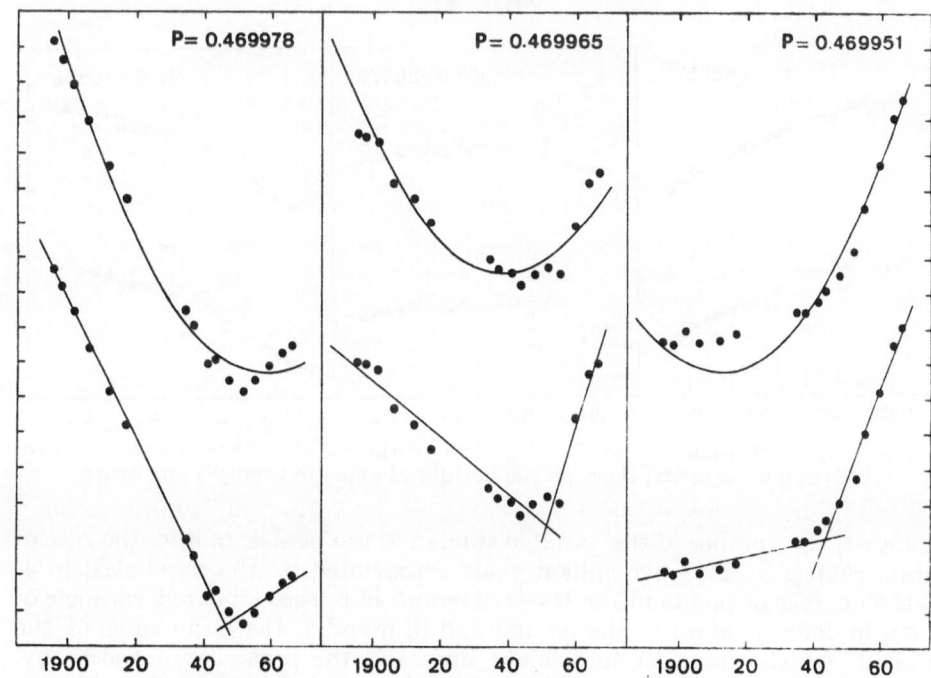

Figure 2d. O—C diagram for Var 19 in M5. The marks on the the vertical axes are two-tenths of a cycle apart; those on the horizontal axes are twenty years apart

VAR 29

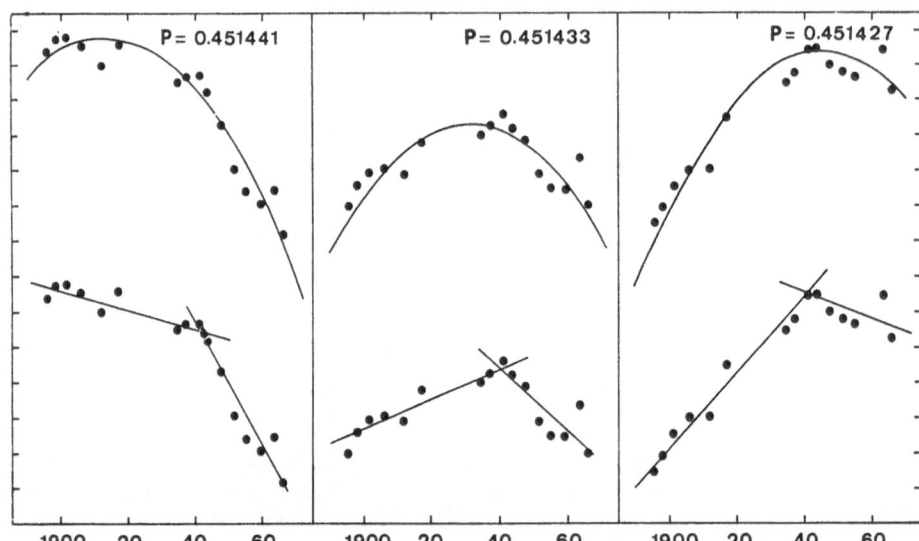

Figure 2e. O—C diagram for Var 29 in M5. The marks on the vertical axes are two-tenths of a cycle apart; those on the horizontal axes are twenty years apart

VAR 41

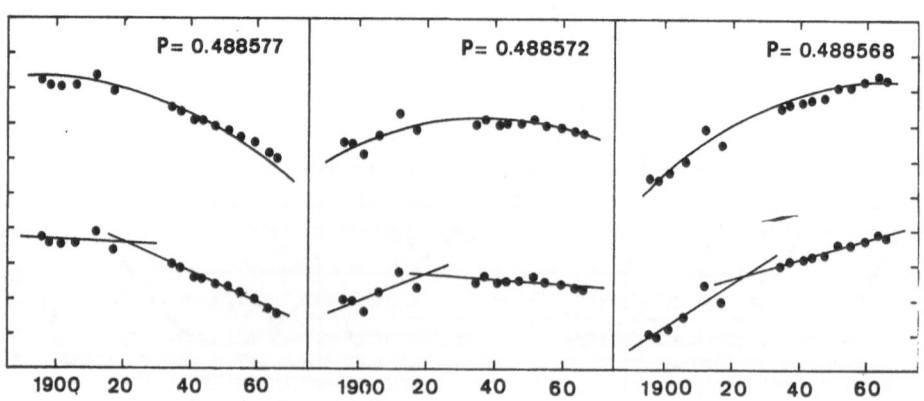

Figure 2f. O—C diagram for Var 41 in M5. The marks on the vertical axes are two-tenths of a cycle apart; those on the horizontal axes are twenty years apart

column 1, the number of the variable star; in 2, the period; in 3, β, the rate of period change in days per million years represented by the parabolas; in 4, N, the number of points in the O—C diagram; in 5, the estimated variance of errors in determination of the points; and in 6 and 7, the mean sums of the residuals for the intersecting straight lines and the parabolas, respectively.

In this investigation, the curves have been fitted to the points by eye, and so the results can only be considered tentative. However, for 5 of the 6 stars, the mean sum of the residuals is less when we consider the hypothesis of the intersecting straight lines than when we consider parabolas. (In the case of variable 29, the residuals are large in both cases and so we are probably not justified to assume that either hypothesis is reasonable.)

This investigation is presently being extended to all the stars for which O—C diagrams have been constructed in M5. The curves will be fitted to the points by the method of least squares and the results will thus be more certain. However, it is interesting to note that these preliminary results appear to indicate that most period changes probably occur in an abrupt rather than gradual manner. The final result will be important for our understanding of period changes in RR Lyrae stars and their connection with the evolution of the stars along the horizontal branch.

I am grateful to Professor Mannino of Bologna for arranging for me to use the IBM1620 computer at the Faculty of Engineering, University of Bologna to make this investigation. Dr. Emma Nasi kindly discussed the computer programs with me.

REFERENCES

Faulkner, J. and Iben, I., 1966, Astrophys. J., **144**, 995.
Sawyer-Hogg, H. and Coutts, C. 1969, Publ. David Dunlap Obs. (in press).

DISCUSSION

Plagemann: Is there any significance to the fact that the break in your 2 straight lines seem to occur between 1917 and 1935 when there were no observations?

Coutts: No, I don't think so.

Geyer: What type of RR Lyrae-stars are covered by your investigation? Are there differences of the O—C diagrams of RRc and RRab-stars?

Coutts: Of the 6 stars I have presented here, only one, variable 15 is an RRc type. In general, in M5, the O—C diagrams of the RRc types stars do not seem to differ from those of the RRab types.

Detre: 1) From which years have you observations? Lovas has taken about 500 plates of M5 in the years 1952—1966. at the Konkoly Observatory

2) Photoelectric observations (e. g. RR Leo) reveal even in the case of parabolic O—C diagrams some irregularities superposed on the parabolas.

3) Can you separate the two branches of RRab variables in the P—A relation first shown by Belserene in M3 and ω Cen? In M3 the stars located on the brighter branch have all positive parabolas in the O—C diagrams.

Coutts: 1) The David Dunlap Collection for M5 includes plates between the years 1936 and 1966. However, the material is somewhat weak between 1959 and 1964. Lovas' material will therefore be very important.

2) This is also true for most of the variables in M5. The deviation of the points from the curves often exceeds their errors in determination. However, χ^2 tests indicate that it is reasonable to accept the parabola or straight line hypothesis for all the stars discussed, except wariable 29. The χ^2 test also rejects the parabola hypothesis for variable 19.

3) The periods of variables in M5 with longer periods which show period changes are increasing. However, many are constant and this may be due to the fact that period changes are harder to detect for longer period stars. The shape of the parabola is related to $\beta/2P^2$. Thus the lower limit for detectable period changes is higher for stars with longer periods.

INSTABILITY OF LIGHT CURVES AND PERIODS OF LONGPERIOD CEPHEIDS BELONGING TO THE SPHERICAL COMPONENT OF THE GALAXY

O. P. VASILJANOVSKAJA and G. E. ERLEKSOVA

Astrophysical Institute, Academy of Sciences, Tadjik SSR, Dushanbe
(Read by Shakhovsky)

The longperiod Cepheids belonging to the spherical component of the Galaxy represent a special group of Cepheids like the cluster variables. From our investigation of the light curves of these Cepheids we notice that they are divided into two groups having light curves (Vasiljanovskaja et al. 1966, 1968) deviating from the standard light curves of classical Cepheids. The first group is characterized by assymmetrical light curves, in most cases with an hump on the descending branch. The second group is characterized by smooth symmetrical or almost symmetrical light curves. The division of Cepheids into two groups was confirmed by the investigations of period changes (Vasiljanovskaja and Erleksova, 1969a) and the statistical relations between light amplitude and logarithm of period, instrinsic colour index and logarithm of period, period change and light amplitude (Vasiljanovskaja, and Erleksova, 1969b).

One of the peculiarities of the spherical component Cepheids is the instability of the shape of their light curves. The first group with periods longer than 17^d shows variability of the light curve from cycle to cycle on the descending branch, in the region of the hump. Real changes of the shape and the size of the hump are shown by the Cepheids W Vir ($P = 17^d$), MZ Cyg ($P = 21^d$) and TW Cap ($P = 29^d$). These changes were traced by photoelectric or photographic observations taken by one author only. For example, let us examine the variations of the light curve of TW Cap. A variation of the hump on the descending branch was found by Soloviev (1955) from his photographic observations. Wallerstein (1958) noted a large scatter in the radial velocity curve. Wallerstein considered the scatter as consequence of irregular variations in the light curve from cycle to cycle. The normal light curve of TW Cap is represented in Figure 1, expressed in fractions of the amplitude (Max = 0, Min = 100). One can see that the whole descending branch of the light curve is changing. A reliable change of the light curves in the first group with periods shorter than 17^d is not observed. However, we note that the photographic light curves of ST Tau ($P = 4^d$) obtained by various authors have different descending branches. We were not able to reduce all observations into one photometric system and to take into account the period changes. Therefore the observed discrepancies may not be real.

Cepheids of the second group with periods longer than 20^d show variations of the light curve more frequently than Cepheids of group I. It is established from photoelectric observations that the light curves of the Cepheids RS Pav ($P = 20^d$), RU Cam ($P = 22^d$) before the sudden fall in the light amplitude in the year 1964, RX Lib ($P = 25^d$) show alternating cycles

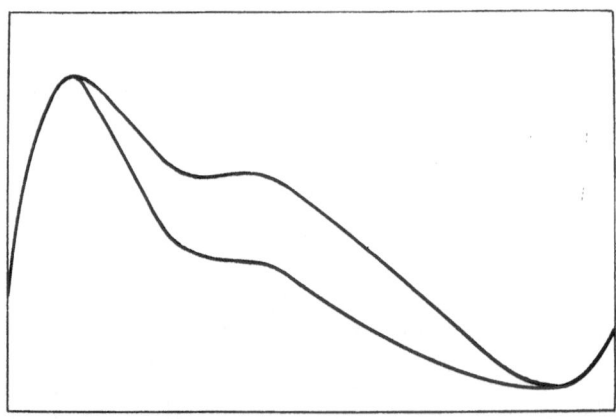

Fig. 1

reminding of RV Tau light curves. The maximum magnitude remains almost constant. The shape of the light curve and the amplitudes change insignificantly from cycle to cycle. The normal light curve near the primary maximum is sharper and narrower (Figure 2 — broken line) than near the secondary maximum (Figure 2 — solid line). We take RU Cam as an example, because RU Cam behaved like a Cepheid of Group II in the course of 60 years.

The other peculiarity of the spherical component Cepheids is the instability of their periods. Sudden changes of the period are typical features of all investigated Cepheids with periods longer than 2^d4. In the cases of SZ Cas, MZ Cyg and CC Lyr sudden changes of the period are accompanied by progressive changes. The investigation of the period changes of long period variables and the comparison with Cepheids show that progressive and sudden changes of the periods take place at the same time. The causes of the progressive and sudden period changes are different and independent. The interpretation of the progressive period change as an evolutionary effect is very probable.

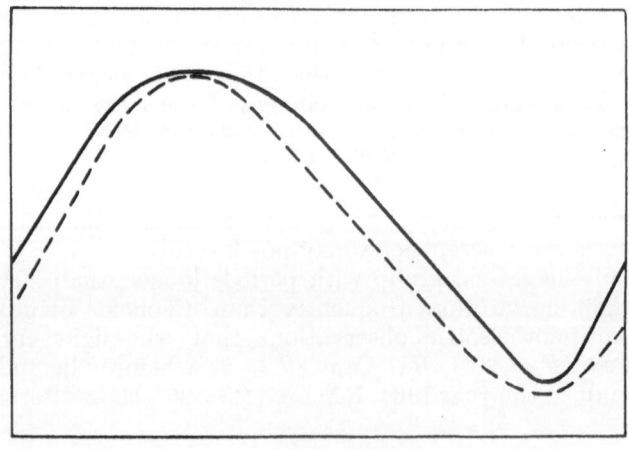

Fig. 2

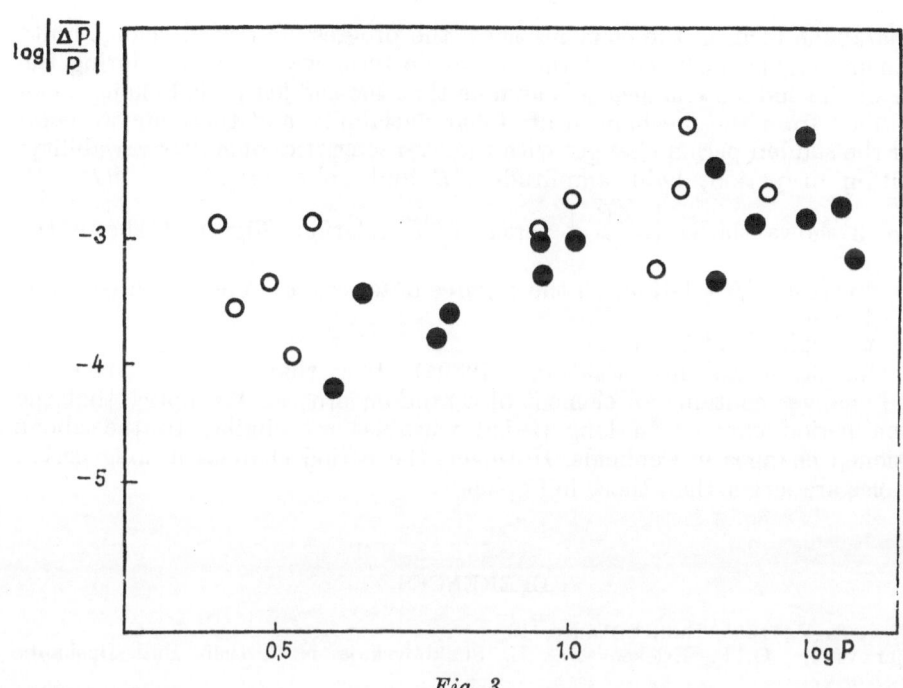

Fig. 3

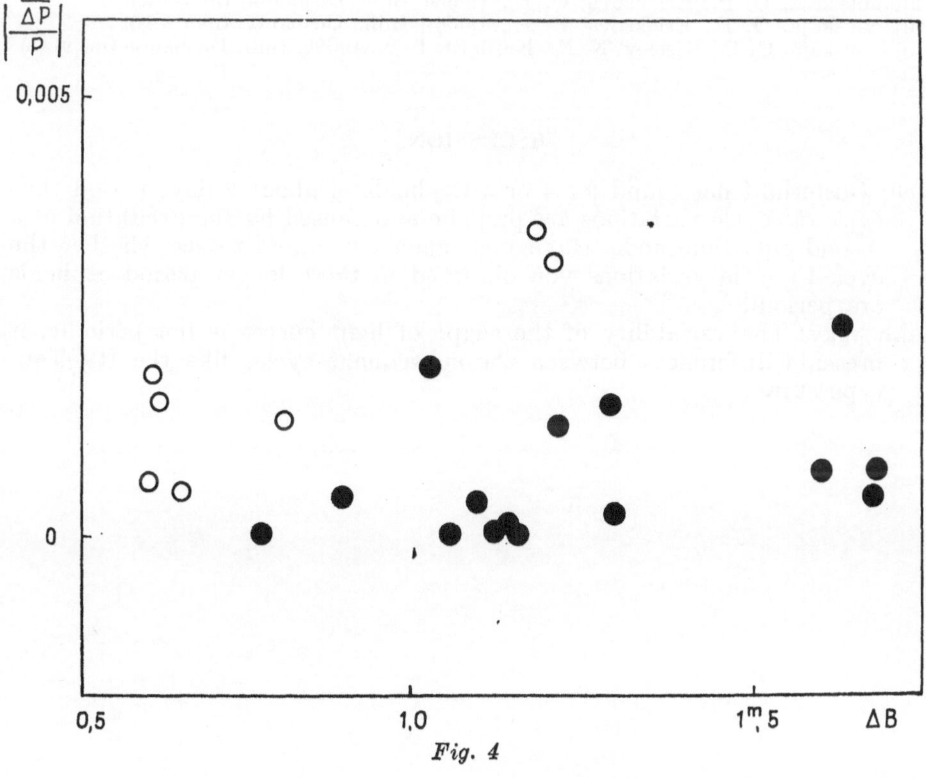

Fig. 4

The parabolic term in the elements show the progressive period change quite convincingly. The influence of the parabolic term was excluded during the study of the sudden changes. We assume that the sudden period changes can be derived from the mechanism of stellar variability and therefore we compared the sudden period changes with the characteristics of stellar variability: logarithm of period, light amplitude ΔB and color amplitude $\Delta(B-V)$.

Figure 3 shows the lg $\left| \dfrac{\overline{\Delta P}}{P} \right|$ versus lg P relation. Figure 4 shows the $\left| \dfrac{\overline{\Delta P}}{P} \right|$ versus ΔB relation. In the Figures dots refer to Cepheids of Group I, circles to Cepheids of Group II.

Our statistical investigations (1969a) show that the mechanism of period-changes contains an element of a random process. We notice that the sudden period changes in long period variables are similar to the above mentioned changes in Cepheids. However, the period changes in long period variables are larger than those in Cepheids.

REFERENCES

Soloviev, A. V., 1955, Bull. SAO No. 13.
Vasiljanovskaja, O. P., Erleksova, G. E., Shakhovskaja, N. I. 1966. Bull. Dushanbe No. 48.
Vasiljanovskaja, O. P., Erleksova, G. E., 1968. AC No. 469.
Vasiljanovskaja, O. P., Erleksova, G. E., 1969a. Bull. Dushanbe (in press).
Vasiljanovskaja, O. P., Erleksova, G. E., 1969b. Bull. Dushanbe (in press).
Vasiljanovskaja, O. P., Kiselev, N. N., Kiseleva, T. K. 1969c. Bull. Dushanbe (in press).

DISCUSSION

Fitch: Oosterhoff has found for 4 or 5 Cepheids of about 2 days period that cycle -to-cycle variations are periodic and caused by the excitation of a second pulsation-mode. Have you made any effort to see whether the cycle-to-cycle variations you observed in these longer period cepheids are periodic?

Shakhovsky: The variability of the shape of light curves is not periodic, it presents differences between the subsequent cycles like the RV Tau-type stars.

Non-Periodic Phenomena in Variable Stars
IAU Colloquium, Budapest, 1968

CORRELATIONS BETWEEN THE IRREGULARITIES OF THE PERIOD AND RADIAL VELOCITY, LENGTH OF PERIOD, AMPLITUDE AND SPECTRUM OF MIRA-VARIABLES

P. AHNERT

Institut für Sternphysik, Sternwarte Sonneberg, GDR

There are several well-known correlations between the abovementioned properties. However, because there are remarkable differences between slowly and rapidly moving Mira-stars (population I and II), I supposed that there may also be different correlations for these two groups. I have taken all Mira-stars of spectrum Me with radial velocities > 80 km/s ($n = 23$) and those with < 20 km/s ($n = 42$) from Merrill's Catalogue (ApJ 94, 171; 1941), which are also mentioned in the "Studies of Long Period Variables" by Leon Campbell (AAVSO 1955).

The comparing shows the following result:

Table 1

V	$\bar{P}$	$\overline{(M-m)/P}$	$\overline{Sp}$	σ_p	$\bar{A}$
20 km/s	325^d	0.436	M5. 42e	$\pm 2.69\%$	4^{m}85
80	236	0.450	M3. 89e	± 3.33	4.24

(σ_p = scattering of cycles in percents of the period.)

In the "Studies of Long Period Variables" Leon Campbell gives the times and magnitudes of the maxima of these stars from 1920 to 1950 with the single cycles, the mean period, the proportion $(M-m)/P$ and the spectrum. From these data I have computed the scattering of the cycles in days and percents of the period and the scattering of the magnitudes of the maxima.

Let us regard the slowly moving group. As seen on Figure 1 there is a loose relation between σ_p and the spectrum in the sense that the larger the scattering the earlier the sub-grioup of Me. The relation is rather loose; it becomes more evident if we divide the spectra into some groups (see the crosses). A corresponding relation exists between the scattering and the length of the periods (Figure 2). But this coincidence is trivial because it is a well-known fact that the longer the period the later the sub-group of Me. And with regard to the relation between the scattering of the cycles and the length of the period we must take into consideration that the percentage of the scattering must be higher at short periods, even if the scattering in days is the same for long and short periods. Nevertheless the relation seems to be at least partially real. (See also Table at the right side of the diagram.)

List of the stars

(Other properties see L. Campbell, "Studies of Long Period Variables")

1. 23 Mira-stars v > 80 km/s

			σ_p					σ_p	
			d	%				d	%
004533	RR	And	±8.8	±2.66	155823	RZ	Sco	±7.4	±4.61
004746a	RV	Cas	±8.5	±2.56	180531	T	Her	±5.7	±3.45
052036	W	Aur	±7.3	±2.68	183308	X	Oph	±10.2	±3.04
065208	X	Mon	±9.1	±5.86	193509	RV	Aql	±6.6	±3.03
073723	S	Gem	±6.9	±2.35	194048	RT	Cyg	±7.0	±3.64
083350	X	UMa	±6.8	±2.71	194348	TU	Cyg	±6.2	±2.81
090024	S	Pyx	±7.4	±3.60	200212	SY	Aql	±10.3	±2.91
110506	S	Leo	±6.9	±3.63	200715a	S	Aql	±6.7	±4.52
123160	T	UMa	±6.6	±2.58	215934	RT	Peg	±8.2	±3.81
153620a	U	Lib	±8.05	±3.55	220613	Y	Peg	±6.4	±3.10
154536	X	CrB	±6.1	±2.53	225914	RW	Peg	±9.3	±4.43
155229	Z	CrB	±6.6	±2.64					

2. 42 Mira-stars v < 20 km/s

			σ_p					σ_p	
			d	%				d	%
010102	Z	Cet	±8.4	±4.6	154715	R	Lib	±7.7	±3.0
010940	U	And	±9.1	±2.6	161122a	R	Sco	±6.5	±2.9
050953	R	Aur	±11.7	±2.5	162816	S	Oph	±7.7	±3.3
052404	S	Ori	±11.7	±2.85	164844	RS	Sco	±7.15	±2.25
053531	U	Aur	±10.2	±2.5	182133	RV	Sgr	±8.9	±2.8
072820b	Z	Pup	±13.3	±2.6	183149	SV	Dra	±6.5	±2.5
075612	U	Pup	±7.8	±2.2	185634	Z	Lyr	±9.7	±3.4
081112	R	Cnc	±9.7	±2.7	190941	RU	Lyr	±9.0	±2.4
094211	R	Leo	±7.3	±2.3	190967	U	Dra	±7.0	±2.2
094735	S	LMi	±6.2	±2.65	194604	X	Aql	±8.0	±2.3
102900	S	Sex	±6.7	±2.55	194632	X	Cyg	±6.65	±1.6
115919	R	Com	±6.8	±1.9	195202	RR	Aql	±9.0	±2.3
120012	SU	Vir	±7.3	±3.5	195308	RS	Aql	±9.25	±2.2
120905	T	Vir	±10.3	±3.0	200812	RU	Aql	±8.2	±3.0
122803	Y	Vir	±7.7	±3.5	203429	R	Mic	±5.2	±3.8
133273	T	UMi	±8.1	±2.6	204016	T	Del	±6.1	±1.8
134536	RX	Cen	±8.4	±2.6	205030a	UX	Cyg	±13.1	±2.35
140528	RU	Hya	±9.8	±2.9	210382	X	Cep	±10.7	±2.00
143417	V	Lib	±7.7	±3.0	221938	T	Gru	±6.3	±4.6
144918	U	Boo	±9.1	±4.55	230110	R	Peg	±7.8	±2.1
154615	R	Ser	±5.2	±1.7	235350	R	Cas	±8.1	±1.9

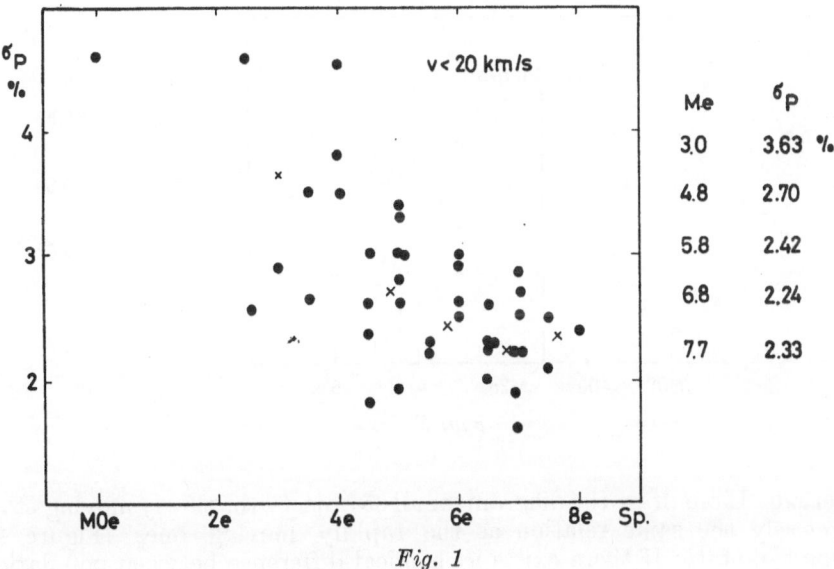

Me	σ_P
3.0	3.63 %
4.8	2.70
5.8	2.42
6.8	2.24
7.7	2.33

Fig. 1

The maxima of the short period stars are more pointed than those of the long period stars and therefore we should expect that the observational part of the scattering is smaller for short period stars than for long period objects. Another argument is the following: From 125 maxima obtained by two experienced observers (by Nijland and me and by Loreta and me) I found a mean error of only $\pm 2^d5$ or roughly 1% of the period for each maximum, if the mean value has been regarded to be correct. Therefore a rest of true physical relation remains in the sense that the shorter the period and the earlier the spectrum the larger the scattering of the cycles.

The rapidly moving stars show the same behaviour (Figure 3). The correlation seems to be steeper but this is only an effect of the shortness of

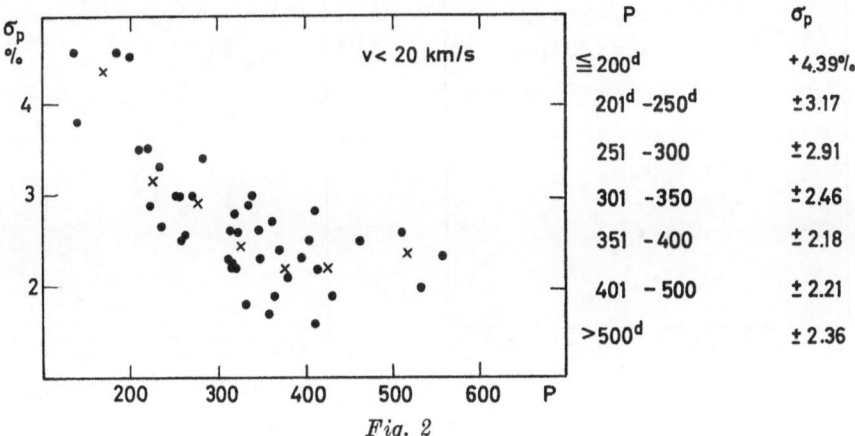

P	σ_p
$\leqq 200^d$	+4.39‰
$201^d - 250^d$	±3.17
$251 - 300$	±2.91
$301 - 350$	±2.46
$351 - 400$	±2.18
$401 - 500$	±2.21
$>500^d$	±2.36

Fig. 2

22*

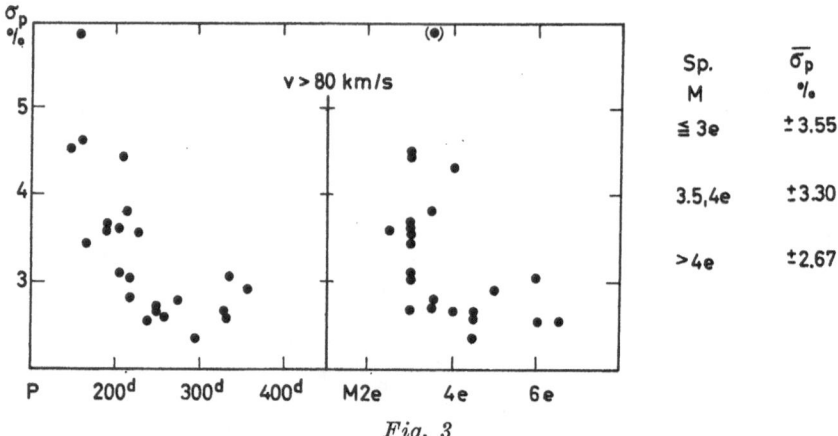

Fig. 3

these periods. If we draw the diagram for all 65 objects, the slowly moving stars give precisely the same relation as the rapidly moving ones (Figure 4).

One can state: If there exists a chemical difference between population I and population II Me-stars, the mechanism of the variability is obviously not influenced. Both groups behave much more in the same way than the classical Cepheids and the W Vir-stars.

Only the relation between the amplitudes and the scattering of the cycles shows an insignificant difference between the two groups (Figure 5). The scattering of the cycles is slightly greater for rapidly moving stars than for slowly moving objects of the same amplitude. If we form groups of about seven to ten stars, the difference becomes evident, though it remains small.

The greater scattering of the cycles of small amplitude variables in both groups is caused partially by the fact that their maxima are mostly very flat and therefore difficult to observe. But only for stars with shorter periods than 200^d are the amplitudes of the rapidly moving stars significantly smaller.

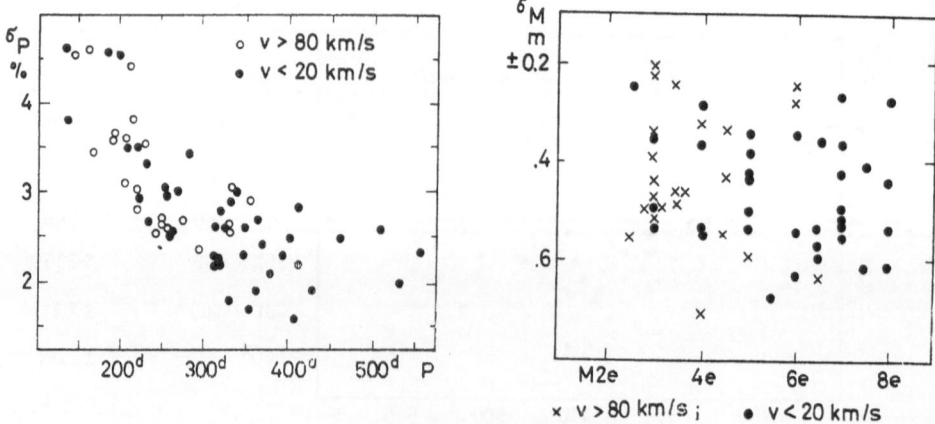

Fig. 4

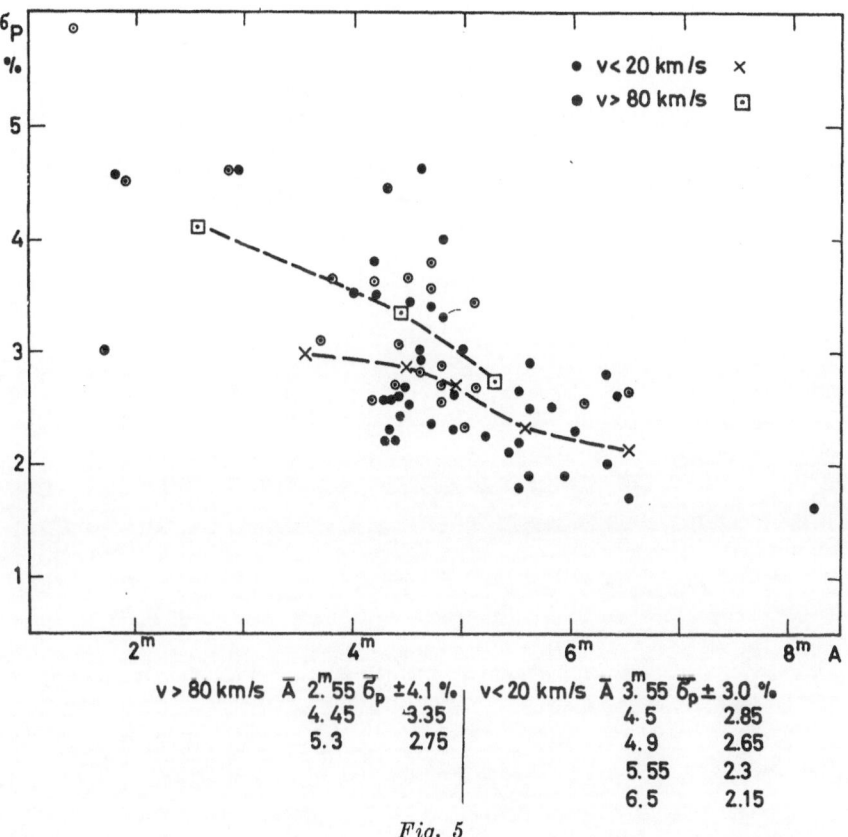

$$v > 80 \text{ km/s} \quad \overline{A} \ 2\overset{m}{.}55 \ \overline{\delta}_p \ \pm 4.1 \% \quad v < 20 \text{ km/s} \quad \overline{A} \ 3\overset{m}{.}55 \ \overline{\delta}_p \pm 3.0 \%$$

4.45	3.35	4.5	2.85
5.3	2.75	4.9	2.65
		5.55	2.3
		6.5	2.15

Fig. 5

For periods longer than 200ᵈ the amplitudes of the two groups are nearly equal (see Table 2).

Table 2

Period	< 200ᵈ	200ᵈ—299ᵈ	300ᵈ—399ᵈ	> 400ᵈ
v > 80 km/s	$\overline{A} = 2^m67$	4^m53	4^m78	—
v < 20 km/s	$\overline{A} = 3.83$	4.40	5.08	5^m88
all	$\overline{A} = 3.05$	4.47	5.02	5.88

Finally it may be of interest to notice that no relation could be found between the scattering of the maximum magnitudes and any other property. Diagram 6 gives an example of this fact. Each sub-group of Me contains stars with high and with small scattering of the maximum brightness.

THE IRREGULARITIES IN THE LIGHT-CHANGES OF MIRA CETI

P. L. FISCHER

Wien, Freyung 6, Austria

The author has collected the whole material he was able to obtain about the visual light-changes of Mira Ceti and about 500 individual light curves of maxima and minima as well. A Monograph will be published next year in the Annals of the University Observatory in Vienna.

Although Mira was discovered in 1596 by Fabricius and since Hevelius it was systematically observed, the observations for serious researches — especially for the study of the period-length — are not useful before 1839 (ep. 268 according to Prager. Gesch. u. Lit. 1934).

The statement by Eddington and Plakidis (M.N. *90*, 65, 1929) "The light fluctuations of long-period variables have a well-marked periodicity complicated by superposed irregularities" is particularly confirmed by Mira Ceti: not only the dates and magnitudes of the maxima and minima show irregularities, but also the shape of the light curve is different from cycle to cycle. In order to measure these changes, *numerical quantities* have been introduced, which together with other essential notions, can be seen in Fig. 1. In addition to the minimum and the subsequent maximum — a natural unit of one cycle — there is another point of special astrophysical interest, the eruption-point: this is the beginning of the nearly linear, very steep slope ES of the ascending branch; in plotting the observations, the magnitudes begin to vary strongly at this point, until a clear ascent is to be seen. The steepness of the eruption ES is η (in days pro 1^m). Between the points A and B the star is 0.4^m brighter than in the minimum and between C and D 0.4^m fainter than in the maximum; thus t can be considered as the duration of the minimum-light, and h as that of the maximum-light. The steepness of B_1C_1 and D_1A_2 defines σ and τ (in days pro 1^m), the mean slope of the ascending and descending branch, respectively.

Which are the irregularities in the light-changes of Mira Ceti? The most important phenomena are as follow:

1. *The individual period-lengths*, defined as time differences between two subsequent observed maxima (P) or minima (Q) or eruption-points (R; see Fig. 1), have a wide range; Table 1 shows the frequency distribution and the mean values. The good agreement of the P-, Q- and R-values proves that the eruption-point is a reliable feature, to which the spectroscopists should pay more attention. Figure 2 shows an everlasting fluctuation of the individual period-lengths up to 30^d and more, which is interrupted only occasionally, and fully irregularly, by quiet times. If a power spectrum analysis is made as Blackman and Tukey (Dover Publ. 190, 1958) have done for climatological time-series, significant waves are to be seen (Fig. 2, below): a wave of 2 cycles

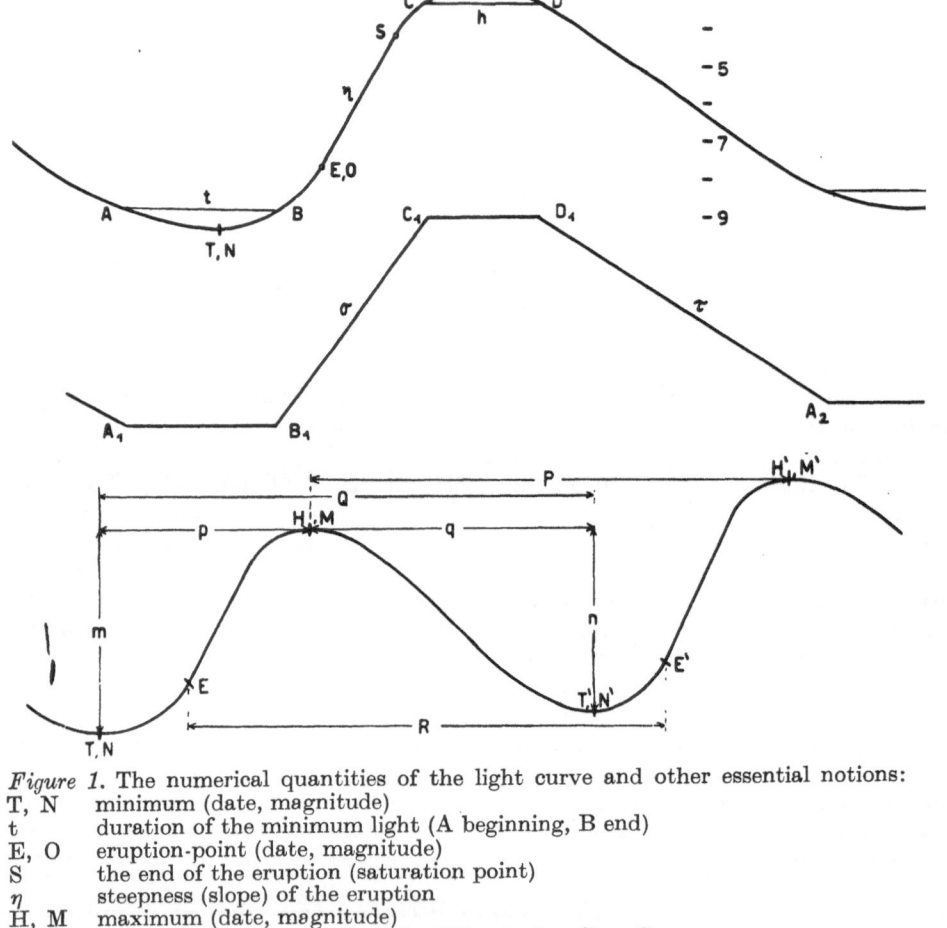

Figure 1. The numerical quantities of the light curve and other essential notions:

T, N   minimum (date, magnitude)
t   duration of the minimum light (A beginning, B end)
E, O   eruption-point (date, magnitude)
S   the end of the eruption (saturation point)
η steepness (slope) of the eruption
H, M   maximum (date, magnitude)
h   duration of the maximum light (C beginning, D end)
σ mean slope of the ascending branch
τ mean slope of the descending branch
P, Q, R   individual period length, found out of 2 subsequent observed maxima or minima. or eruptions respectively
p, q   partial period length of the ascending or descending branch
m, n   partial amplitude of the ascending or descending branch

length in the P's; a less clear wave in the Q's; surprisingly a 10 cycles-wave in the R's; a Markov-persistence has not been observed.

II) *Punctualities of the events (maxima, minima, eruptions):* It is usual to use a formula of the form $C = T + n \cdot X$ for a certain observational series and compare the C's with the observed dates O; the (O—C)-values can be taken for "punctualities" and are denoted in the maxima with U, in the minima with V and in the eruption-points with W; Z is the average of them. The method of least squares furnishes the following formulas for the time after 1938 (ep. 268—408):

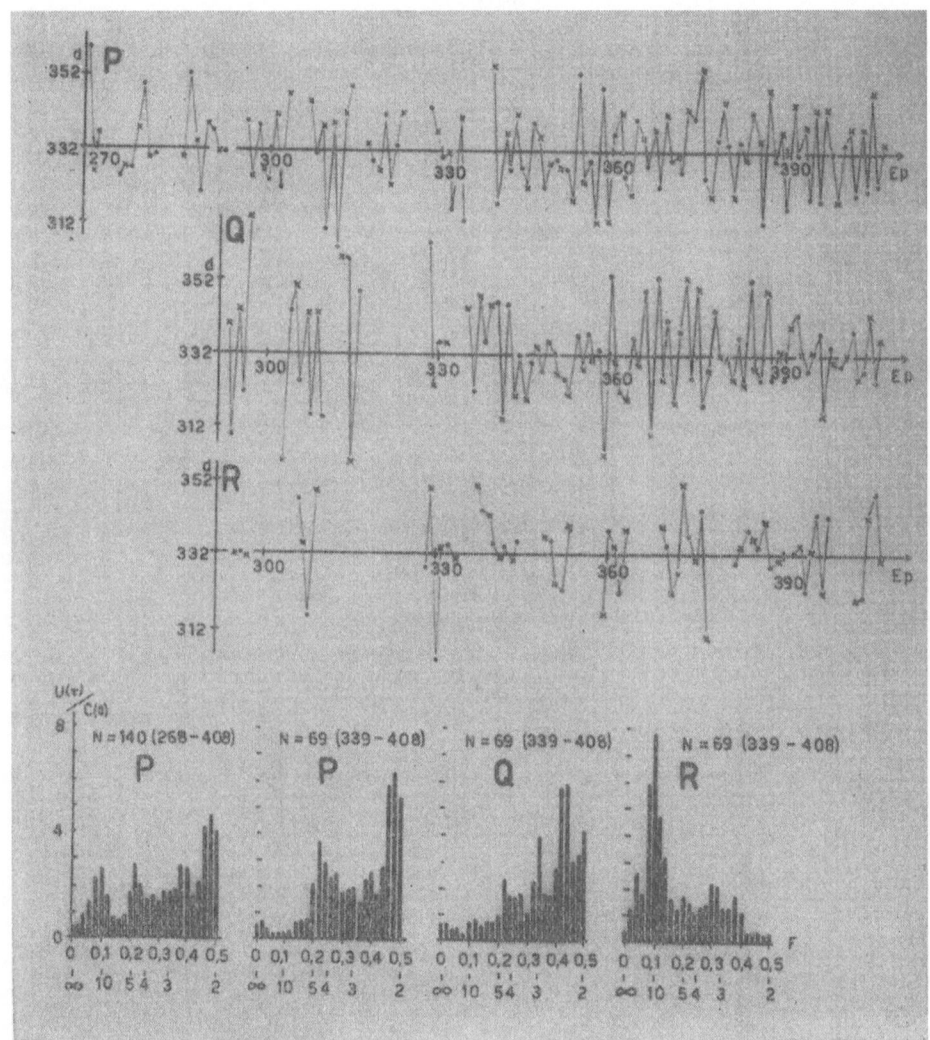

Figure 2. The individual period lengths P, Q, R and their power spectra: the numbers below are the frequencies and the wave lengths (in cycles)

from 123 observed maxima:

$$C_{max} = 239 \ 3019.5 \ (\pm 4.2^d) + n \cdot 331.5545 \ (\pm 0.045^d);$$

from 91 observed minima:

$$C_{min} = 240 \ 1500.8 \ (\pm 6.1^d) + n \cdot 331.7762 \ (\pm 0.075^d);$$

from observed eruption-points:

$$C_{eru} = 240 \ 1916.8 \ (\pm 6.3^d) + n \cdot 331.4809 \ (\pm 0.088^d).$$

Table 1

Mira Ceti. The individual period-lengths P, Q, and R: frequency distribution, mean value, $\pm$ standard deviation, time and number of events

X	From all the material													From the better material
	frequency distribution												$X \pm s$	$X \pm s$
	288	296	304	312	320	328	336	344	352	360	368	376^d		
P	2	2	8	20	47	55	42	24	10	2	1		332.15 $\pm$ 12.7 ep 38—407 *213*	*332.0389* $\pm$ 10.4 ep 71—407 *131*
Q	0	2	3	8	22	23	17	12	5	2	0		332.92 $\pm$ 13.0 ep 293—407 *94*	*332.6256* $\pm$ 11.8 ep. 293—407 *86*
R	0	1	1	3	8	25	15	7	0	0	0		333.28 $\pm$ 9.8 ep 295—407 *60*	*333.0357* $\pm$ 10.0 ep 295—407 *56*
Σ	2	5	12	31	77	*103*	74	43	15	4	1		*367*	*273*

Table 2

Mira Ceti, prognosis of the maxima, minima and eruptions

Events	Ep.	Method of Sterne (H. C. 386) T_S			Method of Fischer T_F			T_S-T_F
maxima	408	243 9459	= 1966	Nov. 29	9453	= 1966	Nov. 23	6^d
	409	9791	1967	Oct. 27	9783	1967	Oct. 19	8
	410	244 0122	1968	Spt. 22	0113	1968	Spt. 13	9
	411	0454	1969	Aug. 20	0443	1969	Aug. 9	11
	412	0785	1970	July 17	0772	1970	July 4	13
	413	1117	1971	Jun. 14	1102	1971	May 30	15
	414	1449	1972	May 11	1431	1972	Apr. 23	18
	415	1780	1973	Apr. 7	1760	1973	Mar. 18	20
	416	2112	1974	Mar. 5	2090	1974	Feb. 11	22
minima	408	243 9340	= 1966	Aug. 2	9339	= 1966	Aug. 1	1^d
	409	9672	1967	Jun. 30	9670	1967	Jun. 28	2
	410	244 0004	1968	May 27	9999	1968	May 22	5
	411	0336	1969	Apr. 24	0329	1969	Apr. 17	7
	412	0668	1970	Mar. 22	0659	1970	Mar. 13	9
	413	1000	1971	Feb. 17	0989	1971	Feb. 6	11
	414	1331	1972	Jan. 14	1318	1972	Jan. 1	13
	415	1663	1972	Dec. 11	1648	1972	Nov. 26	15
	416	1995	1973	Nov. 8	1978	1973	Oct. 22	17
eruptions	408	243 9402	= 1966	Oct. 3	9390	= 1966	Spt. 21	12^d
	409	9734	1967	Aug. 31	9720	1967	Aug. 17	14
	410	244 0066	1968	Jul. 28	0050	1968	Jul. 12	16
	411	0398	1969	Jun. 25	0379	1969	Jun. 6	19
	412	0731	1970	May 24	0709	1970	May 2	22
	413	1063	1971	Apr. 21	1038	1971	Mar. 27	25
	414	1395	1972	Mar. 18	1367	1972	Feb. 19	28
	415	1727	1973	Feb. 13	1697	1973	Jan. 14	30
	416	2059	1974	Jan. 11	2026	1973	Dec. 9	33

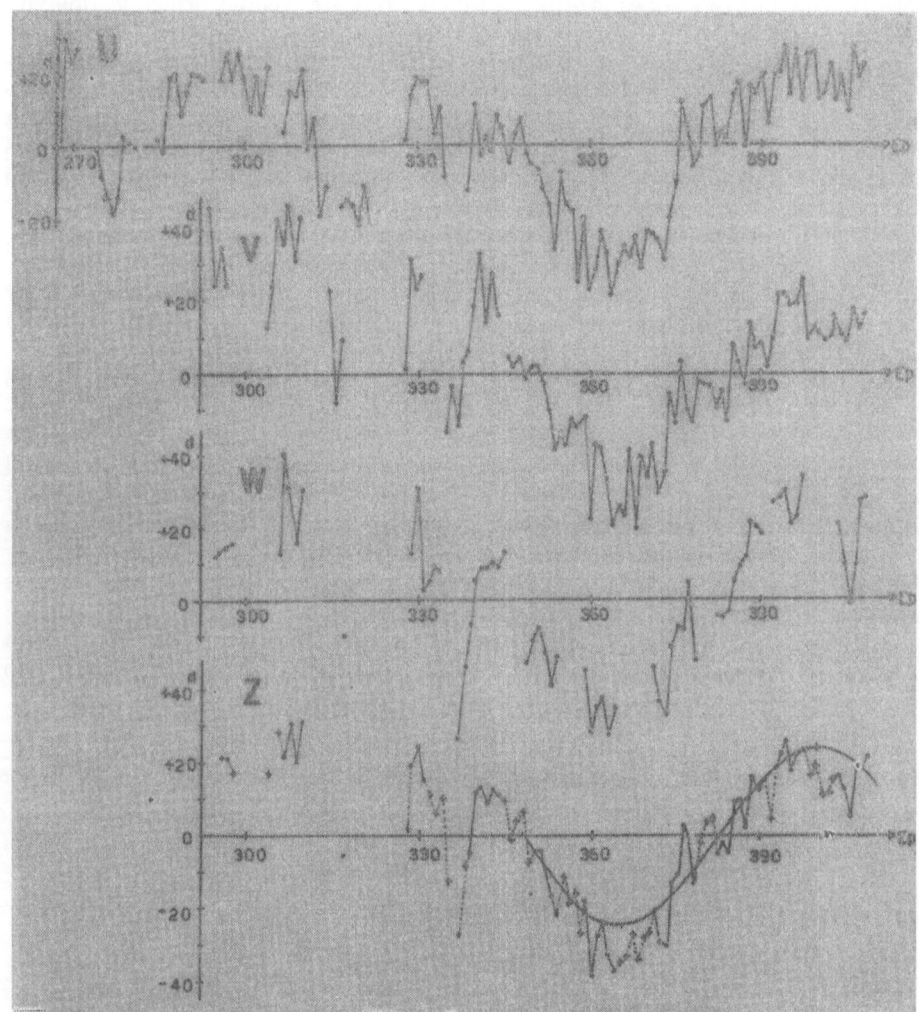

Figure 3. The punctualities (O−C) of the events: U maxima, V minima, W eruptions;
Z average

Figure 3 shows the punctualities U, V, W and the average Z as a function of time: their similar appearance is conspicuous; the sine curve of the
last Z-values (from ep. 348) makes a prognosis possible up to the epoch 416,
regardless of the way of its coming into being: see Table 2.

Two publications have furnished an essential contribution to the correct
understanding of the O—C diagrams:

1. A. S. Eddington and S. Plakidis, "Irregularities of Period of Long-
period Variable Stars" (M. N. *90*, 65, 1929): They distinguish the following
3 kinds of irregularities: a) *permanent i.:* if the individual periods differ from the
mean period by a purely accidental fluctuation; b) *temporary i.:* if the primary

cause of the light-changes works regularly, but its visible effect is delayed or accelerated by casual circumstances (so that the maximum-date differs from the ephemeris-date by an accidental fluctuation); a side remark: observational errors operate in the same manner; c) *repeat i.*: if the phase differencies increase with time — in this case the period-length changes. In the first case there is no tendency to return to the original ephemeris-dates: the delays and accelerations of successive periods build up in the same way as accidental errors; if the star is late, the time lost is written off as irretrievable.

2. T. E. Sterne, in "The Errors of Period of Variable Stars" (H. C. 386, 1934) has further developed these ideas as *"cumulative"* and *"non-cumulative"* errors. About O—C diagrams he remarks: They indicate, particularly after the lapse of many epochs, the accumulated errors of the star; and although changes of true period will be reflected in such diagrams as curvatures, curvatures will also be caused, in the absence of such changes, by the accumulation of errors" (p. 14). Further in his work Sterne demonstrates a way for the calculation of the standard deviations s and e of the cumulative and non-cumulative errors respectively, and the "best approximation" X' to the true period over a certain series of observations, a so-called "run". The mean error (m.e.) of the period determined in this way is greater than that obtained by using the method of least squares. This m.e. is of great importance in ascertaining whether the difference between the periods found from 2 subsequent runs is significant or not; in this way it is possible to find a change of the period-length. Table 3 shows the application of Sterne's method to the rich material of Mira Ceti: The entire time of useful observations has been divided in runs and then

Table 3

Proof of the constancy of the period-length in Mira Ceti after the method of T. E. Sterne (HC. 386)

	Run	Number of epoch	e	s	X' $\pm$ m. e.	Differencies r_1-r_{14}	r_1-r_{34}
maxima	r_1	268—303	4.5d	7.4d	331.7315 $\pm$ 1.26d	+0,14d	—
	r_2	303—335	6.6	6.1	331.0984 $\pm$ 1.11	—0.49	—
	r_3	339—373	6.8	2.7	330.7194 $\pm$ 0.51	—0,87	—1.19d
	r_4	373—408	7.5	2.0	332.4328 $\pm$ 0.39	+0.84	+0.52
	r_{34}	339—408	6.8	4.1	331.9100 $\pm$ 0.50		← ⌐
	r_{14}	268—408	6.5	5.0	331.5885 $\pm$ 0.42	← ⌐	
	r_3	338—373	7.5	4.1	330.7091 $\pm$ 0.73	—	—1.21
	r_4	373—408	5.9	2.8	332.5321 $\pm$ 0.51	—	+0.61
	r_{34}	338—408	6.3	4.8	331.9174 $\pm$ 0.58		←⌐
	r_3	338—373	2.2	7.4	331.1143 $\pm$ 1.25	—	—1.04
	r_4	373—408	6.5	2.9	332.4532 $\pm$ 0.53	—	+0.30
	r_{34}	338—408	4.1	6.7	332.1572 $\pm$ 0.81		←⌐

Table 4

Mira Ceti. The numerical quantities

Quantity			Mean ± st. dev.	Frequency distribution (middle of the class, number)
magnitude of	maximum	M	3.45 ± 0.64m	2.1/4, 2.3/6, 2.6/15, 2.9/37, 3.2/36, 3.5/31, 3.8/41, 4.1/29, 4.4/13, 4.7/3, 5.0m/5 — Σ 220
	minimum	N	9.07 ± 0.27m	8.5/9, 8.8/23, 9.1/41, 9.4/25, 9.7m/3 — Σ 101
	eruption	O	7.7 ± 0.6m	6.6/3, 6.8/5, 7.1/6, 7.4/10, 7.7/15, 8.0/13, 8.3/12, 8.6/6, 8.8m/0 — Σ 70
duration of	max. light	h	50 ± 12 d	15/0, 22/2, 29/6, 36/11, 43/26, 50/30, 57/25, 64/14, 71/4, 78/1, 85d/1 — Σ 120
	min. light	t	76 ± 12d	41/1, 48/1, 55/1, 62/6, 69/20, 76/21, 83/18, 90/2, 97/4, 104/3, 111d/1 — Σ 78
ascending branch	partial period	p	121.16 ± 11.2d	89/8, 100/8, 111/16, 122/44, 133/21, 144/4, 155d/1 — Σ 95
	part. amplitude	m	5.56 ± 0.51m	4.2/1, 4.7/8, 5.2/10, 5.7/34, 6.2/17, 6.7/2, 7.2m/1 — Σ 90
	mean slope	σ	13.5 ± 2.3 d/m	7.5/1, 9.5/4, 11.5/17, 13.5/26, 15.5/14, 17.5/3, 19.5d/m/3 — Σ 68
Mean slope of eruption		η	9.1 ± 2.2 d/m	6.5/22, 10.5/33, 14.5d/m/3 — Σ 58
descending branch	partial period	q	210.62 ± 12.7d	178/2, 189/7, 200/21, 211/36, 222/17, 233/9, 244d/1 — Σ 93
	part. amplitude	n	5.55 ± 0.64m	4.2/21, 4.7/18, 5.2/22, 5.7/27, 6.2/18, 6.7/6, 7.2m/2 — Σ 89
	mean slope	τ	29.7 ± 3.9 d/m	21/1, 24/6, 27/17, 30/30, 33/33, 36/36, 39d/m/3 — Σ 65

the "best approximations" X' together with their m.e. for each part and the whole time has been calculated; as the differences are of the order of the magnitude of the m.e. the main question is solved: the period of Mira Ceti is constant, although the fluctuations of the individual lengths of the period are very large.

III) *The numerical quantities:* their mean values and the frequency distributions are shown in Table 4. Only the maximum magnitudes show a tendency to alternate, not the magnitudes of the minima or eruptions.

Neither the sequence of the partial-periods (p. q, see Fig. 1), nor the sequence of the partial-amplitudes (m, n) have any regularity; it is interesting that the partial period p of the ascending branch is nearly always 37% ($\pm$3) of the time difference (Q) between the correspondent minima.

DISCUSSION

Fernie: I was not quite clear as to what you meant by "eruption". Is this the moment at which the emission lines appear in the spectrum?

Fischer: No. The eruption-point is defined as the beginning of the nearly linear, very steep slope of the ascending branch of the light curve.

Detre: It was a pleasure to hear about an application of power spectrum analysis to a Mira variable. So far I know, this is the first study of this kind.

Fischer: Yes.

ON THE INTRINSIC LIGHT-POLARIZATION IN-SOME LATE TYPE STARS

R. A. VARDANIAN

Byurakan Astrophysical Observatory, Armenia, USSR

The variability of polarimetric parameters of μ Cephei was discovered in 1957 at the Byurakan Observatory (Grigorian 1958). So it became clear, that sometimes, besides the interstellar polarization, a rather strong intrinsic polarization can exist as well. After this discovery the interest in polarimetric observations of late type stars strongly increased. During the past ten years the variability of polarimetric parameters of a number of stars was discovered. (Grigorian 1959; Serkowski 1966; Shakhovski 1964; Vardanian 1966—67; Zappala 1967.)

At present the existence of intrinsic light-polarization can be considered as established for a group of late type stars. Among them, however, not more than 15 stars show a great amount of light-polarization ($P \geq 2\%$).

In this paper we wish to present the results of our polarimetric observations of some late type stars, in which the degree of light polarization according to our observations was higher than 2% in the blue part of their spectra.

175 late type stars (spectral types M, N, R, S) were observed during 1967 and 1968. 65 of them are variable stars. Only six variables showed a considerable amount of light-polarization. The relevant data are presented in Table 1. Its columns give respectively: the name of the star, the galactic coordinates, the maximum and minimum magnitudes, the period, the spectrum, the mean values of maximum and minimum degrees of polarization, the mean value of the positional angle of polarization, the difference of magnitudes corresponding to the maximum and minimum polarization degree and the mean value of the ratio of the polarization degrees measured in the blue and yellow parts of the spectrum.

Table 1.

Star	l^{II}	b^{II}	Type	Magnitude		Period	Spectrum	P_{max} %	P_{min} %	Pos. angl.	$\frac{m_{P_{max}}}{m_{P_{min}}}$	$\frac{P_{blue}}{P_{yellow}}$
				max	min							
RX Boo	001°	+68°	SRb	8^m6	11^m3	78^d	M7e–M8e	2.3	1.7	62°	$+0^m8$	1.7
AB Cyg	050	—15	SRb	9.5	10.1	520	M4III	4.0	3.2	51	+0.1	1.4
AK Peg	054	—44	SRa	8.9	10.8	195	M5e	3.0	1.2	50	+0 7	—
R Gem	162	+15		6.0	14.0	369.93	S 3.9e—S6.9e	3.2	0.3	80	+1.5	1.8
Z Cnc	177	+28	SRb	9.4	10.7	104	M6III	2.0	1.0	45	+0.25	2.0
CD Ser	350	+46	Ib	10.0	11.0	80?	M4	3.4	3.2	70	+0.1	1.7

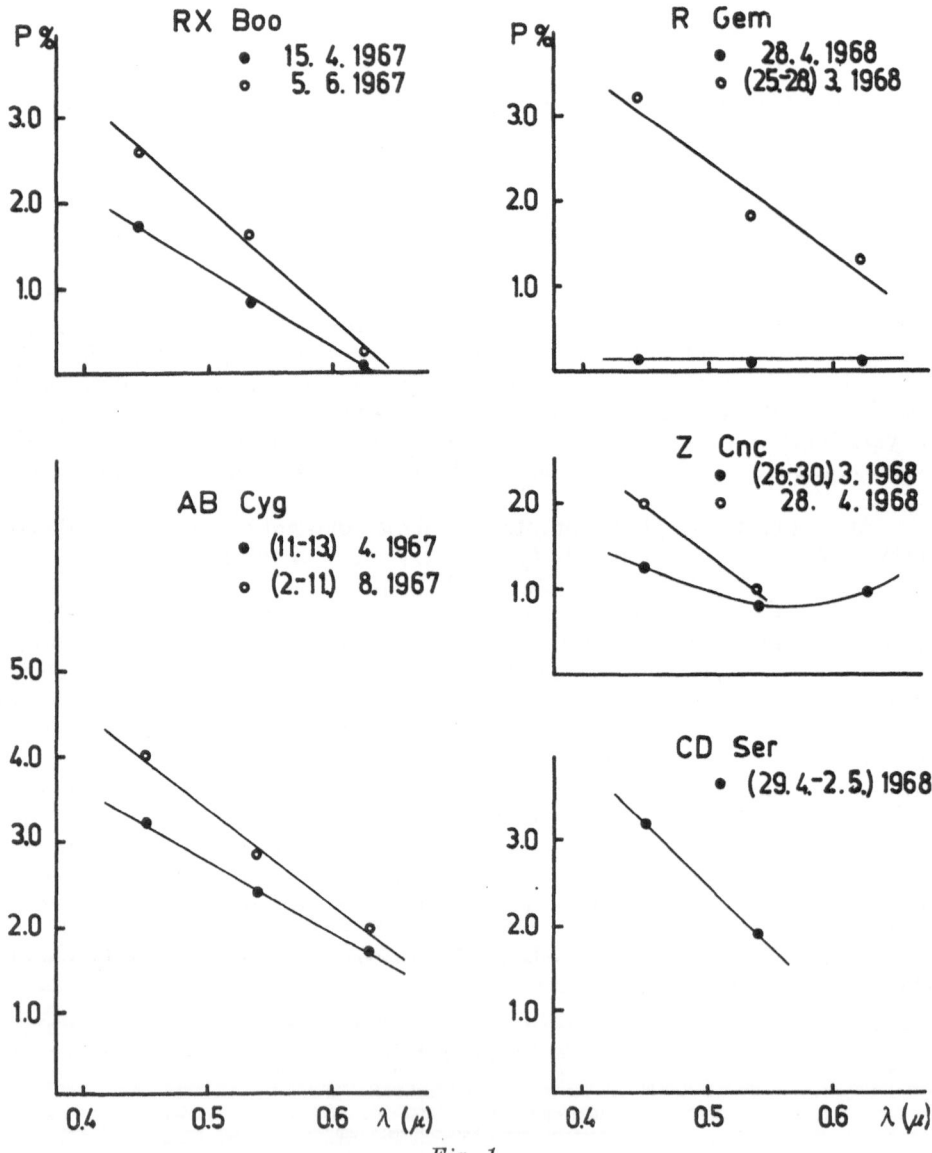

Fig. 1

From the polarimetric data gathered in Table 1 it follows:

a) the variations of polarization degree of these stars are greater than the observational errors $(\overline{\Delta P} = \pm\ 0.2\%)$.

b) it is very probable that at minimum polarization the star is brighter than at maximum polarization.

c) the degree of polarization in the blue part of the spectrum $(\lambda_{\mathrm{eff}} = 4500\ \text{A})$ is on the average 70% higher than in the yellow part $(\lambda_{\mathrm{eff}} = 5400\ \text{A})$.

Fig. 1. presents the relation between the mean values of maximum and minimum degrees of polarization and the wave lengths. The open circles correspond to the state of fainter radiation of the star.

As it was already shown (Vardanian 1967; Goyne and Kruszovski 1968) in the case of cold supergiants and specially in the case of μ Cephei such a correlation was not observed.

Although the number of observed stars is not very large, it is, nevertheless, possible to make some preliminary quantitative analysis.

In fact as only six of 65 randomly chosen variable stars show in the blue part of the spectrum a polarization degree greater than 2%, it can be supposed, that at least 9% of late type stars possess an intrinsic polarization higher than 2%. It should be noted that the polarization degree of stars surrounding the variables was within the limits of observational errors. Moreover, none of the 110 late type stars, situated in the neighbourhood of the 65 variables, showed a remarkable degree of polarization. It follows therefrom, that the presence and the changes of light-polarization of late type stars are connected with the variability of their brightness.

The search for late type star having an intrinsic light polarization is still in progress. The observational data together with their detailed analysis will be published after the programme is finished.

Note added October 24, 1968: The polarization of five more variable stars, TW Peg, AE Cap, Z Eri, T Aqr and Z Psc was discovered by the author.

REFERENCES

Goyne, G., and Kruszovski, A., 1968, Astr. J., **73**. 29.
Grigorian, K. A., 1958, Byurakan Obs. Contr. **25**, 45.
Grigorian, K. A., 1959, Byurakan Obs. Contr. **27**, 43.
Serkowski, K., 1966, Inf. Bull. Var. Stars. No. 141. Budapest.
Shakhovski, N. M., 1964, Astr. Zu. **40**, 6.
Vardanian, R. A., 1966, Byurakan Obs. Contr. **37**, 23.
Vardanian, R. A., 1967. Rus. Astron. Zhirk. No. 433.
Zappala, R., 1967, Astrophys. J. **148**, 81.

PART IV

NON-PERIODIC PHENOMENA IN BINARY SYSTEMS

A. HOT, VERY SHORT-PERIOD ERUPTIVE BINARIES: OLD NOVAE, U Gem STARS etc.

B. SYMBIOTIC STARS

C. CONTACT BINARIES AND W UMa STARS

D. CONVENTIONAL BINARIES: PERIOD VARIATIONS, INTERPRETATION IN TERMS OF GAS STREAMS, MASS EXCHANGE etc.

HOT, VERY SHORT-PERIOD ERUPTIVE BINARIES

Introductory Report by

J. SMAK

Institute of Astronomy, Polish Academy of Sciences Warsaw, Poland

ABSTRACT

Some problems connected with the interpretation of the photometric and spectroscopic observations of the U Geminorum type stars are discussed.

1. INTRODUCTION

In order to discuss the non-periodic or irregular phenomena displayed by a given group of stars one should have a good knowledge of all regular properties and basic physical processes known to occur in these objects. In the case of hot, very short-period eruptive binaries the situation is still quite unsatisfactory in this respect. Within each group we observe a considerable diversity of basic physical characteristics and a large variety of occurring phenomena. Several excellent review articles on these objects have appeared recently (e.g. Kraft 1963, 1966, Mumford 1967) and therefore there seems to be no reason for presenting another review of this type. Instead an attempt will be made here to discuss only some selected problems connected with the interpretation of the apparently most homogeneous group of objects, namely of the U Geminorum type stars.

2. SOME BASIC PROPERTIES

Due to the work of Kraft (1962) it is now well established that all U Geminorum type variables are binaries. Their orbital periods range from about 3 hours to about 9 hours. At minimum light, i.e. between the outbursts, the spectrum of U Geminorum type variables consists of the following components: *(a)* the blue, hot continuum usually associated with the blue component of the system; however, no spectral features belonging to this component are seen in the spectrum and it is difficult to tell what fraction of the hot continuum may belong to it; *(b)* strong, emission lines, primarily those of H, He I, and Ca II, which are interpreted as being due to a rotating gaseous ring around the blue component; as a rule these emission lines are observed to be double when the orbital inclination is close to 90° (e.g. in U Gem); *(c)* absorption lines belonging to the red component; these lines are visible only in some systems, being undetectable in the others (at least in the blue region of the spectrum), thus indicating that the relative luminosity of the red star is quite different in different systems. In order to explain the existence of the ring it is necessary to postulate (Kraft 1962, 1963, Krzemiński 1965) that the red component is in contact with its Roche lobe and is loosing mass through the inner Lagrangian point; this mass, together with the momentum it carries, is responsible — directly or indirectly — for the formation of the ring.

It is really unfortunate that among all U Geminorum type variables studied until now there is no single object that would simultaneously be a double-line spectroscopic binary and an eclipsing variable. This leaves us with some crude estimates only of stellar masses and radii. Two stars, namely U Gem and SS Cyg, can be considered in order to illustrate the existing situation. In U Gem no lines of the red component are visible and therefore no direct estimate is available for the mass-ratio $\mathfrak{M}_b/\mathfrak{M}_r$ (b and r stand for the blue and red components, respectively). On the other hand, the star is known to be an eclipsing variable (Krzemiński 1965) and therefore the orbital inclination is close to $i = 90°$. Taking the mass-ratio as an unknown parameter we can obtain a set of solutions for $\mathfrak{M}_b$ and $\mathfrak{M}_r$ (Krzemiński 1965, Paczyński 1965). With $\mathfrak{M}_b/\mathfrak{M}_r = 0.9$, which is the average value for the double-line U Geminorum type binaries, we get $\mathfrak{M}_b = 1.1\mathfrak{M}\odot$ and $\mathfrak{M}_r = 1.2\ \mathfrak{M}\odot$. Larger mass-ratios would lead to larger masses but there is a limit set up by the resulting radius of the gaseous ring; namely, if we require that the ring should be within the Roche lobe of the blue component, then we get that the mass-ratio cannot be larger than about 1.0. On the other hand, however, it is possible that the mass-ratio is smaller than 0.9 and that the masses are smaller than those given above. SS Cyg is a double-line spectroscopic binary with $\mathfrak{M}_b/\mathfrak{M}_r = 0.9$ (Joy 1956). No eclipse has been detected, however, in spite of many searches and the orbital inclination is an unknown parameter. With $i = 70°$ we get $\mathfrak{M}_b = 0.22\ \mathfrak{M}\odot$ and $\mathfrak{M}_r = 0.24\ \mathfrak{M}\odot$ while with $i = 30°$ we get $\mathfrak{M}_b = 1.44\ \mathfrak{M}\odot$ and $\mathfrak{M}_r = 1.60\mathfrak{M}\odot$.

3. PERIOD CHANGES

Walker and Chincarini (1968) have recently found that the orbital period of SS Cyg is increasing with time. The observed rate of the period variation is $A = +9.^{d}54 \times 10^{-11}$, where A is the coefficient in the usual formula

$$\text{Zero Phase} = T_0 + PE + AE^2. \tag{1}$$

An inspection of the O—C diagram for U Gem (Fig. 1) shows that the orbital period of this star is also increasing; the data presented in Fig. 1 lead to $A \approx +2.^{d}0 \times 10^{-11}$. Since SS Cyg and U Gem are rather typical examples of the U Geminorum type variables, it is tempting to conclude that the lengthening of the orbital period may be a common characteristic of the entire class.

Period changes in close binary systems are usually interpreted as a result of the exchange of mass or of the loss of mass from the system. Much work has been done recently in this field (cf., for example, Kruszewski 1966). The rate of the period variation can be related to the rate of the mass transfer (or mass loss) provided the mechanism of the mass transfer (or mass loss) can be properly identified. Let us see then what conclusions can be obtained in this respect for SS Cyg and U Gem under different assumptions concerning the mechanisms involved.

Case I. Mass exchange between the components with no exchange between the orbital and rotational momenta. Fixing our attention on the red

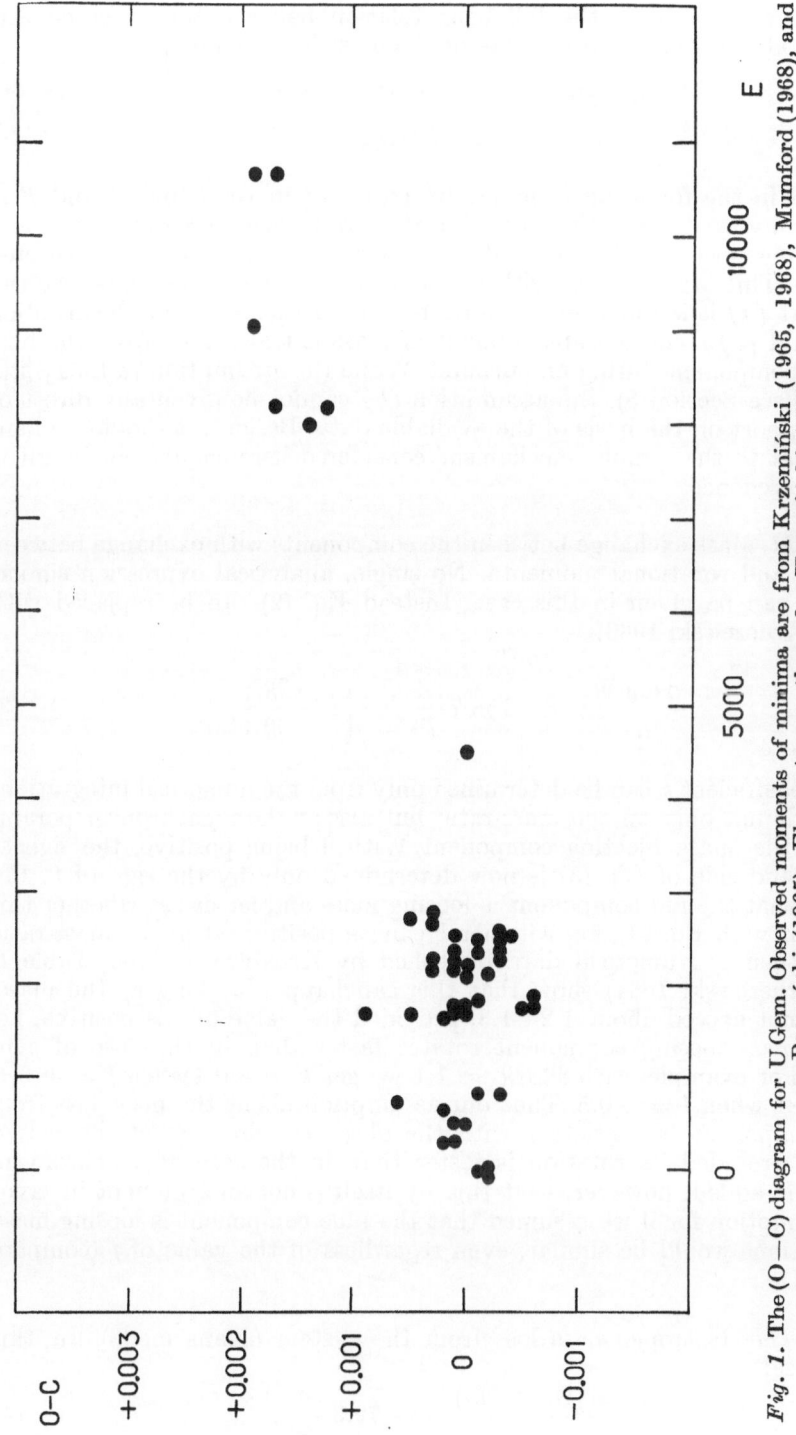

Fig. 1. The (O—C) diagram for U Gem. Observed moments of minima are from Krzemiński (1965, 1968), Mumford (1968), and Paczyński (1965). Elements are those by Krzemiński (1965).

component we can write the following relation between the observed rate
(A) of period variations and the rate of the mass loss (or mass gain):

$$\frac{d \log \mathfrak{M}_r}{d\,t} = 52.9\,\frac{A}{P^2}\left(\frac{\mathfrak{M}_r}{\mathfrak{M}_b} - 1\right)^{-1};$$ (2)

in this and in the following formulae we express t in years and A and P in
days. We may easily note that with the mass-ratio being greater than 1 and
with A being positive, the mass of the red component should be expected
to increase. This is in conflict with the indication given above unless we can
assume that *(a)* it is the blue component that is responsible for the eruption
activity, and *(b)* a considerable amount of mass is transferred from the blue
to the red component during an outburst. While the assumption *(a)* may still
be correct (see Section 5), the assumption *(b)* cannot be given any direct or
indirect support on the basis of the available data. Besides, it should be kept
in mind that to the simple mechanism considered here should not be given
too much credence.

Case II. Mass exchange between the components with exchange between
the orbital and rotational momenta. No single, analytical expression similar
to Eq. (2) can be given in this case. Instead Eq. (2) can be replaced with
(see, e.g., Kruszewski 1966):

$$\frac{d \log \mathfrak{M}_r}{d\,t} = -158.7\,\frac{A}{P^2}\,k^{-1}\left(1 + \frac{\mathfrak{M}_r}{\mathfrak{M}_b}\right),$$ (3)

where the coefficient k can be determined only from the numerical integrations
and depends not only on the mass-ratio but also on the asynchronism parameter, f, of the mass ejecting component. With A being positive, the sign of
the right-hand side of Eq. (3) is now determined only by the sign of k. Let
us assume that the red component is loosing mass and let us see whether this
is consistent with Eq. (3), i.e. whether k can be positive when the mass-ratio
is greater then 1. Numerical data published by Kruszewski (1966, Table I;
see also Kruszewski 1964) show that this can happen as long as the mass-
ratio does not exceed about 1.2—1.3, provided the value of f is positive, i.e.
when the mass loosing component rotates faster than in the case of syn-
chronism. For example, with $\mathfrak{M}_b/\mathfrak{M}_r \simeq 1.1$ we get $k \simeq +0.4$ when $f = +\,0.1$,
and $k \simeq +1$ when $f = +0.5$. Thus our assumption about the mass loss from
the red component is consistent with the observed direction of the period
variations, provided its rotation is faster than in the case of synchronism.
It should be added, however, that this by itself is not an argument in favor
of our assumption for if we assumed that the blue component is loosing mass
our conclusions would be similar, even regardless of the value of f (compare
with Case I).

Case III. Isotropic mass loss from the system (Jeans mode). In this
case we have:

$$\frac{d \log (\mathfrak{M}_r + \mathfrak{M}_b)}{d\,t} = -79.3\,\frac{A}{P^2}.$$ (4)

It is irrelevant here which of the two components is loosing mass, the effect being always the same — namely the increase of period. Situation becomes more complicated, however, when we wish to include the angular momentum effects. A second term appears then in the right-hand side of Eq. (4), which is proportional to $d\log \bar{h}_0/dt$, where h_0 is the angular momentum per unit reduced mass. Therefore depending on the amount of momentum carried away by the ejected matter the period may either increase or decrease. In other words, without going into the details of the angular momentum effects it is impossible to decide whether the observed increase of period in the case of SS Cyg and U Gem is, or is not consistent with the pure mass loss hypothesis. In particular, no much weight should be given to the rate of mass loss computed with the help of Eq. (4). Finally, adopting such a possibility would require neglecting the contribution to the observed period variations due to the mass exchange which is certainly taking place in these systems.

We are not going to consider here the most complicated case of simultaneous mass exchange and mass loss. No simple description can be given in such a case mostly because of the difficulties connected with the angular momentum effects. It should be noted that all the analytical considerations published in the past (e.g. Huang 1956) were necessarily of rather restricted and/or approximate character and their results should never be blindly applied to any specific case.

To summarize, the observed increase of period in SS Cyg and U Gem cannot be used as an argument in favor, or against any of the possible mechanisms of the mass exchange and/or mass loss. Situation becomes slightly better, however, when we try to interpret the numerical results obtained from Eqs. (2—4). First, it turns out that the rates of mass exchange (or mass loss), as computed from Eqs. (2), (3), or (4) for either SS Cyg or U Gem, are practically of the same order, namely between 10^{-7} and 10^{-6} $\mathfrak{M}\odot$ per year. These estimates depend on the adopted values of the masses themselves which were taken to be of the order of 1 $\mathfrak{M}\odot$. At least one possibility can now be shown incompatible with the estimates given above. It is the possibility of the sudden mass loss from the system during the outbursts. The amount of mass loss per outburst would be of the order of 10^{-7} $\mathfrak{M}\odot$, that is only one order of magnitude less than in the case of some novae (see McLaughlin 1960), which is clearly incompatible with the lack of any direct observational evidence for mass ejection during an outburst. Therefore if the mass loss is to be responsible for the observed period variations it should mostly be of continuous character (Walker and Chincarini 1968). Unfortunately, no convincing conclusions can be obtained at the present moment with respect to the mass transfer phenomena.

4. COLORS, HOT SPOTS, ETC.

The aim of this section is to mention some unexplained photometric properties of the U Geminorum type variables. Fig. 2 shows the light and color curves of U Gem at minimum light (i.e. between the outbursts). The following features of these curves are worth mentioning (Krzemiński 1965, Mumford 1964, Paczyński 1965): (a) a total primary eclipse, with no major variations in (B—V), but with a strong excess in (U—B); (b) nearly constant light and colors between phases $0\overset{P}{.}1$—$0\overset{P}{.}6$; (c) a shoulder in the light curves

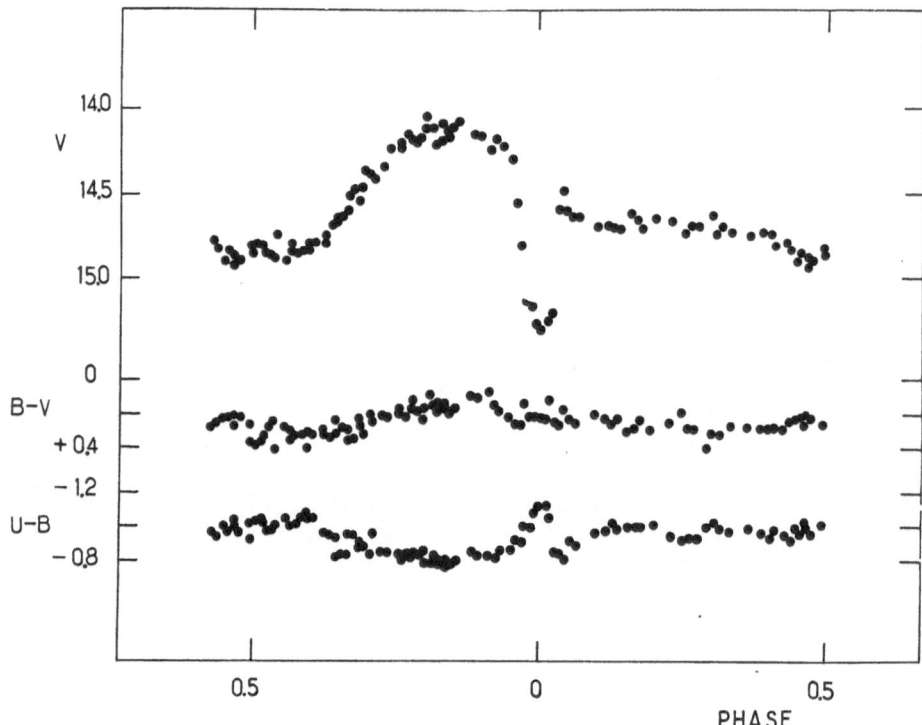

Fig. 2. The V, B—V, and U—B curves of U Gem on JD 2438056 (after Paczyński 1965).

between phases $0^{P}6$—$0^{P}1$; a similar shoulder is present in the (B—V) curve (i.e. the amplitude of the shoulder is larger in B than in V), while the (U—B) curve shows a minimum in this interval of phases. Other examples of stars showing a shoulder in their light curves are: Z Cam (Kraft, Krzemiński, and Mumford 1968), RR Pic which is a nova (van Houten 1966), and VV Pup which is an extremely short-period nova-like object (Thackeray, Wesselink, and Oosterhoff 1950, Herbig 1960, Krzemiński 1965, Walker 1965). Following Herbig's (1960) interpretation of VV Pup, it is now commonly accepted that the shoulder is due to a hot spot located on the surface of the hotter component symmetrically with respect to the line joining the components.

Using the UBV data available for different stars we can easily determine the photometric properties of such hot spots. First, we can assume that the observed brightness and colors outside of the shoulder are due to the combined light of the two components including the gaseous ring around the blue star. Second, from the observed brightness and colors on top of the shoulder we can determine the relative luminosity and colors of the hot spot. The results are given in Table 1 and shown in Fig. 3. Let us discuss them briefly.

(a) Colors of the combined light of the two components (plus the ring), as seen at phases $0^{P}2$—$0^{P}5$, are peculiar; all three objects shown in Fig. 3 are situated 0.3—0.5 mag. above the black body line in the two-color diagram. No quantitative estimates of the influence of the hydrogen emission lines and continuum on colors of these objects are available but it seems almost

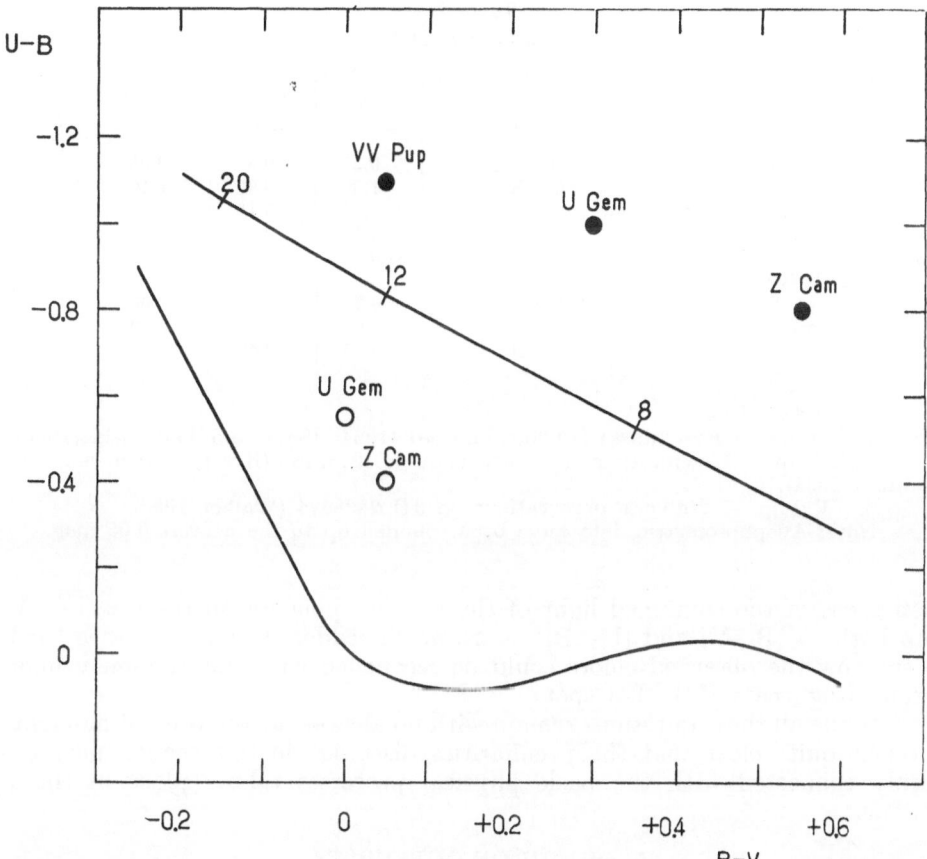

Fig. 3. Colors of U Gem, Z Cam, and VV Pup at phases 0ᴾ2—0ᴾ5 are shown as filled circles. Colors of hot spots in U Gem and Z Cam are shown as open circles. The standard main sequence relation and the black body line (with the temperature in 1000°K units marked) are shown for comparison.

improbable that they could be responsible for the entire ultraviolet excess, unless we could assume that the bound-free and free-free emission of hydrogen dominates in the continuous spectrum. But even this explanation could hardly be accepted for Z Cam where the absorption lines of the red component are seen in the combined spectrum what indicates rather high contribution of light from that component.

(b) Relative luminosities of hot spots (expressed in units of light at phases 0ᴾ2—0ᴾ5) are very large. In the most extreme case of VV Pup the hot spot would seem to radiate much more energy than the "rest" of the system (i.e. the two stars plus the ring). And if such a hot spot is confined to a small region on the surface of the blue star we are forced to conclude that the surface brightness of the spot is at least one order of magnitude higher than that of the star itself.

(c) Colors of the hot spots are peculiar. In the case of U Gem and Z Cam (see Fig. 3) the (B—V) colors are bluer while the (U—B) colors — redder

Table 1
Photometric Data

		U Gem[1]	Z Cam[2]	VV Pup[3]
Colors at $\varphi = 0\overset{P}{.}2 - 0\overset{P}{.}5$,	B—V	+0.30	+0.55	+0.05
	U—B	—1.00	—0.80	—1.10
Amplitude of the shoulder,	A_v	0.60	0.15	1.40
Colors at $\varphi = 0\overset{P}{.}8$,	B—V	+0.15	+0.45	+0.30
	U—B	—0.80	—0.80	—0.80
Relative luminosity of the hot spot,	I_v	0.7	0.2	2.60
Colors of the hot spot,	B—V	0.00	+0.05	+0.40
	U—B	—0.55	—0.40	+0.60

Photometric data from:
 1. U Gem — Krzemiński (1965), Mumford (1964, 1967), and Paczyński (1965).
 2. Z Cam — Krzemiński's observations on JD 2439138 (Kraft, Krzemiński, and Mumford 1968).
 3. VV Pup — Walker's observations on JD 2438474 (Walker 1965).
 Note: All photometric data have been rounded up to the nearest 0.05 mag.

than those of the combined light of the two components; in the case of VV Pup both — (B—V) and (U—B) — are much redder. In all cases it is hard to see how the observed colors could be reconciled with the extremely high surface brightness of the hot spot.

While all these questions remain with no answers at the present moment it seems quite clear that the peculiarities discussed in this section may be deeply connected with the basic physical processes taking place in these systems.

5. ORIGIN OF OUTBURSTS

Although the problem of the origin of outbursts in the U Geminorum type variables is rather of theoretical nature, there is one basic question to which an answer should be given on the basis of the observational data. The question is: which of the two components is responsible for the outburst activity. Photometric observations of U Gem during its outbursts seemed to leave no doubt that the eruption occurs in the red component (Krzemiński 1965, see also Paczyński 1965). Indeed, the eclipses become shallower during the rising branch and disappear almost completely at maximum; their width increases sharply during an outburst and decreases slowly afterwards. All this can be explained if we assume that the outburst consist of a large increase in the surface brightness of the red star and a moderate increase of its radius.

Recent spectroscopic observations of SS Cyg made by Walker and Chincarini (1968) seem, however, to contradict the photometric arguments collected for U Gem. Their observations made during the rise to maximum, some 3 mag. above the minimum level, show that the absorption lines characteristic of that phase display radial velocity variations which are in phase with the radial velocity of the blue component observed at minimum. This means that the relative luminosity contributed by this component (or rather by its gaseous envelope) must be at least comparable with that of the red star. Thus Walker and Chincarini conclude that the outburst is associated with the blue star.

Both pieces of evidence are very convincing. It seems, however, that there is no need to postulate that two, completely different mechanisms operate in U Gem and SS Cyg. It is very likely that an increase in the surface brightness and of the radius of the red star, as postulated in Krzemiński's model, is accompanied by an increase in brightness of the blue star (for example due to accretion of mass transferred from the red star). It may happen, then, that at some specific phase of the outburst the ratio of luminosities is already much different from that at minimum — this would account for shallower eclipses — but the luminosity of the blue star (plus its envelope) is still high enough to contribute to the combined spectrum. Needless to say an extensive series of spectroscopic observations made during the outburst are badly needed to solve this problem.

In conclusion, the author wishes to thank Drs. W. Krzemiński and B. Paczyński for many day-to-day discussions; many of their ideas have been incorporated into the present report. Specific thanks go to Dr. Krzemiński for his unpublished observational data.

REFERENCES

Herbig, G. H., 1960, Astrophys. J., **132**, 76.
Huang, S.-S., 1956, Astr. J., **61**, 49.
Joy, A. H., 1956, Astrophys. J., **124**, 317.
Kraft, R. P., 1962, Astrophys. J., **135**, 408.
Kraft, R. P., 1963, Adv. Astr. Astrophys., **2**, 43.
Kraft, R. P., 1966, Trans. I. A. U., XIIB, 519.
Kraft, R. P., Krzemiński, W., Mumford, G. S., 1968, Astr. J., **73**, S21.
Kruszewski, A., 1964, Acta Astr., **14**, 241.
Kruszewski, A., 1966, Adv. Astr. Astrophys., **4**, 233.
Krzemiński, W., 1965, Astrophys. J., **142**, 1051.
Krzemiński, W., 1968, Unpublished.
McLaughlin, D. B., 1960, Stellar Atmospheres, ed. J. L. Greenstein (Chicago: University of Chicago Press), 585.
Mumford, G. S., 1964, Astrophys. J., **139**, 476.
Mumford, G. S., 1967, Publ. astr. Soc. Pacific, **79**, 283.
Mumford, G. S., 1968, Astr. J., **73**, S 110.
Paczyński, B., 1965, Acta Astr., **15**, 305.
Thackeray, A. D., Wesselink, A. J., Oosterhoff, P. Th., 1950, Bull. astr. Inst. Netherl., **11**, 193.
van Houten, C. J., 1966, Bull. astr. Inst. Netherl., **18**, 439.
Walker, M. F., 1965, Budapest Mitt., No. 57.
Walker, M. F., Chincarini, G., 1968, Astrophys. J.,

DISCUSSION

Feast: Could you tell us what is the evidence for mass exchange in one direction between the stars rather than in the other direction?

Smak: The presence of the rotating gaseous ring around the blue component implies that there must be a transfer of mass and of momentum from the red to the blue star. We do not quite know, however, what is happening during an outburst. If one can assume that the outburst is associated with the blue component, then it is very likely that some amount of mass could be transferred in another direction.

Lortet-Zuckermann: What are the exposure time and the linear dispersion of
   Kraft's spectra showing the doubling of the emission lines?
Bakos: Why do you think that a mass loss of 10^{-7} m$\odot$/year is too high?
   It is less than the mass of the earth and I believe that the blue star could
   absorb it without appreciable changes in its spectrum.
Smak: I think this figure is acceptable as long as we assume that this is the
   amount of mass transferred from one component to the other, without
   any considerable sudden mass loss from the system, or that this is a
   continuous outflow of mass from the system. But I do not think it could
   be associated with a sudden mass loss from the system during an outburst
   only, because there is no direct spectroscopic indication pointing to such
   a large outflow of mass.

Non-Periodic Phenomena in Variable Stars
IAU Colloquium, Budapest, 1968

PHOTOMETRIC OBSERVATIONS
OF THE PECULIAR BLUE VARIABLE TT ARI = BD + 14°341

J. SMAK

Institute of Astronomy, Polish Academy of Sciences, Warsaw, Poland

and

K. STĘPIEŃ

Warsaw University Observatory, Warsaw, Poland

ABSTRACT

Light variations of BD +14°341 can be resolved into three, apparently independent activities: (1) periodic variations with $P = 0^d2658$ and an amplitude of about 0.15 mag.; (2) quasi-periodic fluctuations with periods between 14 and 20 minutes and variable amplitude; and (3) small-scale, apparently irregular fluctuations ("flickering") with a time-scale of the order of 1 min. It is suggested that the star is a hot, subluminous close binary system.

BD + 14°341 = BV 150 Ari = TT Ari was discovered to be a variable star by Strohmeier, Kippenhahn, and Geyer (1957). Their photographic observations, together with those by Huth (1960), seemed to indicate that the light variations were of irregular character. Herbig (1961) took several spectrograms of the star (dispersions from 430 A/mm to 16 A/mm) and found that its spectrum was essentially featureless with very weak and diffuse emission lines of the Balmer series. This indicated that the star could be a hot, subluminous object of the nova-like type. Following Dr. Herbig's suggestion the star was observed photoelectrically (by J. S.) with the Crossley reflector of the Lick Observatory during the 1961/62 season. Later on additional photoelectric data were collected in 1966 with the 60 cm reflector of the Haute

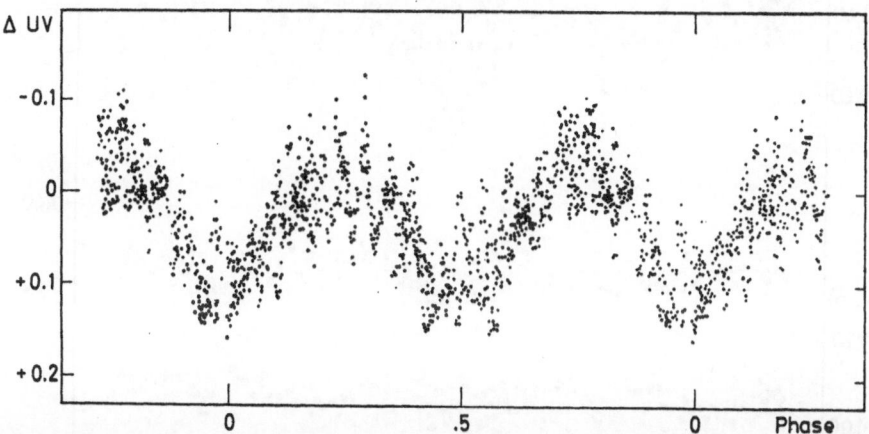

Fig. 1. Composite light curve of BD +14°341 based on measurements made in ultraviolet light on four consecutive nights in September 1966 with the 60 cm reflector of the Haute Provence Observatory. Phases were computed with $P = 0^d2658$. Magnitude differences are plotted with an arbitrary zero-point. The large scatter of points is due to quasi-periodic fluctuations discussed in the text.

Provence Observatory (by J. S.) and in 1966—68 with the 24-inch reflector of the Lick Observatory (by K. S.). Altogether over 3000 measurements were made (mostly in ultraviolet). The aim of the present note is to discuss the results of these observations.

Three, apparently independent components are present in the observed light variations:

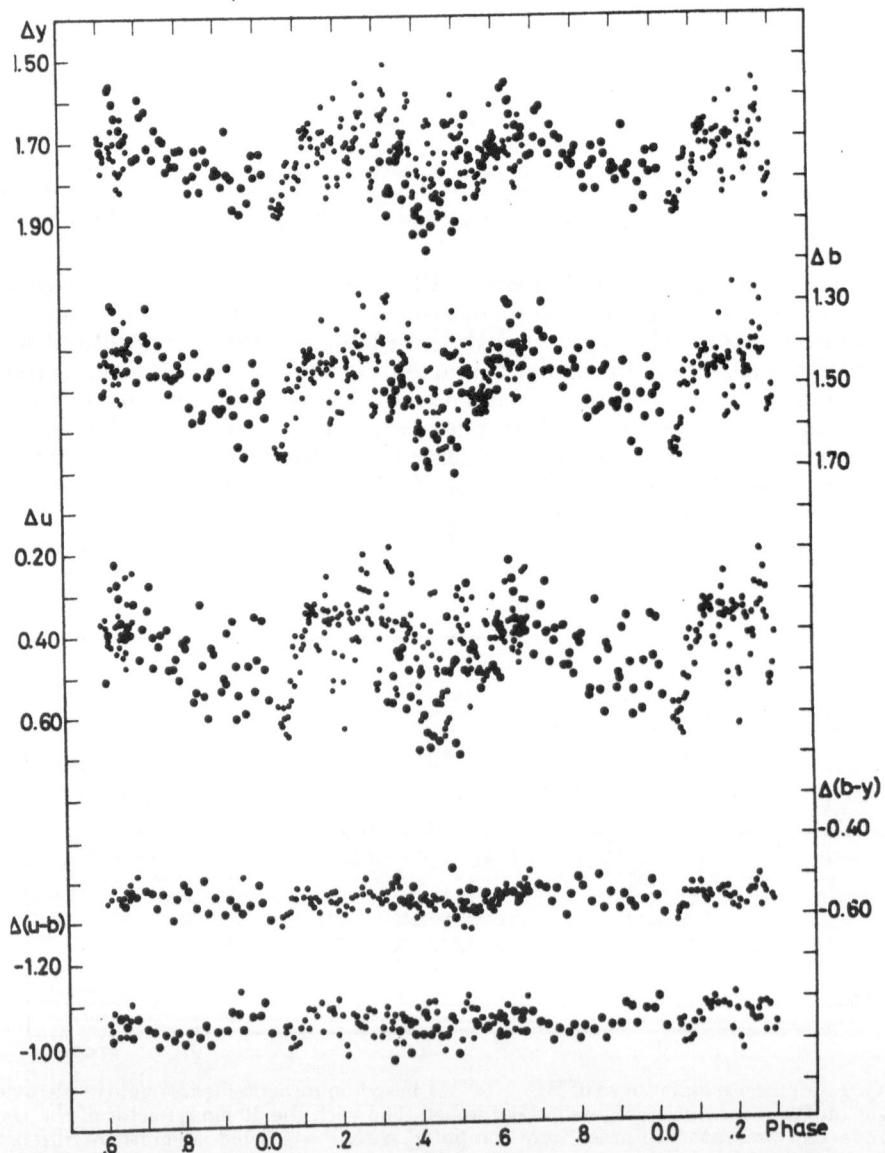

Fig. 2. Composite light and color curves of BD +14°341 based on measurements made on two nights in October 1967 with the 24-inch reflector of the Lick Observatory. All differences are in the sense: BD +14°341 — BD +14°336, and are left in the instrumental system.

(1) Periodic variations with $P = 0^d1329$ (i.e. about 3^h12^m). The shape of the light curve is approximately that of a sine-wave. For reasons given below we shall assume that the fundamental period is twice that given above, i.e. $P = 0^d2658$. Under this assumption we have a double sine-wave light curve (Fig. 1) with two nearly identical minima and two only slightly different maxima; it should be made clear, however, that these differences may not be real. The mean brightness of the star varies from night to night by a few tenths of a magnitude (mean $V = 10.6$), but the period and the shape of the

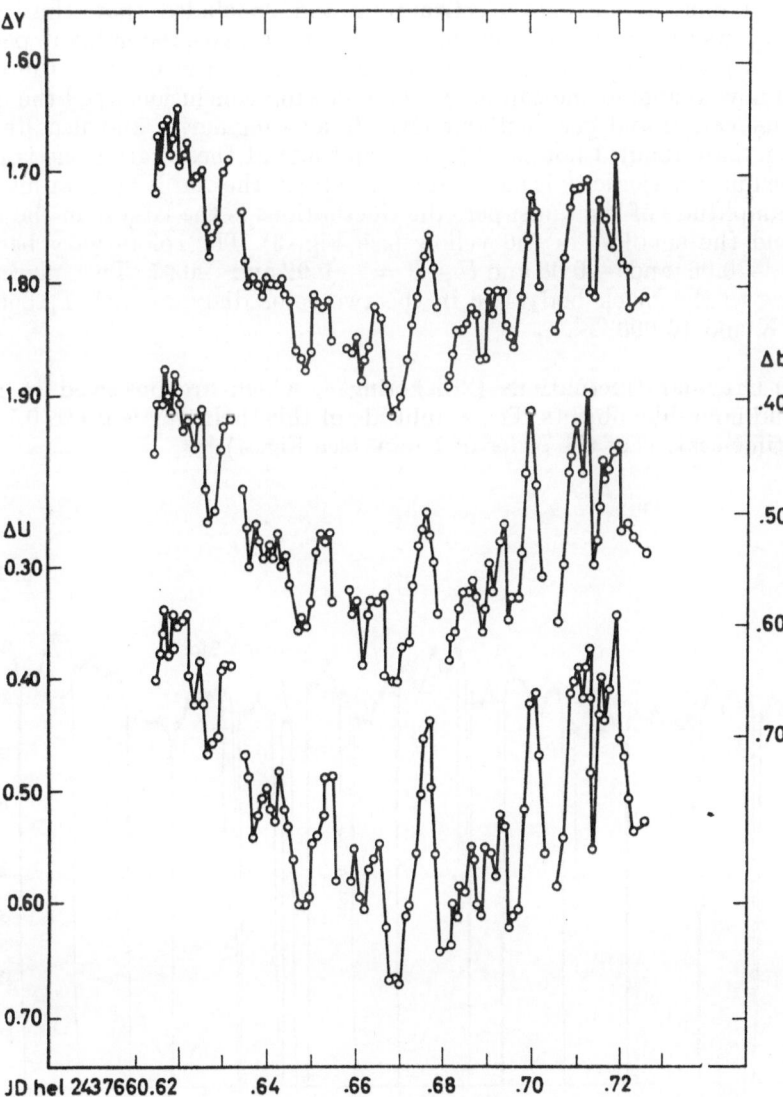

Fig. 3. Three-color observations of BD $+ 14°341$ made on Dec. 28 UT, 1961 with the Crossley reflector. The magnitude differences are in the sense: BD $+14°341 -$ BD $+$ $+ 14°336$, and are left in the instrumental system.

light curve are very stable. The amplitude of the periodic variations is of about 0.15 mag. Three color observations (Fig. 2) seem to suggest that the star is slightly bluer at maxima. Unfortunately, there is a large scatter of observations connected with the quasi-periodic fluctuations discussed under (2). Because of that scatter it is practically impossible to discuss the periodic variations in a more quantitative way.

(2) Quasi-periodic fluctuations with periods between 14 and 20 minutes. Their amplitude can be as large as 0.2 mag. Williams (1966) made a statistical analysis of one of our photoelectric runs and concluded that this quasi-periodicity was a transient phenomenon and that two predominant periodicities were present of about 14 and 18 minutes. On the basis of the entire material now available one can make the following conclusions: *(a)* the quasi-periodicity comes and goes without any obvious regularity and usually lasts no longer than about 3 hours; *(b)* the amplitude of these variations is larger when the quasi-periodicity is more pronounced; *(c)* the star is bluer at maxima, i.e. the amplitude of the quasi-periodic fluctuations is the largest in the ultraviolet and the smallest in the yellow (see Fig. 3). The colors vary between $B—V = —0.06$ and $—0.08$ and $U—B = — 0.92$ and $—0.97$. This places the star close to the black body line in the two-color diagram with T_e between 15 000 °K and 16 000 °K.

(3) Irregular fluctuations ("flickering"), which are observed in many novae and nova-like objects. The amplitude of this flickering is up to 0.1 mag. and its time-scale is of the order of 1 min. (see Fig. 4).

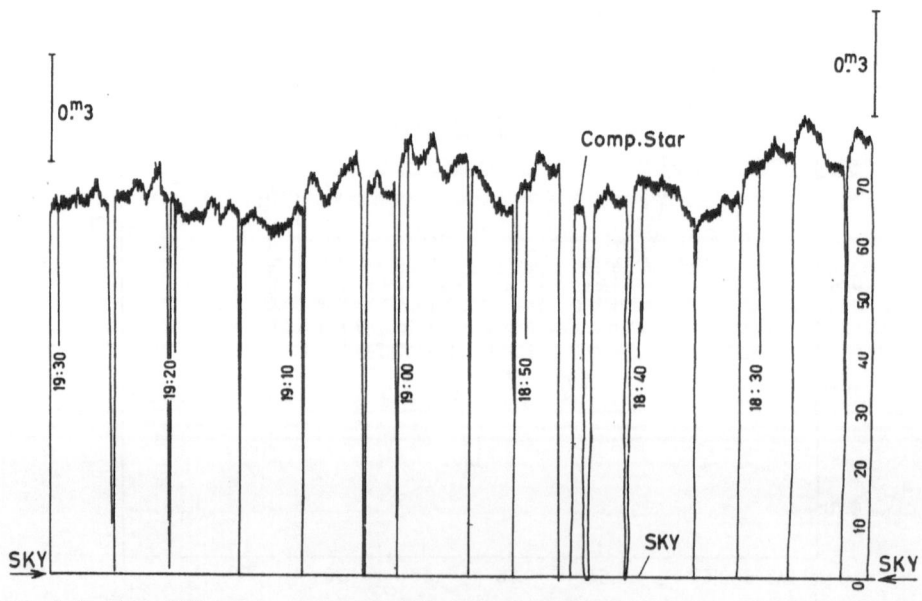

Fig. 4. Section of Brown recorder sheet showing observations of BD + 14°341 in ultraviolet light made on Dec. 31 UT, 1961 with the Crossley reflector. Short-period fluctuations ("flickering") are clearly visible.

As a working hypothesis we suggest the following interpretation of the observational data. BD + 14°341 is a close binary system with the orbital period $P = 0\overset{d}{.}2658$. Regular light variations connected with this period are due to the aspherical shape of either one or both components. Quasi-periodic variations with periods between 14 and 20 minutes are due to pulsations of either one or both components, as suggested by Williams (1966). We believe, however, that in the case of a three-axial ellipsoid the situation should be much more complicated than that considered by Williams; in particular, non-radial pulsations should be present and since no appropriate theory is now available we can only conclude that the resulting light variations could indeed be rather complicated.

REFERENCES

Herbig, G. H., 1961, Private information.
Huth, H., 1960, Mitt. veränderl. Sterne, No. 454.
Strohmeier, W., Kippenhahn, R., and Geyer, E., 1957, Kl. Veröff. Bamberg, No. 18.
Williams, J. O., 1966, Publ. astr. Soc. Pacific **78,** 279.

DISCUSSION

Lortet-Zuckermann: Did you obtain profiles of the hydrogen lines?

Herbig: If I understand the question, the Lick coudé- plates of +14°341 show only very weak emission components at the H lines; the spectrum is otherwise continuous.

PHOTOMETRIC EFFECTS FOR HIGHLY DISTORTED WHITE DWARF SECONDARIES IN CLOSE BINARY SYSTEMS

S. M. RUCIŃSKI

Astronomical Observatory, University of Warsaw, Warsaw, Poland

As a by-product of an analysis of the light curves of the early-type close binary systems (Ruciński 1969a), similar computations for the highly distorted white dwarf hypothetical secondaries of certain peculiar systems were made. The computations were carried out numerically by integrating the monochromatic fluxes emerging from the atmosphere over the visible surface of the star. The effects of eclipses were not taken into account; the reflection effect was also excluded at this step of analysis. A slightly different model atmosphere was used at each point of the star's surface depending on the local effective temperature (with the assumption of von Zeipel proportionality $T_e \sim g^{1/4}$). The shape, variations of the effective gravitaty g, and the cosine of the angle between the local normal to the surface and the direction to the observer, μ, were described using the first order perturbation theory for close binary systems (Chandrasekhar 1933). In that theory the Legendre polynomial P_2 gives the ellipticity of the star; the next P_3 and P_4 polynomials describe respectively the non-symmetric and symmetric deviations from ellipticity.

By integrating over the visible surface S of the star we obtain a quantity proportional to the total monochromatic energy radiated by the star in a given direction:

$$L_\nu = \int\!\!\int_S I_\nu(\mu, T_e) \cdot \mu \cdot dS .\tag{1}$$

The product $L_\nu A^2$, where A is the separation of the binary components, is the total energy radiated by the star in the observer's direction per unit of frequency and per unit of time. L_ν evaluated as a function of the phase angle describes the light changes caused by the distortion of the star. More detailed description of the methods of computation will be published elsewhere.

The emerging intensities I_ν at $\lambda = 5000$ A were computed on the basis of the white dwarf model atmospheres (Terashita, Matsushima 1966) for the effective temperatures 8000° and 10700° and for the gravity $g = 10^7$. These intensities were represented by the series:

$$I_\nu(\mu, T_e) = \sum_{n=0}^{3} I_{\nu n}(T_e) \cdot \mu^n .\tag{2}$$

As the mean effective temperature of the white dwarf secondary was taken near 7800° in computing the integral (1) the values of $I_\nu(\mu, T_e)$ in particular points of the star's surface were found either by interpolation or by extra-

polation of both the model atmospheres depending whether the effective
temperature was in or out of the range 8000°—10700°.

The following assumptions were made: the inclination of the orbit
$i = 90°$, the synchronism of the star's rotation and the orbital revolution,
and the Roche model for the concentration of mass towards the centre.

Two cases were computed with two different mass ratios. In each case
the dimensions of the stars were chosen in such a way as to fit the corresponding
Roche lobe (Plavec, Krotochvil 1964; Kuiper, Johnson 1956) in the direction
of the y-axis which lies in the orbital plane and is perpendicular to the x-axis
joining the centres of the stars. The mass ratios and the respective dimensions
(in units of the separation of the components, A) are:

Case	M_{prl}/M_{WD}	x_a	x_b	y	z
1	10	.226	.219	.197	.187
2	19	.187	.182	.162	.154

Because of the simplified description of the stars the conical parts directed
towards the Lagrangian L_1 point are obviously not filled, but elsewhere the
approximation of the Roche lobes is relatively good (Fig. 1); in any case we
might expect that the photometric effects caused by the distortion of the star
will be rather under-estimated than over-estimated in this way. The changes
in L_v are given by the formula:

$$L_v = C_0 \cdot \left(1 + \sum_{n=1}^{3} C_n \cdot \cos\left(n \cdot \text{Phase}\right)\right) \qquad (3)$$

where the Fourier coefficients C_n for the computed cases (the accuracy being
approximately 0.0005) are:

Case	C_0	C_1	C_2	C_3
1	7.62×10^{-6}	—0.0132	—0.2278	—0.0112
2	5.16×10^{-6}	—0.0122	—0.2467	—0.0099

In both cases the full amplitude of the light variations amounts to almost
three quarters of a magnitude. This implies, therefore, that a strongly distorted
white dwarf secondary whose contribution to the total brightness of the system
may be quite low, might easily account for the quasi-sinusoidal variations
observed in such systems as WZ Sge (Krzemiński, Kraft 1964) and, perhaps,
BD +14°341 (Smak, Stępień 1968).

There is no theory which could be used to evaluate the expected values
of the coefficients C_1 and C_3, both the coefficients giving the asymmetry of the
light curve. The largest coefficient C_2, however, can be estimated on the basis
of the geometric parameters of the star and the limb (u) and gravity (y)

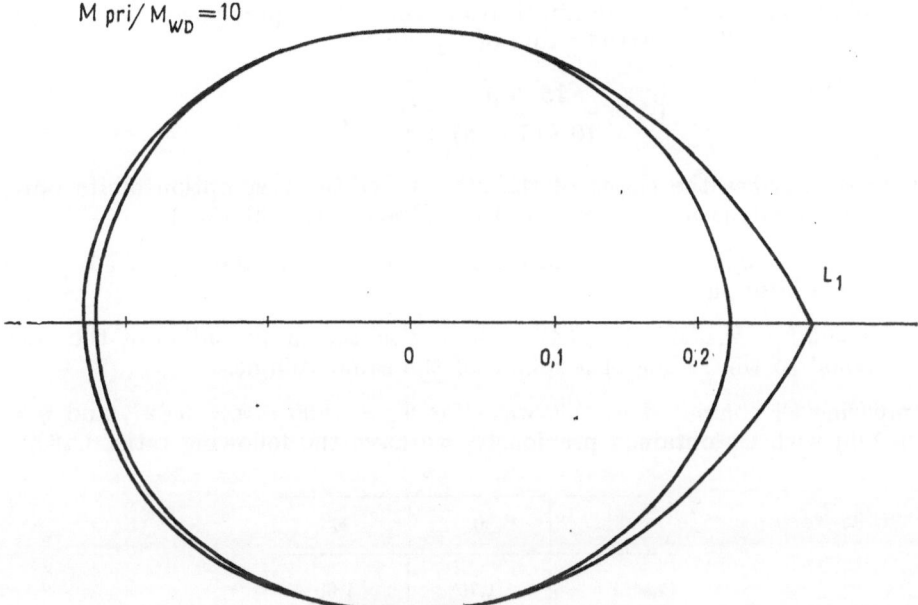

Fig. 1a. The shape of the WD secondary for Case 1 compared with its Roche lobe (orbital plane section).

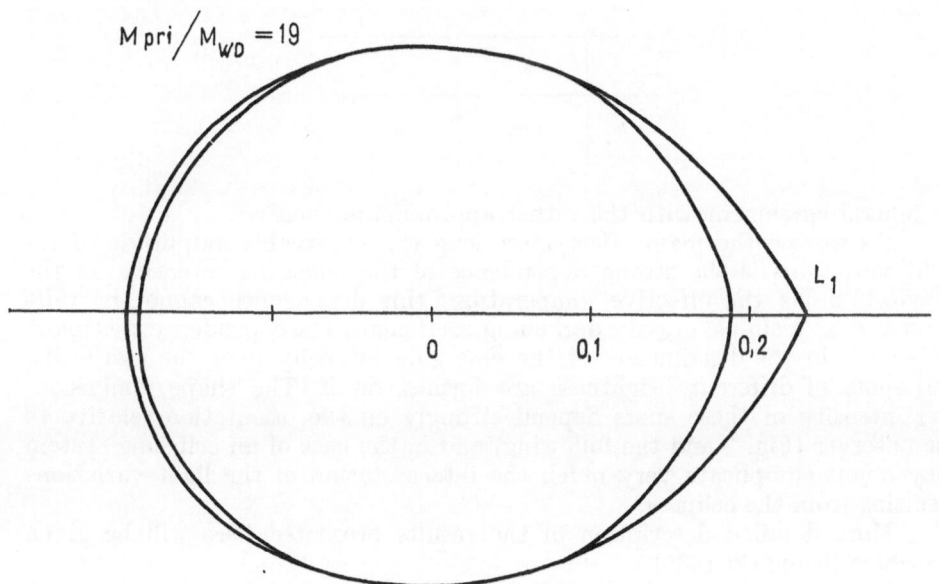

Fig. 1b. Same as in Fig. 1a. for Case 2.

darkening coefficients determined from the same previously used model atmospheres. The theoretical value of C_2' will be:

$$C_2' = \frac{15 + u}{10 \cdot (3 - u)} \cdot (1 + y) \cdot \varepsilon,$$

where ε describes the shape of the star under the assumption of its pure ellipticity. ε might be approximated by (Russell, Merrill 1952):

a) $\varepsilon = (a - b)/a$, where a and b are the two axes of the ellipsoid lying in the orbital plane,

b) $\varepsilon = 3/2 \cdot (M_{prl}/M_{WD}) \cdot (r/A)^3$, where r is the mean radius of the star, equal to the radius of a sphere of the same volume.

Comparing C_2' computed in this way (for $T_e = 7800°$, $u = 0.627$, and $y = 0.785$) with C_2 obtained previously, we have the following ratios C_2/C_2':

	a)	b)
Case 1	1.70	1.60
Case 2	1.74	1.62

The simpler theory leads therefore to the value of C_2 which is 60—70 per cent smaller than that obtained from the model described above.

As a check a modification of Case 1 was computed by assuming $T_e =$ constant over the whole surface of the star; that is, only the geometrical and limb darkening effects were taken into account. The ratio C_2/C_2' was then:

	a)	b)
c_2/c_2'	1.05	0.98

in general agreement with the rather approximate theory.

As we see the main effect increasing the observable amplitude of the light variations is the strong dependence of the emerging intensity on the gravitation via the effective temperature; this dependence cannot be fully described by a simple gravity darkening coefficient. The dependence also manifests itself in the distribution of the emerging intensity over the star's disc and spots of different brightness are formed on it. The shape, dimensions and intensity of these spots depend strongly on the orientation relative to the observer (Fig. 2 and the following) and in the case of an eclipsing system they might complicate very much the interpretation of the light variations resulting from the eclipses.

More detailed description of the results presented here will be given elsewhere (Ruciński 1969b).

The author is indebted to Dr. J. Smak for arousing his interest in the subject.

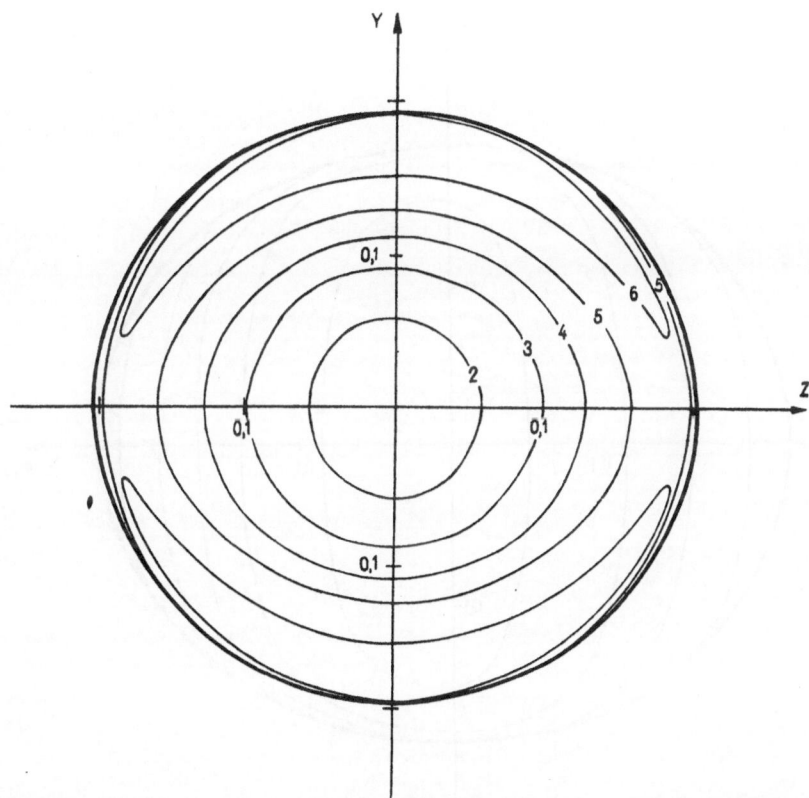

Fig. 2. The distribution of the emerging intensity over the visible disc of the WD secondary. The mass ratio is 10 (Case 1). The isophotes are labelled with numbers which should be multiplied by 10^{-5} to give the intensity I_ν in cgs units. The X-axis is directed towards the observer and the rotation axis lies in the XZ-plane. The inclination of the orbit is $i = 90°$; the phase $\varphi = 0°$, i.e. the star is viewed from the primary component.

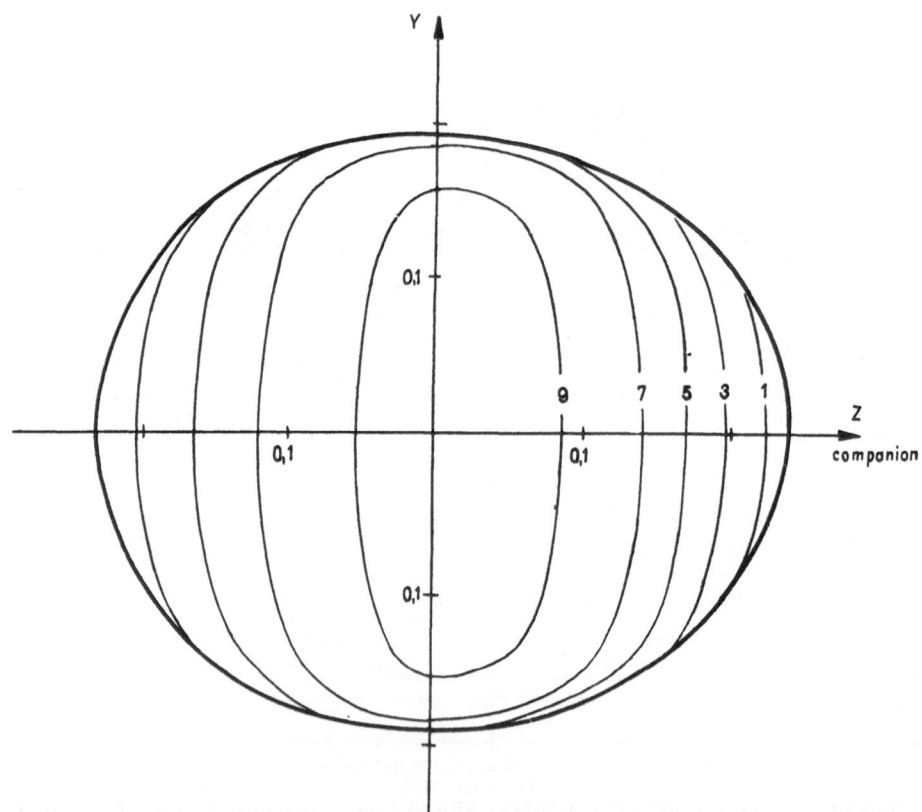

Fig. 3. Same as in Fig. 2, for the phase $\varphi = 90°$, i.e. the star is viewed side-on.

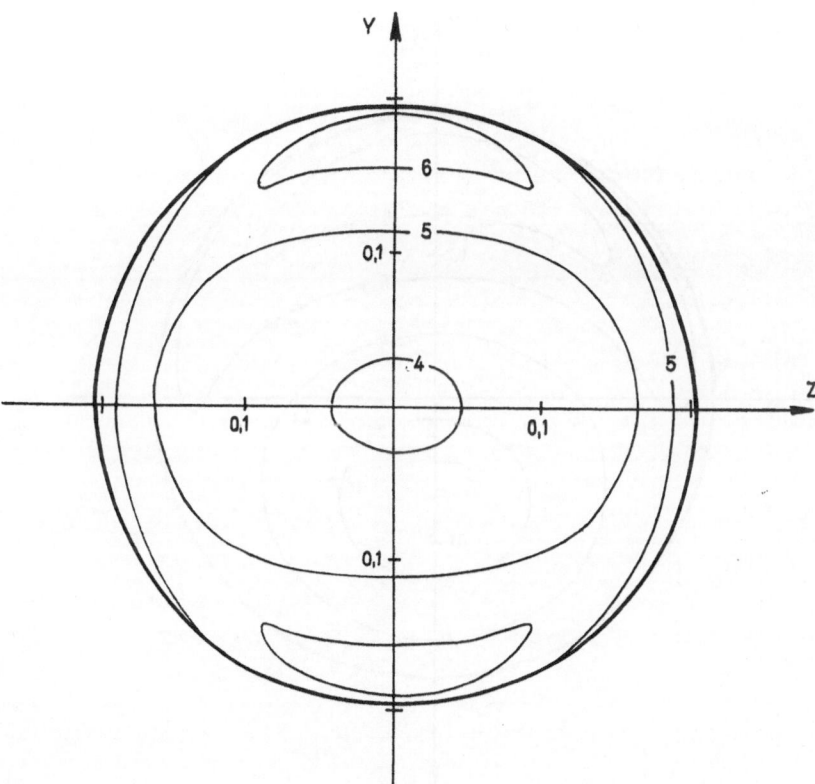

Fig. 4. Same as in Fig. 2, for the phase $\varphi = 180°$, i.e. the opposite side of the star is viewed.

S. M. RUCINSKI

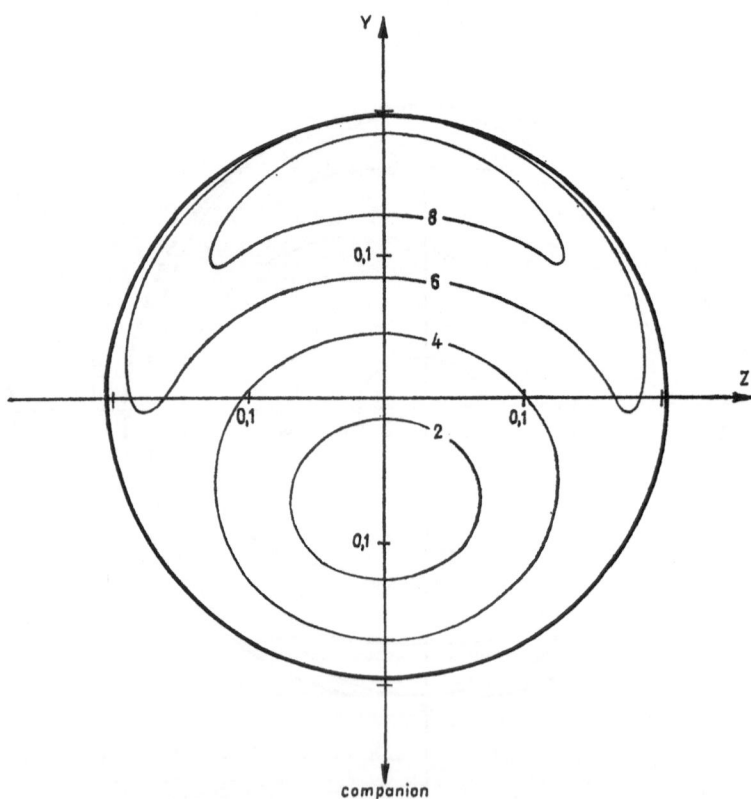

Fig. 5. Same as in Fig. 2, for the phase $\varphi = 0°$ but for the inclination $i = 75°$.

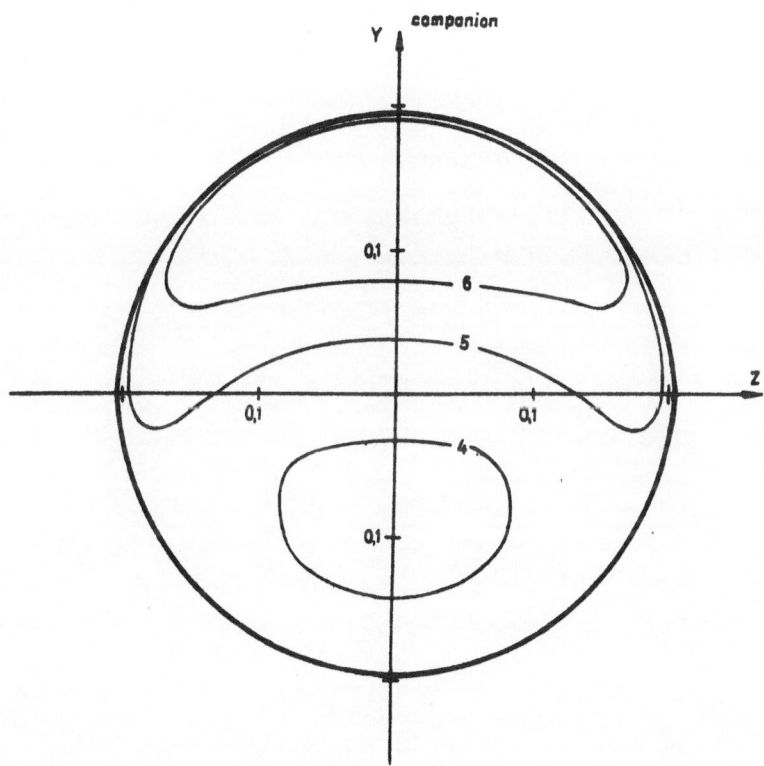

Fig. 6. Same as in Fig. 2, for the phase $\varphi = 180°$ and the inclination $i = 75°$.

REFERENCES

Chandrasekhar, S., 1933, Mon. Not. R. astr. Soc. **93,** 462 and 539.
Krzemiński, W., Kraft, R. P., 1964, Astrophys. J. **140,** 921.
Kuiper, G. P., Johnson, J. R., 1956, Astrophys. J. **123,** 90.
Plavec, M., Krotochvil, P 1964, Bull. astr. Inst. Csl. **5,** 165.
Ruciński, S. M., 1969a, Acta Astr. in preparation.
Ruciński, S. M., 1969b, Acta Astr. in preparation.
Russell, H. N., Merrill, J. E., 1952, Contr. Princ. Univ. Obs., **26,** 32.
Smak, J., Stępień, K., 1968, this Colloquium.
Terashita, Y., Matsushima, S., 1966, Astrophys. J. Suppl., **13,** 461.

PHOTOMETRIC OBSERVATIONS OF NOVA WZ SAGITTAE AND THEIR INTERPRETATION

W. KRZEMIŃSKI and J. SMAK

Institute of Astronomy, Polish Academy of Sciences, Warsaw, Poland

ABSTRACT

A new model of the binary system WZ Sge is proposed, in which the secondary component contributes about 20 percent to the total light. The W UMa-type light curve (except for the primary eclipse) is explained as a result of the aspherical shape of the secondary. Both components are degenerate stars. Their effective temperatures are approximately 20000 °K and 8000 °K.

The binary system Nova WZ Sagittae has several unique properties and differs much from other known close binaries. Four years ago a model of this system was published (Krzemiński and Kraft 1964) based on spectroscopic and photometric observations available at that time. According to that model the masses of the components are about $0.6\ \mathfrak{m}\odot$ and $0.03\ \mathfrak{m}\odot$, giving an unusually small mass-ratio of about 0.05. The primary, more massive component, whose stationary absorption lines of hydrogen are observed, is a white dwarf with $T_e = 13600$ °K and $M_{bol} = +10.4$. The secondary is a dM star of very low luminosity and fills its Roche limit. The primary is surrounded by a gaseous ring rotating with the velocity of 720 km/sec. The characteristic S-wave component of the hydrogen emission lines, which is visible on single-trail spectrograms approximately 90° out of phase with respect to the light curve, is attributed to a strong stream ejected from the secondary toward the primary component. The primary eclipse consists of a partial eclipse of the white dwarf and of covering of the rotating ring and the stream by the dM component. The secondary eclipse (shifted to the phase 0.54) is interpreted as an annular eclipse of the stream by the white dwarf; no photometric effects of the eclipse fo the dM star itself are observed since it is fainter than the primary by about 5 mag. (in V).

In 1964—1967 new photoelectric observations (in the UBV system) were made with the 120-inch, 100-inch, and 193-cm telescopes of the Lick, Mount Wilson, and Haute Provence observatories, respectively. Several objections to the previous model together with the new photometric results raised the need for revising the model. Figs. 1 and 2 show the composite light and color curves based on new data; one can recall here that the photometric data available prior to 1964 consisted of the ultraviolet light curves only. The following points should be raised as being inconsistent with the previous model: (a) the W UMa-type light curve outside of minima; in addition one may note that the first photometric elongation (i.e. near the phase 0.25) is usually slightly brighter than the second (i.e. the phase 0.75); (b) the new U—B color is much different from that given by Walker (1957) and used in the previous model; the new measurements give $\langle(U—B)\rangle = -0.93$, and $\langle(B—V)\rangle = +0.10$; (c) the secondary eclipse in the V-curve

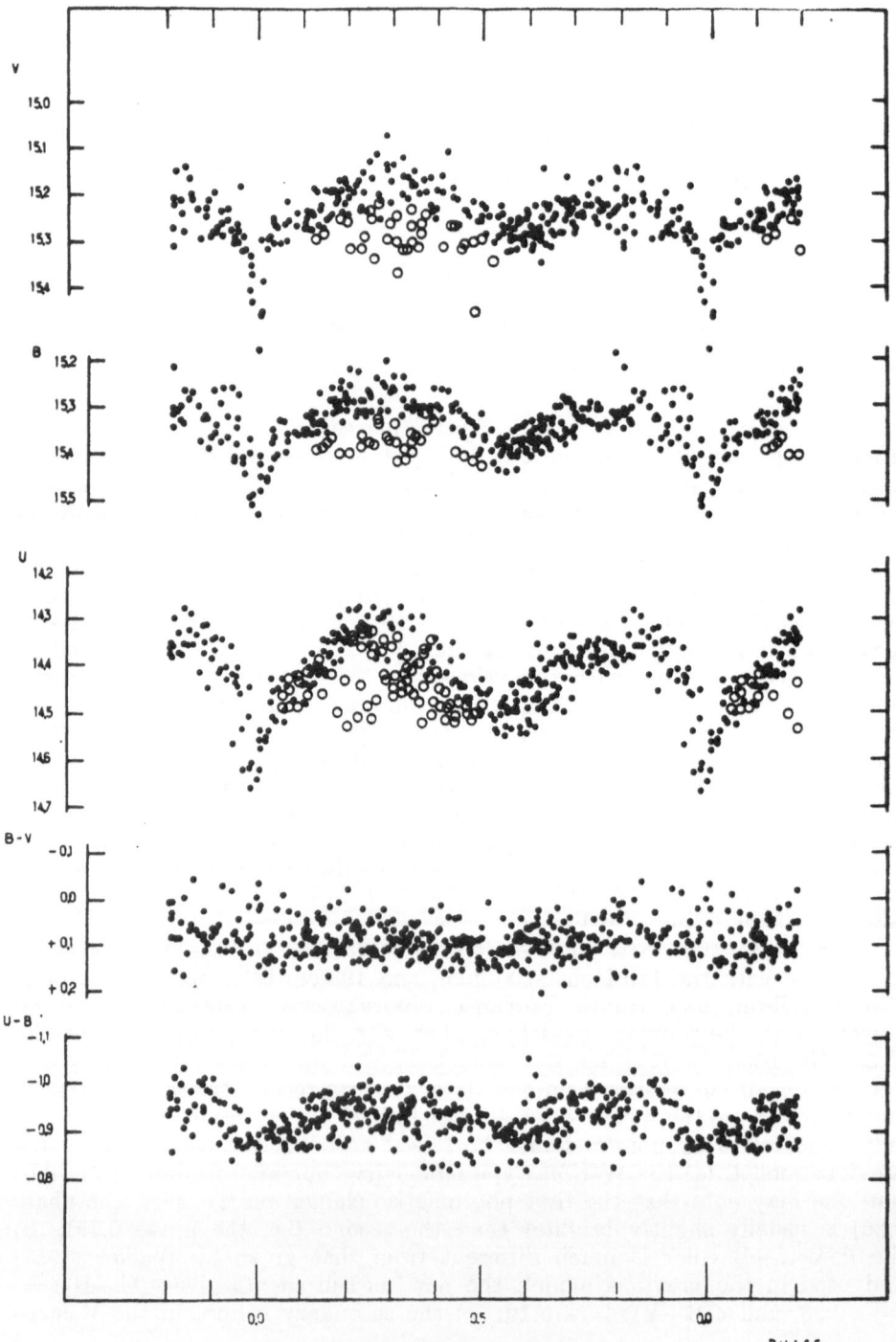

Fig. 1. Light and color curves of WZ Sge based on observations made with the 120-inch telescope in 1964.

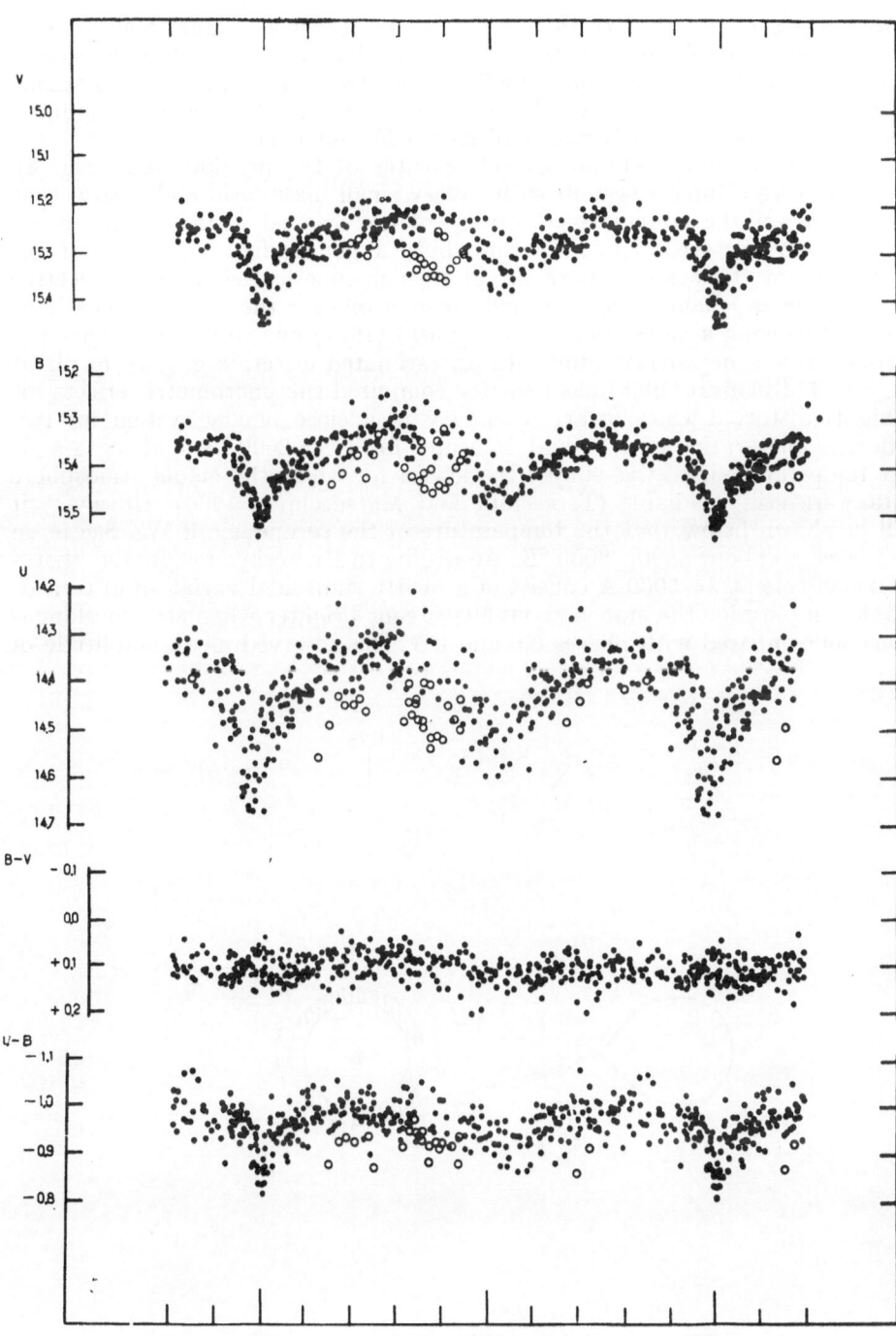

Fig. 2. Light and color curves of WZ Sge based on observations made with the 193-cm telescope in 1964.

is of comparable depth with that in the B-curve; if it were due to an annular eclipse of the stream by the white dwarf it should be almost undetectable in the yellow region where no light from the stream (except for the Paschen continuum) could be present; (d) the position of the secondary minimum is not constant but varies between phases 0.50 and 0.56.

Our new model retains several features of the previous one (Fig. 3). Thus we have a binary system with a very small mass-ratio and with a very small mass of the secondary. Contrary to the original model, however, we assume that the secondary may contribute a non-negligible fraction of the total light of the system. Because of its non-spherical shape and effective gravity effects it could then be made responsible for the W UMa-type light curve. Following a suggestion by Paczyński (1967) one can assume that the secondary is a degenerate star with an estimated effective gravity of about $\log g = 7$. Ruciński (1968) has recently computed the photometric effects for a highly distorted white dwarf secondary in a close binary system for two different mass-ratios of $1/10$ and $1/19$ and for $T_e = 8000\ ^\circ K$ and $\log g = 7$. The temperature used was simply the lowest for which the model atmosphere data were still available (Teraschita and Matsuschima 1966). However, it will be shown below, that the temperature of the secondary of WZ Sge is, in fact, close to about 6500—8000 °K. According to Ruciński's results the photometric effects at $\lambda = 5000$ A consist of a nearly sinusoidal variation of considerable amplitude: the star is about 60 per cent brighter when seen at elongations, as compared with phases 0.0 and 0.5. The observed mean amplitude of

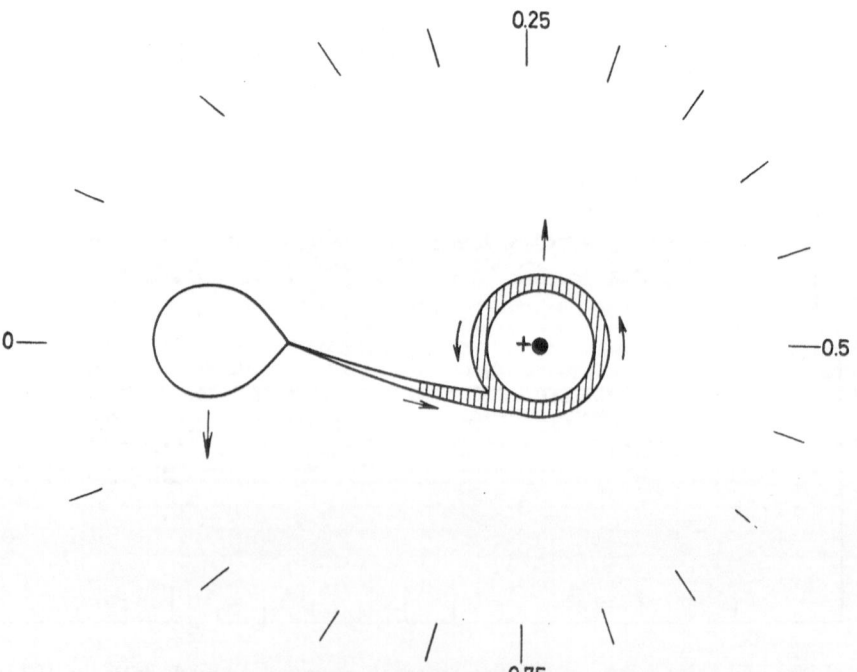

Fig. 3. Schematic model of WZ Sge. The plus sign marks the center of mass.

the W UMa-type variations of WZ Sge (in B and V), including the secondary but excluding the deep part of the primary minimum, amounts to about 0.1 mag. Therefore the fractional luminosity of the secondary component should be equal to about $L_2 = 0.20$, in order to account for the observed amplitude. The ratio of luminosities should be $L_2/L_1 = 0.25$.

The observed phase shift of the secondary minimum cannot be explained by our model. However, since the position of the secondary minimum is not constant, one can assume that the phase shift could be due to an extra source of light of variable intensity located on the surface of the secondary component asymmetrically with respect to the line joining the two components; this extra light could contribute to the light curve between phases (approximately) 0.2—0.6, producing not only the observed shift of the secondary minimum but also the slight excess of luminosity at the phase 0.25. A number of mechanisms connected either with a non-synchronous rotation or mass ejection could be responsible for the existence of such a "hot spot".

Fig. 4 shows the observed position of WZ Sge in the two-color diagram. Also shown is a grid of lines based on model atmospheres computed by Teraschita and Matsuschima (1966). To obtain U—B and B—V of the primary component one has to correct the observed colors for the effect of the hydrogen emission and for the secondary component. Both effects were evaluated in an approximate way. In particular, it was assumed that the ratio of luminosities (mean for B and V) is $L_2/L_1 = 0.25$, as given above, and that the colors of the secondary should be close to those of other more typical white dwarfs with $T_e = 6000$—$8000\,°K$ and $\log g = 7$, i.e. should lie within the rectangle in the lower right part of the two-color diagram (Fig. 4). It turns out that

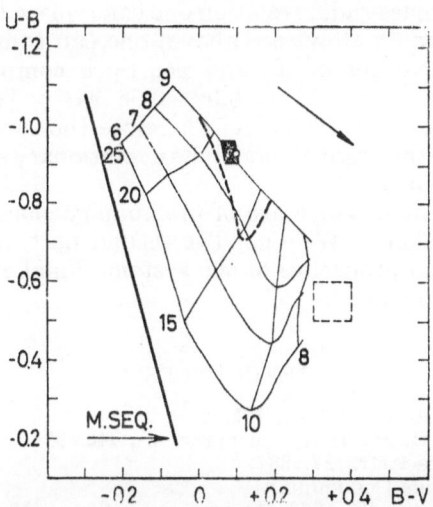

Fig. 4. The two-color diagram. An open circle marks the observed colors of WZ Sge. A rectangle in the lower right part of the diagram is assumed to contain the colors of the secondary component. The colors of the primary should lie in the hatched area located near the observed point. A grid of lines is based on model atmospheres (Teraschita and Matsuschima 1966) with $T_e = 8000$, 10000, 15000, 20000, and 25000 °K, and log $g = 6$, 7, 8, and 9. Finally, the broken line is the "locus" based on the H-gamma absorption profile belonging to the primary component.

the resulting colors of the primary are not very sensitive to these assumptions and should lie within a small region, close to the observed point, as shown in Fig. 4. A comparison with the theoretical data suggests the following characteristics of the primary: $T_{e,1} \simeq 18000$ °K, $\log g_1 \simeq 9$, provided no reddening corrections need to be introduced. An independent estimate of these parameters is possible via the observed profile of the H-gamma absorption line (Greenstein 1957) and the theoretical profiles published by Teraschita and Matsuschima (1966). Because of the emission component in the center of line only the absorption wings could be taken into account and for that reason it was impossible to determine T_e and g in a unique way; instead a "family" of solutions corresponding to the different values of T_e and g was obtained. A "locus" corresponding to these solutions is shown also in the two-color diagram (Fig. 4). It can be seen that the colors of the primary component lie very closely to this locus and with a very small correction for intersteller reddening the agreement would be even better. Taking into account both sets of data (i.e. from colors and from the H-gamma line) one can state that the effective temperature of the primary is between 18000 °K and 20000 °K (instead of 13600 °K in the original model) and its surface gravity about $\log g = 8.5$. The value of log g, together with the theoretical mass-radius relation for white dwarfs would imply that the mass of the primary should be larger than 1 solar mass, i.e. much larger than that obtained by Krzemiński and Kraft (1964) within the previous model. It is clear, however, that the accuracy of this determination is quite low.

Finally one can estimate the effective temperature of the secondary component. The ratio of luminosities, which is known (see above), can be written down as a function of radii and effective temperatures of the components. By using the mass-radius relation one can replace the radii with masses With $T_{e,1}$ known (approximately, see above) one can compute $T_{e,2}$ as a function of various combinations of $\mathfrak{M}_1$ and $\mathfrak{M}_2$. Such computations were made with $\mathfrak{M}_1 = 0.4—1.4 \, \mathfrak{M}\odot$ and $\mathfrak{M}_2 = 0.02—0.06 \, \mathfrak{M}\odot$. The resulting values of $T_{e,2}$, lie in the range 6500—8000 °K. Therefore the secondary turns out to be much hotter than in the original model; its evolutionary significance remains, however, equally obscure.

The results presented above are of preliminary character and form only a part of our rediscussion of WZ Sge. The second part, now under way, will deal with the dynamical properties of the system. Final results will be published in "Acta Astronomica".

REFERENCES

Greenstein, J. L., 1957, Astrophys. J., **126**, 23.
Krzemiński, W., Kraft, R. P., 1964, Astrophys. J., **140**, 921.
Paczyński, B., 1967, Acta Astr., **17**, 287.
Ruciński, S. M., 1968, This Colloquium.
Teraschita, Y., Matsuschima, S., 1966, Astrophys. J. Suppl, **13**, 461.
Walker, M. F., 1957, I. A. U. Symposium No. 3, Ed. G. H. Herbig (Cambridge, Cambridge University Press), p. 46.

Non-Periodic Phenomena in Variable Stars
IAU Colloquium, Budapest, 1968

ECLIPSES OF U GEMINORUM

MARGARET W. MAYALL

A. A. V. S. O., Cambridge, USA

During the last 4 observing seasons, 1964—65 to 1967—68, the AAVSO has received many visual observations of the eclipses of U Geminorum. Some are too fragmentary to use in a preliminary analysis, but we have 97 eclipses which were observed over a long enough period of time to give a value of the width of the eclipse curve.

Most of these 97 eclipses were observed by 4 very careful and experienced observes, and 75% were by Leslie Peltier with his 12-inch Clark refractor. About $^1/_4$ of these were made in collaboration with Carolyn Hurless. The two observers made alternate estimates about one minute or less apart.

The other 25% were made by Clinton B. Ford, Carolyn Hurless and Thomas Cragg, with a few by Vicki Schmitz and Diane Lucas.

Visual observers find it too difficult to make rapid accurate estimates of one star for more than about one hour at a time. Consequently, it is often impossible to determine the time of eclipse by the method used by Dr. Krzeminski, although the observations do confirm the change in period found by the photoelectric observers.

In this preliminary discussion I have used the width of the eclipse curve at magnitudes 14.2 and 14.6. The normal minimum magnitude of U Geminorum between eruptions is 13.8 to 14.0. At times the eclipses reach 15th magnitude and at other times they barely reach 14.2.

One of our observers, Ron Thomas (AAVSO Abstracts, 1968 June) arranged the eclipse curves into different types, and attempted to find a correlation between type of curve and the time elapsed since an eruption. In a modified form, the types are:

Type I — similar to the schematic curve used by photoelectric observers (Krzeminski Ap. J. *142*, 1053, 1965) to determine the time of minimum, with an increase before the sharp drop to minimum, followed by a rapid rise $^3/_4$ of the way to normal magnitude.

Type II — steep decrease and slow rise.

Type III — slow decrease and rise.

Type IV — slow shallow decrease, followed by an interval of nearly constant brightness.

Type V — erratic variations of a few tenths to three-quarters of a magnitude before and after predicted time of minimum.

Table I lists the eruptions of U Geminorum which were observed during 1964—1968. The numbering is a continuation of the systems started by Leon Campbell. The data for Epochs 1 to 3 is published in JRASC Vol 51, 2, March 1957.

Table I
Eclipses of U Geminorum

J. D. 2,430,000+	Type	Min. Mag.	Width		Observer	Days from eruption
			14.2	14.6		
8799	II	14.9	.009	.004	CR	+15
8802	V	14.9	—	.027	CR	+18
8810	IV	14.9	.010	.007	CR	+27
8811	I	14.8	.013	.008	CR, BM	+27
8812	?	14.7	?	.002	P	+28
8818	I	14.6	.011	.003	CR	+34
8820	III	—	.012	—	P	+36
8842	II	14.4	.010	—	FD	+58
8847	II	14.5	.011	—	CR	+63
8851	III	14.7	.016	.005	FD	+67, —38
8853	III	14.6	>.015	.003	P	+69, —36
8855	I	14.75	.019	.007	FD	+71, —34
8856	I	14.7	.016	.004	FD	+72, —33
8870	I	14.6	.012	.001	P	+86, —19
8873	I	14.8	.013	.003	P, HR	+89, —16
8878	II	14.8	.011	.004	P, HR	+94, —11
8879	I	14.75	>.027	.005	P	+95, —10
8881	III	14.8	>.035	.007	P, HR	+97, —8
8884	I	14.8	>.018	.007	P, HR	+100, —5
8897	II	14.7	.016	.004	CR	+8
9024	II	14.7	.013	.002	P	+25
9033	II	14.75	.019	.008	P	+34
9035	II	14.7	.014	.003	P	+36
9036	III	14.5	.016	—	P	+37
9038	II	14.7	.015	.003	P	+39
9058	III	14.8	>.017	.006	P	+59
9059	III	14.5	.016?	—	CR	+60
9060	III	14.7	.014	.007	FD	+61
9062.7	I	15.0	.016	.007	FD	+63
9062.9	III	14.8	>.022	.006	P	+63
9063	III	14.7	>.024	.007	P	+64
9064	III	14.7	>.022	.006	P	+65, —45
9065	II	14.75	>.018	.008	P	+66, —44
9069	II	14.6	.011	.001	FD	+70, —40
9084	I	14.8	>.014	.005	P	+85, —25
9087	I	14.4	.007	—	P	+88, —22
9094	I	14.3	.002	—	FD	+95, —15
9095	III	14.4	.008	—	P	+96, —14
9113	IV	12.2	—	—	F, BM	+4
9120	II	14.4	.010	—	P	+11
9123	III	14.5	.016	—	P	+14
9129	IV	14.4	.019	—	P	+20
9137	IV	14.5	.014	—	P	+28
9139	II	14.8	.012	.003	CR	+30
9140	I	14.8	.017	.007	FD	+31
9141	III	14.5	.015	—	P	+32
9143	I	14.5	.012	—	P	+34
9144	II	14.6	.015	.002	P, HR	+35
9150	III	14.5	.008	—	FD	+41
9154	III	14.9	.016	.009	FD	+45
9168.5	—	14.65	>.020	.008	P	+59
9168.7	—	14.55	>.012	.001	P	+59
9170	IV	15.0	>.032	.030	CR	+61
9173	III	14.6?	.019	.003	LS	+64
9174.5	?	14.8:	?	?	P	+65, —40

Table I (Cont.)

J. D. 2,430,000+	Type	Min. Mag.	Width 14.2	Width 14.6	Observer	Days from eruption
9174.7	I	15.0	.018	.007	FD	+65, —40
9176	IV	14.6	>.020	.001	P	+67, —38
9177	III	14.8	.018	.003	P, LS	+68, —37
9178	III	14.6	.018	.001	P, HR	+69, —36
9179	III	14.75	.018	.007	P, HR	+70, —35
9198	III	14.8	.019	.007	P	+89, —16
9200	I	14.9	>.020	.010	P	+81, —14
9201	I	14.75	>.026	.006	P, LS	+92, —13
9202	I	14.9	.017	.010	P, HR	+93, —12
9207	?	14.6	.018	.001	P, HR	+98, — 7
9229	III	14.3	.008	—	P, HR	+15
9238	III	14.4	.014	—	P, HR	+24
9240	I	14.2	.006	—	FD	+26
9241	III	14.5	.011	—	P, HR	+27
9256	III	14.6	.017	.001	P, HR	+42
9262	IV	14.5	.013?	—	P, HR	+48
9264	III	14.5	.015?	—	P, HR	+50
9452	II	14.6	.019?	.003	P	+69, —17
9453	II	14.9	>.015	.007	P	+71, —16
9478	III	14.6	.017	.002	P	+9
9493	III	14.9	.026	.013	P	+24
9500	III	14.7	?	.006	P, HR	+31
9503	III	14.8	.031	.008	P	+34
9506	IV	14.7	>.038	.014	P	+37
9508	I	14.8	>.021	.007	P	+39
9524	II	14.8	>.017	.009	P	+55
9531	III	14.9	>.026	.007	P	+62
9533	III	14.9	>.031	.010	P, HR, SV	+64
9535	III	14.9	>.106	.007	P	+66, —47
9558	III	14.8	>.022	.002	P	+89, —24
9559	II	14.9	>.017	.008	P, HR	+90, —23
9567	III	14.85	>.021	.010	P, HR	+98, —15
9596	III	14.5	.016	—	P	+14
9940	IV	14.8	>.021	.010?	P, HR, SV	+43
9941	IV	`14.7	>.021	.013?	P	+44
9944	II	14,9	>.014	.008	P	+47
9945	II	14.8	>.023	.008	P	+48
9946	III	14.8	>.035	.007	P	+49
9948	III	14.8	>.046	.008	P, HR	+51
9972	IV	14.8	>.015	.002	P	+75

BM	= Baldwin		HR	= Hurless
BOR	= Bornhurst		LS	= Lucas
CR	= Cragg		P	= Peltier
FD	= Ford		SV	= Schmitz

Table II gives the data for the eclipses used in this discussion.

Table II
U Geminorum Eruptions

No.	Class	J. D. 11.0I	Diff.	J. D. Max	Mag	Diff.	J. D. 11.0D	Diff.
388	n	8467	108	8468	9.0	106	8472	99
391	w	8781	314	8784	8.8	316	8792	320
392	n	8888	107	8889	9.5	105	8891	99
394	n	9108	220	9109	9.5	220	9112	221
395	w	9212	104	9214	9.3	105	9223	111
397	w	9377	165	9382	9.2	168	9387	164
398	n	9468	91	9469	9.8	87	9472	85
399	w	9579	111	9582	8.9	113	9591	119
401	w	9768	189	9771	8.6	189	9776	185
402	n	9893	125	9897	9.5	126	9898	122

The diagram shows the time before or after an eruption when various types of eclipse light curves were observed. I find no correlation between the number of days from an eruption and the type of light curve, depth of mini-mum, or width of the curve at magnitude 14.2 or 14.6.

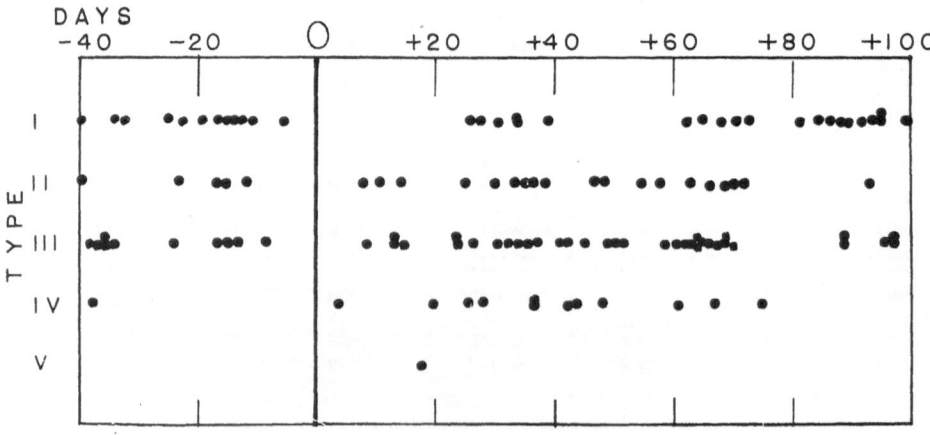

Fig. 1.
Type of Minimum vs. Days from Eruption

I have great confidence in these visual light curves, and feel sure that a large number of minimum observations give good evidence that U Geminorum eclipses are irregular in shape and there is no difference in the shape before or after the outburst.

Non-Periodic Phenomena in Variable Stars
IAU Colloquium, Budapest, 1968

THE STANDSTILLS OF LIGHT OF Z CAM STARS PHOTOELECTRIC AND SPECTROGRAPHIC OBSERVATIONS

MARIE-CLAIRE LORTET

Institut D'Astrophysique, Paris

This paper is a summary of a longer paper to be published in the new "European Journal of Astronomy and Astrophysics"; a detailed paper will also appear in the Supplement Series of this Journal.

There is curiously an almost complete lack of photoelectric or spectroscopic observations of Z Cam stars during their standstills of light. The study of standstills may be however quite important for the understanding of the physical nature of U Gem stars. In particular, it is not understood why certain U Gem stars undergo standstills and not the others, although the physical characteristics (masses, dimensions, separation) do not appear to be different.

Photoelectric observations with B and V filters (and three observations with a U filter) were obtained for two standstills of light of Z Cam and for one of RX And. The results allow two important conclusions to be drawn:

1) During a standstill, the rate of outflow of energy in the U, B, V region is *the same* as the mean rate of emission during normal periods (that is, during periods when outbursts occur without standstills), within 20 or 25 per cent.

2) Rapid brightness fluctuations are present, probably of the same nature as those observed for ordinary U Gem stars, at all stages of activity (outside and during outbursts).

A spectrum of Z Cam in the middle of a 480 days long standstill was obtained in December 1966. The phase of the orbital period extended from 0.11 to 0.56 during the exposure time. Thus, according to the photoelectric light curve of Z Cam over the orbital period (Kraft and al. 1968; Smak, 1968), the observation was made entirely outside the part of the cycle when the light-curve shows a bright shoulder, so that the G-star was never seen through the thickest part of the ring around the other star.

The spectrum has been briefly described previously (Lortet-Zuckermann 1967). It is fairly similar to the spectra obtained for Z Cam and other U Gem stars during the decline after an outburst.

The profiles of the hydrogen lines exhibit central emission (about 20 A in total width) and broad shallow absorption wings extending to about ± 80 A on both sides of the central emission.

No evidence is found for a strong increase in mass loss from the red star during the standstill, but small features would be lost, both due to the lack of time resolution and spectral resolution (about 2 A).

The absorption lines may be explained by Stark broadening at the surface of the G star, while the emission lines arise probably in the ring.

REFERENCES

1) Kraft, R. P., Krzemiński, W., Mumford, G. S., 1968. Astr. J. **73,** S21.
2) Lortet-Zuckermann, M. C., 1967. C. r. hebd. Seanc. Acad. Sci. Paris, **265,** 826
3) Smak, J., 1968, Colloquium on Variable Stars, Budapest, September 1968.

ANALYSE DES COURBES DE LUMIÈRE DES ÉTOILES DU TYPE U GEMINORUM

MICHEL PETIT

Depuis 1963, j'ai cherché, en collaboration avec Mr Léon Ménager, à analyser les courbes de lumière de plusieurs étoiles du type U Geminorum suffisamment observées. De nombreux résultats ont été publiés, ou le seront ultérieurement. Nous n'envisageons ici que les méthodes applicables à l'étude des liaisons et à l'analyse chronologique des courbes de lumière. Les résultats concernant la classification des maxima et l'étude statistique des critères de classification seront publiés par ailleurs.

ÉTUDE DES CORRELATIONS

Rappelons brièvement le principe de la méthode du calcul des corrélations pouvant exister entre deux données, par exemple entre deux éléments de la courbe de lumière.

Soient n données de la variable x_i et n' données de la variable y_i; on nomme m_x et m_y leurs moyennes. Le coefficient de corrélation r se définit comme suit:

$$r = \frac{\sum\limits_{i=1}^{n} (x_i - m_x)(y_i - m_y)}{\sqrt{\Sigma(x_i - m_x)^2 \; \Sigma(y_i - m_y)^2}}$$

où $x_i - m_x$ et $y_i - m_y$ sont les écarts par rapport aux moyennes m_x et m_y.
Les écarts-types se définissent par:

$$\sigma(x) = \sqrt{\frac{\Sigma(x_i - m_x)^2}{n-1}}$$

$$\sigma(y) = \sqrt{\frac{\Sigma(y_i - m_y)^1}{n-1}}$$

On peut démontrer que

$$-1 \leq r \leq +1$$

Pour juger de la valeur d'une corrélation, on peut calculer l'écart à la valeur vraie:

$$\varepsilon = \frac{1 - r^2}{\sqrt{n}}$$

et considerer que la corrélation est bonne, ou pas, selon des valeurs arbitraires de ε. Nous avons préféré uliliser la méthode du coefficient de Pearson. Il s'écrit:

$$z = \frac{1}{2} \log (1 + r) - \frac{1}{2} \log (1 - r)$$

L'écart réduit t_i^2 se calcule par la méthode des χ^2, t_i^2 correspondant aux différences $\Sigma(x_i - x_{th})^2$.

Dans le cas d'une fausse liaison, on a $t^2 = 0$, aux fluctuations aléatoires près, et la distribution est la même que celle des χ^2. Comme t n'est pas lié à la moyenne des z_i, il est possible d'apprécier la signification d'un groupe de corrélation.

Ménager et moi-même avons calculé des coefficients de corrélation entre la magnitude, au maximum et au minimum, la largeur (1) des maxima et des minima et deux autres données:

— l'énergie totale E, calculée en prenant pour unité l'énergie dégagée en un jour au minimum d'éclat (et exprimée en intensité lumineuse)

$$E = \int_{\text{explosion}} \left(I_\nu(t) - I_\nu \min \right) dt$$

— la vitesse de croissance de l'éclat pendant la montée au maximum

$$v = \frac{\varDelta m}{h \text{ phase montée}}$$

Le tableau suivant résume l'analyse faite pour quelques corrélations (n est le nombre d'étoiles étudiées)

Corrélations	n	χ^2	$p(\chi^2 > \chi_0^2)$	Valeur
1 max—V	9	8,23	0,50	faible
1 max—E	9	84,79	$<$0,01	bonne
E — V	9	22,86	0,06	moyenne
1 max—mv max	15	74,04	$\leq$0,001	bonne
1 min—mv min	5	24,52	0,04	moyenne

La liaison entre 1 min et mv min, signalée par Campbell (1933) sur SS Cygni, confirmée par Martel (1961) a été retrouvée par l'auteur sur cinq autres étoiles observées au minimum. Sa valeur moyenne, relativement faible, est probablement due à l'incertitude des observations utilisées.

ANALYSE DES SUITES DE MAXIMA

Soient m_x et m_y les valeurs moyennes des intervalles et des largeurs des maxima, n_x et n_x le nombre des échantillons, σ_x et σ_y les dispersions correspondantes. Si ces valeurs sont distribuées au hasard, nous aurons, puisque les moyennes sont indépendantes:

$$m_x - m_y = 0$$

L'écart réduit t vaut:

$$t = \frac{(m_x - m_y)}{S \sqrt{\dfrac{1}{n_x - 1} + \dfrac{1}{n_x + 1}}}$$

où S, degré de liberté totale, s'exprime par

$$S = \sqrt{\frac{(n_x - 1)\,\sigma_x^2 + (n_y - 1)\,\sigma_y^2}{n_x + n_y \cdot 2}}$$

Les écarts-types sont

$$\sigma_x = \frac{\Sigma\,(x - m_x)^2}{n_x - 1}$$

$$\sigma_y = \frac{\Sigma\,(y - m_y)^2}{n_y - 1}$$

On démontre ainsi que les familles de maxima se groupent en deux classes, selon les valeurs de t, l'une comprenant les maxima d'ordre court (C ou SC), l'autre les maxima d'ordre long (L ou SL).*

Appelons évènement A les maxima d'ordre court, et B les maxima d'ordre long des observations fournissent de A et B disposées comme la suivante:

$$^{3A}_{1B}\ ^{5A}_{1B}\ ^{2A}_{2B}\ ^{1A}_{1B}\ ^{3A}_{1B}\ \text{etc...}$$

Soit une série chronologique comprenant m fois A et n fois B. Cet ensemble peut être considéré comme une suite de tirages indépendants d'éléments distincts. Un schéma d'urne montre que le nombre de variations d'une telle suite est une variable aléatoire de moyenne R_{th}, dans le cas d'une variable gaussienne, telle que:

$$R_{th} = \frac{2\,mn}{m + n} + 1$$

de variance

$$\sigma(R) = \left[\frac{2\,mn\,(2\,mn - m - n)}{(m + n)^2\,(m + n - 1)}\right]^{1/2}$$

L'écart réduit t, qui suit la loi de Gauss, s'écrit:

$$t = \frac{R_{th} - R_0}{\sigma(R)}.$$

* La classification adoptée par l'auteur, très proche de celle établie par Mme Lortet-Zuckermann (1961), (1964) est la suivante:

maxima normaux longs	L
maxima normaux courts	C
maxima symétriques longs	SL
maxima symétriques courts	SC
maxima faibles	F

Si la série est aléatoire, la probabilité de trouver $t > t_0$ s'écrit:

$$P(t > t_0) = \frac{2}{\sqrt{2n}} \int_0^\infty e^{\frac{-u^2}{2}} \, du .$$

Une telle analyse a été faite pour six étoiles (tableau suivant); Lorsque est positif, les évènements A et B ont tendance à se *grouper*, quand t est négativ, ils ont tendance à alterner

Étoiles	R_0	$\sigma (R)$	t	$P(t > t_0)$	Auteur
SS Aur	32,1	3,3	−3,7	0,0001	Ménager
Z Cam	23	2,4	4,1	0,00002	Ménager
U Gem	53,2	5,5	3,3	0,0001	Petit
VW Hyi	42,9	4,1	2,6	0,009	Ménager
CN Ori	97	8,5	3,2	0,0001	Petit
TZ Per	45	4,1	3,6	0,0001	Petit

La même méthode permet de mettre en évidence la tendance au groupement des maxima symétriques, des maxima faibles des oscillations (Lortet, 1964) et d'autres époques erratiques.

Cette méthode a été aussi utilisée pour l'étude des variations non normales de deux étoiles du groupe Z Camelopardalis; nous appelons évènement A les suites variations normales–variations erratiques, évènement B les suites variations normales–paliers. On obtient:

	R_0	$\sigma (R)$	t	$P(t_0 > t_0)$
Z Cam	13,4	1,9	3,9	0,0001
TZ Per	23,5	3,4	3,3	0,001

Ménager et Petit (5) ont montré, pour VW Hyi, que pour les séries de maxima d'ordre court, les largeurs des minima *sont d'autant plus petites que le nombre de maxima courts successifs est plus élevé*. J'ai vérifié sur une douzaine d'étoiles que ce phénomène existe dans tous les cas. Il semble donc que, lorsque des maxima d'ordre courts se succèdent, *leur processus de formation a tendance à s'accélérer*. L'étude approfondie de ce phénomène reste à faire, mais on note que les distributions de mv max, l max et E sont sensiblement les mêmes pour les maxima d'ordre C alternés avec des maxima d'ordre L et pour des maxima C formant une série.

ANALYSE CHRONOLOGIQUE DES VARIATIONS LUMINEUSES

Nous avons utilisé quatre méthodes d'étude, que voici:

1. *Les chaînes de Markov*

Considérons une suite chronologique d'états E, observés à des instants t, tels que l'on ait $E(i_1, t_1)$, $E(i_2, t_2)$ ... $E(i_n, t_n)$ où $i = 1, 2, \ldots n$. Par définition, une telle suite obéit à une chaîne de Markov simple si, quels que soient n et j, la probabilité d'observer un état donné $E(j)$ à l'instant $tn + 1$ ne dépend que de l'état précédent $E(i_n, t_n)$ et non pas des états antérieurs.

L'observation permet d'établir une matrice carrée d'ordre r, dite matrice de transition $|P^K|$

La probabilité $P[E(j, t_{n+1})/E(i_n, t_n)]$ est appelée probabilité de transition en une épreuve. Elle est représentée par une matrice $[P1]$(ou $k = 1$).

On peut aussi calculer des matrices de transition en p épreuves, pour lesquelles:

$$P[E(j, t_{n+1})/E(i, t_{n-p+1})] = [P1]_j^p$$

Dans la pratique on calcule des matrices de transition en deux et trois épreuves, $[P2]$ et $[P3]$ et on les compare avec le carré et le cube de la matrice $[P1]$ observée.

Si l'on a:

$$[P1]^2 \simeq [P2]$$

$$[P1]^2 \simeq [P3]$$

(relation de Chapman—Kolmogorov) c'est que les maxima évoluent suivant une chaîne de Markov simple: la nature d'un maximum dépend alors du maximum précédent *et seulement de celui-ci.*

Martel (1961) avait signalé que c'est le cas de SS Cygni, mais Mme Lortet (1964, 1966) appliquant divers tests sur le caractère aléatoire d'une série de deux ou trois paramètres, arrive aux conclusions suivantes:

— la succession des maxima de caractère long et non-long est assez bien représentée par une chaîne de Markov d'ordre 1, mais notons que SS Cygni a une nette tendance à l'alternance régulière des deux types, ce qui n'est pas le cas de toutes les étoiles que nous avons étudié.

— la succession des maxima faibles (F) et non faibles peut se représenter par une chaîne de Markov d'ordre 2.

— un schéma markovien d'ordre 1, 2 ou plus n'explique pas l'existence des séries de maxima courts, ou des périodes erratiques, riches en variations irrégulières.

Mme Lortet a donc proposé un *schéma d'évènement récurrent* où les structures successives sont indépendantes, ont chacune une certaine probabilité, et où l'évènement est constitué par un changement de structure. Ces structures cycliques peuvent être, soit les suites long-court, les suites court-court et les variations non normales (oscillations, paliers, etc . . .).

Pour plusieurs étoiles, Ménager et moi avons appliqué deux tests, l'un en calculant des matrices théoriques normalisées, et en appliquant à ces matrices normalisées $|Pk|$ un test en χ^2 des transitions en une épreuve, l'autre en calculant des matrices $[P1]^2$ et $[P1]^3$, déduites de $[P1]$, malgré la difficulté de leur attribuer un intervalle de confiance, et en les comparant aux matrices $[P2]$ et $[P3]$.

Les résultats obtenus sont résumés par le tableau suivant:

Variables	m	$\chi\,|\,P2\,|$	$P(\chi^2 > \chi_0^2)$	$\chi[P3]$	$P(\chi^2 > \chi_0^2)$	Auteur
SS Aur	4	11,54	0,02	5,19	0,25	(7)
Z Cam	1	5,30	0,025			Ménager
U Gem	1	10,8	0,01	4,7	0,16	Petit
VW Hyi	1	3,44	0,07	3,2	0,08	Ménager
CN Ori	1	4,7	0,04			Petit

Dans tous les cas *il n'apparaît pas que l'on soit devant une chaîne de Markov simple.*

2. La fonction d'autocorrélation

Il faut remarquer que la suite des intervalles peut être markovienne sans que la suite des maxima le soit. Ménager a repris l'analyse, en calculant une fonction d'autocorrélation, où le coefficient d'autocorrélation $r(p)$ est défini par:

$$r(p) = \frac{\Sigma\,(x_i - x)^2(\nu_i - y)}{\sqrt{\Sigma\,(x_i - x)^2\,(y_i - y)^2}}$$

où p est l'ordre du maximum suivant, et où

$$y_1 = x_{i+p}$$

$$x = \frac{\sum_{i=1}^{m} x_i}{m}$$

$$y = \frac{\Sigma\,x_{i+p}}{m}$$

On a

$$r(p) = \frac{[M\Sigma\,x_i(x_{i+p}) - \Sigma\,x_i\,\Sigma\,x_{i+p}]}{\sqrt{M\,\Sigma\,x_i - [\Sigma\,(x_i)]^2}\,\sqrt{M\,\Sigma^2\,x_{i+p} - [\Sigma\,x_{i+p}]^2}}$$

m étant le nombre de termes et N le nombre de couples observés $(M = N + p)$. On calcule ensuite les variances en chaque point par

$$\varepsilon = \frac{1 - r^2\,p}{\sqrt{M}}$$

Pour SS Aur et pour Z Cam l'autocorrélogramme des intervalles suggère une chaîne de Markov d'ordre 4 ou plus; pour VW Hyi, la suite des intervalles semble suivre un processus markovien simple, tandis que la suite des maxima obéit à un processus d'ordre $\geqslant 4$.

3. *Calcul du temps de retour*

Le temps de retour est le nombre d'épreuves nécessaires pour le retour à un état spécifié, c'est à dire le nombre théorique de suites long–non long que l'on compare ensuite aux nombres observés.

Pour Z Cam on observe, par rapport au calcul d'une matrice théorique, un déficit de transition pour les suites de maxima courts d'ordre 1, ou d'ordre > 4, et un excès de transition pour les valeurs moyennes.

On conclut que les suites de maxima longs suivent un schéma conforme à celui de la matrice $[P1]$ en une épreuve, tandis que les suites de maxima courts se rapprochent du schéma d'un tirage exhaustif d'evènements A et B absolument aléatoires.

4. *Test de récurrence*

Cette méthode a été utilisée pour l'analyse des l min. On pose

$$q^2 = \frac{1}{2(N-1)} \sum_1^{N-1} (x_{i+1} - x_i)^2$$

$$S^2 = \frac{1}{N-1} \sum_i^n (x_i - \bar{x})^2$$

q^2 étant la variance de 2 valeurs successives de la variable x et S^2 la variance de x. Soit

$$\gamma = \frac{q^2}{s^2}$$

$$\sigma^2(\gamma) = \frac{1}{N+1}\left(1 - \frac{1}{N-1}\right) \simeq \frac{1}{N+1}$$

Pour Z Cam on obtient $q^2 = 25$, $S^2 = 36{,}2$ $\gamma = 0{,}69$. Le test de Student nous donne:

$$t = \frac{1-\gamma}{\sigma\gamma} = 5.1 \qquad p(t > t_0) < 10^{-6}$$

ce qui indique une *tendance à la récurrence*. Les calculs effectués sur SS Aur et U Gem confirment ce phénomène; dans ces deux cas $p(t > to) < 10^{-5}$.

Ainsi notre analyse confirme qu'un schéma de chaîne markovienne, s'il suffit dans certains cas, à l'étude de la succession des intervalles n'explique ni la tendance au groupement des maxima faibles ou symétriques, ni l'existence de périodes erratiques diverses. Il semble donc préférable d'adopter un schéma d'évènement récurrent, mais une analyse complète demandera de disposer de séries de maxima longues et continues et d'une bonne classification des phénomènes observés.

26

REFERENCES

Campbell, L., 1933, Ann. Harv. Coll. Obs., **90**, No. 3.
Lortet-Zuckermann, M. C., 1964, Ann. Astrophys. **27**, 65.
Lortet-Zuckermann, M. C., 1966, Ann. Astrophys. **29**, 205.
Martel, L., 1961, Ann. Astrophys. **24**, 267.
Ménager, L. et Petit, M., (à paraître).
Petit, M. et Ménager, L., 1963, Ciel Terre **79**, 407.
Zuckermann, M. C., 1961, Ann. Astrophys. **24**, 431.

ON A POSSIBLE CAUSE OF BRIGHTNESS FLUCTUATIONS IN CLOSE BINARY SYSTEMS OF DWARF STARS

by V. G. GORBATZKY

Leningrad University Observatory, USSR.

The observational data on novae, recurrent novae and U Gem stars recently obtained (Kraft, 1963) show that most of these stars and very likely all of them are binary systems. They consist of dwarf stars and some of these systems are observed as eclipsing ones. In this case the light curve of the system has the following peculiarities:

1. Short periods, usually near to 6^h.
2. The primary light minimum is asymmetrical.
3. The secondary light minimum is very shallow or unobservable.
4. Brightness fluctuations, superposed on the eclipse light curve, are observed. Their amplitude is about $0^m.05$ and their time scale is of the order of 1 min. The fluctuations are more outstanding near the phase 0.25 P and before the primary light minimum. They are smallest on the descending branch to the light minimum.

A possible explanation of these brightness fluctuations is proposed in this note. As it is well known from spectroscopic observations, the primaries of the close binaries of the type considered here have disk-like gaseous envelopes. It was found from theoretical study of gas motions in a close system of two gravitating points (Prendergast, 1960), that in the vicinity of each point these motions differ only a little from the circular Keplerian ones. Consequently, an azimuthal velocity gradient must exist in the disklike envelope.

The flow of gas in the envelope towards the stellar surface is caused by turbulent viscosity. If there is no permanent supply of gas, the envelope will vanish. All the matter it contains will join the star. Estimates show (Gorbatzky, 1968) that this will occur within one day or even at a shorter time. However, we observe the envelopes over long time intervals, and this must be considered as an evidence of the supply of the envelopes with gas. This supply may be realized only by gaseous streams flowing from the other star of the binary system. In the case of WZ Sge the stream ("jet") that transfers the matter to the envelope of the primary star, can be detected directly (Krzeminsky and Kraft, 1964) by spectroscopic observations.

The gaseous jet must be accelerated by the star's gravitational field. If the initial velocity of the gas is small enough, its radial velocity v near the envelope is of the order

$$v \approx \sqrt{\frac{2GM_*}{r}} \tag{1}$$

Taking the mass of the star $M_* \approx M_\odot$, and the radius of the envelope $r \approx 10^{10}$ cm, as found from observations, we get $v \approx 5 \cdot 10^7$ cm sec^{-1}. A more precise calculation, making allowance for the gas pressure as well as for the effects caused by the rotation of the stars around their common center of gravity, leads to values of V of the same order.

The disk-like envelope emits the radiation not only in discrete frequencies but also in the continuum. As it may be inferred from photometric data, the brightness of the system diminishes subtantially even during the eclipse of the envelope. Consequently, the observable continuous radiation of the envelope is significant, sometimes it may give the main contribution to the total radiation of the system.

The kinetic energy of the gas flowing into the envelope must be one of the main sources of the radiation emitted by the envelope. The heated region is formed at the place where the gas flow encounters the envelope. The radiation of this region is more intense than that of the other parts of the envelope.

Apparently just in this place we observe the so called "hot spot", the presence of which has been supposed in some studies to explain the peculiarities of the light curve. The "shoulder" on the light curve that precedes the primary minimum cQrresponds to phases of best visibility of the spot. During this light minimum the hot spot is eclipsed by the cold secondary star.

The calculations show that almost half of the total kinetic energy of the gaseous streams will be transformed into the radiation of the hot spot region (the rest of the energy dissipates in other regions of the envelope later). Hence the energy E radiated by the envelope is

$$E = K \varrho v^3 S \tag{2}$$

Here ϱ is the gas density, v the gas velocity, K is a coefficient of the order of unity.

In theoretical studies the gaseous stream is usually considered as a continuous jet flowing from point L_1 and having constant velocity and constant capacity. The envelope is assumed to be a homogeneous disk. There are no observational data which confirm the validity of such a simple model. On the contrary, the observed differences in the depths of light minima in different cycles are evidences of rapid changes in the dimensions of the envelope. Hence the capacity of the gaseous flow also changes. Variations in the velocity of the jet are observed in case of WZ Sge.

A variation of not more than 10 per cent in the density of the jet, or a change of a few per cent in the velocity of the gaseous streams is sufficient to change the luminosity of the envelope by about ten per cent and, correspondingly, the brightness of the system changes by several hundredths of a magnitude. Apparently, inhomogeneity of the jet on such a scale is quite possible and it may be the main cause of the observed brightness fluctuations.

The stream seems to flow out from its envelope and not from the secondary star. There are spectroscopic evidences of the presence of envelopes around the cool secondaries (Greenstein, 1960). Since well developed turbulence must exist in such an envelope, the stream flowing out from it cannot be fully homogeneous. The scale of inhomogeneities in the stream and in the envelope must be of the same order.

If the brightness fluctuations are caused by imhomogeneity of the gaseous stream, one can estimate the size d of the largest inhomogeneities in the envelope of the secondary star. Using data on the duration (Δt) of the brightness fluctuations and taking into account that $d \sim v\Delta t$, we find that d is of the order of 10^9 cm. On the other hand, we have earlier estimated (Gorbatzky, 1965) the scale (l) of the turbulence in a disk-like envelope. It can be found from the equation, that

$$l \approx \frac{r\,\Delta u}{v_\varphi} \tag{3}$$

where Δu is the velocity of the turbulent pulsations and v_φ is the azimuthal velocity. The value of Δu cannot be higher than some tens of km/sec and $v_\varphi \approx 5.10^7$ cm/sec. Therefore, taking $r \approx 10^{10}$ cm we have $l \leq 10^9$ cm. In this way we found that the size of inhomogeneities in the envelope of the secondary star and correspondingly the scale of non-uniformity in the stream are of the same order as the main scale of turbulent motions in the envelope. The turbulence in the envelope of the secondary star can result in inhomogeneity of the gaseous jet flowing into the envelope of the primary. The fact, that $d \approx l$, seems to confirm our assumption on the cause of brightness fluctuations in close binary systems of dwarf stars.

The inhomogeneity of the stream combined with the rotation of the envelope makes the envelope structure rather complex. Densities in the envelope in the same distance from the primary star may be different. This must result in different emissivity. As the envelope rotates, this may also cause brightness fluctuations. In both cases changes in the capacity of a gaseous flow give rise to brightness fluctuations in the systems considered.

REFERENCES

Gorbatzky, V. G., 1965, Trudy astr. Obs. Leningr. gos. Univ. **23**, 15.
Gorbatzky, V. G., 1968, Report Trieste Symp. (in press).
Greenstein, J. L., 1960, Stellar Atmospheres, ed. by J. L. Greenstein (Univ. of Chicago Press. Ch. 19).
Kraft, R., 1963, Cataclysmic Variables as Binary Stars. Adv. Astr. Astrophys. **2**, p. 43.
Krzeminsky, W. and Kraft, R., 1964, Astrophys. J., **140**, 921.
Prendergast, K., 1960, Astrophys. J., **132**, 162.

SYMBIOTIC STARS

Introductory Paper by

A. A. BOYARCHUK

Crimean Astrophysical Observatory, USSR

The term "symbiotic stars" was first introduced by P. W. Merill and now is widely used for the designation of astronomical objects, whose spectra represent a combination of absorption features of a low temperature star with emission lines of high excitation. Several dozens of such objects are known today. Bidelman's (1954) list contains 23 "stars with combination spectra". Mrs. Payne-Gaposchkin (1957) attributed 32 objects to symbiotic stars. Most stars are common in both lists. Four reviews of symbiotic stars were published for the las 10 years (Merrill, 1958; Payne-Gaposchkin, 1957; Sahade 1960, 1965) and many problems, which have been mentioned, are still very important.

As far as this Colloquium is concerned with non-periodic phenomena, I will concentrate your attention on such phenomena and will touch upon other sides of the problem of symbiotic stars only when it is necessary.

First of all it is necessary to note that the criteria of "symbiotic stars" above mentioned are rather rough. Many of long-period variables as well as classical symbiotic stars like Z And satisfy such criteria. As a result the lists of symbiotic stars given by Bidelman and by Mrs. Payne-Gaposchkin are not homogeneous.

The detailed investigations of the typical symbiotic stars Z And, AC Peg, AX Per, CI Cyg and BF Cyg give a possibility to propose the following criteria for symbiotic stars. I. The absorption lines of late-spectral type (TiO bands, CaI, CaII and al.) must be seen. II. The emission lines of HeII, OIII or higher ionized atoms must bee seen. The widths of emission lines do not exceeded $\sim$ 100 km/sec. III. The stellar brightness can vary with an amplitude up to 3 magn. and with a period of several years.

If the informations about a star correspond to the criteria mentioned, then the star belongs with high probability to the group of symbiotic stars. The list of that stars are found in Table 1. Table 2 lists the stars, for which we have not all necessary informations and the known data indicate only that they may be symbiotic stars.

LIGHT VARIATIONS

The irregular variations of brightness is one of the most characteristic features of symbiotic stars. Their light curves can be considered as a whole complex of small simultaneous flares. This circumstance has given a possibility to name these stars as "novalike stars".

Table 1

N	Name	R. A. 1900	Dec. 1900	m_{max}	m_{min}	S_p	em	V_r	P_{rv}	P_m	Ref.
1	2	3	4	5	6	7	8	9	10	11	12
1	AX Per	$01^h29^m57^s$	+53°44.9	9^m7	13^m4	M5III	[FeVII]	−110		630^d	Boyarchuk, 1968
2	VV 8	01 52 13	+52 24.8		14.3	G5III	[OIII]	−6			O'Dell, 1966
3	RX Pup	08 10 42	−41 24.0	11.1	14.1	M5III	[FeVII]				Swings and Struve 1941.
4	SY Mus	11 27 36	−64 52.0	11.3	12.3	M3III	[OIII]			623	Henize, 1952
5	RW Hya	13 28 47	−24 52.1	10.0	11.2	M2III	[OIII]	+10	370^d	370	Merrill, 1950a
6	AG Dra	16 01 07	+67 04.7	9.1	11.2	K3III	HeII	−140			Boyarchuk, 1966
7	HZ 177	16 39 34	−62 25.4		13.1	M	[OIII]				Webster, 1966
8	YY Her	18 10 18	+20 57.4	11.7	>13.2	M2III	[OIII]				Herbig, 1950
9	AR Pav	18 10 20	−66 07.0	10.2	12.7	M	[OIII]	−50		605	Sahade 1949
10	FR Sct	18 17 46	−12 44.0	11.7	12.5	M2III	[OIII]				Bidelman, Stephenson, 1956
11	V443 Her	18 19 00	+23 24.0	12.4	12.6	M3III	[OIII]	−55			Tift, Greenstein, 1958
12	FN Sgr	18 48 01	−19 07.1	9	13.9	Pec.	[OIII]	−51			Herbig, 1950
13	CM Aql	18 58 22	−03 12.2	13.2	16.5	M4III	HeII				Herbig, 1960
14	BF Cyg	19 19 57	+29 28.8	9.3	13.5	M5III	[OIII]	+5	750^d		Boyarchuk, 1968a
15	My 129	19 33 18	−68 22.3	9.4	13.7	M3	[OIII]				Thackeray, 1954
16	CI Cyg	19 46 30	+35 25.9	13.3	>16.5	M5III	[FeVII]	+15			Boyarchuk, 1968a
17	V407 Cyg	20 58 41	+45 22.8		>15.5	Mep		+15			Merrill, Burwell 1950
18	MHα 328−116	19 55 12	+39 33.0	10		M3III	[OIII]	−58			Boyarchuk, 1968b
19	AG Peg	21 46 11	+12 09.5	6.8	8.2	M3III	[OIII]	−16	800^d	800^d	Boyarchuk, 1967a
20	Z And	23 28 51	+48 16.0	8.0	12.4	M2III	[FeVII]	−5		714^d	Boyarchuk, 1967b
21	R Aqr	23 38 39	−15 50.3	5.8	11.5	M7e	[OIII]	+15	26y, 3y	387^d	Merrill, 1950b

Table 2

N	Name	R.A. 1900	Dec. 1900	m_{max}	m_{min}	S_p	em.	V_r	Ref.
1	2	3	4	5	6	7	8	9	10
1	HD 4174	00h41m54s	40°24'.0	7m5		M2III	[OIII]	—101	Wilson, 1950
2	DV Aur	05 15 18	+32 24.7	8·2	10·0	C5	[OIII]		Sanford, 1944
3	Hz 134	15 41 04	—66 18.5		15		[OIII]		Webster, 1966
4	HD 330 036	15 47 15	—48 42				[OIII]		Webster, 1966
5	Hz172	16 29 45	—55 38	11·7	12·9		[OIII]		Webster, 1966
6	MH$_\alpha$ 276—52	16 45 06	—25 49	11·5	15·8	Pec.	[FeVII]		Merrill, Burwell, 1950
7	HK Sco	16 48 17	—30 13.7	13·1		Com.	HeII		Elvey, 1941
8	V 455 Sco	17 00 47	—33 57.9	12·8	>16·5		H		Merrill, Burwell, 1950
9	MH$_\alpha$ 276—12	17 05 42	—32 28		17		[FeVII]		Merrill, Burwell, 1950
10	MH$_\alpha$ 79—52	17 37 24	—22 43	12			[FeVII]		Merrill, Burwell, 1950
11	MH$_\alpha$ 359—110	17 45 00	—22 17	11			[FeVII]		Merrill, Burwell, 1950
12	KW Sco	17 45 42	—27 59.8	11·0	13·2	Mp	H		Swope, 1940
13	F 6—7	17 59 36	—20 21	11		M3			Merrill, Burwell, 1950
14	Y CrA	18 07 12	—42 52.3	12·0	12·9	Pec.			Bidelman, 1954
15	MH$_\alpha$ 305—6	18 59 18	+16 18	11·5			[FeVII]		Merrill, Burwell, 1950
16	MH$_\alpha$ 80—5	19 41 24	+18 22	11		Com.	H		Merrill, Burwell, 1950

It is necessary to mention that different stars have rather different light curves. Moreover, the character of the light curve of the same star varies strongly from time to time. Many astronomers have observed light variations of symbiotic stars for many years. Now we shall briefly consider the results of the observations of several symbiotic stars.

Fig. 1 shows the light curve of Z And. The papers by Prager (1941), Payne-Gaposchkin (1964), Erleksova (1964), Beljakina (1968), were used for the construction of the curve.

The light variations of Z And have a rather complicated character. There are periods, when the stellar brightness changes very little, for instance in 1905—1913. On the other hand the star flared up to 4^m in 1914 and 1939. Quite a few flares with smaller amplitude were observed in other years. In 1920—1931 the light variations had a quasi-periodic character. According to photoelectric observation (Belyakina, 1968) the light variations were very complicated after 1960. Mrs. Payne-Gaposchkin (1945) found light maxima following one after another in 714 days. But departures from the average period can reach hundred days in several cases. On the whole one can say that non-periodic processes play a very important role in the case of Z And.

Fig. 2 shows the light curve of BF Cyg, which was constructed on the basis of the papers by Jacchia (1941), Aller (1954), Wachmann (1961) and Romano (1966). We can see that the light curve of BF Cyg is similar to that of Z And. Here we also have periods for small variations, for instance for 1929—1931. A rather large flare was observed in 1955. The quasi-periodic variations of the brightness of BF Cyg are smaller than those of Z And, the non-periodic part is very important here.

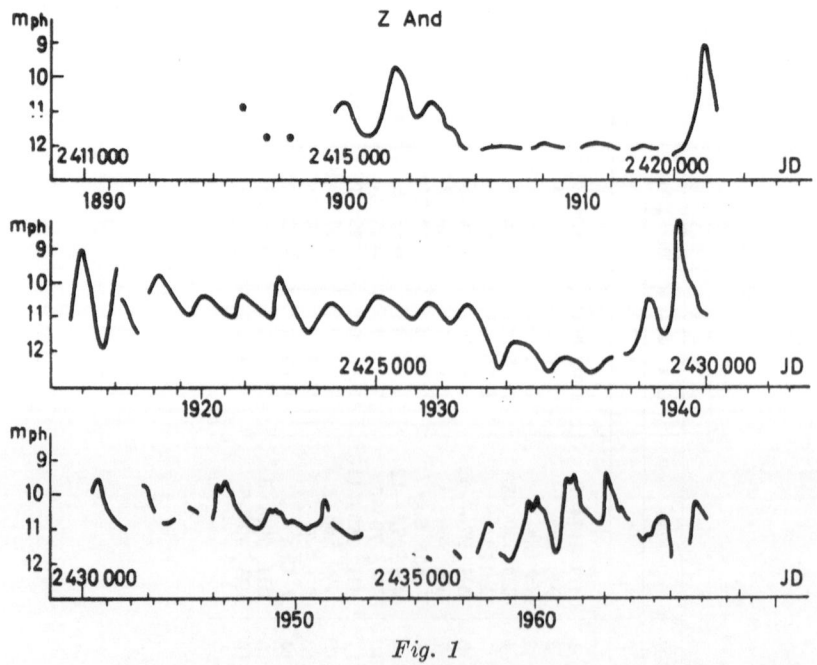

Fig. 1

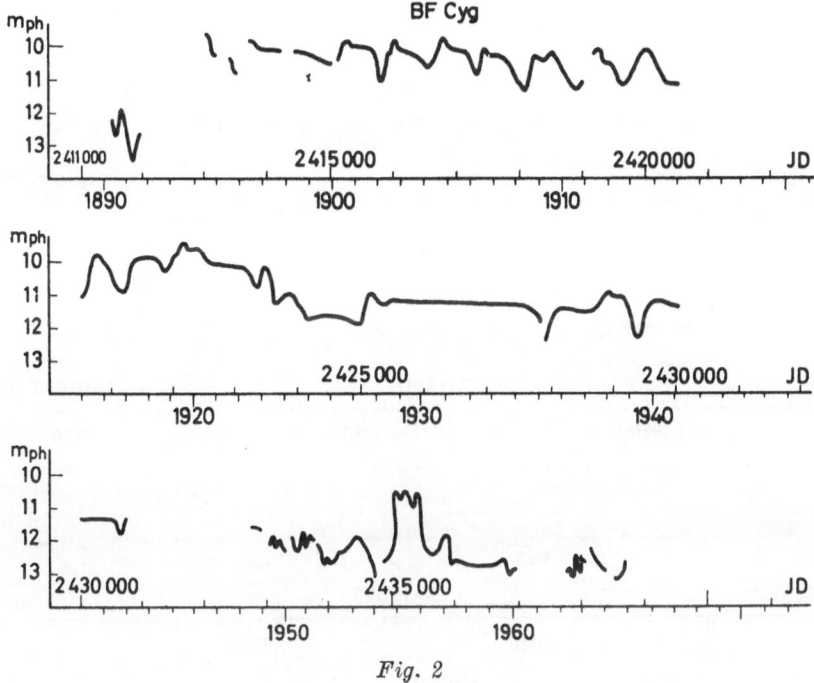

Fig. 2

Fig. 3 shows the light curve of CI Cyg, which was constructed on the basis of papers by Greenstein (1937), Aller (1954) and Hoffleit (1968). The brightness of the star varied generally with small amplitude. Rather large flares were observed only twice, in 1911 and in 1958. Mrs. Hoffleit has found that quasi-periodic variations have a period of 815^d. We can see the form of those variations in Fig. 4, which was adopted from Hoffleit's paper (1968). The scattering on that figure is large. It means that non-periodic variations are important.

Fig. 5 shows the light curve of AX Per, which was constructed on the basis of papers by Lindsay (1932), Payne-Gaposchkin (1946), Wenzel (1956), Sieder (1956) and Romano (1960). On the whole the light variations of AX Per are similar to those of Z And.

The light variations of AG Peg have different character. Fig. 6 shows the light curve of AG Peg from 1825 on. The symbols are as follow ●—m_{vis}, ×—m_{ph}, △—u, □—B, o—V. The original data were published by Lundmark (1921), Himpel, (1942), Sanding (1950), Payne-Gaposchkin (1950), Mayal (1964) and Belyakina (1965).

In general terms the light curve of AG Peg resembles that of a slow developed flare. Using more detailed observations in the last years (Mayall, 1964, Belyakina, 1965) Miss Belyakina discovered the periodical variation of brightness with $P = 800^d$ and $\varDelta m \sim 0.^m3$. This is easy to see in Fig. 7. The symbols are: O—visual, ●— phootelectric V.

Special photoelectric observations of the light variation of symbiotic stars (Belyakina 1965, 1967) have shown, that many of them show rapid light

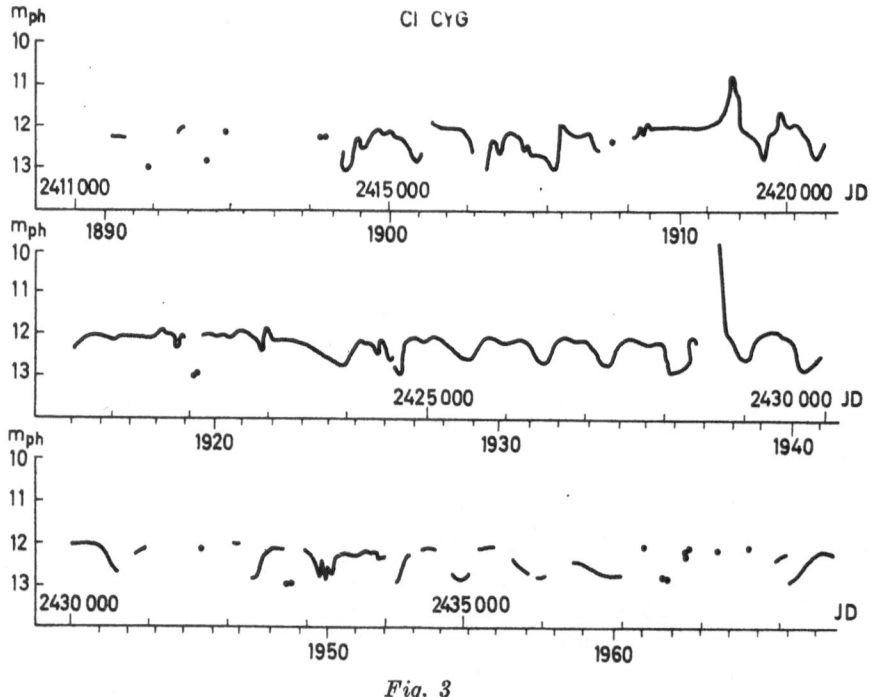

Fig. 3

variations with small amplitudes and such variations have a non-periodic character. Some examples are shown in Fig. 8 (Z And) and in Fig. 9 (AG Dra). In these figures the dots correspond to brightness differences between the symbiotic star and the comparison stars, and the crosses correspond to those between two standard stars. The largest short-time variations have been observed in the ultraviolet.

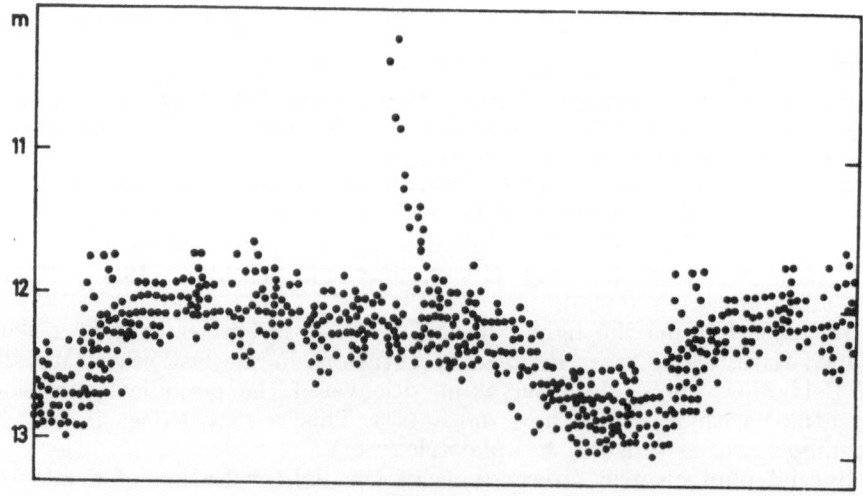

Fig. 4

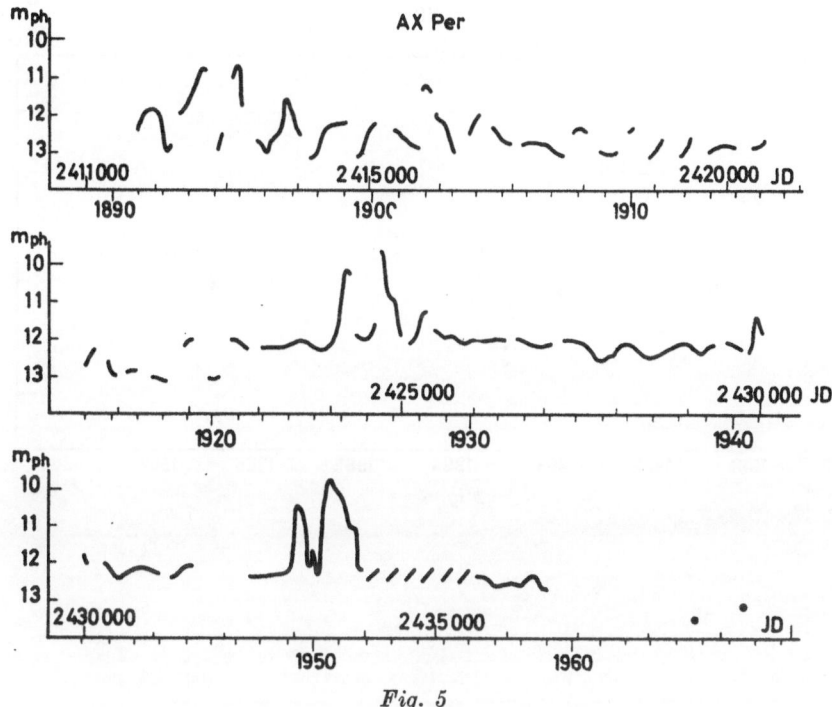

Fig. 5

Thus the available observations show that non-periodic processes play a very important role in the light variation of symbiotic stars.

The brightness variations of symbiotic stars are accompanied by color variations. As it was noted by Jacchia (1941), Himpel (1941), Payne-Gaposch-kin (1946) and al., the value $m_{ph}-m_v$ increases with decreasing brightness, i.e. star becomes redder. Belyakina's photoelectric observations confirm this statement, moreover, she has shown, that the U—B color decreases with decreasing stellar brightness, i.e. the ultraviolet excess increases.

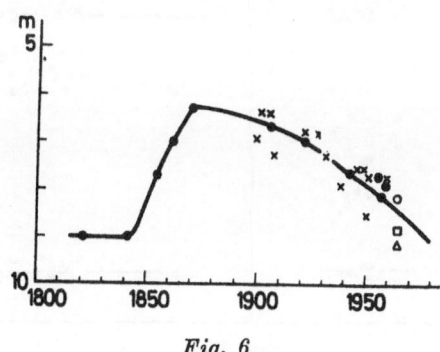

Fig. 6

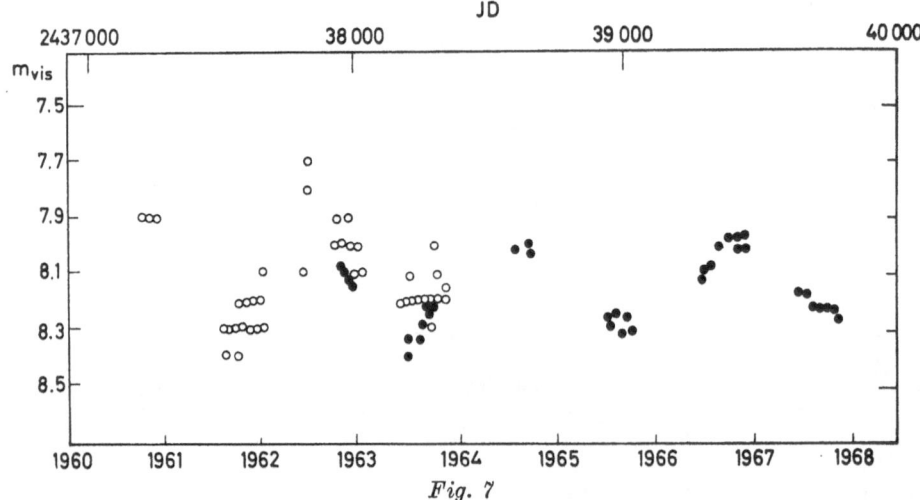

Fig. 7

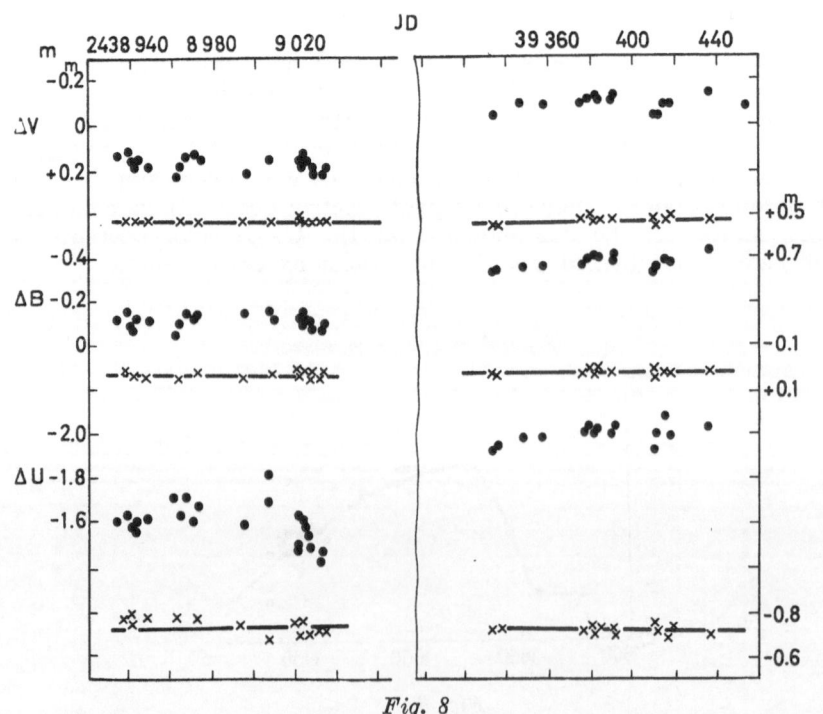

Fig. 8

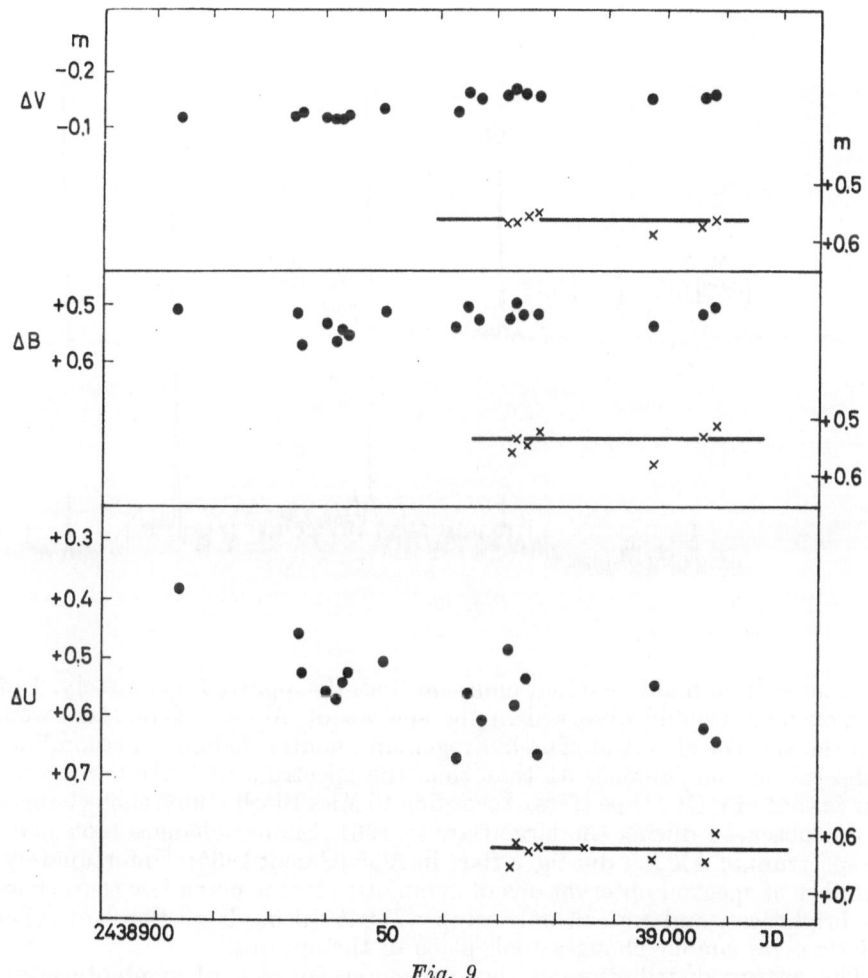

Fig. 9

SPECTRAL VARIATIONS

The character of spectra of symbiotic stars also changes together with the light variations. Already in 1934 Hogg (1934) noted that features of a late type spectrum strengthened when the brightness of Z And decreased. At the same time the excitation degree of the emission spectrum increased.

Later similar changes of the spectrum were found for other symbiotic stars.

Fig. 10 shows the intensity tracing of spectra of AX Per taken in 1964, at $m_{ph} = 13^m5$, and in 1965, at $m_{ph} = 12^m8$. We can see that the intensities of the [FeVII] and He II lines appreciably decreased in comparison with those of hydrogen when the stellar brightness increased by 1 mag. At the same time the intensity of the TiO bands strongly decreased.

Swings and Struve (1941a) have recognized that during the large flare of Z And in 1939 its spectrum underwent great changes. Late-type absorption

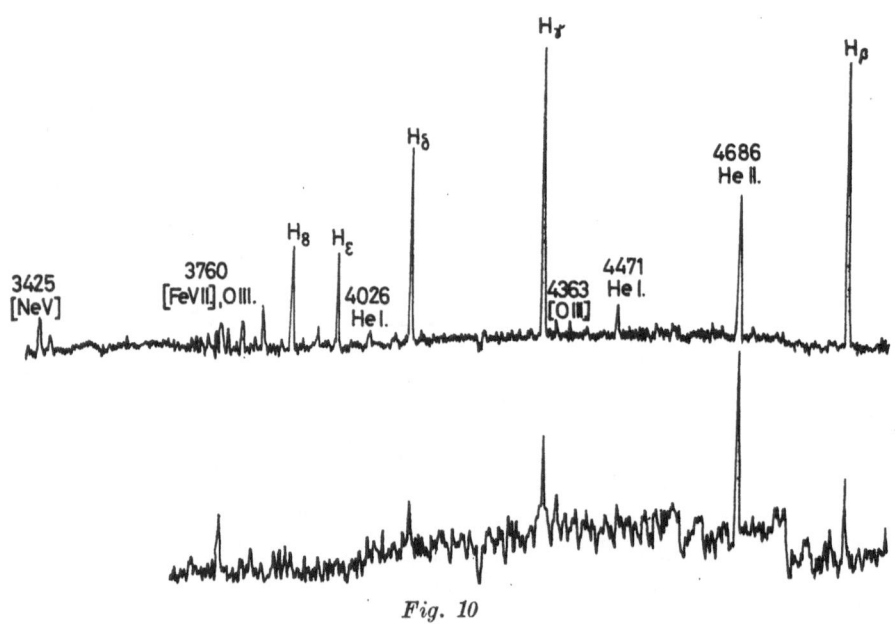

Fig. 10

features as well as highly excited emission lines disappeared completely. But absorption lines usually observed in the spectra of A0—A2 type stars were seen in the spectra of Z And. The hydrogen and neutral helium emission lines had absorption companions. At that time the spectrum of Z And was very similar to that of P Cyg type stars. According to Miss Bloch (1963) such changes were also observed during the large flare in 1961. Similar changes took place in the spectrum of AX Per during a flare in 1955 (Gauzit 1955). Unfortunately, the number of spectral observations of symbiotic stars is much less than those of the brightness, and we can only suppose that during large flares of other symbiotic stars similar changes took place in the spectra.

   The energy distribution in the continuous spectra of symbiotic stars have been investigated by many authors (Tcheng Mao Lin, Bloch 1952, 1954, Ivanova 1960, Boyarchuk 1967 and al.). The results of these investigations are in good agreement with those of color variation. In Fig. 11 the solid lines represent the observed energy distribution in the spectrum of Z And for different data. It is seen that the energy distribution becomes steeper and the Balmer jump increases when the star becomes fainter.

   The displacement of different lines, or the radial velocities can give important informations about the processes in symbiotic stars. At present a rather large number of measurements of radial velocities for five stars: AG Peg, BF Cyg, Z And, RW Hya and R Aqr have been published mainly by Merrill and by Swings and Struve (1941—1943). The variations of radial velocities have periodic character for four stars: AG Peg, BF Cyg, RW Hya and R Aqr. The velocity curves of the stars mentioned are represented in Fig. 12. This figure brings into evidence the binary nature of symbiotic stars. It should be noted that AG Peg shows progressive changes of radial velocity as well as periodic variations. Table 3, which was compiled by using Merrill's data (1929,

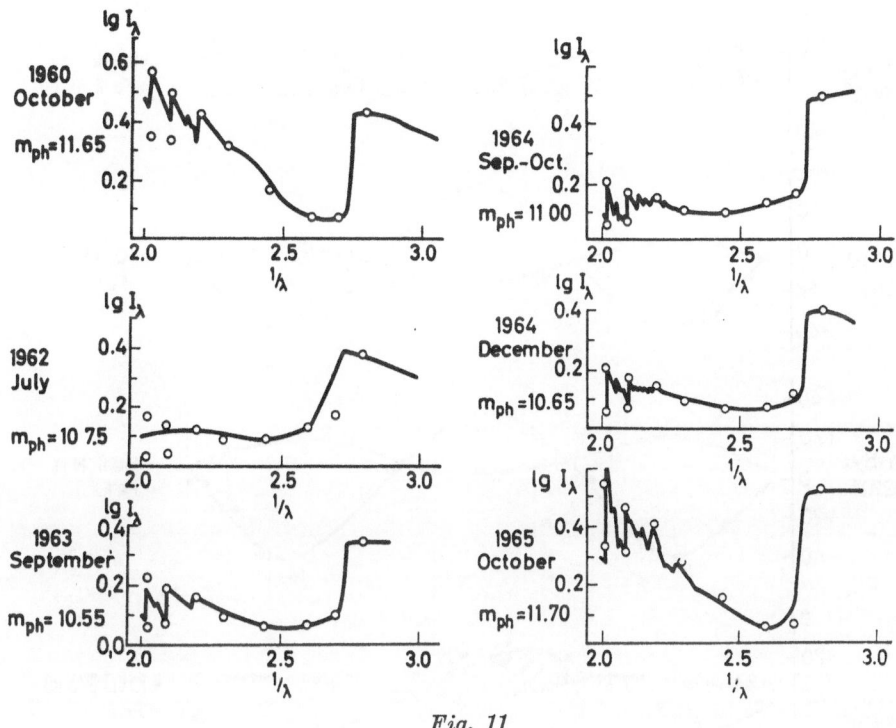

Fig. 11

1942, 1951a, b, 1959) contains data about the average velocities determined from the displacements of different lines. Boyarchuk (1967) has explained this phenomenon by the decreasing opacity of an envelope and by the changing excitation conditions.

The influence of a change of the excitation conditions on radial velocities can especially be seen for Z And, Fig. 13 represents variations of radial velocities determined from the ionized metal lines, A, and from the lines of highly ionized elements, B. The symbol o means that the star had $m_{ph} < 9^m$, × that

Table 3.

Element	V_r km/sec				
	1915	1926	1939	1946	1952
H	+16	+12	—4	—8	—27
HeI		+6	+1	—8	
NII		+6	—12	—20	
HeII			—13	—6	
NIII			(—8)	—20	
OIII5007⎤ 4959⎦				—17	
OIII 4363				—55	
FeII, TiII, SiII		—22	—14	—18	

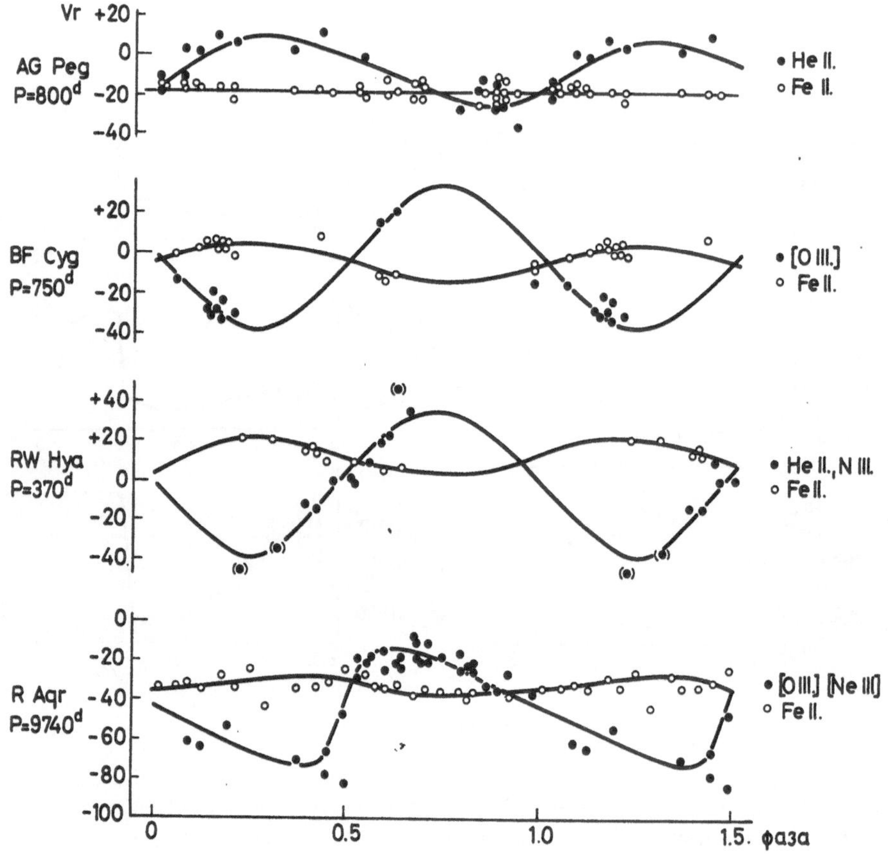

Vr +20
0
-20
-40
AG Peg
P=800d
• He II.
○ Fe II.
+20
0
-20
-40
BF Cyg
P=750d
• [O III.]
○ Fe II.
+40
+20
0
-20
-40
RW Hya
P=370d
• He II., N III.
○ Fe II.
0
-20
-40
-60
-80
-100
R Aqr
P=9740d
• [O III.] [Ne III]
○ Fe II.
0
0.5
1.0
1.5.
фаза

Fig. 12

Vr
+40
+30
+20
+10
0
-10
-20
A
+10
0
-10
B
0d
250d
500d
750d
ФАЗА

Fig. 13

mph
8
9
10
11
20
0
20
40
60
Vr

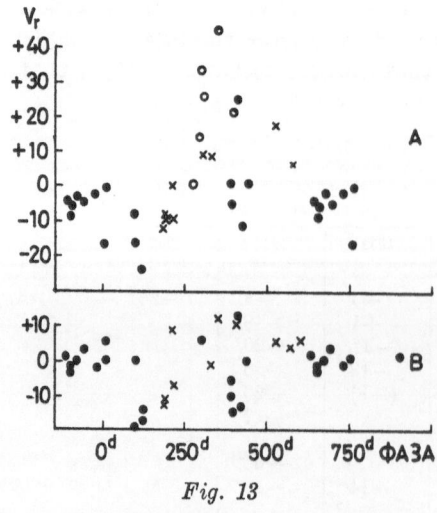

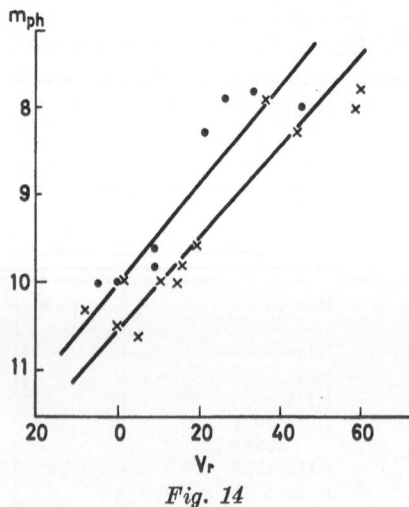

Fig. 14

Table 4

		1960 October	1962 July	1963 September	1964 Sept.—Oct.	1964 December	1965 October
m_{ph}	Z And	11.65	10.75	10.55	11.00	10.65	11.70
	M2III	12.11	11.98	11.49	12.20	10.74	12.17
	Nebula	14.00	13.25	12.90	12.86	12.78	13.68
	Hot Comp.	13.29	11.34	11.37	11.77	11.47	13.50
Hot comp.	$T \cdot 10^{-3}$ °K	108	70	79	96	88	141
	M_V	2.82	1.07	1.57	1.32	1.55	3.05
	$R/R\odot$	0.30	0.70	0.55	0.60	0.54	0.26
Nebula	$\lg n_e$			7.1	7.3	7.5	
	$M \cdot 10^{-29}$g	0.6	1.2	1.0	2.1	1.3	0.9
	$R \cdot 10^{-15}$ cm	0.8	0.9	0.9	1.1	1.0	0.9

$10^m > m_{ph} > 9^m$ and ● that $m_{ph} > 10^m$. The scattering is very large and is connected with change of stellar brightness. Fig. 14 represents the dependence of the value of radial velocity upon the magnitude of the star for phases $300^d—400^d$.

The symbol o corresponds to FeII lines, the symbol × to hydrogen lines. Thus, in the case of Z And, the radial velocity variations reflect the variations of exitation conditions in a higher degree than an orbital motion.

Summing up the review of basic data about the variation of symbiotic star characteristics, the conclusion should be drawn that non-periodic processes take a very important place. Naturally the question arises which physical process is responsible for these variations and what is its nature. Some information about it can be obtained from an investigation of the continuous spectra of symbiotic stars.

We will suppose that the continuous spectrum is formed by three sources of energy (Boyarchuk 1967) I) an M-type giant, II) a hot component with $T \sim 10^5$ °K, III) a nebula with $T_e \sim 17\,000$ °K and $n_e \gtrsim 10^6$ cm^{-3}. The contributions of the three hypothetical sources of energy to the combined spectrum of Z And are shown in Fig. 11 by the symbol o. The agreement between the observed and theoretical distributions is quite satisfactory. This comparison gives us a possibility to determine the part of radiation contributed by each sources at any wave length for different times.

These results together with the light curve of Z And give us a possibility in turn to determine the brightness variation of each component. Table 4 contains some results. One can see that the brightness variations of the cool component were negligible. On the contrary, the brightness variations of the hot component were very large. They cause the brightness variations of Z And on the whole. The temperature of the hot component which was determined by Zanstra's modified method (Boyarchuk 1967b) changed significantly. The temperature increased simultaneously with the increase of m_{ph}. Such behaviour of the hot component is in good agreement with the observed variations of the spectral and color characteristics of symbiotic stars. As it

27*

follows from the calculations, the variations of the visible magnitude of the hot component and of its temperature occur in such a way that the bolometric luminosity does not change significantly. There are at least two possibilities for the origin of such variations. First, the hot component is affected by a pulsation. Second, an optically thick envelope surrounding the hot component is formed as a result of mass outflow from the cool component. We need further investigations for the solution of this problem.

If we suppose that the cool components of symbiotic stars are normal giants, then their hot components are located below the main sequence as it can be seen from Fig. 15. The central stars of planetary nebulae, hot components of SS Cyg-type stars, novae a.o. are located near this place too. Perhaps the nonstability of all these objects has a common nature.

In conclusion we briefly consider the hypotheses for the nature of the symbiotic stars. All hypotheses can be divided into three groups.

I. The symbiotic stars consist of a hot star surrounded by an optical envelope. The cool absorption spectrum is formed in the outer part of the envelope and the hot emission spectrum is formed in the inner part of the envelope. This hypothesis was suggested by Sobolev (1945), Menzel (1946) and Aller (1953). But many observations do not agree with this hypothesis. For instance, the theory predicts that the relative intensity of emission lines

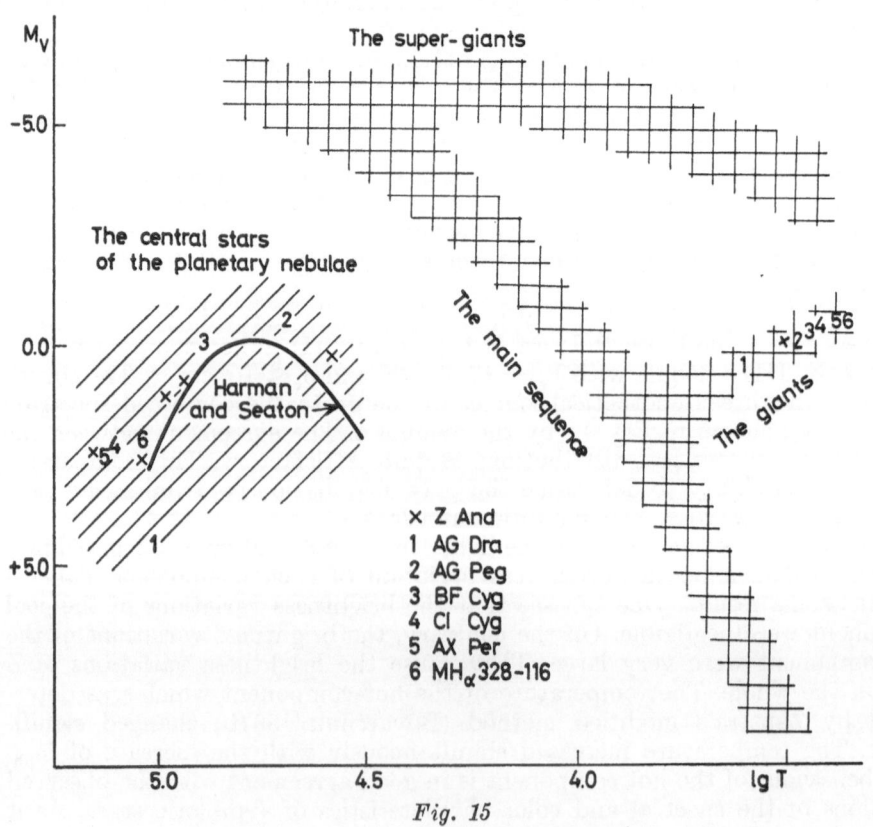

Fig. 15

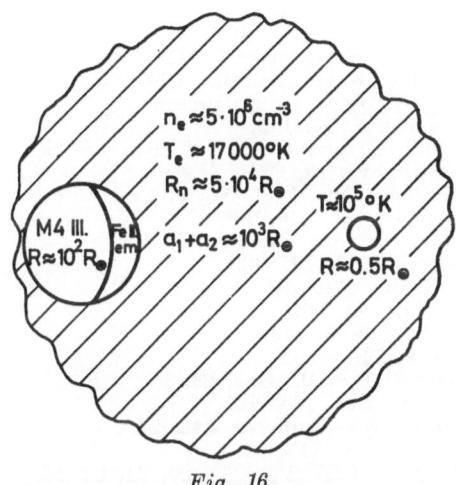

Fig. 16

must have such disturbances as that in the spectra of longperiod variables. But the observations have not detected them.

II. A symbiotic star is a cool star with very extended dense corona (Aller 1953, Gauzit 1955). Then the main difficulty is in the heating problem. It is impossible to get such a large flux of mechanical energy without heating the whole stellar photosphere.

III. The simplest hypothesis is that we are concerned with a binary, one of the components is a late-type giant and the other one is a hot small star being the source of the excitation of a nebula surrounding both components. The hypothesis was first suggested by Hogg (1943) and then by Swings and Struve (1941), Aller (1954), Sahade (1960) a.o.

The binary nature of symbiotic stars is confirmed by radial velocity variations (see Fig. 12) and the light curve of AG Peg. All basic results of observations can be explained from the point of view of binary hypothesis.

The quantitative investigations of symbiotic stars (Boyarchuk 1966, 1967) renders it possible to construct a rough hypothetical model of symbiotic stars (see Fig 16). Of course, we need much more observations and theoretical investigations to improve this model.

REFERENCES

Aller, L. H., 1954a, Astrophysics. New York.
Aller, L. H., 1954b, Publ. Dom. astrophys. Obs. **9.**, 321.
Belyakina, T. S., 1965, Izv. Krym. astrofiz. Obs. **33,** 226.
Belyakina, T. S., 1966, Astrophys., **2,** 115.
Belyakina, T. S., 1967, Izv. Krym astrofiz. Obs. **38,** 171.
Belyakina, T. S., 1968, Astr. Zu. **45.**
Bidelman, W. P., 1954, Astrophys. J. Suppl. **1,** No. 7.
Bidelman, W. P., Stephenson, Ch. B., 1956, Publ. astr. Soc. Pacific **68,** 152.
Bloch, M., 1952, Ann. Astrophys.
Boyarchuk, A. A., 1966, Astr. Zu. **43,** 976.
Boyarchuk, A. A., 1967a, Astr. Zu. **44,** 12.

Boyarchuk, A. A., 1967b, Astrophys. **3**, 203.
Boyarchuk, A. A., 1967c, Astr. Zu. **44**, 1016.
Boyarchuk, A. A., 1967d, Izv. Krym. astrofiz. Obs. **38**, 155.
Boyarchuk, A. A., 1968a, Izv. Krym. astrofiz. Obs. **39**, (in press).
Boyarchuk, A. A., 1968b, Astrophys. **4**, (in press).
Elvey, C. T., 1941, Astrophys. J. **94**, 140.
Erleksova, G. E., 1964, Bull. astrophys. Inst. of Acad. of Science of Tad. SSR, **37**, 43.
Gauzit, J., 1955, Ann. Astrophys. **18**, 354.
Greenstein, N. K., 1937, Harv. Obs. Bull. No. 906.
Henize, K. G., 1952, Astrophys. J. **115**, 133.
Herbig, G. H., 1950, Publ. astr. Soc. Pacific, **62**, 211.
Herbig, G. H., 1960, Astrophys. J. **131**, 632.
Himpel, K., 1940, Astr. Nachr. **270**, 184.
Himpel, K., 1942, Beob. Zirk. **24**, 53.
Hoffleit, D., 1968, Irish. Astr. J. **8**. 149.
Hogg, F. S., 1934, PAAS, **8**, 14.
Ivanova, N. L., 1960, Soobšč. Byurak. Obs. **28**, 17.
Jacchia, L., 1941, Harv. Obs. Bull. No. **912**.
Lindsay, E. M., 1932, Harv. Obs. Bull. No. **888**, 22.
Lundmark, K., 1921, Astr. Nachr. **113**, 94.
Mayall, M., 1963, Quart, Bull. A. A. V. S. O. 20, 21, 23, 25.
Menzei, D., 1946, Physica, **12**, 768.
Merrill, P. W., 1929, Astrophys. J. **69**, 330.
Merrill, P. W., 1942, Astrophys. J. **95**, 386.
Merrill, P. W., 1950a, Astrophys. J. **111**, 484.
Merrill, P. W., 1950b, Astrophys. J. **112**, 514.
Merrill, P. W., 1951a, Astrophys. J. **113**, 605.
Merrill, P. W., 1951b, Astrophys. J. **114**, 338.
Merrill, P. W., 1958, Etoiles à raies d'émission, Université de Liège, p. 436.
Merrill, P. W., 1959, Astrophys. J. **129**, 44.
D'Dell, C. R., 1966, Astrophys. J. **145**, 487.
Payne-Gaposchkin, C., 1946, Astrophys. J. **104**, 362.
Payne-Gaposchkin, C., 1950, Astrophys. J. **115**.
Payne-Gaposchkin, C., 1957, The Galactic Novae, Amsterdam North-Holland Pub. Co.
Prager, R., 1941, Pop. Astr. **45**, 445.
Romano, G., 1960, Publ. Osser. astr. Padova, **119**.
Romano, G., 1966, Mem. Soc. astr. Ital. **37**, N3.
Sahade, G., 1949, Astrophys. J. **109**, 541.
Sahade, G., 1960, Stellar Atmospheres, J. L. Greenstein ed., Chicago p. 494.
Sahade, G., 1965, 3rd Colloquium on Variable Stars, Bamberg, p. 140.
Sandig, H. U., 1950, Astr. Nachr. **278**, 187.
Sanford, R. F., 1944, Publ. astr. Soc. Pacific. **56**, 122.
Sieder, Th., 1956, Mitt. veränderl. Sterne 238.
Sobolev, V. V., 1947, Moving Envelopes of Stars, Leningrad.
Sobolev, V. V., 1940, Astrophys. J. **91**, 601.
Swings, P., Struve, O., 1941a, Astrophys. J. **93**, 356.
Swings, P., Struve, O., 1941b, Astrophys. J. **94**, 291.
Swings, P., Struve, O., 1942a, Astrophys. J. **95**, 152.
Swings, P., Struve, O., 1942b, Astrophys. J. **96**, 254.
Swings, P., Struve, O., 1943a, Astrophys. J. **97**, 194.
Swings, P., Struve, O., 1943b, Astrophys. J. **98**, 91.
Swope, H. H., 1940, Harv. Obs. Ann. **109**, No. 1.
Tcheng Mao Lin, Bloch, M., 1952, Ann. Astrophys. **15**, 104.
Tcheng Mao Lin, Bloch, M., 1954, Ann. Astrophys. **17**, 6.
Thackeray, A. D., 1954, Observatory, **74**, 257.
Tift, W. G. Greenstein, J. L., 1958, Astrophys. J. **127**, 160.
Wachmann, A. A., 1961. Astr. Abhand. Hamburger Sternw. **4**, No. 1.
Webster, L. B. 1966. Publ. astr. Soc. Pacific. **78**, 136.
Wenzel, W., 1956. Mitt. Veränderl. Sterne 227.
Wilson, R. E. 1950. Publ. astr. Soc. Pacific **62**, 14.

SPECTRAL EVOLUTION OF THE PECULIAR STAR MHα 328—116 (V 1016 CYG) FROM 1965 TO 1967

A. MAMMANO, L. ROSINO

Astrophysical Observatory of Asiago, Italy

ABSTRACT

Some informations are given on the spectral evolution of the peculiar star MHα 328—116 from 1965 to the end of 1967. It is shown that the degree of excitation is increasing. Some considerations on the nature of the object follow.

A general study of the spectrum of MHα 328—116 in 1965 has been published two years ago by the Authors (1966); a second detailed paper on the spectrum 1966—67 is in print and will appear very soon (1968). Only short informations on the spectral evolution of this interesting object will therefore be given here.

The star was known as a faint emission object ($m_{pg} = 15.5$) since 1950, when Merrill and Burwell (1950) included it in a list of 519 objects with a bright Hα. The spectrum was estimated near type M. The general attention on this star was called, however, only in 1965 after the announcement of McCuskey (1965) that the star had risen to magnitude 12 developing at the same time a very rich emission spectrum. Since then, the variable has slowly increased in luminosity of about one magnitude in two years.

The 1965 spectrum of MHα 328—116, as observed at Asiago on a series of 16 spectrograms with dispersions between 40 and 180 A/mm at Hγ, was characterized by the presence of relatively sharp emission lines of H (Balmer series from Hα to H_{18} and some infrared components of the Paschen series), FeII, NIII, CIII, CIV and forbidden lines of [NII], [OI], [OII], [OIII], [FeII], [FeIII], [SII], [SIII], [NeIII], [AIII], etc. The N_1—N_2 nebular lines and the 4363 line of [OIII] were particularly strong. The mean radial velocity was found to be —65 ± 3 km/s. The spectrum maintained more or less the same characteristics during the first months of 1966, but in November 1966 a decisive increase of the degree of ionization was apparent. The excitation further increased in 1967. Twenty five spectrograms were obtained at Asiago with the 122 cm telescope during this period, some of them, as that reproduced in Fig. 1, with the Carnegie intensifier applied to camera VI of the cassegrain spectrograph (60 A/mm at Hγ).

The spectral evolution in 1966—67 was chiefly characterized by a gradual increase in the state of ionization, demonstrated by the strengthening of the HeI and HeII lines, particularly HeII 4686, and by the emergence of new lines of high ionization as OIII, [NeIV], [NeV], [AIV], [AV], [FeV], [FeVI] which were not recorded or were faintly visible in the spectra of 1965*. The new high ionization lines observed in 1967 are reported in Table I.

---

* Boyarchuk (1968) claims that the excitation in MHα 328—116 was decreasing in 1966 (October) as compared with 1965. The Asiago spectrograms obtained in 1966 do not support his conclusions. It is true, however, that a definite increase of excitation was only observed in the second half of November.

Another striking point has been the fading of the FeII permitted lines, which disappeared or became very weak in 1967, while the forbidden lines of [FeII] maintained their intensities or even strengthened. The fading of the FeII permitted lines relative to the forbidden ones was also observed by Thackeray (1955) in RR Tel.

Table I

New high ionization lines observed in the spectrum 1967 of MHα 328—116.

3426	[NeV]	1F	3924	HeII, [FeV]	4.3F	5060	[FeIII]	1F
3444	OIII	15	4181	[FeV]	1F	5146	[FeVI]	2F
3715	OIII	14	4230	[FeV]	1F	5176	[FeVI]	2F
3735	[FeV]	3F	4330	NIII	10	5192	[AIII]	3F
3757	OIII	2	4511	NIII	3	5237	[FeVI]	1F
3760	OIII	2	4626	[AV]	2F	5271	[FeIII]	1F
3774	OIII	2	4634	NIII	2	5309	[CaV]	1F
3790	OIII	2	4711	[AIV]	1F	5677	[FeVI]	1F
3813	HeII	4	4724	[NeIV]	1F	6087	[CaV]	1F
3820	[FeV]	3F	4740	[AIV]	1F	6102	[KIV]	1F
3839	[FeV]	3F	4754	[FeIII]	3F	6435	[AV]	1F
3892	[FeV]	3F	4769	[FeIII]	3F	7006	[AV]	1F
3896	[FeV]	1F	4778	[FeIII]	3F			
3911	[FeV]	1F	4972	[FeVl]	2F			

Fig. 2 illustrates the most conspicuous changes observed in the spectrum 1967 of MHα 328—116 when compared with the spectra obtained in 1965. The simultaneous presence in the spectra of lines emitted by atoms in very different stages of ionization suggests stratification of the emitting ions in an extended envelope, excited by a hot central star. The increasing degree of excitation can be attributed partly to the increasing temperature of the exciting source and partly to the decreasing density of the envelope in slow expansion.

There still remain many uncertainties on the real physical nature of MHα 328—116. There are three possibilities: *a)* that the star is a symbiotic object, as suggested by the writers in 1966, and also by McCuskey and *ass.* (1966) and by Boyarchuk (1968). Although the light curve is not typical of a symbiotic, this hypothesis remains the most sounded. *b)* That MHα 328—116 is a slow nova, like RR Tel. However, the late pre-outburst spectral type, the absence of an absorption spectrum near maximum and the appearance of a nebular spectrum while the star is still increasing in brightness, are in disagreement with this hypothesis. *c)* Finally, a third possibility is that the object may represent an initial stage in the formation of a planetary nebula.

Further observations in the next years will probably give the possibility of making a choice between the various hypotheses.

REFERENCES

Boyarchuk, A. A., 1968, Astrophysica, 4, 289.
Fitzgerald, M. P., Houk, N. McCuskey, S. 1966, 1966, Astrophys. J. **144,** 1135.
McCuskey, S. W., 1965, Circ. IAU 1916, 1917.

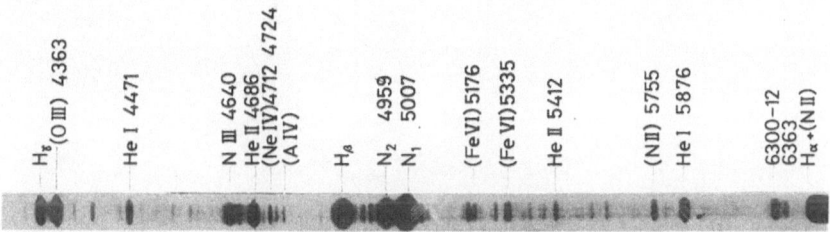

Fig. 1. The spectrum of MHα 328−116. Carnegie Intensifier, Oct. 23, 1967.

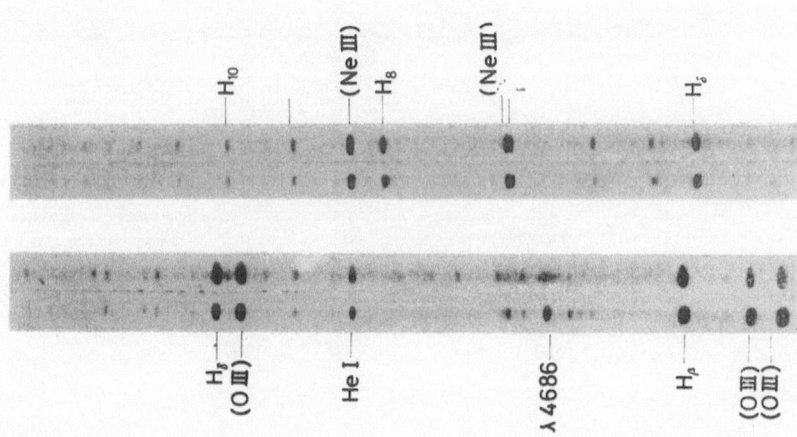

Fig. 2. Comparison of spectrum No. 6524 (August 8, 1965) with spectrum No. 7440 (Apr. 4, 1967) below. See the strengthening of HeII 4686.

Mammano, A., Rosino, L., 1966, Mem. Soc. astr. ital. XXXVII, 493.
Mammano, A., Rosino, L., 1968, Mem. Soc. astr. ital., XXXIX, 471.
Merrill, P. W. and Burwell, C. G. 1950, Astrophys. J. **112,** 72.
Thackeray, A. D. 1955, Mon. Not. R. astr. Soc. **115,** 236.

COMMENT

Fernie: You may be interested to know that we have made some photoelectric observations of this object at Toronto during the summer of 1968. During this time it remained quiescent at about $V = 11.06$, with fluctuations of no more than about ± 0.03 mag.

SPECTROSCOPIC VARIATIONS IN WY GEM, W CEP AND HD 4174

A. MAMMANO

Astrophysical Observatory, Asiago

A. MARTINI

Astrophysical Laboratory, Frascati

Some late-type stars with forbidden lines, suspected to be binaries or to have combination spectra, are under observation at Asiago. In this paper the preliminary results for WY Gem, W Cep and HD 4174 are presented.

WY Gem (= HD 42474) was classified by Bidelman (1954) as VV Cep type star. The available spectral classifications are: Harvard (HD): K5; Adams et al. (1935): M3ep; Swings and Struve (1940): M3ep + B3; Bidelman (1954): M2ep Iab + B.

The best description of the WY Gem spectrum is that of Swings and Struve (1940) who found an ultraviolet region "free from the late-type spectrum" and showing broad Balmer absorption lines, down to H8, with sharp centers. Moreover, they found the CaII (K) line "sharp and narrow" and a strong line at 3819 A (probably HeI).

At present, the late-type absorption spectrum is that of a normal M0 supergiant whose strong continuous spectrum masks the emissions in the visual region and falls off below 4000 A. The Balmer absorption lines (from $H\beta$ to H9 on our plates) are narrow and, with the exception of $H\beta$, relatively strong. The CaII (H and K) lines are strong and very broad. Finally, the presence of the HeI lines is seriously doubtful. The singlet lines are absent while the triplet lines, present at $\lambda\lambda$ 4471 and 4026, must be identified (at least as major contributors) with the blend FeI (2), TiI (146) and with TiI (12) respectively, whose multiplets are well represented in the spectrum. The HeI at 3819 A, which Swings and Struve found "strong", is absent. Therefore, the absorption spectrum shows well marked changes.

The emission spectrum of WY Gem consists of 12 forbidden lines, 11 of [FeII] and 1 of [SII], compared with 6 observed by previous authors. The identification of the lines and their estimated intensities are given in Table IV. The radial velocities have been measured on two spectrograms (dispersion 40 A/mm at $H\gamma$). The displacements of the lines of different elements are listed in Table I.

It is interesting to note that all the lines, both M spectrum absorption lines and forbidden emission lines, exhibit the same displacement. The mean radial velocity of M spectrum absorption lines seems to be slightly variable by considering the previous determinations: +18.2 km/s (Mt. Wilson) and +10 ± 2 km/s (Redman, 1938).

W Cep (= HD 214369) was classified by Bidelman (1954) as VV Cep type star. The available spectral classifications are: Harvard (HD): Kp; Mt. Wilson: Mep; Bidelman (1954): K0epIa.

Table I

Radial velocities in WY Gem

	Dec. 12, 1967 (km/s)	Mar. 12, 1968 (km/s)
Selected lines of M spectrum (absorption)	$+26 \pm 1$	$+33 \pm 2$
FeI ,,	$+29 \pm 1$	$+34 \pm 1$
V I ,,	$+28 \pm 1$	$+36 \pm 3$
CrI ,,	$+27 \pm 2$	$+35 \pm 3$
TiI ,,	$+27 \pm 1$	$+35 \pm 3$
CoI ,,	$+28 \pm 3$	$+31 \pm 3$
MnI ,,	$+29 \pm 4$	$+32 \pm 3$
Ionized metals ,,	$+29 \pm 3$	$+31 \pm 2$
Balmer lines (H_β. H_γ, H_δ) ,,	$+28 \pm 4$	$+32 \pm 4$
Forbidden lines (emission)	$+27 \pm 2$	$+30 \pm 3$
Mean	$+27 \pm 1$	$+34 \pm 2$

W Cep was observed by Merrill et al. (1932). They found H_α in emission, variable in intensity. The only description of W Cep spectrum is that of Swings and Struve (1940): the star was similar to WY Gem, though the [FeII] emission lines were weaker than in WY Gem. Moreover, they found a continuous spectrum extending far into the ultraviolet and "a strong, but sharp, line at CaII K". At present, the late-type absorption spectrum is that of a normal K3 supergiant with the continuous spectrum very weak below 3900 A. As for WY Gem, the examined plates give no evidence of HeI spectrum. In agreement with Swings and Struve's observations, the [FeII] emission lines are weaker than in WY Gem. The identification of the lines and their estimated intensities are given in Table IV. Of greatest interest is the Balmer series structure: we find $H\beta$ $H\gamma$, $H\delta$ with a weak sharp emission component displaced toward the red by about 30 Km/s. $H\alpha$ is present strong in emission and variable in intensity. Down from $H\delta$ the Balmer lines are in absorption.

The radial velocities have been measured on the best spectrogram (dispersion 40 A/mm at $H\gamma$). The displacements of the lines of different elements are listed in Table II.

Table II

Radial velocities in W Cep

	Dec. 12, 1967 (km/s)
Selected lines of K spectrum (absorption)	-39 ± 2
FeI ,,	-37 ± 2
CrI ,,	-40 ± 4
TiI ,,	-40 ± 3
V I ,,	-39 ± 3
Ionized metals ,,	-37 ± 3
Forbidden lines (emission)	-37 ± 2
Mean	-38 ± 2

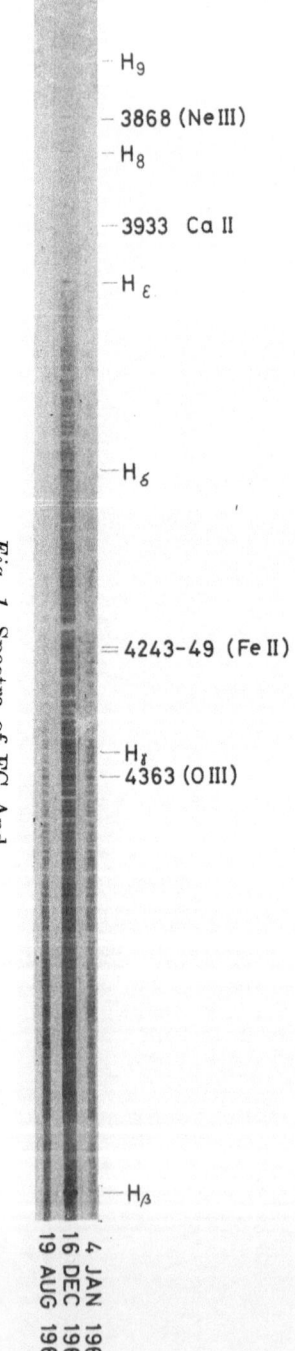

Fig. 1. Spectra of EG And.

The spectrum of HD 4174 (EG And) was described by Wilson (1950). He found a gM2 background absorption spectrum and an emission spectrum consisting of the Balmer series from Hβ to H18 and of the three nebular lines $\lambda\lambda$ 3868 and 3967 of [Ne III] and 4363 of [OIII]. Babcock (1950) found Hβ variable "in intensity and character". Photometric and magnetic variations have been found by Jarzebowsky (1965).

Our plates (taken during the years 1966 to 1968) show very marked variations mainly in the emission spectrum (Fig. 1).

In August 1966 Balmer emission lines (from H$_\alpha$ to H11), and the nebular lines of 4363 [OIII] and [Ne III] 3868 are present. Moreover, we find in emission also CaII (K) at 3933A; the line is weak and superimposed on a broad and diffuse absorption. All of the emission lines but [OIII] 4363 are narrow. In December 1966, [NeIII] 3868 becomes stronger. CaII K emission line is absent while HeI 4026 appears in emission. In Jan 1967, that is in less than one month, all the emission lines but OIII [4363] and [NeIII] 3868 disappear. In Dec 1967, H$_\alpha$ through H$_{10}$ reappear in emission together with [OIII] 4363, [NeIII] 3868, 3967, CaII (K) and two lines at $\lambda\lambda$ 4243, 4248, identified as [FeII] (21F) and [FeII] (36F) respectively. HeI 4026 is absent. In Aug 1968, the hidrogen and CaII emissions lines disappear while [OIII] 4363, [NeIII] 3868 and [FeII] 4243—4248 persist. The estimated intensities give:

	Aug. 10 1966	Dec. 3 1966	Jan. 4 1967	Dec. 16 1967	Aug. 19 1968
Hγ : [OIII] 4363	2	3	—	4	—
[NeIII] 3868 : H8	0.7	2	—	1	—
[NeIII] 3868 : [NeIII] 3967	4	5	—	5	—

The radial velocities have been measured on the best spectrogram (dispersion 40 A/mm at Hγ). The displacements of the lines of different elements are listed in Table III.

Table III

Radial velocities in HD 4174

	Dec. 3, 1966 (km/s)
Selected lines of M spectrum (absorption)	---104 $\pm$ 2
FeI ,,	—102 $\pm$ 1
V I ,,	104 $\pm$ 2
MnI ,,	—102 $\pm$ 4
CoI ,,	—100 $\pm$ 4
TiI ,,	—101 $\pm$ 3
CrI ,,	—106 $\pm$ 4
CaI ,,	—105 $\pm$ 4
Ionized metals ,,	—101 $\pm$ 4
Forbidden lines (emission)	—103 $\pm$ 4
Balmer lines (Hγ, Hδ, Hε, H$_8$) ,,	—101 $\pm$ 3
Mean	—103 $\pm$ 3

Summarizing, we have found that: 1) the intensity variations previously noted in HD 4174 are not confined to Hβ. At least on two occasions (Jan 4, 1967 and August 19, 1968) all of the emissions, except [OIII] 4363, [NeIII] 3868 and [FeII] 4243, 4249 disappeared. 2) No significant differences between radial velocities inferred from absorptions and emissions (both permitted and forbidden) have been revealed for the three stars (the approaching velocity of HD 4174 is remarkably high). 3) The shell lines noted by Swings and Struve in WY Gem are no longer present, and the radial velocity is variable. We may be confronted with a huge binary system, like VV Cep, Boss 1985 and Boss 5481, whose late-type components are surrounded by extended atmospheres.

Table IV

Forbidden lines in the spectra of WY Gem, W Cep, HD 4174

λ_0	Element	Estimated intensities		
		WY Gem	W Cep	HD 4174
4457.95	FeII (6F)	3s	—	—
4452.11	FeII (7F)	4n	—	—
4416.27	FeII (6F)	3s	—	—
4413.78	FeII (7F)	3d	5b	—
4402.60	FeII (36F)	3D	4D	—
4363.21	OIII (2F)	—	—	5D
4359.34	FeII (7F)	5d	—	—
4319.62	FeII (21F)	3s	3s	—
4287.91	FeII (7F)	8n	5n	—
4276.83	FeII (21F)	3d	4D	—
4249.07	FeII (36F)	—	2d	?1s
4243.98	FeII (21F)	6D	7n	?2s
4231.56	FeII (21F)	2D	3d	—
4178.95	FeII (23F)	—	3s	—
4068.62	SII (1F)	2s	—	—
3967.51	NeIII (1F)	—	—	?1s
3868.74	NeIII (1F)	—	—	5n

Notes: s = sharp;   d = diffuse;   D = very diffuse;   n = narrow;   b = broad

REFERENCES

Adams, W. S., Joy, A. H., Humason, M. L. and Brayton, A. M., 1935, Astrophys. J. **81**, 187.
Babcock, H. W., 1950, Publ. astr. Soc. Pacific. **62**, 277.
Bidelman, W. P., 1954, Astrophys. J. Suppl. **1**, 175.
Jarzebowsky, T., 1965, I. A. U. Bamberg Coll. on Variable Stars.
Merrill, P. W., Humason, M. L. and Burwell, C. G., 1932, Astrophys. J. **76**, 156.
Redman, R. O., 1938, Publ. Dom. Obs. Victoria **6**, 27.
Swings, P. and Struve, O., 1940, Astrophys. J. **91**, 546.
Wilson, R. E., 1950, **62**, 14. Publ. astr. Soc. Pacific 6.

Non-Periodic Phenomena in Variable Stars
IAU Colloquium, Budapest, 1968

THEORETICAL LIGHT CHANGES OF W UMa STARS WITH LOW MASS RATIO

H. MAUDER

Remeis Sternwarte, Bamberg, Germany

The main obstacle in calculating reliable elements for W UMa-stars is due to the fact that these stars are highly distorted. Therefore one should carefully take into account the influence of distortion on the light curves of these stars. Usually this is done by a rectification process which transforms the observed light curve to the equivalent light curve of a pair of spherical stars. There are two assumptions that make this process inadequate in the case of very close binaries. The first assumption is that both stars are ellipsoidals of equal shape. The second assumption is following from the first, namely that it should be possible to extrapolate the light changes of the rotating deformed stars which are observable outside the eclipses to the whole cycle and especially to the minima.

At the last Colloquium on variable stars at Bamberg in 1965 Binnendijk presented a paper on W UMa stars. It was impossible to derive elements for the first four systems he had shown. This is due not only to the shallowness of the eclipses but to another effect. One should take attention of the fact that there is a remarkable change of light outside the eclipses and a large period of constant brightness during secondary minimum, especially in the case of AW UMa. This star was observed by Paczińsky and later on by Kalish who confirmed this effect. Any rectification based on the light changes outside the eclipses will lead to a rectified light curve which cannot be attributed to any pair of spheroidal stars. In Fig. 1 one can see a fairly trustworthy primary minimum and a totally unfamiliar secondary minimum. Reaching the secondary minimum the star brightens, than decreases a little in brightness and

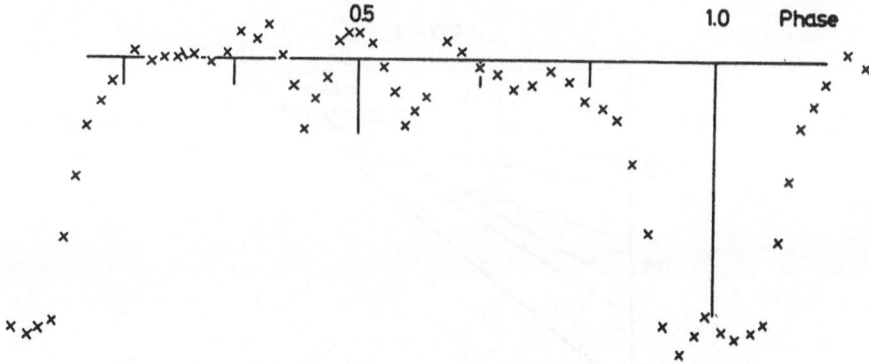

Fig. 1. Rectified light curve of AW UMa, ellipsoidal model

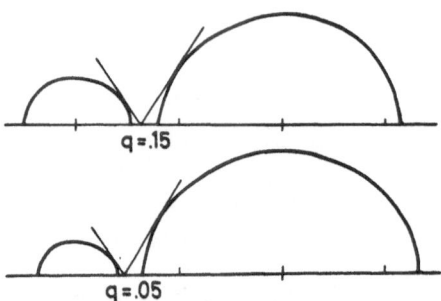

Fig. 2. Model for contact binaries with low mass ratio q

than shows a large hump around the middle of the minimum that makes the system actually brighter than its rectified maximum brightness. This effect is present though less pronounced in the case of the other three stars, too.

To understand the behaviour of these stars theoretical light changes were calculated for a pair of stars which are actually in contact, that means, they are filling their critical Roche lobe. A preliminary analysis had shown that the systems should have low mass ratio, less than 0.20. It was tried to find a model that represents satisfactorily the actual shape of these stars. It was checked numerically that the critical Roche lobe can be approximated by a three-axial ellipsoid combined with a cone whose top is at the inner Lagrangian point and which osculates the ellipsoid. (Fig. 2). The straight line in the figure is not the osculating cone but the tangent to the Roche lobe at the inner lagrangian point. The difference against the osculating cone is too small to be drawn. Numerically the difference between the model and the apparent roche lobe for this small values of the mass ratio is less than 0.2 per cent of the radius for the smaller component and less than 0.5 per cent of the radius for the larger one. Calculating the light changes of this con-figuration, one finds the following (see Fig. 3). There is a large variation in brightness outside the eclipse but this variation is much lower when the

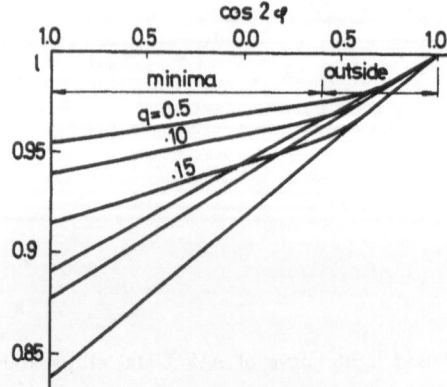

Fig. 3. Light changes for the roche-model with low mass ratio. The straight line is the extrapolation of the light changes outside the eclipses

eclipse begins. This is due to the fact that just at the onset of the eclipse the osculating cone is projected directly on the surface of the eclipsed star and the apparent area of this star — not influenced by eclipse — is now indeed of an ellipsoidal shape. However, in any rectification process one extrapolates the straight line of fig. 3 which leads to a variation in brightness that is much larger than really present. This causes the rectified minima to be much shallower and even narrower than they really are, leading to orbital elements you hardly can trust in. Rectifying according to the postulation that the brightness of the system not influenced by the eclipses remains constant leads to the light-curve of Fig. 4. One can see a remarkable difference in depth and shape of the light curves rectified by a classical method and by applying the Roche model. However, one should be very cautious when deriving elements from this light curve. This curve no longer is due to a spherical model. One can get elements from this curve but these elements still need corrections.

There may be a way out of these difficulties. In a previous paper of the author it was shown that one can get three essentially independent correlations from the Fourier transform of the total light curve. In the case of a contact configuration there are three main parameters, namely the ratio of luminos-ities, the inclination of the orbit and the mass ratio, which determines fully

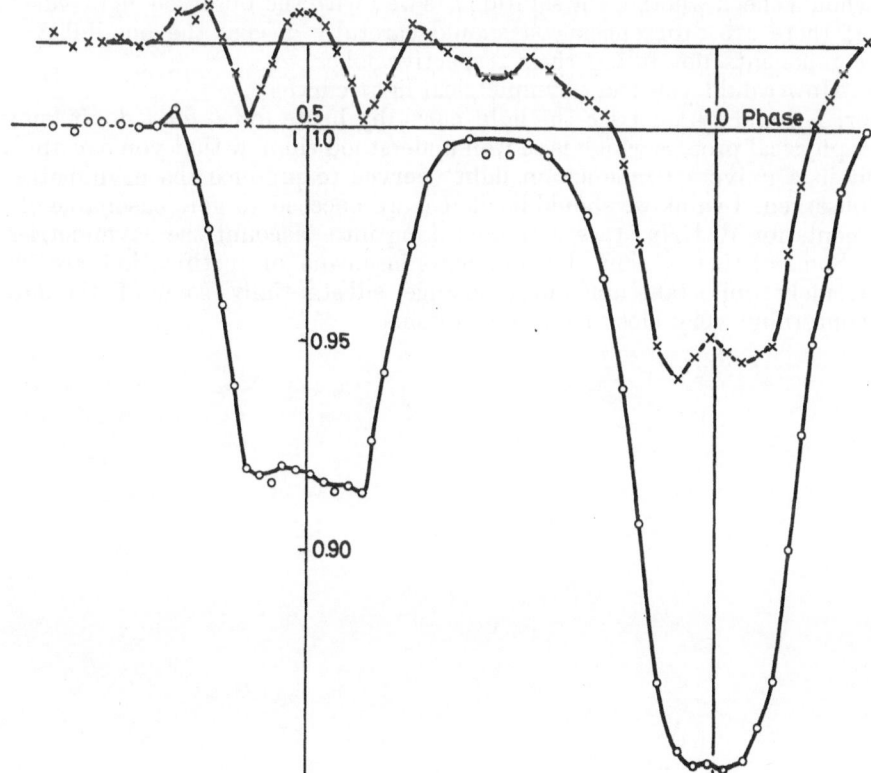

Fig. 4. Rectified light curve of AW UMa according to normal rectification (top) and by applying the Roche-model (bottom)

the geometrical properties of the system. Therefore it is to be expected that one can solve the problem by the Fourier method. However, this possibility is still under numerical examination.

REFERENCES

Binnendijk, L., 1965, Kleine Veröff. Bamberg Nr. 40, p. 36.
Paczynski, B., 1964, Astr. J. **69**, p. 124.
Kalish, M. S., 1965, Publ. astr. Soc. Pacific **77**, p. 36.
Mauder, H., 1966, Kleine Veröff. Bamberg Nr. 38.

DISCUSSION

Avery: What happens, when the stars do not quite fill the Roche lobes?

Mauder: This is a case much more complicated, for you have two additional parameters that describe the relative radii of both components beside the mass ratio. You cannot then solve the problem by the Fourier method but you are forced to use some other method to solve the light curve. However, if you use the assumption of a contact configuration you should check whether the solution fits well with the observed light curve. If there are differences you should carefully discuss the possibility of components not filling their respective lobes.

Bakos: How would you treat symmetrical light curves?

Mauder: I would symmetrize the light curve by brute force, for I don't know a physical process which is so well understood till now that you can apply it in a universal manner on light curves to produce the asymmetries observed. I think we should be glad if we succeed to get reasonable elements for W UMa stars without taking into account the asymmetries.

Wood: Some of these recent developments in means of rectification are extremely important and may change substantially some of the data concerning very close eclipsing systems.

ON THE REFLECTION EFFECT IN CLOSE BINARIES

I. B. PUSTYLNIK

Physical and Astronomical Institute of the Academy of Sciences, Tartu

Among various problems of the theory of close binary systems, investigations of the effects which are due to the gravitational and radiative interactions of the components are of special interest.

It is inevitable in close binary systems that a part of the radiation of either component penetrates into the atmosphere of its mate, suffers some changes and will be subsequently reemitted or scattered. This is a wellknown interaction phenomenon, called the reflection effect. At the same time, the shape of the reflecting surface is governed by tidal perturbations and stellar rotation.

At present we dispose only of somewhat fragmentary observational data, concerning the values of amplitudes and phase functions of the reflected light for several dozens of eclipsing variables. As to the physical theory of the reflection effect, it is based on a number of simplifying assumptions hardly accessible to observational checking. Such a state of affairs is due, on one hand, to the fact that the fractionally reflected light constitutes a too small part of the whole brightness of the binary system to be measured by the direct methods of photometry or spectrophotometry. On the other hand, in the outer layers of the reflecting star which are responsible for re-emission, deviations from LTE and an anisotropy of radiation field can be appreciable.

The present report deals with two different aspects of the problem of radiation transfer in a semi-detached binary system, where a B-type star of the main sequence combines with secondary subgiant component of F—K spectral class. First we discuss the negative O—C values of the amplitude of the reflection effect. Another question concerns some details of the mechanism of the reflection effect.

It is generally known that in close binaries of the afore-mentioned type theoretical estimates of the bolometric amplitude of the reflection effect appear as a rule to be larger than the values obtained from the analysis of light curves of eclipsing variables. Sir Arthur Eddington (1927) was the first to draw attention in his pioneer work to this hardly explicable feature of the reflection effect. Recently Sobieski (1965) has taken into account the nongreyness of stellar matter and calculated the monochromatic amplitudes for several well-studied close binaries. His results reaffirmed the presence of the negative O—C. We still have no comprehensive explanation of this peculiarity.

Recently we examined the following possibility. As long as the reflecting star, usually subgiant, fills its critical Roche lobe, the gravitational darkening on its surface must be important. Indeed, let the mass ratio value be equal to 0.3 (for instance RS Vul or TX UMa). Then the dimensions of critical Roche lobe are such, that on the top of a tidal bulge the value of gravitational acceleration is approximately half as much as in the point diametrically opposite to it. If the mechanism of the reflection effect lies in absorption with subsequent reemission, rather than scattering, then at the maximum of light the top of a tidal bulge will send out in the direction of the observer substantially less energy than it would do for the case of a spherical reflecting star. Therefore the value of the amplitude of the effect will be significantly lower for the distorted star compared to the spherical one.

A quantitative approach to the problem was outlined in our article in "Astrophysics" vol. III, 1. The problem has been reduced to the solution of the radiative transfer equation for re-emitting non-grey, plane-parallel atmosphere. An irradiation flux of given magnitude and spectral distribution is assumed as parallel beam. Next our solution of the radiative transfer equation is to be applied to the idealized binary system, where the point source represents the irradiating star and the reflecting star is identified with its critical Roche lobe. Then the reflecting surface is approximated locally by plane-parallel layer. Thus the entire irradiated area is divided into elemental zones and the problem of computing the brightness of reemitting surface, as viewed by a distant observer, is reduced to the summation of the brightnesses from each visible differential zone, allowing for gravitational darkening and fore-shortening effects. The numerical calculations required have not yet been performed, since a sufficiently powerful computer was not available for the time being.

It goes without saying that a mere confrontation of the predicted values of the bolometric amplitudes with rectification constants for several well-studied binaries would essentially contribute to the full understanding in this question. But it is worth keeping in mind the low accuracy of determination of the reflection effect amplitudes through an analysis of out-of-eclipse light variations. It would be even more interesting to study in detail the influence of gravitational darkening upon the phase law. We anticipate that in presence of the strong gravitational darkening the maximum of the reflected light would not fall any more on the phase π, as usually adopted in the rectification procedure.

In this connection the improbably small values of the ellipticities of the secondary components for the majority of Algol-type binaries may be recalled. Hosokawa (1957; 1958; 1959) has managed to establish that the systems with small ellipticities of the secondary components possess also negative O—C values of the reflection effect amplitudes. Let us examine the system TX UMa as an example. According to the rectification constants of the light curve the ellipticity of its secondary subgiant component equals to 0.03. At the same time the mass of the primary component is three times as much as that of the secondary and orbital elements indicate that the cooler component fills it critical Roche lobe. We state the presence of discrepancy of the observational data in case of TX UMa. This binary has also large negative O—C.

It should be noted also that in all theoretical works on the reflection effect the role of convection has hitherto been neglected. On the other hand, according to contemporary ideas of stellar evolution the subgiant components of binaries possess well-developed convective zones. The effective depths of formation of the latters are highly moderate (of order $\tau = 0.5$). If so, then about a half of the irradiated energy will be absorbed within the convective zone. Moreover, the convective zone apparently stretches out to the boundary of the star here and there, as long as observational evidence exists for mass transfer. The problem, whether all this absorbed energy will be re-emitted in outer layers or any appreciable amount of it will be swept away by the mass loss or even by shock waves, had not ever been studied.

We proceed now to the brief discussion of the mechanism of the reflection effect. It can be shown that Lyman continuum of the B-type star is responsible for high electron pressure ($\sim 10^2$ bars or even more) in the reflecting layers. Normal electron pressure for a single star of G—K spectral class would be 1 to 10 bars. Lyman continuum photons of B-type star ionize atoms of H of its cooler companion and, after recombinations, will be transformed into Balmer continuum and L_α photons. There is some chance to discover this effect through the observations in L_α of the nearest close binaries with large orbital velocities.

L_α photon lives only a split of second in a free state. It will be absorbed by a ground state H-atom. The latter will be ionized in its turn by photons of the Balmer continuum, as long as these constitute a predominant part of irradiative energy. An additional electron density depends on the value of the flux in Lyman continuum.

It would also have been difficult to understand, what induces attenuation of the irradiated flux in the reflecting atmosphere if the influence of L_c continuum had been neglected. Indeed, in a typical Algol-type close binary the value of Balmer continuum flux of the primary star falling on the surface of its mate exceeds the proper flux of the latter. Assuming both stars to be black body radiators of definite effective temperature, we are in position to estimate a mean number of B_c photons falling from outside on each cm^2 of irradiated surface per sec and to compare it to the numbers of scattefings or absorptions. The latter is proportional to $\int \alpha_\nu B_\nu \, d\nu$. Calculations indicate that at a normal electron pressure absorption by H^- ions and scatterings on H atoms are the most significant opacity sources. The former is slightly more effective. At the same time radiation of the primary component in Balmer continuum would penetrate quite deeply into its companion. Calculations give for the depth 10^9 cm, if the mean density equals to 10^{-8} g/cm^3. But on the other hand, we expect that subjected to irradiation, outer layers of the reflecting star would adjust themselves in some way to hamper "strange" radiation. Assuming that all free electrons, originated due to L_c continuum, recombine with H atoms to form H^- ions, a relatively low ionisation degree of H justifies this assumption, we will find that H^- absorption is much more effective than Rayleigh scattering. Furthermore, we obtain that the column of matter one or two scores of kilometers high at a normal density just will do to absorb completely irradiated energy.

We hope that our qualitative approach is valid as a satisfactory first approximation. Of course, the numerical results must be treated with a considerable portion of reservation. A more rigorous approach, allowing for the rates of proceeding of the various processes and permitting at the same time to work out the equation of equilibrium conditions, is highly desirable.

REFERENCES

Eddington, A. S., 1927, Mon. Not. R. astr. Soc. **87**, 43.
Hosokawa, Y., 1957, Sendaj astr. Raportoj, **52**, 208.
Hosokawa, Y., 1958, Sendaj astr. Raportoj, **56**, 226.
Hosokawa, Y., 1959, Sendaj astr. Raportoj, **70**, 207.
Sobieski, S., 1965, Astrophys. J. Suppl. ser. **12**, 263.

NEARLY CONTACT BINARY HD 17514

V. I. BURNASHOV and E. A. VITRICHENKO

Crimean Astrophysical Observatory, USSR

This system was discovered in 1967 at the Crimean Observatory. Light and radial velocity curves were obtained. The star is an eclipsing binary with elliptical components. The masses of the components are 48 and 13.5 $\mathfrak{M}_\odot$, the stars almost fill the Roche's surface. In spite of the extremely large mass of the primary (O8 V) and the extremely small distance between the components (cf. Fig. 1), no irregularity has been detected in the light curve (1777 photoelectric measurements made in 1967—1968 with a mean standard error of the order of $0^{m}005$) and in the spectra (29 spectrograms with a dispersion of 37 A/mm). Hα is in absorption, without any evidence of emission. It is possible that we are dealing with a very young system in which mass exchange has not yet begun.

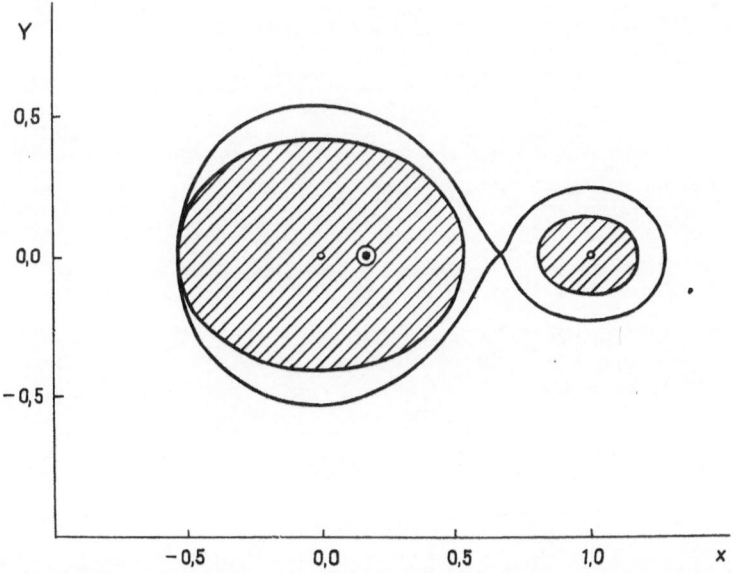

Fig. 1. Cross-section of the system HD 17514 in the equatorial plane. Roche's surface is shown.

NON-PERIODIC PHENOMENA IN BINARY SYSTEMS. CONVENTIONAL BINARIES

Introductory Report by

FRANK BRADSHAW WOOD

Department of Physics and Astronomy, University of Florida
Gainesville, Florida, USA

This paper is primarily for those astronomers who are not specialists in the field of close double stars. It will review some of the evidence for non-periodic phenomena in eclipsing systems, especially those changes connected with the orbital period, and it will attempt to set a background for the more specialized papers on specific topics which follow.

The detection and the interpretation of non-periodic changes in the radiation received from stars present many problems. This is especially true if the changes are small or if they occur only rarely. In the case of small changes in particular, we have the question of whether these have really occurred in the star or whether they have been introduced by the earth's atmosphere or the observational equipment.

The earlier photometric observers of variable stars found various irregular changes chiefly in the form of "humps" or intermittent increases in brightness over normal light. In time, these showed secular changes, decreasing in number and size in direct proportion to the increase in the precision of observing techniques. In other words, they were not changes in the stars themselves. The realization of this, combined with the notorious temperamental behavior of the early photoelectric photometers, led to a tendency to believe that all such observed changes were a result of atmospheric or instrumental effects. In efforts to obtain light curves of the highest possible precision, observers frequently applied arbitrary "corrections" called night errors or seasonal errors. This produced beautiful light curves, but probably prevented the discovery of many non-periodic phenomena of the type with which we are concerned in these discussions.

The study of period changes is made difficult because the observed effects can be detected only long after the changes have taken place. If a change of period of 0.1 seconds occurs, usually several thousand epochs must pass before the differences between observed and computed minima can be detected. For a star with a three day period, this can amount to well over a year. Thus we can never be certain whether we are observing one large period change or a series of smaller ones. Careful photoelectric monitoring of selected systems should help at least partially to answer this question.

In principle, the study of period changes is extremely simple; in practice it is frequently more difficult. We simply plot the (O-C)'s — differences between observed times of minima and those computed from linear light elements— against the epoch of observation. If the period is constant, these scatter about a straight line; if the period is constantly changing, a curved line is needed to fit the observations; if an abrupt change (or closely spaced series of small changes) has occurred, the (O-C)'s before the epoch of change will be fitted by one straight line and those following it by another of different slope.

The difficulties arise in the usual scarcity of observed minima at the critical times which frequently leaves a wide choice in interpretation possible. The situation would be even more serious were it not for the long series of minima observed during the past fifty years at the Cracow Observatory. Working with relatively small instruments the Polish astronomers have observed a large number of times of minima to which modern observers using instruments of high precision turn as a first step in the interpretation of their own observations. When properly used, observations like wine grow more valuable with age, and while it is recognized that visual and photographic estimates — as distinct from measures — are of little use in obtaining accurate solutions, the times of minima from such observations are frequently the only data we have to combine with modern methods for studies of period changes.

Instead of attempting the lengthy task of discussing all types of non-periodic phenomena in eclipsing stars, I shall concentrate on one system as an example of the sort of behavior we at times encounter, and for this purpose, select R Canis Majoris. In one sense, this is not a good choice, because R CMa certainly cannot be considered a "type" star — that is, as an example of some sub-class of eclipsing stars. Any attempt so to use it would have to rest on very superficial comparisons. On the other hand, it is a system which has been studied for more than seventy years and a great deal of observational data exists. The earliest reports of a non-periodic phenomenon came from observations made in 1898—9 when Pickering (1904) and Wendell (1909) reported an unusual increase in brightness immediately after the following shoulder of primary minimum. The plotted normal points of the two observers show a marked discrepancy but fortunately the individual observations were published and it is possible to show that this is a result of different combinations of the observations on the nights when the "hump" was present with those on the nights when it was not. When the individual observations are compared directly, any discrepancy is smaller than the observational scatter. Each observer used a polarizing photometer; while the accidental errors using such an instrument are considerably larger than those from photoelectric observations (probable errors of $\pm$ 0.04 magnitudes or larger for an individual observation were not uncommon) the technique was remarkably free from systematic error, and the magnitude of the hump—when it was present — indicates that it was a real although transient feature of the curve. Several astronomers have studied these observations independently and have reached this same conclusion.

This remarkable discovery of "flaring" (although not in the modern use of the term) attracted little comment and was not detected by other observers; however, it appears that no systematic attempt was made to monitor this part of the light curve. Then, Jordan (1916) observed radial velocity curves in the years about 1912. These showed two features. The first was a large orbital eccentricity when interpreted by conventional methods. The location and duration of secondary minimum on the light curve had indicated a circular orbit in distinct contradiction to this spectrographic data. The second unusual feature of the velocity curves was a large variation in the eccentricity from year to year. At the time there was no known explanation for this, although a number of cases have been found later and work by Struve (1944) and by

Hardie (1950) has shown the cause to be distortion of the spectral lines in circumstellar gas streams.

About this time, Dugan (1924) began a photometric study, possibly in an effort to solve this discrepancy between photometric and spectrographic data. Instead of doing so, his observations not only confirmed the discrepancy, but added another problem. The relative depths of his minima were shallower than those found by Wendell, and the smaller secondary in particular, where the addition of a small amount of extra light would be relatively more noticeable, could not be reconciled with Wendell's observations. Little attention was paid to this problem for twenty years, although Gadomski (1930) did suggest a shortening of the period which was confirmed by other observers. Then photoelectric observations made in 1938—9 (Wood, 1946) agreed with Wendell's rather than Dugan's observed depth of secondary. These also confirmed the changed period, which seemed to occur at about the time of Jordan's work. At that time, the only conclusion was that further work in the infrared might help in reaching some conclusion.

In light of current ideas concerning mass ejection and the frequent existence of circumstellar material in the neighborhood of close double stars, a reasonable explanation of these data is possible. If the intermittent flaring observed in the 1890's be considered an indication of developing instability which culminated in the ejection of material shortly before Jordan's observations, this ejection could be responsible for the period changes and the circumstellar material responsible for the distortion of the spectral lines and for the added light which caused the discrepancy between the earlier and later light curves and the one observed at this epoch. Many problems are solved by this hypothesis; however, I do not want to imply that this system presents no further problems; later work, in particular that by Koch (1960) and by Kitamura and Takahashi (1962), discusses some of these in more detail.

I turn now to a general non-periodic phenomenon of many eclipsing systems — that of sudden and unpredictable period changes, as distinct from those caused by the rotation of the line of apsides of an elliptical orbit or changing light time in a three-body system. The first general study of these was by Dugan and Wright (1939). From period studies of a number of systems, they showed that in nearly all there were period changes which could neither be predicted nor explained by a periodic change. This conclusion has been confirmed for many other systems by later observers. In particular, Plavec (1960) has made a thorough study and has concluded that no reasonable combination of periodic terms can explain the observed changes.

In interpreting these changes, it is natural to turn first to Newtonian mechanics, since relativistic effects such as gravity waves have been shown to be too small to be detected.* If Newtonian mechanics suffice, then by Kepler's third law change of period indicates either change of mass or change of separation of the components or both. Of these, a sudden change of mass roughly analogous to major prominence activity on the sun seems most reasonable and this possibility was postulated in 1950 (Wood, 1950). Earlier workers (e.g., Kuiper 1941) had considered mass loss but, if Kepler's third law alone is considered period changes in one direction only can take place whereas in

---

* However, see comments by L. Detre in the discussion following.

reality changes in either direction are observed to occur. If, however, we consider the effect of the high velocity ejection of matter upon the motion of the star itself, it can be shown that loss of the order of 10^{-7} of the mass of the star per ejection can cause the larger period changes observed, with much smaller values possible if we are really observing the cumulative effects of many small changes. It also was shown that systems showing erratic period changes had one component near the Roche stability limit, whereas those showing long intervals of constant period had both components well outside these limiting surfaces. This rather primitive early work has been carried further by a number of astronomers making detailed computations of particle trajectories; a recent summary of this work has been given by Piotrowski (1967). The figure of 10^{-7} of the mass of the star, which looked uncomfortably large at the time, looks far more reasonable in terms of recent work on mass loss from stars. As early as 1956, Deutsch from spectrographic data found ejection of at least 3×10^{-8} needed in his work on Alpha Herculis and Dadaev showed a series of small changes would not produce photometric effects large enough to be noticeable. Batten's (1964) discussion of pronounced disturbances in the light curve at the time of a sudden period change of U CrB and related spectrographic evidence is also of considerable importance.

We might end this part of the discussion by classifying our knowledge into three degrees of certainty. That irregular and — as far as predictability goes — erratic fluctuations of period occur in close double stars is a well-established fact of modern observational astronomy. Less certain, but still looking reasonably firm, is that as a general rule these changes are found to occur in systems in which one component is near the Roche lobe and will occur only rarely (e.g., AR Lac) when both systems are well separated from these lobes. The old suggestion (Wood, 1950) of a rough separation of close binaries into these two classes seems a reasonable working hypothesis and fits well with much later work on the evolution of close binaries, where the eventual growth of the more massive system to fill this limit seems an almost inescapable consequence of stellar evolution for all but the most widely separated systems. The later suggestion of a third classification of "contact" binaries is difficult to support on present observational evidence and it is not easy to rationalize the creation of such systems in any number from known evolutionary procedures. Finally, the details of the mechanism causing these changes are less certainly established but are a fruitful field for further investigation, and we may hope for an increase in our knowledge of these in the future. Mass loss from at least one component is almost certainly involved, but the precise physical mechanism or mechanisms involved and the ultimate disposition of the mass lost are still matters to be investigated.

In conclusion, the importance of precisely observed light curves in known spectral regions and of times of minima continue to be of considerable importance. Theoretical advances are hindered by lack of enough observational evidence concerning the non-periodic phenomena which we know to occur. Interesting information can come from studies of the association of close double star systems with clusters of various ages, and this should be a profitable field of investigation which should aid in evolutionary studies. The absence of eclipsing systems in globular clusters has long been noted, and if they prove to be absent or very rare in the older galactic clusters, this will be significant information. Even in the younger clusters, the relative number of close double stars compared to that of non-cluster systems will give help in studying their formation and evolution.

The idea of eclipsing systems being important to such studies is not a new one although it is now receiving much emphasis. I should like to close with the statement: "The study of Algol should bring us to the very threshold of the question of stellar evolution . . ." The writer was A. W. Roberts in volume 24 of the Proceedings of the Royal Society of Edinburgh. The year was 1902.

REFERENCES

Batten, A. H., 1964, Q. J. R. astr. Soc. **5,** 145.

Deutsch, A. J., 1956, Astrophys. J. **123,** 210.

Dugan, R. S., 1924, Princeton Contr. No. 6, 49, 1924.

Dugan, R. S., Wright, F. W., 1939, Princeton Contr. No. 19.

Gadomski, J., 1930, Astr. Nachr. **239,** 96.

Hardie, R. H., 1950, Astrophys. J. **112,** 542.

Jordan, F. C., 1916, Publ. Allegheny Obs. **3,** 49.

Kitamura, M., Takahashi, C., 1962, Publ. astr. Soc. Japan **14,** 44.

Koch, R. H., Astr. J. **65,** 326.

Kuiper, G. P., 1941, Astrophys. J. **93,** 133.

Pickering, E. C., 1904, Ann. Harv. Coll. Obs. **46,** 172.

Piotrowski, S. L., 1967, Commun. Obs. R. Belgique, Ser. B, No. 17, 133.

Plavec, M., 1960, Bull. astr. Inst. Csl. **11,** 197.

Struve, O., 1944, Astrophys. J. **99,** 89.

Wendell, O. C., 1909, Ann. Harv. Coll. Obs. **69,** 66.

Wood, F. B., 1946, Princeton Contr. No. 21, 31.

Wood, F. B., 1950, Astrophys. J. **112,** 196.

DISCUSSION

Bakos: It appears that not only globular but also galactic clusters appear to have fewer eclipsing binaries, about a factor of 3, than expected in a general stellar field.

Detre: I have only the same comment I already said in my introductory paper: if at least one component is of high temperature, a lot of ionized matter is flowing between the components and the high temperature star is losing ionized matter. Mass loss of this kind might be more effective in causing period changes than simple "gravitating" matter. I refer to Schatzman's papers on "magnetic breaking".

Wood: I am in complete agreement with Dr. Detre's remarks and had intended to say this in the body of the paper.

PHOTOMETRIC RESEARCH ON RS CVn AT THE CATANIA ASTROPHYSICAL OBSERVATORY

S. CATALANO and M. RODONÒ

Catania Astrophysical Observatory, Italy

SUMMARY

On the ground of extensive photoelectric observations of RS CVn made at Catania since 1963 several photometric peculiarities of this system are analysed.

A good deal of hypotheses have been proposed to explain the peculiarities of the system RS CVn (e.g. Sitterly 1930; Mergentaler 1950; Catalano and Rodonò 1967). However, as the accumulation of photometric and spectroscopic observations has been increasing, none of these hypotheses seems to be completely correct.

Since 1963 photoelectric observations were collected at Catania in order to reach a description as complete as possible for the properties of this system.

We should like to review here these properties as far as concerns:
1. light curve variation
2. colour index outside eclipses
3. variation of the primary minimum depth
4. orbital period variation and displacement of the secondary minimum
5. spectral peculiarities.

LIGHT CURVE VARIATION

The light curve of RS CVn shows fairly regular variations, in contrary to many systems having irregular fluctuations or humps at certain phases.

The luminosity of the system outside eclipses is perturbed by a wave-like distortion (Fig. 1). This distortion, maintaining its shape, moves in the sense of decreasing phases. In the Fourier expansion of rectified light curve, all coefficients are negligible compared to those of the $\cos \varphi$ and $\sin \varphi$ terms (Chisari and Lacona, 1965).

Therefore the observed light curve outside eclipses may be well represented by the following simple equation:

$$L[\varphi, \Theta(t)] = L_0 - \Delta L \cos [\varphi - \Theta(t)]$$

where φ is the phase angle, and Θ, variable in time, gives the phase angle of the minimum of the wave-like distortion relatively to the primary minimum.

The Θ mean values for each year were derived from the Catania observations by the method of least-squares using the above-mentioned equation. Three values were also derived from Keller and Limber's (1951) and Popper's (1961) observations, but they are somewhat uncertain because their light curves are not complete.

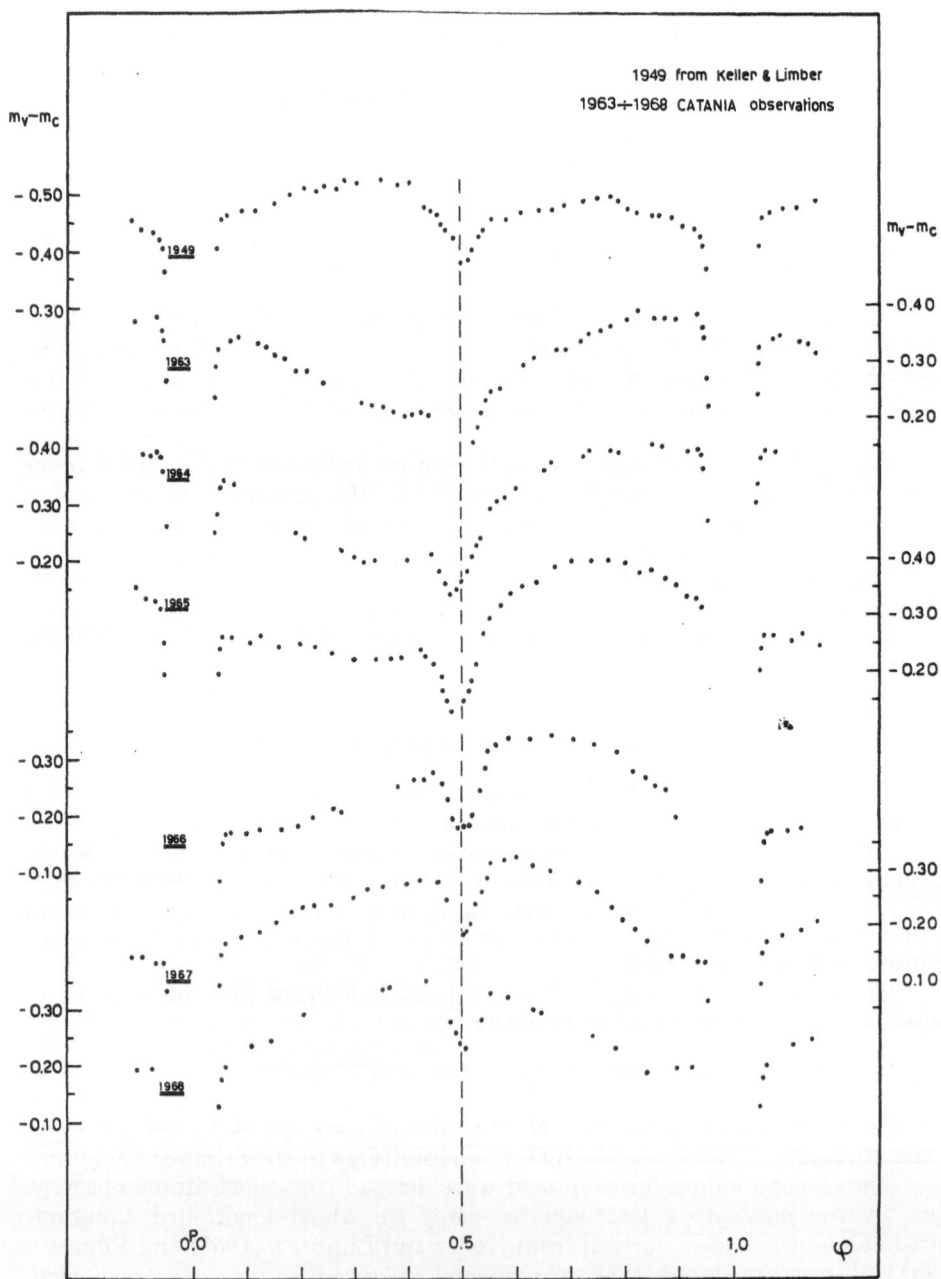

Fig. 1. Observed light curves.

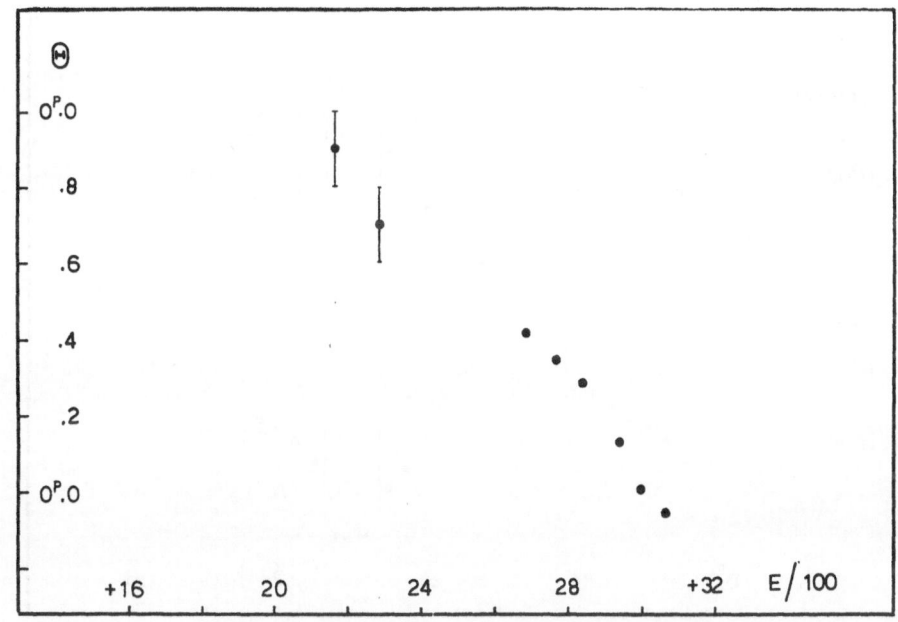

Fig. 2. Position angle of light curve distortion.

Fig. 2 shows the Θ values versus the number of orbital period from an assumed initial epoch (JD 2425249.028; Schneller, 1928). Because of the uncertainty of the first three values of Θ it is difficult to determine the exact period in which Θ reaches the same value. On the hypotheses that Keller and Limber's observations and ours up to 1966 belong to the same cycle, we made a rough estimate of 2400 orbital periods (Catalano and Rodonò 1967). Using new observations obtained during 1967 and 1968, it seems that Keller and Limber's observations belong to a preceding cycle. Therefore a more suitable value might be about 800 orbital periods. Limiting ourselves to the Catania observations this period would be even smaller.

COLOUR INDEX OUTSIDE ECLIPSES

The amplitude of the distortion is larger at longer wave-lengths. The data in the following Table refer to the observations of 1968:

U	B	V
$0.^{m}10$	$0.^{m}14$	$0.^{m}17$

Consequently the light outside eclipses is bluer at its minimum than at its maximum. This appears clearly in Figs. 3a and 3b, where for 1967 and 1968 the mean light curves and the colour index variations $\Delta(B-V)$, both outside eclipses, are reported.

This behaviour of the colour index, as observed in many systems showing distortions of their light curves, has been discussed by Mergentaler (1950). He supposed that gaseous streams of negative ions of hydrogen of different optical thickness could cause the observed distortions in the light curve. If the distortion of RS CVn is due to this reason, the absorbing matter should

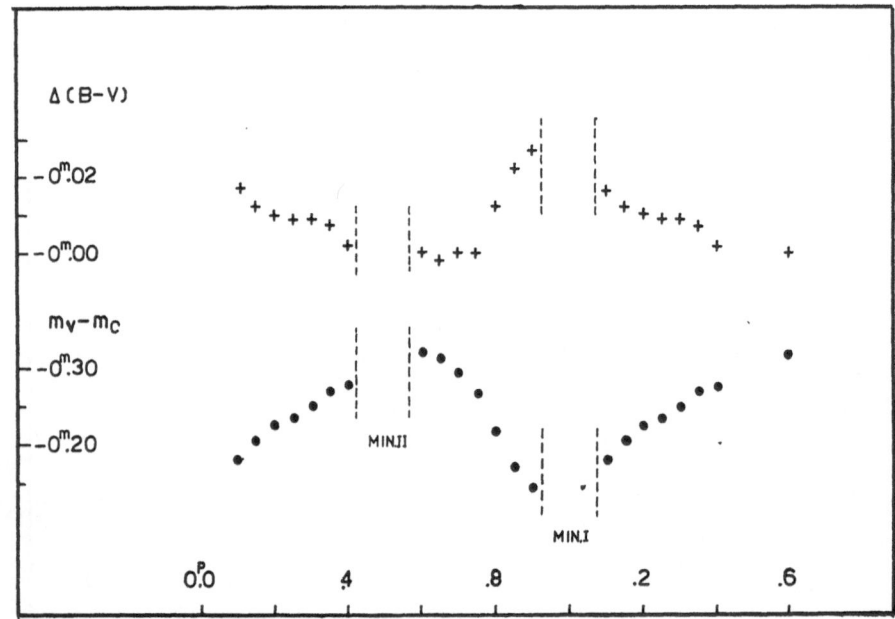

Fig. 3a. Correlation between C. I. variation (+) and the mean V light curve outside eclipses (●) (1967 observations).

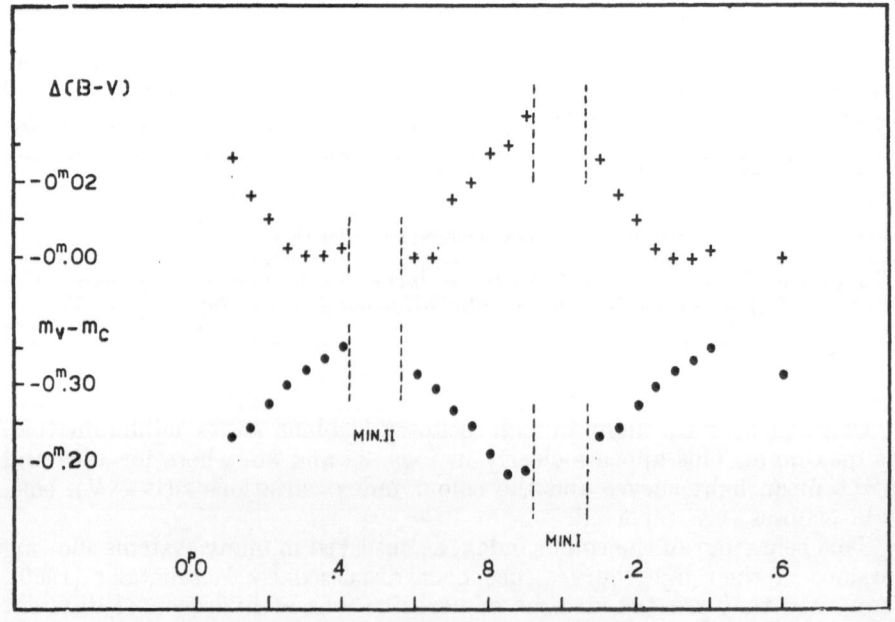

Fig. 3b. Correlation between C. I. variation (+) and the mean V light curve outside eclipses (●) (1968 observations).

have an equilibrium configuration, because the wave-like distortion, as we have seen, maintains its shape with time. We had suggested (Catalano and Rodonò 1967) that a ring around the equatorial plane of the primary component might cause the distortion of the light curve and its shift with time. This idea was supported primarily by the good agreement between the previously determined period of the distortion shift (2400 P) and the theoretical period of precession of the equator, which was assumed to have inclined to the orbital plane. But the new estimate of the period of the light variation, as previously reported, is too short to be compared with the theoretical one.

VARIATION OF DEPTH OF PRIMARY MINIMUM

Many observational evidences give the suspicion that the distortion of the light curve is due to the secondary component. Fluctuations of the depth of the primary minimum were already observed by Keller and Limber and by Popper. These fluctuations are confirmed by our observations. The primary eclipse is total, therefore these variations are due to the secondary component.

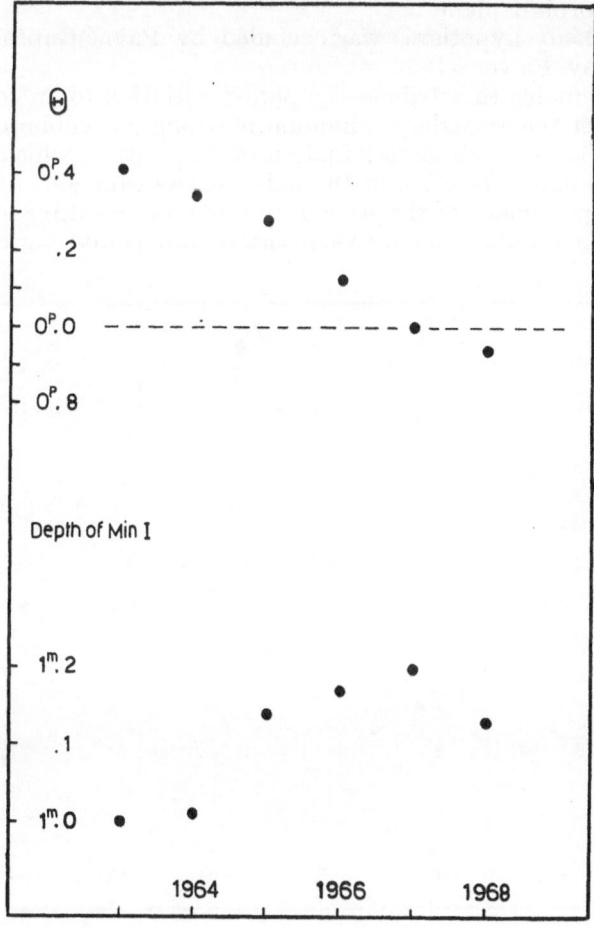

Fig. 4. Correlation between position angle of distortion and depth of primary minimum.

29*

In Fig. 4 (above) the phases Θ of minima of the light curve distortion and (below) the depths of the primary minima are plotted versus time. It is evident that the variation of the depth of primary minimum is clearly connected with the position of the distortion in the light curve. In particular, when the minimum of the distortion falls near the primary eclipse (i.e., the secondary component is fainter at this phase) the eclipse appears deeper.

The fluctuation in the luminosity of the secondary component could be connected in some way with the orbital motion.

ORBITAL PERIOD VARIATION AND DISPLACEMENT OF THE SECONDARY MINIMUM

From Fig. 5 we can deduce that the fluctuations of the orbital period around the mean value $4^{d}797865$ have a cycle of about 4000 orbital periods which is little shorter than Plavec' value (Plavec 1960).

Plavec (1960) pointed out that the observed period variation of RS CVn could not arise from rotation of one or both components around axes inclined to the orbital plane.

The third body hypothesis was excluded by Payne-Gaposchkin (1930) and decisively by Plavec (1960).

It is questionable to attribute the period variation to an eccentricity of the orbit. In fact the secondary minimum is strongly asymmetrical and this does not permit an accurate determination of its position, which seems to be affected by the relative position of the light curve distortion. At present the period of the displacement of the secondary minimum resulting from our observations is incompatibly smaller than the orbital period variation.

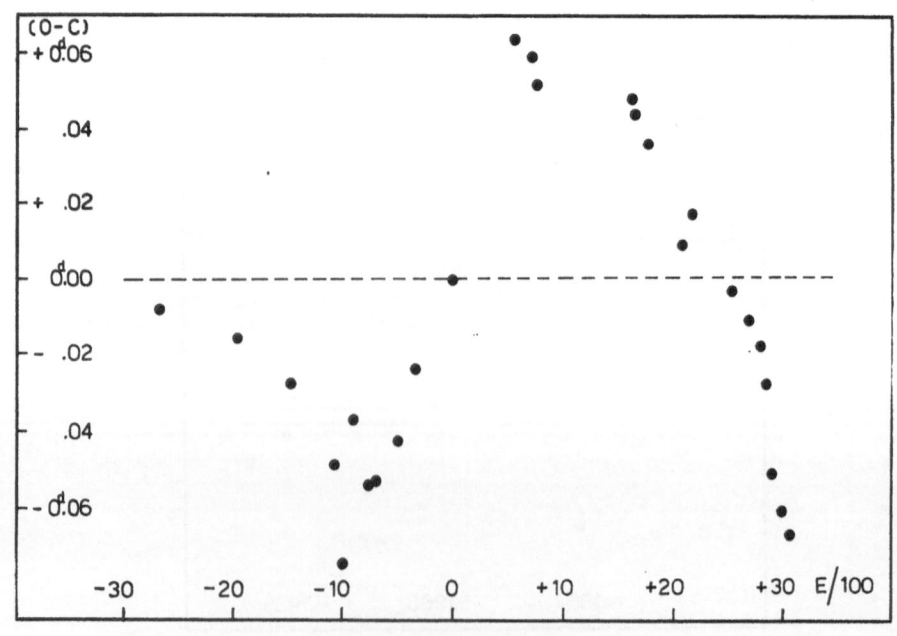

Fig. 5. Period variation.

The radial velocity curves do not show evidence of orbital eccentricity.

Ejection of matter from one or both components seems to be the only mechanism that may explain the orbital period variation.

SPECTRAL PECULIARITIES

Probably variable emission lines, as H_α, H and K lines of ionized calcium, have been observed in the spectrum of the secondary component (Hiltner 1947; Joy 1940; Popper 1961).

Azimov (1965) found that the electron density and the temperature of the secondary component are higher than those of single subgiants of the same spectral type.

Finally, our unpublished spectroscopic observations carried out at the Asiago Astrophysical Observatory confirm the presence of the mentioned emission lines, which do not disappear during the secondary eclipse, as previously reported by Hiltner (1947).

REFERENCES

Azimov, S. M., 1965, Abastumansk. astrofiz. Obs. Bjull., **33**, 81.
Catalano, S., Rodonò, M., 1967, Mem. Soc. astr. Ital., **38**, 395.
Chisari, D., Lacona, M., 1965, Mem. Soc. astr. ital., **36**, 463.
Hiltner, W. A., 1947, Astrophys. J., **106**, 481.
Joy, A. A., 1930, Astrophys. J., **72**, 41.
Keller, G., Limber, D. N., 1951, Astrophys. J., **113**, 637.
Mergentaler, J., 1950, Wrocław Contr. No. 4.
Payne-Gaposchkin, C., 1939, Harvard Repr. No. 170.
Plavec, M., 1960, Bull. astr. Inst. Csl., **11**, 152.
Popper, D. M., 1961, Astrophys. J., **133**, 148.
Schneller, H., 1928, Astr. Nachr., **233**, 361.
Sitterly, B. W., 1930, Princ. Contr., **11**, 21.

Non-Periodic Phenomena in Variable Stars
IAU Colloquium, Budapest, 1968

CHANGES IN THE LIGHT CURVE OF BETA LYRAE 1958—1959

G. LARSSON-LEANDER

Lund Observatory, Lund, Sweden

SUMMARY

Photometric results obtained during the 35 days of the international programme on β Lyrae are compared with observations made in 1958, mainly at the Lick Observatory. From the minimum epochs of the two seasons a period of 12.9355 days is obtained. The total B magnitude is found about 0.10 mag. fainter and the colour about 0.05 mag. redder in 1959 than in 1958. A slow decrease in brightness during the 1959 campaign is indicated. The depths of the minima are about the same as in 1958, but the bottoms of the primary minima show great dissimilarities among themselves. The first minimum, on August 12, 1959, was almost flat, and the nearly constant phase had a duration of about 0.70 days. The other two primary minima, on August 25 and September 7, 1959, had bottom widths of similar duration, but exhibited pronounced dips shortly after mid-eclipse. The rising branches from the primary minima are steeper than in 1958, making the minima almost symmetric. The ensuing maxima and the secondary minima occurred at phases about 0.24 and 0.49, respectively, or somewhat earlier than in 1958. The $B-V$ curve of 1959, showing the reddening at primary eclipse, is suspected to have a narrow "blue top" at mid-eclipse. Another feature of the colour variation is indication of a small "red dip" at about phase 0.37.

1. INTRODUCTION

Following a proposal by the late Dr. O. Struve, Commission 42 of the I.A.U. decided at the 1958 Moscow meeting to make β Lyrae the subject of an international programme of coordinated photometric and spectrographic observations. Mainly two reasons were given for the proposal. Firstly, no recent photoelectric light curve was available. The second and more important reason was the existence of erratic changes, both in the total light and in the intensity of certain spectral features. Some kind of correlation between these secondary variations might be expected. It thus seemed desirable to make a concentrated effort to observe the star more or less continuously during a certain interval of time, to make possible a direct intercomparison of successive cycles. The present writer accepted the task of coordinating the programme.

From subsequent correspondence with Dr. Struve it was learnt that extensive observations of β Lyrae had been made at the Lick Observatory during the interval 1958, June 21 to July 12. Thus, D. B. Wood and M. F. Walker had obtained a large number of photoelectric measurements on the U, B, V system on all nights of this interval, and spectrograms had been obtained by S. N. Svolopoulos on almost all the nights. In addition, Struve had taken a large number of coudé spectrograms at Mount Wilson during nights of the interval. All these observations were part of an international programme that Struve and Walker had organized. Five observatories outside the United States participated in the photometric part of the programme, but only a rather limited number of observations were made by them. All the

photometric observations from this programme have been published by Wood
and Walker (1960) together with a discussion of the light curve and its changes.
The secondary light variations were compared by Struve (1959) with changes
in the intensities of the absorption lines originating in the outer shell.

As β Lyrae had thus already been subjected to an international programme, the question arose if it would be worth-while to proceed with the programme
decided upon by the I.A.U. The interval of the 1958 programme included
two primary minima and one secondary minimum. According to the preliminary results, communicated by Dr. Struve, the magnitude at the four
maxima was the same, while the two primary minima showed some striking
differences. Most obvious was a difference in width: the first minimum,
centred on 1958, June 25, was found considerably wider than the following
one, on July 8. Struve further stated that the differences were clearly correlated with differences in the intensities of the spectral lines produced by the
shell. In view of these results Struve strongly endorsed the plan of organizing
a more extensive photoelectric study in 1959.

2. THE 1959 CAMPAIGN

In consultation with the president of Commission 42 the international
campaign was planned to cover three successive primary minima, viz., those
of 1959, August 12.9, 25.9, and September 7.8. The limiting dates were set
at August 8 and September 11, 1959.

One of the objects of the 1959 campaign was to produce light curves
for β Lyrae that could be directly compared with those obtained in 1958.
For this reason it would be most advantageous to adopt the comparison
stars used by Wood and Walker. Details on these stars, HR 6997, 8 Lyrae,
and 9 Lyrae, were kindly communicated by Mr Wood. According to this
information the colour of HR 6997 was very close to that of β Lyrae, while
8 Lyrae was somewhat bluer and 9 Lyrae somewhat redder. HR 6997 had
therefore been used as primary comparison star in the 1958 programme, but
numerous checks with 8 Lyrae and 9 Lyrae had been made. The preliminary
results indicated that 9 Lyrae might be variable by about 0.02 mag., while
HR 6997 and 9 Lyrae appeared to be reasonably constant. On the basis of
this, it was decided to exclude 8 Lyrae from use in the 1959 campaign and the
observers were recommended to adopt HR 6997 as primary comparison star
and 9 Lyrae as check star. Further, observers using such equipment that
transformation to the U, B, V system might be feasible were asked to observe
a sufficient number of standard stars to make this transformation possible.

Not all observers who had intended to participate in the photometric
part of the campaign were successful in obtaining observations. As a com-
pensation, however, observations were made also by colleagues who had not
in advance announced their participation. The net photometric outcome of
the campaign is sixteen series of observations, made at fourteen observatories.
The total number of observations, from the interval of the campaign, amounts
to 2361. Some observers extended their series outside the campaign limits.

Details on the sixteen series of observations are given in Table 1, where
the participating observatories are listed in longitudinal order. The table is
self-explanatory, except for the penultimate column, giving information on
the colour systems. In cases where the instrument-filter combinations are

Table 1.

Observers and instruments

Series	Observatory	Observers	Instrument	Multiplier	Colour system	Ref.
1	Nanking	Chang, Hong, Mo, Chow	24″ reflector (silvered mirrors)	RCA 1P21	b (no filter)	1
2	Byurakan	Grigoryan	16″ reflector	EMI 6094	u, b, y, r	2
3	Abastumani	Magalashvili, Kumsishvili	13″ reflector		v, y	3
4	Cracow	Szafraniec	8″ refractor	RCA 931-A	b, y	4
5	Budapest	Balázs-Detre	24″ reflector	RCA 1P21	u, B, V, r	4
6	Stockholm	Larrson-Leander	24″ refractor	EMI 5060	B, V	4
7	Capodimonte	Fresa	7″ refractor	RCA 1P21	b (no filter)	5
8	Copenhagen (Brorfelde)	Gyldenkerne, Jaeger	10″ reflector	EMI 5060	B, V	6
9	Hoher List	Herczeg	14″ reflector	RCA 1P21	$u, b, r \rightarrow U, B, V$	7
10	Leiden, I	Kwee	18″ reflector	EMI 6094	U, B, V, r	4
11	Leiden, II	van Agt	10″ refractor	RCA 1P21	v, B, y	4
12	Pic-du-Midi	Bouigue, Pedoussant, Rochette	24″ reflector	Lallemand	U, B, V	4
13	Sidmouth	Archer	7″ astrograph	RCA 931-A	$b, y \rightarrow B - V$	4
14	Flower and Cook, I	Binnendijk	28″ reflector	RCA 1P21	$u, v, b, y \rightarrow U, B, V$	8
15	Flower and Cook, II	Bookmyer	15″ siderostat		$y \rightarrow V$	8
16	Lick	Gordon	22″ reflector	RCA 1P21	B, V	9

1. Chang, Hong, Mo, and Chow (1959); 2. Grigoryan (1961); 3. Magalashvili and Kumsishvili (1960); 4. Larsson-Leander; 5. Fresa (1960); 6. Gyldenkerne and Jaeger (1963); 7. Herczeg (1964); 8. Binnendijk (1960); 9. Gordon (1960).

different from those of the standard U, B, V system, the colour regions have been roughly indicated by the letters u, v, b, y, and r, which stand for ultra-violet, violet, blue, yellow, and red, respectively. Arrows following these symbols and pointing to U, B, V indicate that the observers have also furnished data transformed to the standard system. Particulars concerning filters and transformation formulae are given in the papers referred to in the final column of Table 1. These same papers also contain the original observations. Several of the observers have desisted from publishing their observations, but kindly submitted them to the present writer for inclusion in this final report of the campaign (Larsson-Leander, reference 4 of Table 1).

Photometric observations were made during all 35 days of the campaign. However, because of the unequal longitudinal distribution of the observers, gaps of about 0.5 days are frequent. Table 2 gives for each colour the number of observations furnished by the various observers, with the number of observing nights within parentheses.

The internal mean errors of the various series of observations were estimated by two methods:

Table 2.

Number of observations and observing nights (between parentheses)

Observatory	u	v	b	y	r
Nanking			442 (25)		
Byurakan	53 (21)		56 (21)	55 (21)	35 (12)
Abastumani		59 (6)		59 (6)	
Cracow			28 (6)	28 (6)	
Budapest	44 (22)		43 (22)	42 (22)	45 (22)
Stockholm			93 (16)	93 (16)	
Capodimonte			193 (28)		
Copenhagen			80 (15)	92 (15)	
Hoher List	53 (11)		53 (11)		53 (11)
Leiden, I	14 (6)		14 (6)	13 (6)	13 (6)
Leiden, II		8 (6)	12 (6)	12 (6)	
Pic-du-Midi	12 (7)		12 (7)	12 (7)	
Sidmouth			7 (7)	7 (7)	
Flower and Cook, I	27 (5)	74 (6)	80 (6)	81 (6)	
Flower and Cook, II				34 (3)	
Lick			115 (13)	115 (13)	
All	203	141	1228	643	146

(1) From the scatter shown in magnitude differences obtained for comparison and check stars.

(2) From the scatter in the observations of β Lyrae at epochs when the variable should be almost stationary in light, i.e. at maxima and minima, or from scatter during very short intervals of time.

The pre-requisite for the use of method (1) is, of course, that the comparison and check stars remained constant during the interval of the campaign. As further discussed in Section 3, this was found to be true. In several cases mean errors derived in this way were communicated by the observers themselves.

As regards method (2), many series of closely spaced observations exhibit a rather large scatter, which one would be tempted to interpret as rapid fluctuations. However, intercomparisons of various overlapping series indicate quite clearly that the main part of the "fluctuations" merely reflects observational errors or effects introduced by variable atmospheric extinction. This view is supported by the fact that for series containing a sufficient number of observations, mean errors calculated by means of method (2) agree with those from method (1).

The average values of the internal mean errors are given in Table 3. Two series, Pic-du-Midi and Sidmouth, are missing from this tabulation, because of insufficient data.

We end this section by noting that high-dispersion spectrograms of β Lyrae were obtained by Abt at McDonald Observatory on thirteen nights of the campaign. In addition, K. O. Wright and A. McKellar at Dominion Astrophysical Observatory obtained spectrograms on three nights, one just after the end of the campaign. The entire spectrographic material has been discussed by Abt (1962).

Table 3.

Internal mean error of a single observation

Series	u	v	b	y	r
Nanking			.018		
Byurakan	.02		.02	.02	.02
Abastumani		.030		.034	
Cracow			.05	.04	
Budapest	.020		.018	.020	.015
Stockholm			.008	.010	
Capodimonte			.02		
Copenhagen			.010	.010	
Hoher List	.012		.010		.008
Leiden, I	.017		.010	.008	.010
Leiden, II	.026		.012	.010	
Flower and Cook, I	.013	.015	.018	.015	
Flower and Cook, II				.019	
Lick			.01	.01	

3. THE COMPARISON STARS

As recommended by the coordinator most observers used HR 6997 and 9 Lyrae for comparison purposes, but other stars were adopted by some observers. Table 4 lists the stars actually used in the various series, the primary comparison star being the first one. The observers checked the constancy of their respective primary comparison star by repeated measurements of their second (or third) star in Table 4. From studies of plots of the magnitude differences versus time, it is concluded that all comparison stars remained constant, within observational accuracy. During the 1958 programme Wood

Table 4.

Comparison stars

Series	Stars
Nanking	HR 6997, 9 Lyr
Byurakan	HR 6997, 9 Lyr
Abastumani	9 Lyr, HR 6997
Cracow	HR 6997
Budapest	HR 6997, 9 Lyr, γ Lyr
Stockholm	HR 6997, 9 Lyr
Capodimonte	HR 6997
Copenhagen	HR 6997, 9 Lyr
Hoher List	HR 6997, 9 Lyr
Leiden, I	HR 6997, 9 Lyr
Leiden, II	HR 6997, 9 Lyr
Pic-du-Midi	γ Lyr, φ Lyr A, φ Lyr B
Sidmouth	HR 6997
Flower and Cook, I	γ Lyr, 9 Lyr
Flower and Cook, II	γ Lyr, 9 Lyr
Lick	HR 6997, 9 Lyr

and Walker (1960) obtained occasionally discordant results for the magnitude
difference between HR 6997 and 9 Lyrae, which were ascribed to variations
of HR 6997. No such discordances, beyond the observational errors, were
noted during the 1959 campaign.

The complicated procedure of obtaining the most probable B and V
values for the comparison stars from the observations made by the various
observers is omitted here. Full details are given in the writer's comprehensive
report, which is being printed in *Arkiv för Astronomi*.

The finally adopted standard magnitudes and colours for the three main
comparison stars, HR 6997, 9 Lyr, and γ Lyr, are given in Table 5. Corresponding values from 1958, as obtained by Wood and Walker (1960) are also
listed. It is seen that the 1959 V magnitudes are slightly fainter than the
1958 values. The $B—V$ colour of HR 6997 turned out somewhat redder in
1959 than in 1958, while the colour found for 9 Lyr is the same during both
seasons.

Table 5.

Standard magnitudes and colours for comparison stars

Star	Sp.	1959		1958	
		V	$B—V$	V	$B—V$
HR 6997	B8	5.452	—0.126	5.430	—.0154
9 Lyr	A2	5.279	+0.058	5.254	+0.059
γ Lyr	B9 III	3.250	—0.045		

4. THE DERIVATION OF LIGHT CURVES

It is known that the $B—V$ colour of β Lyrae is only slightly affected
by the light variations. Wood and Walker (1960) found an increase of $B—V$
by about 0.07 mag. at primary eclipse, but no change at secondary eclipse.
It was therefore expected that the reduction of the 1959 observations to a
common system would present no particular difficulties, even though the
various series had been made in different colour regions. However, the problem
turned out to be rather more complicated than anticipated.

Attempts to derive transformation formulae by means of the measurements of the comparison stars proved unsuccessful, probably because of the
small colour differences involved. Instead of using this straightforward method,
it was then necessary to adopt one or several series as standards, and to
reduce the other series by means of empirical corrections. Obviously, the
standard series had to be chosen among those stated by the observers to be
on the B, V system.

With the 1959 magnitudes of the comparison stars, as given in Table
5, it was found that the partly overlapping Stockholm, Pic-du-Midi, and Flower
and Cook I series of both B and V magnitudes agree very satisfactorily. These
series, without any corrections, were taken to define frame-works of standard
magnitudes. They were plotted on large-scale graphs, and fragmentary light
curves were obtained. These contained points at a variety of phases, including
maxima, as well as primary and secondary minima. One by one the other

series were reduced to these frameworks, and more points were successively added to the light curves. For each series the necessary corrections were determined at all epochs where overlaps occurred. The runs of the corrections were studied versus magnitude and colour in order to disclose possible non-constant terms. The order in which the various series were taken and the corrections obtained are shown in Tables 6 and 7.

Table 6.

Reduction of V magnitudes

Series	V $(\beta$ Lyr$)$
Stockholm	$5.452 + \Delta V$
Pic-du-Midi	V
Flower and Cook, I	$3.250 + \Delta V$
Copenhagen	$5.452 + \Delta V -0.047$
Leiden, I	$5.452 + \Delta V -0.071$
Budapest	$5.452 + \Delta y -0.052$
Leiden, II	$5.452 + \Delta y -0.025$
Hoher List	$5.452 + \Delta V -0.053$
Byurakan	$5.28 \ \ + \Delta y -0.07$
Lick	$5.28 \ \ + \Delta y -0.11$
Abastumani	$5.452 + \Delta y -0.047$
Flower and Cook, II	$3.250 + \Delta y$

Table 7.

Reduction of B magnitudes

Series	B $(\beta$ Lyr$)$
Stockholm	$5.326 + \Delta B$
Pic-du-Midi	B
Flower and Cook, I	$3.205 + \Delta B$
Copenhagen	$5.326 + \Delta B -0.010$
Leiden, I	$5.326 + \Delta B -0.041$
Leiden, II	$5.326 + \Delta b +0.004$
Hoher List	$5.326 + \Delta b$
Budapest	$5.326 + \Delta b -0.33(\Delta v - \Delta b) + \\ + 0.010$
Lick	$5.34 \ \ + \Delta b -0.07$
Byurakan	$5.34 \ \ + \Delta b -0.05$
Abastumani	$5.326 + \Delta v +0.030$
Nanking	$5.337 + \Delta b -0.065 \ \Delta b -0.037$

Three series, Cracow, Capodimonte, and Sidmouth, which are missing from these tabulations, had to be omitted. The Cracow and Sidmouth series contain rather few observations and show, when compared with other series, erratic deviations of a considerable amount. For the more numerous Capodimonte observations, obtained with a refractor and no filter, it was not possible to find any single correction formula that could be used consistently for the whole series. Further, because of large deviations, some five observations had to be rejected in each of the Byurakan and Abastumani series.

As shown in Table 6, all the accepted series of yellow magnitudes could be reduced to the framework of standard V magnitudes by means of constant corrections. It is noteworthy that none of these corrections appear with positive sign. The largest correction, that of the Lick series, amounts to −0.11 mag.

On the other hand, for the blue observations it was necessary, as indicated in Table 7, to introduce a colour term for the Budapest observations and a magnitude term for the Nanking series. The corrections applied to the individual observations of these two series vary from −0.007 to −0.040 mag. and from +0.031 to +0.097 mag., respectively. For the Hoher List blue observations the instrumental values Δb were used rather than values of ΔB, as calculated from formulae derived by Herczeg (1964). The reason is that no correction was needed in the former case, while use of ΔB made a correction necessary. It is noted that here again the Lick series requires a rather large negative correction, amounting to 0.07 mag.

The resulting composite light curves, showing the fit of the various series, are not reproduced here, because of their bulky nature. They are, however, included in the more comprehensive report (Larsson-Leander), already referred to.

5. NORMAL POINTS, MINIMUM EPOCHS, AND PHASES

The individual observations, reduced to the B, V system as described in the previous section, were allotted weights in accordance with the mean errors of Table 3. The Pic-du-Midi series, not appearing in this table, was included in the group of series of highest accuracy, the observations of which were given unit weight. Normal points were then formed from the observations accepted. Because of the unequal spacing of the observations and the various degree of accuracy, the weights of the normals vary between 0.2 and 8.4. The total number of normals is 173 in V and 221 in B.

From the B light curve, which is the most complete one, the epochs of the three primary minima are found as

I. J. D. 2436 793.47
II.   806.405
III.   819.36

where the second value has the highest accuracy. Combining this with the minimum epoch J. D. 2436 379.532, derived by Wood and Walker (1960), a period of 12.9355 days is obtained. Phases for the normal points were computed using this value for the period and the observed epoch for primary minimum II.

The B normal points and the $B—V$ colours, derived as differences between the B and V normals, are plotted versus phase in Fig. 1. Different symbols have been used for the various cycles. Some few normals have been omitted, to avoid crowding of symbols. For comparison, the runs of B and $B—V$ according to the 1958 Lick observations (Wood and Walker, 1960) are indicated by line segments. These have been drawn with respect to the phases designated II by Wood and Walker.

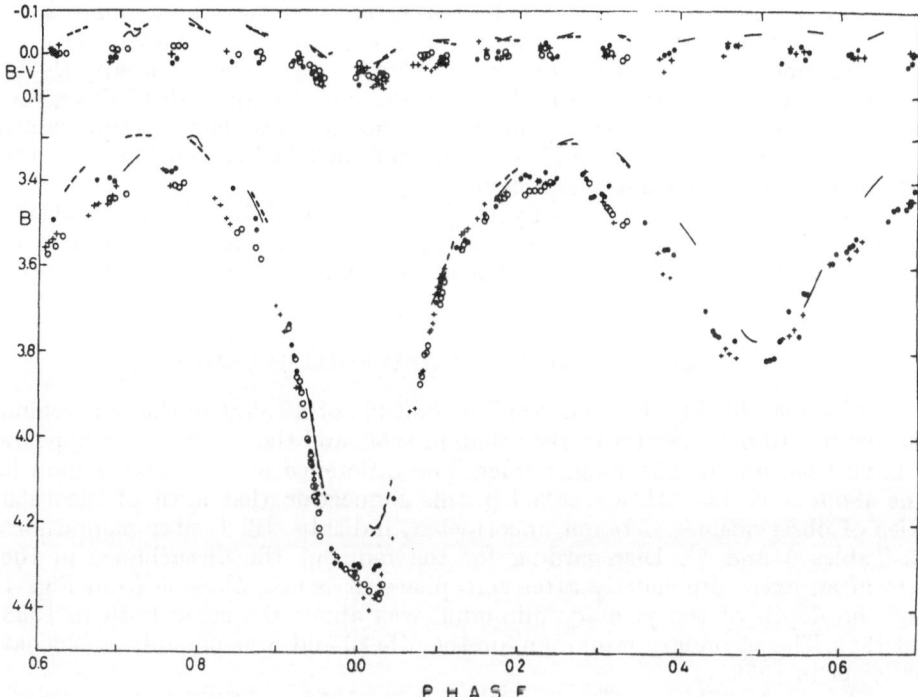

Fig. 1. Light and colour curves for β Lyrae according to normal points determined during the 1959 campaign (symbols), compared with curves (line segments) derived from the 1958 Lick observations (Wood and Walker, 1960). Dots refer to the interval J. D. 2436 788 — 802.5, crosses to 801.3—815.4, open circles to 814.3—823. Line segments drawn in full refer to J. D. 2436 375—387, dashed segments to 387—396.

It may be remarked here that the period of 12.9355 days, derived from the 1958 and 1959 minima, is somewhat longer than expected from ephemerides in current use. Wood and Walker (1960), following J. Sahade, S.-.S Huang, Struve, and V. Zebergs (1959), calculated phases from a formula given by R. Prager (1931) and modified by K. Saidov (1955), namely,

$$\text{Min.} = \text{J. D. } 2398590.57 + 12.908006\,E + 0.3919 \times 10^{-5}\,E^2 - 0.3 \times 10^{-10}\,E^3.$$

The period predicted by this formula, and corresponding to the one quoted above, is 12.9303 days. The residuals, $O-C$ are $+0.266$ day for the 1958 minimum epoch and $+0.439$ day for the 1959 minimum II. The failure of the formula to represent the minima was noted by Wood and Walker, and by adjusting the phases by the corresponding amount, $-0.0206\,P$, they obtained the system designated II (using the period 12.93016 days, predicted for the epoch of observations).

More recently a new ephemeris,

$$\text{Min.} = \text{J. D. } 2433\,289.47185 + 12.928481\,E + 0.3556 \times 10^{-5}\,E^2 - 0.648 \times$$
$$\times 10^{-10}\,E^3$$

has been derived by Wood and J. E. Forbes (1963) from a leastsquare solution of 465 minimum epochs. The period predicted for the mean epoch of the 1958 and 1959 observations is 12.93029 days. The residuals, $O—C$, are —0.049 for the 1958 epoch, but +0.12, +0.125, and +0.15 days, respectively, for the three 1959 minima. Although these latter residuals are smaller than the dispersion, 0.17520 days, obtained in the solution, they are obviously to be regarded as significant.

Of course, the difference between the periods used for calculating phases for the 1958 and the 1959 observations, is much too small to invalidate in any noticeable degree the comparison of light and colour curves, presented in Fig. 1.

6. DISCUSSION OF LIGHT AND COLOUR CURVES

As shown by Fig. 1, the over-all magnitude of β Lyrae in the blue region was about 0.10 mag. fainter in 1959 than in 1958, and the $B—V$ colour appears to have been about 0.05 mag. redder. The difference in the visual region is thus about 0.05 mag. It is recalled in this connection that most of the 1959 series of observations, if taken uncorrected, indicate still fainter magnitudes (cf. Tables 6 and 7). Disregarding for the moment the disturbance in the form of an extra dip shortly after zero phase, it is also obvious from Fig. 1 that the depth of the primary minimum was about the same both in 1958 and 1959. The secondary minimum, on the other hand, was possibly somewhat shallower in 1959.

The most striking difference between the 1958 and 1959 light curves is the change in asymmetry. In 1959 the rise from primary minimum was much steeper than in 1958. This is probably to some extent connected with the phases of the ensuing maximum and of the secondary minimum. In 1958 the maximum, M 1, following primary minimum arrived at about phase 0.27, while in 1959 M 1 came at a phase somewhat earlier than 0.25. A similar phase shift is apparent for the secondary minimum, which in 1958 arrived at phase 0.51 and in 1959 at phase 0.49.

Data on the phases of the two maxima, M 2 and M 1, and of the secondary minimum are given in Table 8, according to independent measurements on both the B and V light curves from 1958 and 1959. The cycles are counted according to the ephemeris of Wood and Forbes (1963). The first two cycles

Table 8.

Phases of maxima and secondary minimum

Cycle* E	M 2		M 1		Sec. min	
	B	V	B	V	B	V
239	−0.219	−0.230	0.276	0.272	0.508	0.512
240	−0.258	−0.258	0.274	0.265		
271	−0.247	−0.253	0.247	0.246	0.492	0.490
272	−0.244	−0.254	0.236	0.228	0.491	0.489
273	−0.240	−0.251	0.235	0.232		

* Cycles are counted according to the ephemerise of Wood and Forbes (1963).

of Table 8 thus correspond to the interval observed by Wood and Walker, while the three later ones are those of the 1959 campaign. Besides the systematic differences mentioned above, we note a large difference in the M 2 phase for the two 1958 cycles. This, of course, is due to the abnormally faint magnitudes found during the first of the 1958 observing nights (cf. Fig. 1).

The asymmetry of the primary minimum was first noted by J. Stebbins (1916) from his observations made at the Lick Observatory in 1915. From a mean light curve, covering three cycles, he found M 2 at phase 0.267 and M 1 at —0.228. As pointed out by Wood and Walker (1960) the difference in slope of the declining and rising branches was at that time found much larger than in 1958. It appears from Fig. 1 that in 1959 the slope of the two branches was almost the same, or even somewhat steeper for the rising branch, if the comparison is extended to phases more distant from the minimum epoch than 0.10 P.

At phases corresponding to the shoulders of the principal minimum, say from phase 0.80 to 0.90 and from 0.10 to about 0.20, the various cycles deviated systematically from each other both in 1958 and 1959. The deviations in 1958, which affected the upper width of the primary minimum, were found by Wood and Walker. As shown in Fig. 1 the 1959 deviations are exaggerated because of a slight progressive change in the total magnitude of the system.

This change is further substantiated by Table 9, giving B and V magnitudes for the maxima and minima observed in 1958 by Wood and Walker (1960) and in 1959 during the international campaign. Smoothed light curves have been used in all cases. Note that the magnitudes of the 1959 primary minima refer to the faintest portion of the minima, the dips around phase 0.02 are thus included. From the magnitudes at the maxima observed in 1959, it appears that during the interval of the campaign the brightness of the system decreased rather regularly by about 0.05 mag. This is a second evidence for the existence of slow magnitude variations, the first one being simply the over-all difference between the 1958 and 1959 magnitudes. Of course, such variations are contributing to the scatter obtained when observations from several cycles are combined to a mean light curve.

The bottoms of the three primary minima observed during the 1959 campaign exhibit striking dissimilarities. The first minimum ($E = 271$) seems to have been almost flat, with perhaps a very small dip in blue light at about phase 0.02. The duration of the constant, or nearly constant phase was about

Table 9.

Magnitudes at maxima and minima

Cycle*	M 2		Prim. min.		M 1		Sec. min.	
E	B	V	B	V	B	V	B	V
239	3.29	3.36	4.21	4.19	3.30	3.35	3.77:	3.85:
240	3.29	3.35	4.22:	4.20:	3.31	3.35		
271	3.37	3.36	4.31	4.26	3.39	3.39	3.80	3.81:
272	3.40	3.41:	4.40	4.34	3.40	3.41	3.82	3.85
273	3.41	3.42	4.37	4.31	3.41	3.41		

* Cycles are counted according to the ephemeris of Wood and Forbes (1963).

0.06 P. If considered flat, the minimum magnitudes are $'B = 4.30$ and $V = 4.26$, and the depths of the minimum are $\Delta B = 0.92$ and $\Delta V = 0.88$. These depths are only slightly larger than the values, $\Delta B = 0.91$ and $\Delta V = 0.84$, found by Wood and Walker (1960) for the two 1958 minima. The other two of the 1959 primary minima show pronounced dips shortly after zero phase. At the second minimum ($E = 272$) the dip amounts to $\Delta B = 0.10$ and is centred at phase 0.015. At the third minimum ($E = 273$) the dip is $\Delta B = 0.07$, and the centre falls at phase 0.025. As shown by the V magnitudes in Table 9, the two dips appear slightly smaller in the yellow light curve. For phases 0.02 to 0.03 these pronounced differences between the first cycle and the two following ones are clearly shown already in the light curve based on the Lick observations alone (Gordon, 1960).

The bottoms of the secondary minima appear rounded and no peculiarities are noted. The depths are about $\Delta B = 0.41$ and $\Delta V = 0.43$, which may be compared with the $\Delta V = 0.47$, given by Wood and Walker (1960).

Turning now to the $B-V$ colours, we note that the 1958 and 1959 mean curves are closely parallel to each other. The reddening during primary minimum is well shown also in the 1959 curve. There is, however, in the 1959 data some indication that the system grew somewhat bluer very close to mid-eclipse, say between phases -0.015 and 0.015. Such a slight top in the colour curve is not incompatible with the 1958 $B-V$ data, and a wider top is certainly apparent in $U-B$ (cf. Fig. 7 in Wood and Walker, 1960). The existence of this top in the colour curve seems to be supported by the six-colour observations by M. J. S. Belton and H. J. Woolf (1965). These observations were made at Lick Observatory on twelve selected nights 1961, spaced over an interval of two months. Segments of light and colour curves were obtained, which were joined to mean curves, using the ephemeris of Wood and Forbes (1963). The resulting $U-V$ curve, as drawn by Belton and Woolf, has a peculiar shape at primary eclipse. In contrast to $V-I$ and other similar, long-wave colours, the minimum of the $U-V$ curve occurs much earlier than mid-eclipse. This is because the $U-V$ was found "redder" around phase 0.94 (Wood and Forbes' system) than at the next phase observed, around 0.02, while no further measurements were made until phase 0.17. It is implied that the measurements around phase 0.02 refer to the colour top found from the 1958 and 1959 observations.

Finally, the $B-V$ colour curves from both 1958 and 1959 indicate a slight dip at about phase 0.37. It would perhaps have remained unnoticed, if it had not been for the much more pronounced dip, at the same phase, that appears in the $V-I$ curve given by Belton and Woolf (1965).

7. CONCLUDING REMARKS

The aim of the present paper has been to report very briefly on the photometric result of the 1959 international campaign, and to compare these results with the photometric data obtained in 1958 (Wood and Walker, 1960). The 1959 data have added to the complexity of the system. As a whole the system appears to be slightly variable in light and colour. The rising branch of the light curve from primary minimum was found much steeper in 1959 than in 1958, and the ensuing maximum and the secondary minimum occurred at somewhat earlier phases. The bottoms of the three primary minima observed in 1959 have different shapes; two of them show pronounced dips just after mid-eclipse. New features of the $B—V$ colour curves are indications that the system exhibits a "blue top" at mid-eclipse and a "red dip" at phase 0.37, i.e. at about the beginning of the secondary eclipse.

The interpretation of the photometric and spectrographic data of β Lyrae, and the changes reported by various observers, in terms of a coherent model, presents formidable problems. Probably the tentative model proposed by Huang (1963) offers the most promising possibilities. The invisible secondary component is assumed imbedded in an opaque semi-stable disk in the orbital plane, inclined to the line of sight. Eclipses, caused by the opaque disk and the B8 star, may be expected to show a variety of changes, depending upon slightly variable size and opacity of the disk.

In any case, β Lyrae is a system in rapid evolution, showing ample evidence of mass loss and mass transfer. Already the star most often studied, except for the Sun, it still represents a challenge. Future cooperative photometric and, if possible, spectrographic programmes would certainly be of the highest value. Much greater emphasis than in 1959 should, however, be placed on the standardization of the photometry.

The writer is indebted to all those who participated in the 1959 programme and for their kindness to make their observations available. He apologizes for his long delay in furnishing the final results.

REFERENCES

Abt, H. A., 1962, Astrophys. J., **135**, 424.
Belton, N. J. S. and Woolf, N. J., 1965, Astrophys. J., **141**, 145.
Binnendijk, L., 1960, Astr. J., **65**, 84.
Chang Chia-hsiang, Hong Hen-jung, Mo Ching-er and Chow Hsin-hai, 1959, Acta astr. Sin., **7**, 198.
Fresa, A., 1960, Mem. Soc. astr. ital., **31**, 365.
Gordon, Katherine C., 1960, Publ. astr. Soc. Pacific, **71**, 1960.
Grigoryan, K. A., 1961, Soobsch. Bjurak. Obs., **29**, 51.
Gyldenkerne, K. and Jaeger, J. R., 1963, Publ. mindre Medd. Kbh. Obs., No. 177, 6.
Herczeg. T., 1964, Veröff. astr. Inst. Univ. Bonn, No. 69.
Huang, S. S., 1963, Astrophys. J., **138**, 342.
Larsson-Leander, G., Ark. Astr., (in print).
Magalahsvili, N. L., and Kumsishvili, Ya. I., 1960, Abastumansk. astrofiz. Obs. Gore Kanobili Bjul., **25**, 91.
Prager, R., 1931, Kl. Veröff. Univ. Sternw. Berlin—Babelsberg, 3, No. 10. 125.
Sahade, J., Huang, S.-S., Struve, O., and Zebergs, V., 1959, Trans. Am. Phil. Soc., **49**, Part 1.
Saidov, K., 1959, Astr. Cirk. Isdav. bjuro astr. Soobshch, Kazan, No. 158, 12.
Stebbins, J., 1916, Lick Obs. Bull. **8**, 186.
Struve, O., 1959, Publ. astr. Soc. Pacific **71**, 441.
Wood, D. B. and Forbes, J. E., 1963, Astr. J., **68**, 257.
Wood, D. B. and Walker, M. F., 1960, Astrophys. J., **131**, 363.

DISCUSSION

Bakos: In your paper you mentioned that the B—V does not change during the secondary minimum. From your light-curve I notice a small change to the blue at the time of secondary minimum.

Larsson—Leander: It is very small, so it may not be real

Sahade: Were there spectra taken during the campaign?

Larsson—Leander: Yes, in the U. S. A.

THE O'CONNELL EFFECT IN SOME ECLIPSING VARIABLES

E. F. MILONE

Gettysburg College, Gettysburg, Pennsylvania
University of Maryland, Astronomy Program, College Park, Maryland
Kitt Peak National Observatory, Tucson, Arizona

INTRODUCTION

The O'Connell effect, a name Dr. Wesselink and I have given to the phenomenon of unequal light maxima in certain eclipsing binary stars, was formerly called the 'periastron effect'. Although there is no definite known cause, in the majority of cases it cannot be due to any periastron effect. This is clear from the negative correlation found by Mergentaler (1950) between the magnitude of the effect and the orbital eccentricity. A more likely origin lies in clouds or streams of matter existing on or around the Lagrangian surfaces of close binaries with nearly circular orbits. UBV observations of two stars in O'Connell's (1951) list do not contradict the latter hypothesis.

PRESENT WORK

The systems RT Lacertae and CG Cygni were selected for observation on these grounds: strong O'Connell effect, lack of previous photoelectric photometry, and brightness.

The data were gathered with more than the usual care. Careful attention was paid to changes in sky transparency in the following ways: 1) a double pair of "U" observations were placed on the outside of the observing sequence, 2) observations of the comparison star inevitably flanked those of the variable; and 3) three times a night high and low standard stars of matching color were observed to provide accurate primary extinction coefficients. Transformations to the standard UBV system were done in the usual way (Hardie, 1962) using coefficients obtained from the low air-mass observations of standards paired according to contrasting spectral types. The reduction technique, described more fully elsewhere (Milone, 1967), produced both differential magnitudes and colors (in the sense: variable-comparison stars) and UBV values for the comparison. Comparison stars (BD + 34°4216 for CG Cygni and BD + 43°4108 for RT Lacertae) were selected because of similarity in colors to the variables and because air mass differences between the variable and comparison stars never exceeded 0.01 within a ten-hour range in hour angle. These conditions minimized the effects of extinction and transformation coefficient changes on the light curve. As a check on the constancy of each comparison star, at least once per night a near-by check star was observed. The magnitudes and colors of the comparison and check stars are given in Table I.

Table I

Comparison and Check Stars

Star	V	B—V	U—B
for RT Lac BD + 43°4108	7.410 ± 0.003	1.355 ± 0.003	1.527 ± 0.005
BD + 43°4109	8.562	0.336	0.112
ε_{mse} in mean difference: ± 0.002		± 0.002	± 0.004
For CG Cyg BD + 34°4216	8.969 ± 0.003	0.744 ± 0.002	0.231 ± 0.003
BD + 34°4213	6.636	1.484	1.811
ε_{mse} in mean difference: ± 0.002		± 0.002	± 0.003

RT LACERTAE

The differential light curve for the $5^{d}07$-period binary RT Lacertae during 1965 is shown in Figure 1. Filled circles are normal points of Kitt Peak observations obtained by the author in October—November, open circles are normal points of Yale observations, x's represent individual Kitt Peak observations contributed by Dr. Douglas Hall (1967b) over a somewhat wider range in time. The light curve is clearly incomplete. Further photoelectric observations are being obtained by Hall (1968), and the author plans simultaneous spectrographic observations. The remarkable features of the present light curve have been reported earlier (Milone, 1967, 1968a, 1968b) and need only be summarized here as:

1) an anomalously blue primary minimum, independently discovered by Hall (1967a) and

2) an apparent change in the magnitude and *sign* of the difference between maxima from Wachmann's (1935) photographic light curve.

A previous radial velocity solution by Joy (1931) yielded masses of 1.90 and 1.00 for the "fainter" and "brighter" components respectively. Entering with the mass ratio of .53 in Kratochvil's (1964) Table II, the limits of the inner contact surface in the direction normal to the line of centers are for the primary: .434 and for the secondary: .318. These exceed by 20% the

Table II

Scatter in the light curve maxima of CG Cygni

Run	Maximum	σ_V*	σ_B*	σ_U*
1965	I	0^m015	0^m015	0^m022
	II	.014	.014	.025
	I & II	.015	.015	.025
1967	I	.011	.010	.023
	II	.012	.010	.021
	I & II	.012	.010	.022
1965 + 1967	I	.020	.020	.026
	II	.016	.019	.027
	I & II	.019	.020	.027

* σ refers to the mean standard error of a single differential observation.

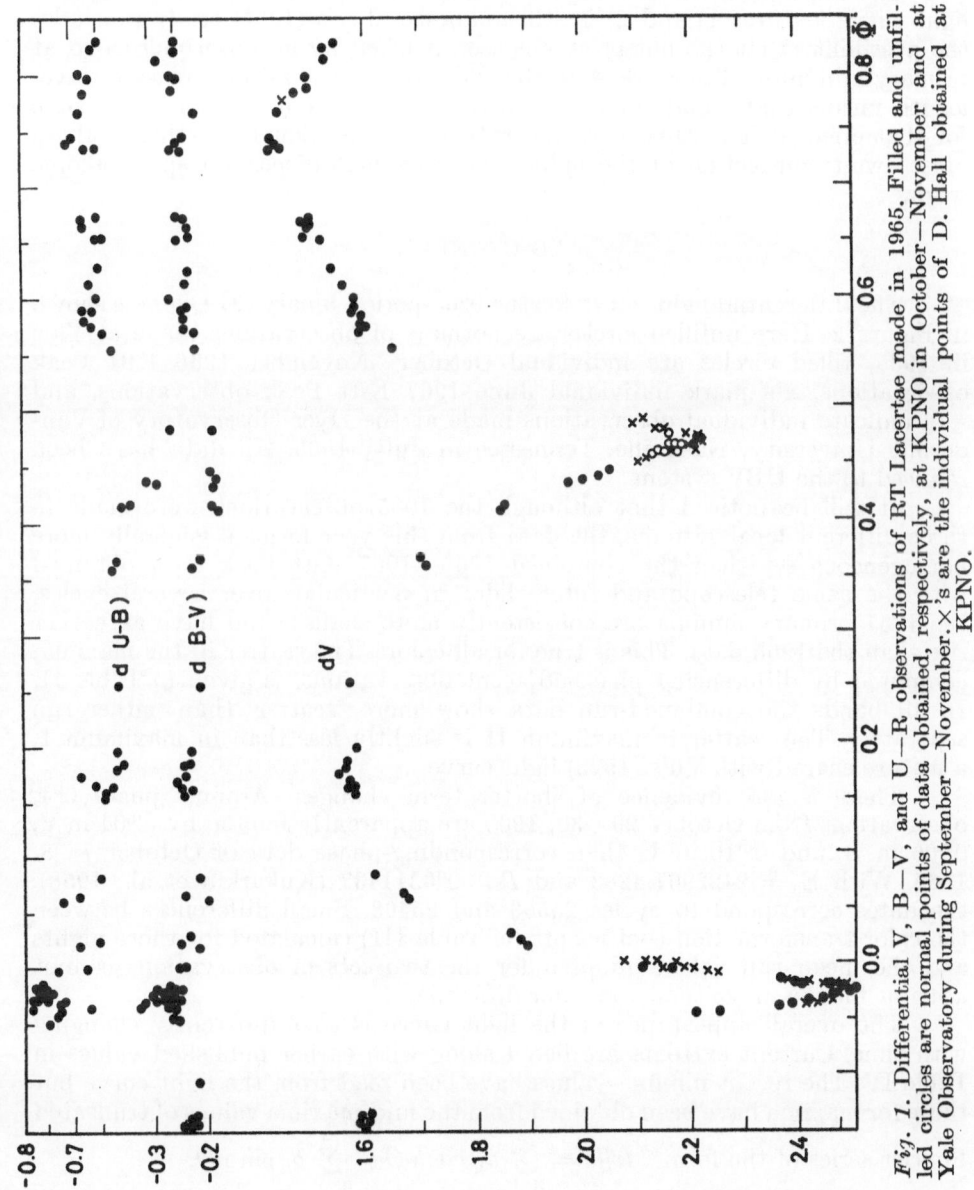

Fig. 1. Differential V, B−V, and U−R observations of RT Lacertae made in 1965. Filled and unfilled circles are normal points of data obtained, respectively, at KPNO in October−November and at Yale Observatory during September−November. ×'s are the individual points of D. Hall obtained at KPNO.

largest radius derived for either component in this direction by previous workers, viz: Fowler (1920) using Luizet's (1910, 1915) data; Krat and Nekrasova (1936) using Wachmann's light curve. Consequently we cannot assume that the system is a contact binary.

The mechanism for causing the O'Connell effect in this star, as well as light curve features 1) and 2) is still unknown. Joy's (1931) study precludes the possibility that the hotter star is seen masked by an absorbing cloud at primary minimum. The increase of the O'Connell effect with decreasing wavelength means that clouds of the negative hydrogen ion are not responsible for that effect if it is caused by absorption at maximum I. Further analysis must await completion of the light curve and high-dispersion spectroscopic work.

CG CYGNI

The differential light curve for the $0\overset{d}{.}63$-period binary CG Cygni is shown in Figure 2. Here unfilled circles are normals of observations made at Yale in 1965, filled circles are individual October—November 1965 Kitt Peak observations, x's mark individual June 1967 Kitt Peak observations, and +'s indicate individual observations made at the Dyer Observatory of Vanderbilt University, Nashville, Tennessee in July, 1965. All data have been reduced to the UBV system.

It will be noticed that although the 1965 observations were made in three different local systems, the data from this year forms a generally more homogeneous set than the combined 1965—1967 Kitt Peak data obtained with the same telescope and filter slide. In particular, over several cycles, the 1967 primary minima are consistently more shallow and have an earlier rise than the 1965 data. This is true for all colors. The scatter in the maxima, computed by differencing phase-adjacent observations, is given in Table II. In all bands the combined-run data show more scatter than either run separately. The scatter in maximum II is slightly less than in maximum I, a feature shared with Yü's (1922) light curve.

There is also evidence of shorter term changes. Around phase $0\overset{P}{.}4$, observations from October 29—30, 1965 are apparently fainter by $0\overset{m}{.}04$ in V, $0\overset{m}{.}06$ in B, and $0\overset{m}{.}10$ in U than corresponding-phase data of October 7—8, 1965. With $E_0 = 2422967.4283$ and $P = 0\overset{d}{.}6311437$ (Kukarkin et al., 1958), the dates correspond to cycles 25503 and 25468. Small differences between the color transformation coefficients (cf Table III) calculated for those nights and the mean run values adopted for the two sets of observations cannot account for the magnitude and color differences.

The overall appearance of the light curve is also apparently changing with time. Current extrema are listed along with earlier published values in Table IV. The recent minima values have been read from the light curve but those for maxima have been obtained from the mid-maxima values of truncated Fourier series of the form $\quad lv/lc = \sum\limits_{n=0}^{3} a_n \cos n\Theta + \sum\limits_{n=1}^{3} b_n \sin n\Theta$.

The purpose in obtaining the 1967 data was to fill in the light curve — particularly on the branches of the minima — so that a preliminary solution could be attempted. The apparent depression of maximum I between fall, 1965 and June, 1967 suggests that it is growing fainter with time and that

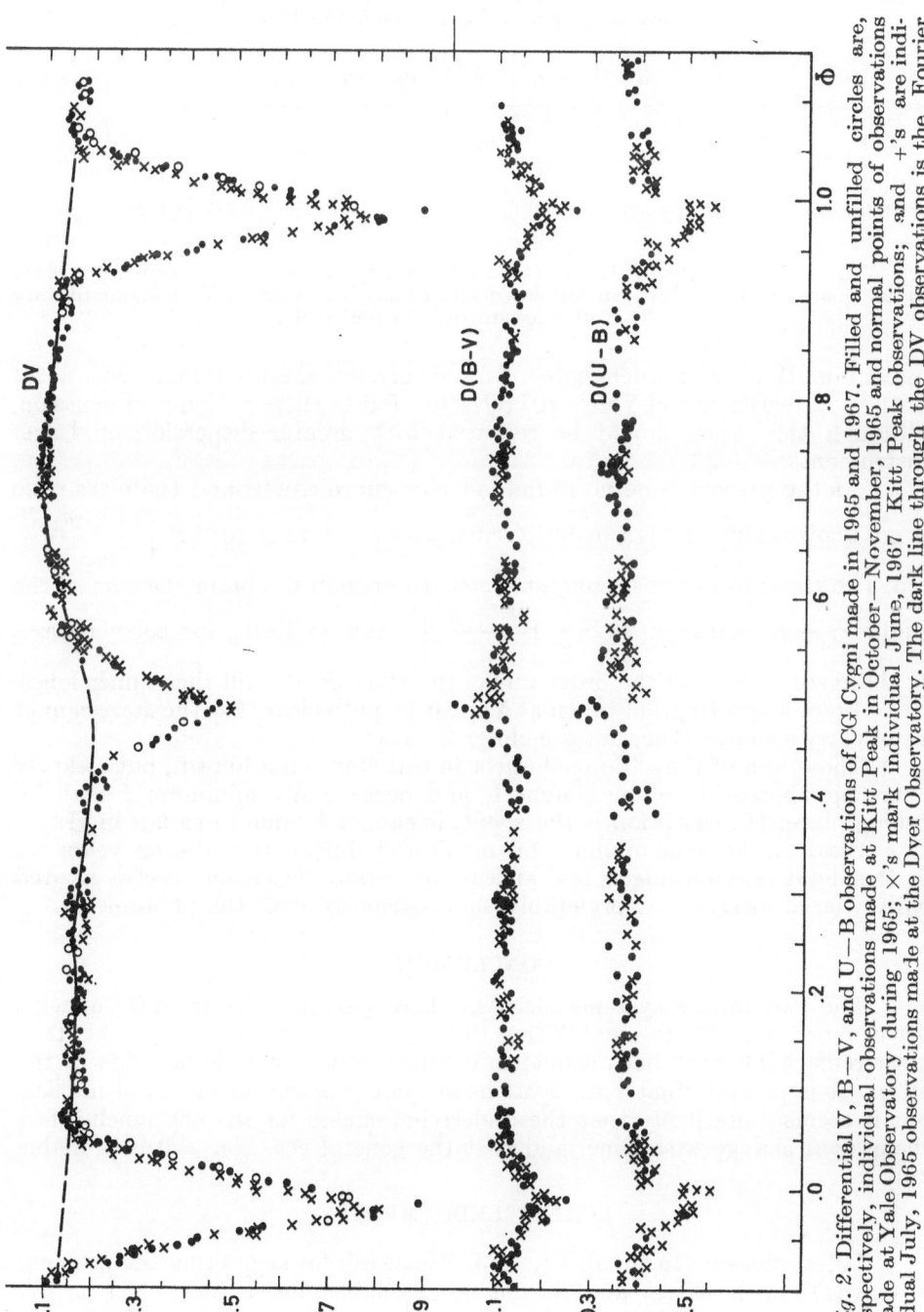

Fig. 2. Differential V, B—V, and U—B observations of CG Cygni made in 1965 and 1967. Filled and unfilled circles are, respectively, individual observations made at Kitt Peak in October—November, 1965 and normal points of observations made at Yale Observatory during 1965; ×'s mark individual June, 1967 Kitt Peak observations; and +'s are individual July, 1965 observations made at the Dyer Observatory. The dark line through the DV observations is the Fourier representation of the combined 1965—67 Kitt Peak observations outside of eclipse.

Table III

Transformation coefficients and fitting errors for Kitt Peak Telescope No. 3 photo‑
metric system during two nights in 1965

Date	ε	$\sigma\varepsilon$	$\sigma_{V \cdot Y_0}$	μ	σ_μ	$\sigma_{(B \cdot V) \cdot (b \cdot y)_0}$	Ψ	$\sigma\Psi$	$\sigma_{(U \cdot B) \cdot (u \cdot b)_0}$
10/7—8	+0.003			1.092			0.989		
Mean for Run 1:	−0.015	±0.008	±0.026	1.088	±0.002	±0.020	0.967	±0.006	±0.017
10/29—30	−0.040			1.118			0.989		
Mean for Run 2:	−0.026	±0.006	±0.029	1.103	±0.004	±0.019	0.977	±0.001	±0.019

The notation is that of Hardie (1962). σ_ε is the m.s.e. of the run mean, $\sigma_{V \cdot Y_0}$ is the mean square deviation between tabulated and calculated values of V for standard stars. The other quantities are analogous.

maximum II may be unchanging. Low dispersion spectra taken with a 36″ Kitt Peak telescope and Yale's 40″ reflector fail to show evidence of emission, although this work should be repeated with greater dispersion on larger instruments.

At the present time no radial velocity curve exists, and the mass ratio is not known. The orbit is sensibly circular with $r = 1.44 \times 10^{-2} \times \left(\dfrac{m_1 + m_2}{m_\odot}\right)^{1/3}$ a.u. The times of external contact are sharp enough to obtain the sum of the radii: $r_1 + r_2 = 0.48 \times 10^{-2} \times \left(\dfrac{m_1 + m_2}{m_\odot}\right)^{1/3}$ a.u. $= 1.2R_\odot$ for solar masses. If the mass ratio is of the order unity, the stars do not fill their inner lobes.

From a spectrogram taken at Yale, it is quite clear that the stars cannot be of early spectral type, but are closer to late G.

The origin of the O'Connell effect in this star is not known, but a slowly changing absorption of maximum I, and occasionally minimum I may be taking place. If absorption is the agent, it cannot be due to clouds of H^-.

Clearly, the system must be monitored during the coming years for further light curve changes, and at least moderate dispersion spectra secured with a large instrument, preferably simultaneously with the photometry.

CONCLUSION

The two binary systems discussed here were selected from O'Connell's list of more than 50 stars exhibiting the asymmetry at maximum light. It is most remarkable that both should have undergone a shift in the sign of the effect. It is possible that both systems are not properly members of his list, but it seems more likely that the underlying causes for the O'Connell effect themselves change with time, and that the general case $\Delta m > 0$ is probable only.

ACKNOWLEDGEMENTS

It is a pleasure to thank Dr. A. J. Wesselink for suggesting the problem of the O'Connell Effect and for many valuable discussions, Mr. E. W. McClurken who provided valuable observing assistance in June, 1967, and Dr. Douglas Hall for his RT Lacertae data.

Table IV

Extrema of the light curves of CG Cygni

Source	Band	Date	max I	max II	d max (II—I)	min I	min II	d min (I—II)
Williams (1922)	visual	1921	9m93	9m94	+0m01	10m42·	(10m15)	(0m27)
Yü (1922)	ptg.	1922	9.219	10.264	+0.045	11.374	10.544	0.830
Milstein and Nicolaev (1940)	ptg.	~1936	11.02	11.02	≤0.00	11.78	11.29	0.49
Milone (1966, 7)	V	1965	10.124	10.060	−0.064	(10.86)	10.429	(0.43)
Milone (1966, 7)	B	1965	10.990	10.918	−0.072	(11.78)	11·253	(0.53)
Milone (1966, 7)	U	1965	11.381	11.301	−0.080	(12.42)	11.604	(0.82)
Milone (unpubl.)	V	1967	10.151	10.069	−0.082	10.737	10.459	0.276
Milone (unpubl.)	B	1967	11.004	10.898	−0.106	11.663	11.255	0.408
Milone (unpubl.)	U	1967	11.409	11.301	−0.108	12.212	(11.61)	(0.60)
Milone (unpubl.)	V	1965+1967	10.134	10.063	−0.071	—*	10.444	—
Milone (unpubl.)	B	1965+1967	10.992	10.917	−0.075	—*	11.258	—
Milone (unpubl.)	U	1965+1967	11.393	11.303	−0.090	—*	11.60	—

* Differences are too extreme.

Brackets indicate uncertain values. In addition, the visual ($\sim$1931) and photographic ($\sim$1951) light curves of Tsesevich (1954) show no discernible O'Connell Effect. The values cited for the early investigations are in local magnitude systems.

This work was begun when the author was a graduate student at Yale, and was carried forth with the help of a Creativity and Research Grant of the Lutheran Church in America and Gettysburg College in 1967—1968, a Gettysburg College Faculty Fellowship in June, 1968, and a summer research participation fellowship at the University of Maryland from June to September, 1968, the help of all of which the author gratefully knowledges.

REFERENCES

Fowler, M., 1920, Astrophys. J. **52,** 257.
Hall, D. S., 1967a, private communication.
Hall, D. S., 1967b, I. A. U. Information Bulletin on Variable Stars, No. 259, Budapest.
Hall, D. S., 1968, private communication.
Hardie, R. H., 1962, Photoelectric Reductions in *Astronomical Techniques,* Stars and Stellar Systems II, 178.
Joy, A. H., 1931, Astrophys. J. **74,** 101.
Krat, W. and Nekrasova, S., 1936, Acta Astron. Ser. C., **2,** 129.
Kratochvil, P., 1964, Bull. Astron. Inst. Czech. **15,** 165.
Kukarkin, B. V., Parenago, P. P., Efremov, Yu. I., and Kholopov, P. N., 1958, General Catalogue of Variable Stars, 2nd ed., Moscow.
Mergentaler, J., 1950, Wrocław Contrib. 4.
Milone, E. F., 1966, Astron. J. **71,** 864.
Milone, E. F., 1967, Thesis, Yale University.
Milone, E. F., 1968a, Astron. J. **73,** S26.
Milone, E. F., 1968b, Astron. J., in press.
Milstein, I. P. and Nicolaev, S. P., 1940, Vsesoiuznoe Astron.-Geodetic Soc. N. 6, 9.
O'Connell, D. J. K., 1951, Riverview Publ. **2,** 85.
Tsesevich, V. P., 1954, Odessa Izvestia **4,** Part I, 255.
Wachmann, A. A., 1935, Astron. Nachr. 255, 367.
Williams, A. S., 1922, MN **82,** 300.
Yü, Ch'ing-Sung, 1923, Astrophys. J. **58,** 75.

SUDDEN CHANGES IN THE PERIOD OF ALGOL

T. HERCZEG

Hamburg—Bergedorf
(Read by W. Seitter)

Orbital motion in binary systems was considered, for a long time, a true paradigm of periodic phenomena. Even its occasional changes were supposed to be of strictly regular, i.e. periodic nature like rotation of the apsidal line or light time effects in multiple systems. Exceptional objects like β Lyrae, with a secular increase of the period, could have been recognized as cases still showing some regular, predictable character. Later on, however, after longer series of observations became available, a sort of eclipsing systems (like XZ And) were found showing obviously non-periodic changes of the period: an erratic but continuous up and down in the O—C diagrams for the times of minima. Other well observed photometric binaries exhibited a different type of non-periodic changes: the eclipsing period was, from time to time, subject to sharp sudden changes ("jumps") while between two consecutive jumps it remained practically constant. AR Lacertae or AH Virginis are important examples of this type.

Schneller (1962) in a lecture at the Budapest Observatory advocated his thesis that, while "regular" changes of the eclipsing period are relatively infrequent, these sudden, erratic changes seem to be the rule, at least in the case of the contact and semi-detached systems. Among other typical objects he mentioned Algol (β Persei) itself — certainly one of the best observed variables — as indicating period changes of this nature; a similar proposal concerning Algol's period, as a matter of fact, was already put forward by Sterne.

In the case of Algol, Schneller tried to represent the O—C curve of the times of minima by a polygon having two angles, two "sharp bends". They correspond to sudden changes, discontinuities of the period, having occurred in 1840 and 1924. This proposal has been essentially substantiated by a recent analysis of all photoelectrically determined epochs of minima, carried out at the Hamburg observatory (cf. a forthcoming paper by Frieboes-Conde, Herczeg and Hög). In particular, the occurrence of "period jumps" seems to be definitely established now though numerical details (like the data of these changes) had to be revised.

This interpretation of the O—C curve offers us an explanation of the much discussed "great inequality" of Algol, too. This is a hypothetical long period light time effect requiring the existence of an additional unseen companion (Algol D) in the system. Up to now there are three observed and, at least, two hypothetical members listed as belonging to the complicated Algol system. Those observed, definitely recognized components are:

Algol AB the eclipsing pair, a semi-detached system;

Algol C, a distant companion with a period of 1^a862. This orbital motion in the triple system gives rise to a light time effect of the times of minima; having a small amplitude ($\approx$ 6 min.), it is rather difficult to handle. On the other hand, there exists an accurate and reliable set of spectroscopic elements for this 1.9 year orbit, derived by Ebbighausen (1958).

The hypothetical members are:

Algol E, introduced for explaining by light time effect a clearly distinguishable 32-year periodicity of the O—C values by light time effect. This periodicity can be, however, much more convincingly explained by the aid of apsidal motion in the eclipsing system;

Algol D, a component proposed already in the last century in order to find an explication for the above mentioned great inequality, dominant feature of the O—C diagram, with a suggested period of about 170—180 years. But an interpretation of the great inequality as a light time effect faces also serious difficulties. Until now, the periodicity itself defied exact representation: predictions based upon a recent, very elaborate discussion of the period changes by Kopal, Plavec and Reilly (1960) yielded residuals amounting to half an hour. Besides, the mass of this hypothetical unseen companion turns out to be unexpectedly high, about 4 $M_\odot$. A further important question is the following: a possible light time effect would require an orbit large enough to cause detectable changes ($>$ 1 sec of arc) in the position of Algol AB, a much observed 2nd magnitude star. Our discussion of all meridian observation reaching back to about A. D. 1750 indicated, however, that Algol's proper motion remained sensibly constant — no traces of the orbital motion in the hypothetical quadruple system Algol ABC-D could be found. This seems to be a direct and decisive argument against the existence of the component Algol D.

We are now obliged to find an alternative explanation for the great inequality. Let us make the basic assumption that — besides the eclipsing period, the 1.9-year light time effect and the 32-year period — no further periodicities exist in the O—C diagram. (This assumption, which might be opposed by some observers, has turned out a very useful and also successful working hypothesis.) Then a rather sensitive test can be carried out, based on the best determined, photoelectrically observed epochs of minima, about 60 in number since 1920. We treated this rather accurate observational material in the following way:

We subtracted the effect of 32-years periodicity This correction was based on data by Hellerich (1919);

We accepted as a definite representation of the 1$\overset{d}{.}$862 orbit the spectroscopic elements given by Ebbighausen (1958, 1962). Thereafter, the corresponding light time effect can easily be taken into consideration.

Then the residuals have shown with considerable accuracy a polygon indicating two sudden changes of the period (in 1944 and 1952) and three different values of a practically constant eclipsing period outside these "jumps" (Fig. 1). The sudden changes amount to $\Delta P = +3\overset{s}{.}5$ and $\Delta P = -2\overset{s}{.}0$, respectively. The constancy of the period within the time intervals, say 1920—1942 and 1955—1965 can be judged from the good representation of the 1.9-year light time effect, the remaining scatter of the O—C values being but slightly higher than the usual error of photoelectrically determined epochs of minima. It seems very improbable that any further periodicities could be accomodated within this small margin.

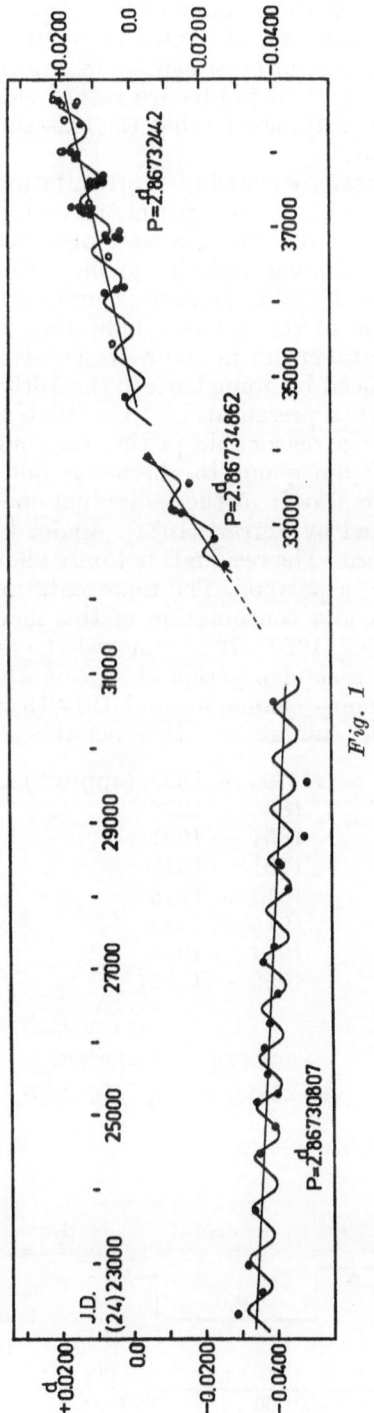

Fig. 1

The reliability of Hellerich's numerical representation of the 22-year period is, of course, a basic question. Though small systematic deviations are not to be entirely excluded, no serious error can, — in my opinion — arise hereby. Especially the fact that the two jumps are rather close together, separated by a quarter of the period only, means that the existence of two sudden changes remains incontestable.

This somewhat perhaps surprising representation of the photoelectric times of minima encourages me to an experiment: to apply a similar procedure to the earlier measurements, too. Minima were observed since 1784 in a very great number. The overwhelming majority of these times of minima is based on visual estimates, and their rather modest accuracy (of the order of $\pm 0.02^d$), doesn't permit a detection of the 1.9-year light time effect. Besides, the influence of minor systematic errors in the representation of the 32-year period will be considerably enhanced by going back to the 18th century observations. This makes our proposed representation to a tentative one; on the other hand the deviations from a reasonable picture turn out relatively small, thus suggesting the proposal I am going to discuss is not completely arbitrary.

Because of the large scatter of the individual epochs I used the normal points meticulously derived by Ferrari (1934). Again, I subtracted Hellerich's formula for the 32-year term. The residuals not only allow, they clearly suggest a representation given by a polygon. The representation derived earlier from the photoelectric minima is a continuation of this new polygon — the long interval of constant period 1920—1944 can well be extended back to 1915. There is no indication of a sudden period change in 1924 as it was proposed by Schneller; the short sharp change around 1914/15 seems to be real (Fig. 2).

The following table summarizes this possible sequence of "events":

$P_1 = 2^d\!8673442$	(1784) — 1835 (approx.)	$\Delta P \approx -3^s\!7$
$P_2 = 2.8673012$	1835 — 1854	≈ -2.3
$P_3 = 2.8672775$	1854 — 1901	$\approx +1.7$
$P_4 = 2.8672967$	1901 — 1913	≈ -4.0
$P_5 = 2.8672506$	1913 — 1915	$\approx +5.0$
$P_6 = 2.86730807$	1915 — 1944	$\approx +3.5$
$P_7 = 2.86734862$	1944 — 1952	≈ -2.0
$P_8 = 2.86732442$	1952 —(1965)	

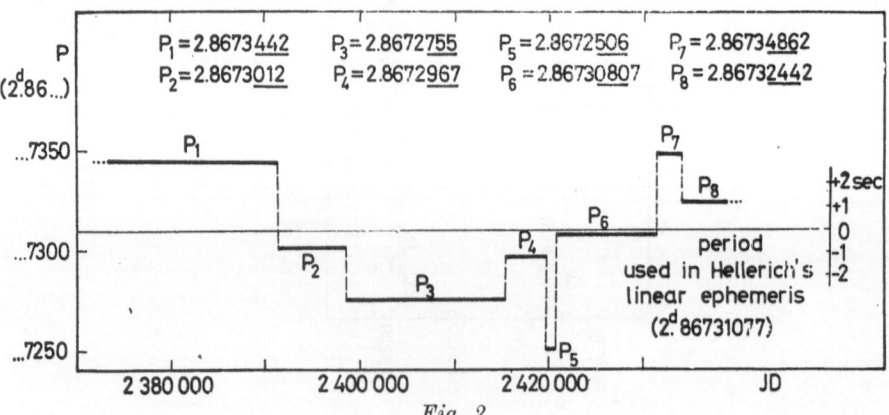

Fig. 2

This interpretation decomposes the great inequality into a series of sudden period changes. Apparently they are forming a random set, the changes occurring on an average once in 25—30 years. This is certainly a non-periodic phenomenon, probably caused by discontinuities of the mass exchange between the components or in the mass loss from the whole system. Perhaps, the "overflowing" of the secondary component of its Roche-limit is not a smooth phenomenon but it causes from time to time directional outbursts of mass from the secondary, hereby changing the eclipsing period in a sudden way.

The obviously spurious periodicity of about 160—180 years length was an understandable suggestion of the early observers: it was favoured by the existence of two relatively long intervals of constant period, P_1 and P_3. This led to the use of a mean period in the linear ephemeris formula which, in its turn, gave rise to the characteristic wedge-shaped figure in the O—C diagram, simulating the "first half" of a sine curve (Fig. 3).

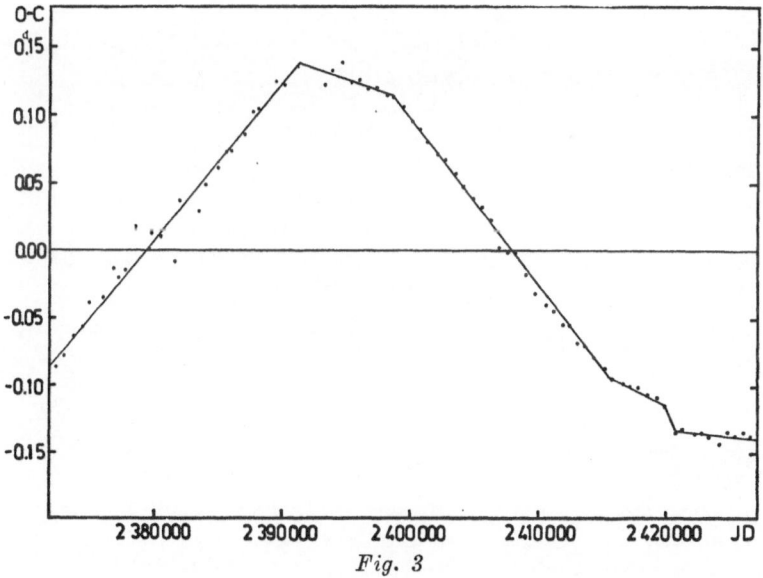

Fig. 3

DISCUSSION

Detre: Algol's O—C diagram looks rather like to one resulting from cumulative effects of random period variations: on long cycles there are superposed shorter cycles. How far is the 32 year period established?

Seitter: Irrespective of smaller errors having an effect on the period length and the zero phase (and Hellerich's representation of the 32 year period seems to be a very reliable one!), most of the jumps, especially those rather thoroughly observed in 1944 and 1952, are quite well established. This is mainly based on the fact that these two jumps occurred within a time interval short enough not to be seriously affected by any erroneous assumption concerning the 32 year periodicity. Numerical data could certainly be shifted to some extent but the existence of sudden period changes remains beyond doubt.

REFERENCES

Ebbighausen, E. G., 1958, Astrophys. J. **128,** 598.
Ebbighausen, E. G., and Gange, J. J., 1962. Publ. Dominion Astrophys. Obs. Victoria **12,** 151.
Ferrari, K., 1934, Astr. Nachr. **253,** 225.
Hellerich, J., 1919, Astr. Nachr. **209,** 227.
Kopal, Z., Plavec, M. and Reilly, Edith, 1960, Jodrell Bank Ann. **1,** 374.
Schneller, H., 1962, Mitt. Sternwarte Budapest No. 53.

ON THE RESEARCH PROGRAM CONCERNING ECLIPSING VARIABLES AT THE OBSERVATORIES NÜRNBERG AND IZMIR

E. POHL

Nürnberg Observatory, Germany

The observatory of Nürnberg in Germany was opened in 1930 and has had above all the task to promote the astronomical education of the public. Besides, an astronomical research program concerning eclipsing variables has been carried out. Since the summer of 1963 a photoelectric photometer constructed in the workshop of the observatory was used for this observation program. We use a photomultiplier 1P21 attached to a 34 cm Cassegrain-reflector. Since the spring of 1967 the measurements were registered with a Siemens-compensograph.

The observation program covers a large part of the eclipsing variables brighter than magnitude 10. The measurements are carried out by myself and some members of the Nürnberg Astronomical Amateur Association. Unfortunately, the observatory is situated in the area of the city of Nürnberg, so the background in the case of faint stars often amounts to 30 per cent of the intensity of the star. The centre of the city is in the west, and during many clear nights we can only measure up to a zenith distance of 45 degrees in this direction.

From 1963 to 1968 about 120 minima of 38 various eclipsing variables could be determined in Nürnberg photoelectrically. For some stars minima were derived every year. Some of these frequently observed objects shows large O—C's against the elements of the General Catalogue 1958 and also of the last edition of the Krakow Catalogue. The star TX UMa shows finally an O—C value of more than 1 hour, and SW Lac a deviation of even $1\frac{1}{4}$ hours from the elements of the Krakow Catalogue. For TX UMa, SW Lac, V 477 Cyg and i Boo new elements were determined basing on our results and published in the Bulletins of Com. 27 of the IAU in the year 1967.

During several travels, first to the University Observatory of Ankara and than to the University of Izmir 1966, a cooperation for this observation program was agreed upon with Turkish astronomers. A particularly friendly and close cooperation with the director of the University Observatory of Izmir, Prof. Kizilirmak, and his staff has been existing for three years. In the spring of 1966, Prof. Kizilirmak and I agreed upon the erection of a common instrument for the photometry of eclipsing variables. This 48 cm Cassegrain reflector is fitted out with a photoelectric equipment similar to that existing already in Nürnberg. The telescope and the electronic parts of the photometer, wich were all constructed in Western Germany, were shipped to Izmir and set up at the University Observatory of Izmir in October last year. The main observational program of this instrument, which is to be carried out by the astronomers of the University Observatory of Izmir and by myself, is the

31*

regular control of a large number of eclipsing variables up to the 11th magnitude, in order to revise the periods. The University Observatory of Izmir is situated on a mountain, 650 m above sea-level, about 20 km away from the city. It was founded 1964/65 under the direction of Prof. Kizilirmak, and its enlargement is still in progress. The future observation program — of the observatories in Nürnberg and Izmir, which will be essentially broadened — will be carried out commonly, and all results will also be published together. During the last 4 years already visual results have been achieved at the University Observatory of Izmir according to the Argelander method. The minima derived so far by the Nürnberg and Izmir observatories have been published in the Astronomische Nachrichten.

On the 30th April 1968 the first photoelectric measurements of i Boo and SV Cam could be made with our new 48 cm telescope in Izmir. In the last 4 months about 40 minima were derived under the favourable climatic conditions in Izmir. Between the Nürnberg and Izmir minima there are no systematic time-differences in the dates. Till today, at both observatories we have obtained photoelectric minima for more than 60 different eclipsing binaries.

We hope, we shall be able to make a valuable contribution to the old but not yet clarified problem of the period changes of eclipsing binaries.

PART V

MISCELLANEOUS PROBLEMS

THE THIRD CATALOGUE OF VARIABLE STARS IN GLOBULAR CLUSTERS

HELEN SAWYER HOGG

David Dunlap Observatory, Ontario, Canada

A subtitle of this paper might be "Some Notes on the Observational Program on these stars at the David Dunlap Observatory." These two topics are so interwoven that it is hard to separate one from the other. The observational program has been running since the 74-inch telescope went into operation in 1935, and the first catalogue was published in 1939.

At first glance the variables in globular clusters seem to have little to do with the topic of this colloquium, Non-Periodic Variables, for most of the variables in globular clusters are definitely periodic. Nevertheless there are a few which are not, two or three U Geminorum stars which might be cluster members, and of particular interest three novae, two of which almost certainly are members. The first was Nova T Sco, discovered visually by Pogson in M 80 in 1860. It changed the whole appearance of the cluster, as it was 7th magnitude at maximum. Then in 1949 Mrs. Margaret Mayall discovered the spectrum of a nova near NGC 6553, but there is doubt as to its cluster membership. The third was discovered five years ago by Dr. Amelia Wehlau in M 14 on plates I had taken with the 74-inch David Dunlap reflector in 1938. This nova hovered near mag. 16.0 on eight plates taken during one week in June, 1938, but it may very well have been several magnitudes brighter earlier. As this nova has already been reported at a former IAU meeting, it will not be discussed further now.

However, the discovery of this nova prompted us to start a systematic search by blink microscope of hundreds of globular cluster plates in our collection. Before the discovery of the nova our philosophy had been that once a cluster has been thoroughly blinked for variables, with plates over a considerable interval of time, there was nothing further to be gained by searching. Now it seems important to try to ascertain the frequency with which novae do occur in globular clusters. To do this, a careful search (probably blink microscope is the easiest) is necessary on thousands of plates. With the help of assistants we have now spent some hundreds of hours blinking hundreds of plates. So far the search has yielded no more novae, but a few new variables have resulted from it. We hope to continue with the checking of all the globular cluster plates at the David Dunlap Observatory, bearing in mind that the nova we did discover was not of spectacular appearance, — a 16th magnitude star in the central region of a globular cluster, no brighter than some of the variables. To get a statistical frequency for novae in globular clusters, it will be necessary to search thousands of plates over long periods of time.

In the 33 years since the program with the 74-inch reflector began, we have missed only 4 cluster observing seasons. We have plates on about 50

Table I

Variable Stars in Globular Clusters
(September 1, 1968)

NGC	No. Vars.	No. Periods	No. RR Lyr	No. Others	All Types
104	11	5	2	3+6a	M, I, RR
288	1	1	0	1	SR
362	14	10	7	3	RR, C, M
1261	6	0			—
1851	9+1s	0			—
1904	5+2s	3	3	1	RR, SR
2298	0+6u	0			—
2419	36	2	2+28a	0+5a	RR, I
2808	0+4u	0			—
Pal 3	1	0	1	0	RR
3201	77+5u	59	58	1+1a	RR, E, M
Pal 4	2	2	0	2	M
4147	16	15	15	0	RR
4372	0				O
4590	38	36	35	1	RR, M (field)
4833	10	9	6	3	RR, M (field)
5024	44	33	33	0+1a	RR, C
5053	10	10	10	0	RR
5139	171+2s+4f	154	138	16+5a	RR, C, RV, M, I, SR
5272	189	176	173	3	RR, SR, C, EW
5286	7	0			—
5466	22+22s	18	18	0	RR
5634	7	1	1+6a		RR
5694	0	0			O
Pal 5	5	5	5		RR
IC4499	6	0			—
5824	27	9	9+16u		RR
5897	4	0	0+4a		RR
5904	97	93+1a	90+2a	3+1a	RR, C, SR, UG
5927	2+1s+12f	0			—
5986	5	0	3	1	RR
6093	8	3+1a	0	4	C, N, M (field)
6121	43	41	41	1	RR, SR
6144	1	0			—
6171	24+16f	18	18	0+1a	RR, M
6205	10+2s	6	2	4+2a	RR, C, I
6218	1	1	0	1	C
6229	22	16	15	1	RR, C
6235	2	0			—
6254	3	2	0	2	C, SR
6266	83+6u	74+4u	74	1	RR
6273	4	0			—
6284	6	0			—
6287	3	0			—
6293	5	0			—
6304	7+4f	0	3a		RR
6333	13	11	11		RR
6341	15	13	12	1	RR, EW (field)
6352	3+9f	0			—
6356	5	0			—
HP1	4	0			—
6362	32	0			—
6366	2	0+1u	1u		RR

Table I. (cont.)

NGC	No. Vars.	No. Periods	No. RR Lyr	No. Others	All Types
6397	3	3	1	2	RR, SR, M (field)
6402	77	40	34	6	RR, C, M, N
6426	12	10	10		RR
6522	9	8	8	1	RR, I (field)
6528	0	0			—
6535	1	0			—
6539	0+1u	0			—
6541	1	0	0	0+1a	M
6553	6	3+1a	3	0+3a	RR, M, N
6558	9+14f	0	4a		RR
I1276	5	1	1	4	RR, SR, M
6569	5+3f	0			—
6584	0	0			O
6624	3+10f	0			—
6626	17+1u	2+3u	3u	3	RR, C, RV, UG
6637	5+5f	0	1a	1	RR
6656	27+2u+1f	22+4u	18	9	RR, M, SR
6681	12+4f	0	2a		RR
6712	19+1s	15	9	6+3a	RR, SR, M, E (field) UG
6715	80	37	34	3+2a	RR, C, SR, E (field)
6723	19+5u	19	24	0	RR
6752	1	0			—
6760	4	0			—
6779	12	4	2	4	C, SR, RV, RR (field)
6809	6	5	5		RR
6838	4	0+1u	0	2	E, SR
6864	11	0			—
6934	51	0+44u	0+44u		RR
6981	39	28	28		RR
7006	72+3s	27	26+20a	2	RR, SR
7078	102+1f	77	74	3	RR, C
7089	21	21	17	4	RR, C, RV
7099	4	3	3	1	RR, UG
Pal 12	3	0	2a	1	RR
Pal 13	4	4	4	0	RR
7492	2	2	2	0	RR

Notes: a = assumed; f = field; u = unpublished.

globular clusters, but by no means have all of these been observed every season. In addition a 19-inch telescope has been used, and four additional seasons had been obtained earlier, from 1931 through 1934, with the 72-inch telescope of the Dominion Astrophysical Observatory, Victoria, B.C. We have about 4800 globular cluster plates at the David Dunlap Observatory. An instrumental development this year may be of interest. On our new building on the University of Toronto campus in the heart of the city, on the 16th floor a new 16-inch Boller and Chivens reflector is in operation. We are taking plates on globular clusters at the Cassegrain focus with a scale of 0″.5 to the millimeter, which compares quite favorably with the 22″ per millimeter scale of the 74-

inch. The 16-inch has an advantage in that it is located only a mile from the northern shore of Lake Ontario and hence can photograph in the southern sky looking across 25 miles of dark lake. To see the region toward the galactic center our 74-inch now has to look across many miles of brilliant city lights. It is hoped to continue these programs with both instruments for many years to come.

Of the variables in the third catalogue, there is no doubt that most of them are the RR Lyrae types recognized in these clusters by Professor S. I. Bailey of Harvard almost 70 years ago. However, there are about 30 Type II Cepheids, and a few long period stars, but no actual members so far with period greater than 220^d, as pointed out by Dr. M. W. Feast who has made a major contribution in indentifying with his radial velocity work some of these stars as actual cluster members, such as CH Scuti in NGC 6712.

Since Bailey's time, however, and particularly in recent years, we have gained a rather different impression of the RR Lyrae stars in respect to their period changes. As the interval of time spanned by plates of different observatories increases, more and more period changes are being detected, and their importance in the scheme of stellar evolution is recognized. Period changes are becoming one of the most fascinating aspects of variables in clusters. They also make the publication of the catalogue more complicated as to choice of period and epoch for listing because not all the values can be published in such a catalogue.

The third catalogue of variables in globular clusters has been in preparation for some time and it is expected to deliver the manuscript to the printer this winter. It will be similar in format to the second catalogue of 1955 and will be a David Dunlap Observatory publication. It is no coincidence that a substantial proportion of the new information has been contributed by astronomers who are in this gathering here today. Strongholds of research in this field have been at the Konkoly Observatory of our host institution here under Professor Detre, with the work of Dr. J. Balázs-Detre, Dr. Szeidl, Mr. Lovas, and Mrs. Barlai; and at Asiago under Professor Rosino, whose contribution is prodigious, with Dr. Margoni and Dr. Mammano, and Dr. Christine Coutts who is a link between Asiago and David Dunlap. I wish to take this opportunity of thanking publicly those workers in this field who are present here today for the wonderful cooperation they have given me in supplying me with their important material, both published and unpublished.

In all, 89 clusters have now been searched for variables, with a total of 1754 variables published, and 61 unpublished or suspected. Of these clusters, 4 have no variables, and in 3 others the variables are unpublished. In 49 clusters 1155 periods have been published, and 60 additional periods remain unpublished in 3 more clusters. Of these periods, 1086 are of RR Lyrae stars in 47 clusters, and there are 132 more such periods unpublished or assumed for a total of 53 clusters. In 36 clusters there are 105 other types of variables with periods determined, and 31 more periods assumed for a total of 41 clusters. A problem of increasing importance is to distinguish between cluster members and field stars, and it is not yet possible to treat this problem in a uniform way from cluster to cluster. Table I gives a brief summary of present data. References which form the basis of this table will be published in the catalogue.

DISCUSSION

Bakos: Are you considering a patrol of globular clusters for novae?

Sawyer Hogg: Not really. Fairly large scale plates would be necessary for an efficient survey. We are trying to get series of plates on different clusters, which we will examine.

Herbig: Is there any hope of detecting the remnant of T Sco in M 80?

Sawyer Hogg: I think that years ago some of the Mt. Wilson astronomers, perhaps Dr. Baade, looked into it, and thought not. It would be very difficult now to identify the nova in the center of this compact cluster.

Feast: It seems to me extremely important to try to determine periods or quasi-periods for semiregular red variables in globular clusters. In 47 Tuc beside the 200 day Mira stars there are sveral semiregular stars for which Arp derived quasi-periods of the order of 50 to 150 days. It would be valuable to have more data on these stars as well as similar stars in the clusters. This should help in understanding the evolution of red variables.

Sawyer Hogg: I certainly agree with this. There are some variables of this type which we are trying to follow.

THE CEPHEIDS ON THE COLOUR-COLOUR PLOT

N. S. NIKOLOV and P. Z. KUNCHEV

Department of Astronomy, University of Sofia

A study of the behaviour of cepheids on the colour-colour plot provides significant information about the radiation of these stars in different regions of the spectrum and, about the physical conditions in their atmospheres, which are of importance for the explanation of the physical nature and evolution of these objects.

The behaviour of more than 200 cepheids were investigated on the colour-colour plot using the mean values of the colours U—B and B—V at intervals of $0^{m}05$ from Nikolov's catalogue of the light and colour curves of the cepheids (1968). At first we tried to classify the curves. In spite of the considerable variety of their shape these curves can be ordered in a few groups:

1. *Linear or almost linear curves.* Their colour-changes on the ascending and descending branches are almost parallel, and the differences in the U—B values in both branches do not exceed 0.03—0.05 magn. (Fig. 1)

2. *"Open" curves.* They represent closed contour which includes a considerable part of the area of the colour-colour plot. More than 50 per cent of the cepheids belong to group 2. (Fig. 2)

3. *Curves exhibiting a loop on the colour-colour plot.* In this group are included those cepheids whose clearly defined loops are not consequences of uncertain data. (Fig. 3)

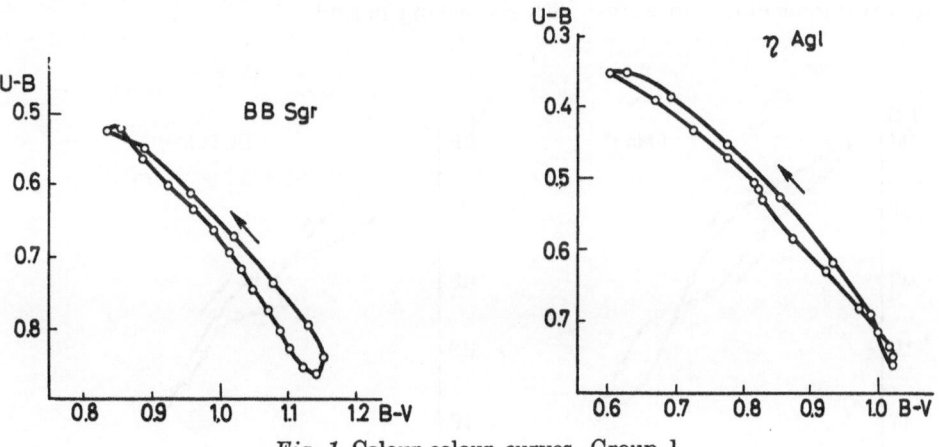

Fig. 1. Colour-colour curves. Group 1.

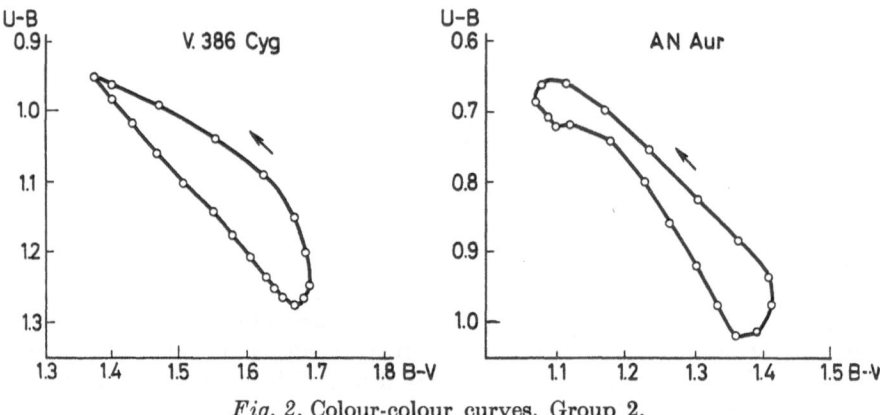

Fig. 2. Colour-colour curves. Group 2.

Our data do not confirm Mianes' opinion (1963) that the presence of a loop on the colour-colour plot may be considered as a criterion for the star belonging to Population II. We find among the stars of group 3 CW-type cepheids and also Cδ-type stars. On the other hand, a considerable number of cepheids classified as CW in the General Catalogue as well as by Petit (1960) do not show a loop on the colour-colour plot at all.

It is interesting that W Vir itself, as a typical representative of the Population II cepheids, does not exhibit a loop according to our data, as well as according to other authors (Kwee 1965, Oosterhoff and Walraven 1966).

We tried to find some relations between the characteristics of the cepheid curves on the colour-colour plot and the logarithm of the period of these stars. It was found that the tangent of the angle by which the curve for each cepheid on the $(U—B)_0/(B—V)_0$ plot is inclined towards the $(B—V)_0$ axis, increased with the period (Fig. 4). This dependence shows that for a given increase of the temperature during the pulsation the relative change in the ultraviolet emission grows with increasing period.

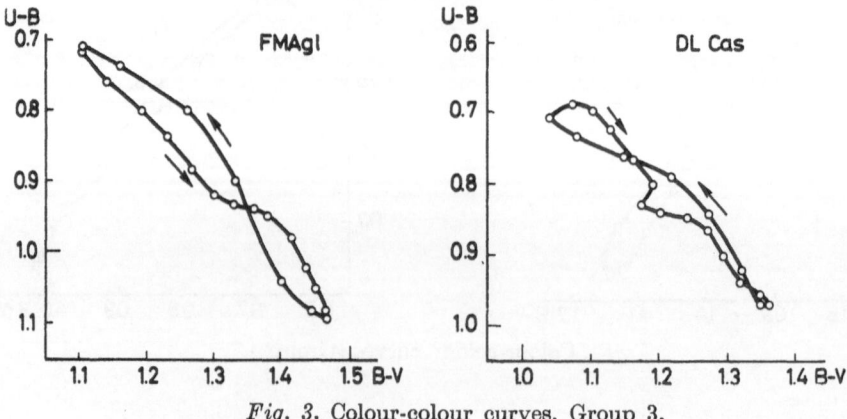

Fig. 3. Colour-colour curves. Group 3.

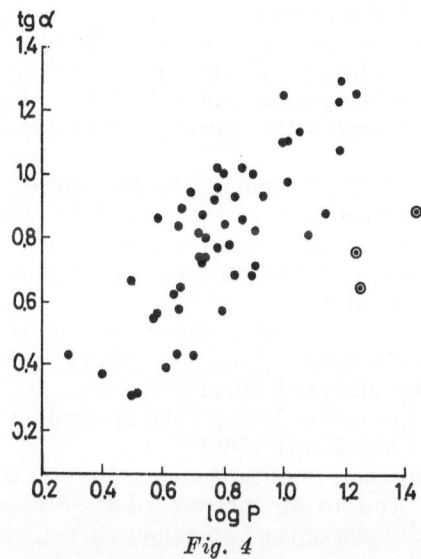

Fig. 4

Since the areas limited by the cepheid curves on the $(U—B)_0/(B—V)_0$ plot do not differ essentially from those on the $(U—B)/(B—V)$ plot, the areas limited by the twocolour curves of about 160 cepheids were determined and plotted against the logarithm of the periods (Fig. 5). It is evident that for all stars with periods from 1^d to about 10^d the surface occupied by the twocolour curves practically does not increase, or it increases only slightly with increasing period. For periods longer than about 10 days this surface increases

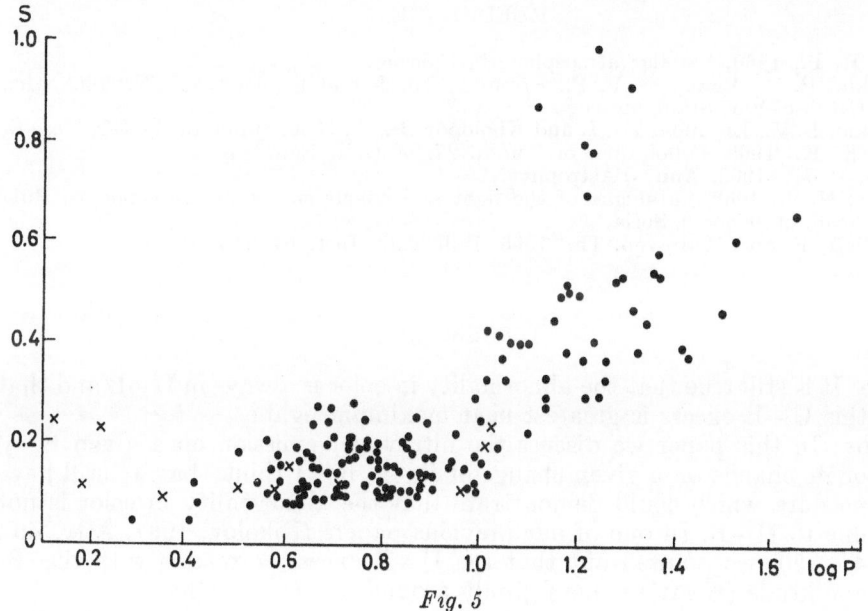

Fig. 5

comparatively quickly. On the basis of some additional investigations we concluded that the major part of the observed effect is due to the increasing differences between the values of U—B for the ascending and descending branches of the curve, although the increase of the range B—V and U—B with the period also influences the surfaces determined by the two-colour curves.

Thus the increase of the "opening" of the curves (U—B)/(B—V) correlates strongly with the logarithm of the period. So, for a given B—V the ultraviolet emission for the ascending branch is stronger than for the descending branch of the two-colour curve. The difference in this emission is greater for the periods longer than about 10^d.

It is difficult to arrive at definitive conclusions about the physical causes which are responsible for the observed relations. Three causes can be mentioned, which may influence the observed effects.

1. The presence of emission lines in the spectrum of the cepheids with relatively longer periods (see Kraft 1960).

The presence of emission lines in the spectrum on the ascending branch of the light curve will lead to an increase of the ultraviolet emission and therefore the ascending branch of the two-colour curves will be shifted towards smaller U—B values. The surface mentioned above will therefore increase.

2. Probably the influence on the ultraviolet emission will also cause a decrease of the effective gravity at the descending branch of the two-colour curve. The longer the cepheid pulsation period, the larger the pulsation range, and the effective gravity will become still smaller on the descending branch. In accordance with this, one observes with the increase of the period a still smaller ultraviolet emission on the descending branch.

3. The blanketing effect may also influence the observed phenomena.

REFERENCES

Kraft, R. P., 1960, "Stellar atmospheres", Chicago.
Kukarkin, B. V., Parenago, P. P., Efremov, Yu. I. and Kholopov, P. N., 1958, Gen. Catal. of Var. Stars, Moscow.
Kukarkin, B. V., Efremov, Yu. I. and Kholopov, P. N., 1960, Suppl. of the GC, Moscow.
Kwee, K. K., 1965, Colloquium of Comm. 27. of IAU, Bamberg.
Mianes, P. M., 1963, Ann. d'Astrophys. **26**, 1.
Nikolov, N. S., 1968, Catalogue of the light and colour curves of the cepheids, Bulg. Acad. of Sciences, Sofia.
Oosterhoff, P. and Walraven, Th., 1966, Bull. astr. Inst. Netherl. **18**, 387.

DISCUSSION

Fernie: It is still true that the abnormality in color is always in U—B and that this U—B excess is greatest near maximum light?

Nikolov: In this paper we discuss the ultraviolet emission on a given B—V or its change on a given change of B—V, but I think that we still have no data which could demonstrate that the abnormality in color is not due to U—B. In one of our previous papers (Nikolov, 1967, Astr. Zu., **44**, 120) we pointed out that the U—B excess correlates with the B-amplitude (Kraft's value f_B) only around maximum light.

IRREGULAR VARIATIONS OF RADIO SOURCES

M. V. PENSTON and R. D. CANNON

Royal Greenwich Observatory, Hailsham, Sussex, England

At the Royal Greenwich Observatory, we have been carrying out a program to study the optical variations of quasars and other radio sources. For the past three years we have been using our 26″ f10.5 refractor for photographic photometry of these objects. This seems an opportune time to give a general account of the situation to variable star observers since if one is set up to observe variable stars, one is then just as well set up to study quasars and it may be of interest to state what the current position is.

We have already (Penston and Cannon, 1968) given an account of our observations of quasars before 1967 November 5 = JD 2439799.5. Then we included an extensive review of the literature on optical variations of quasars. Since then other papers on this topic have appeared notably from Kinman et al. (1968) on the detailed behaviour of 3C345 and from Barbieri and Erculani (1968) on the observations of quasar variability made at Asiago.

At Herstmonceux we have been observing all quasars with V or m_{pg} less than 17^m in an attempt to survey such objects for variability, and have been trying to obtain one plate per month of each object. We feel this is a better use of our resources, given our relatively poor weather at Herstmonceux, than trying to obtain detailed light curves in a few cases as Kinman

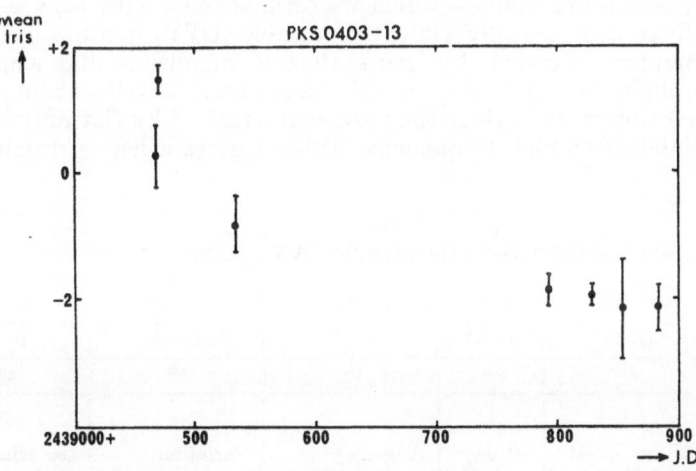

Fig. 1. Light curve for the quasar PKS 0403−13. There is no magnitude sequence for this object and the brightness is presented in terms of mean iris readings. A difference of one in mean iris reading corresponds to between 0^{m}1 and 0^{m}2.

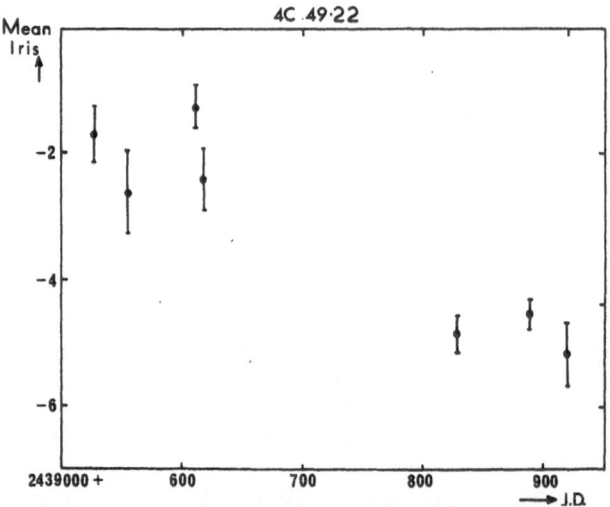

Fig. 2. As Fig. 1 but for the quasar 4C49.22

and his collaborators have been doing at Lick. Presently we have about 20 objects with enough data to say something about the presence or absence of variability. We find that all quasars vary, most by about 0^m1 to 0^m2 in a year. Our results to date (1968 Sept. 29 = JD 2440128.5) are illustrated for PKS 0403—13 and 4C49.22 in Figures 1 and 2. These slow rises and falls could be similar to the 13 year quasi-periodic variations (Smith and Hoffleit, 1963) of 3C273 but it is too early to draw firm conclusions as yet. Further light-curves have been given elsewhere (Penston and Cannon, 1968). On the other hand a small proportion — which we estimate at 10 per cent — of the quasars, which we call optically violently variable (OVV) quasars, behave in a different manner, varying by more than a magnitude and sometimes changing from night to night by appreciable amounts. As well as being distinguished by their optical behaviour they are characterized by flat radio spectra and radio variability at high frequencies. Table I gives a list of bright OVV

Table I

Properties of the brighter OVV quasars

Name	3C 273	3C 345	3C 454 · 3	3C 446	3C 279
B Magnitude	12—13	15—17	15—17	15—18	16—18
Amplitude	$0^m_.7$	2^m	2^m	3^m	2^m
Comments	13-year quasi-period[1]. May not be OVV.	80 day period[2]	Angione[3] reports 2m flare in 1953.	Variation characterized by "anti-flares"[4,5]	see Kinman's light curve given by Schorn et al.[6]

[1]Smith and Hoffleit (1963); [2]Kinman et al. (1968); [3]Angione (1967); [4]Penston and Cannon (1968); [5]Cannon and Penston (1967); [6]Schorn et al. (1968).

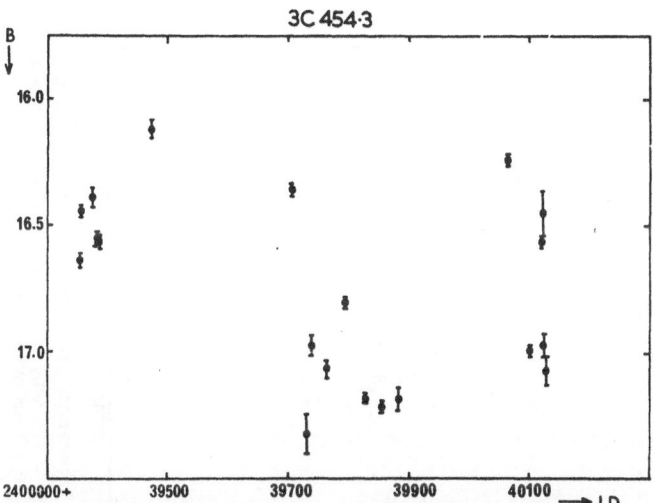

Fig. 3. Light curve for the OVV quasar 3C454.3

quasars with approximate magnitudes, amplitudes of variation and comments. Fig. 3 illustrates the variation of a typical OVV quasar 3C 454.3.

We turn now to the variation of other radio sources first found independently by Oke (1967) and Sandage (1967a) for the N-galaxy 3C 371. At Herstmonceux we have been taking plates of this and other radio galaxies in search of optical variation. In order that our photometry should not be influenced by the non-stellar nature of these objects, we have been taking our plates deliberately out of focus to spread out the star-images to be the same size as the galaxies. In fact it is the nuclei of the galaxies that vary and in the case of the N-galaxies these appear stellar on our plates and the correction is not important. Subsequent investigation of old plates at Herstmonceux and Harvard has produced a light curve from 1890 to date. The Harvard results have already been published by Usher and Manley (1968) and in Fig. 4 we show their results plotted with our results from Herstmonceux, complete to date. Another case for which old plates have been examined to show variability is the similar N-galaxy 3C390.3; we have already published our results for both old and modern plates (Cannon et al. 1968) and Sandage (1967b) has demonstrated the variability of 3C390.3 independently.

The fact that 3C390.3 was brighter in the past provides a "continuity" argument between quasars and N-galaxies indicating that quasars are at cosmological distances. In fact 3C371 was also brighter in the past and sor the same argument applies. In fact in the period 1890—1900 3C371 was within half a magnitude of being as intrinsically bright (assuming the Hubble law) as the quasar 3C48. If 3C371 had been observed in 1890—1900 it seems probable that it would have classified as a quasar.

The fact that these two objects were bright and variable prompted a search by one of us (Penston, M. V., 1967) for similar objects that were known variables and were listed as such in the General Catalogue of Variable Stars (Kukarkin, 1958). It appeared that the interesting Seyfert galaxy 3C120 was discovered to be variable by Hanley and Shapley (1940) in 1940 and had

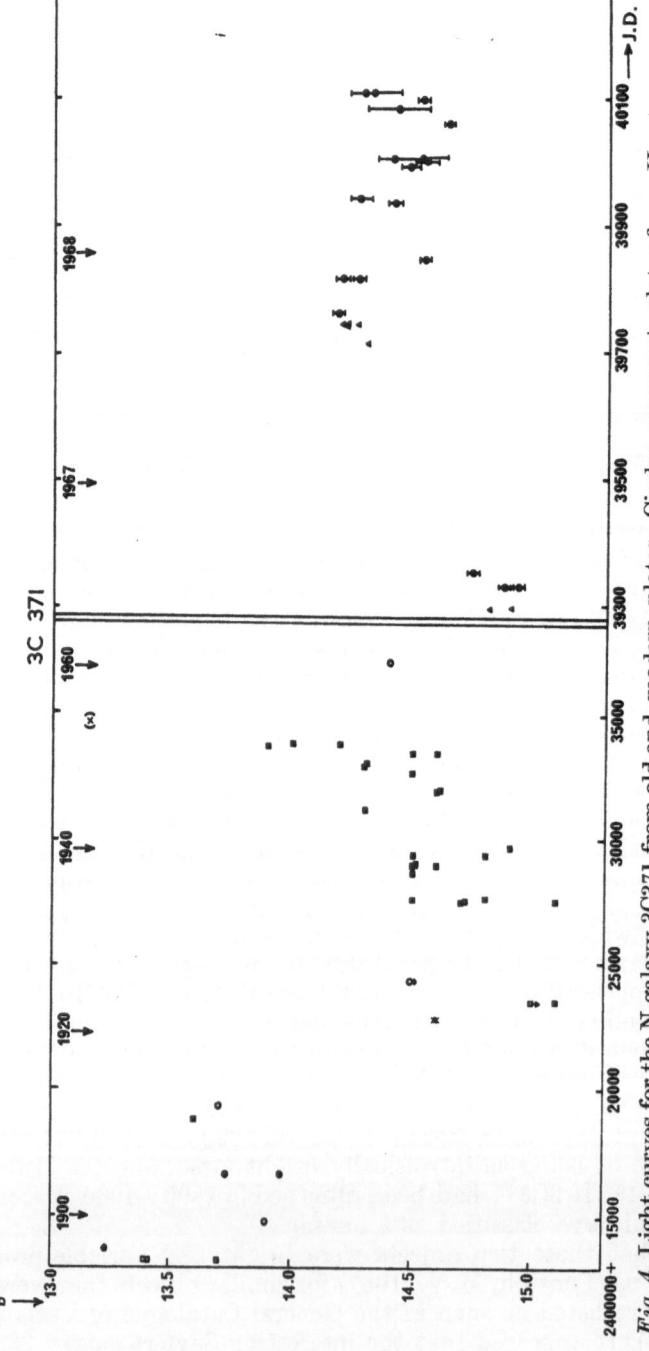

Fig. 4. Light curves for the N-galaxy 3C371 from old and modern plates. Circles represent plates from Herstmonceux or Greenwich, squares plates from Harvard[10], triangles photoelectric observations by Sandage[12] and crosses other plates. Open circles or squares are eye-estimates, while filled symbols show the plates were photometered.

been designated BW Tau. Since then Kinman (1968) has confirmed that this object is still varying today — our few plates at Herstmonceux fit his light-curve well. Another radio source was found to be identified with a known variable star by J. Schmitt (1968). He showed that the radio source VRO 42.22.01 was the variable star BL Lac (discovered by Hoffmeister in 1930). On the Palomar charts this object appears as a blue slightly fuzzy object and a plate taken at the prime focus of the 98″ Isaac Newton Telescope shows that it has an ultra-violet excess, typical of quasars and N-galaxies.

The fact that both BW Tau and BL Lac are radio galaxies and are listed as irregular variables in the General Catalogue of Variable Stars suggests that all 2000 irregular variables in that catalogue should be examined to see if any of them are nebulous on the Palomar Charts. A further radiogalaxy that varies appreciably in the optical region is the Seyfert galaxy NCG 4151 as was shown by Fitch, Pacholczky and Weymann (1967). We have a few plates at Herstmonceux which also show variability. The galaxies mentioned above show large variations and form a group similar to the OVV quasars in many ways but other cases of smaller variability have been found, in particular 3C109 and III Zw 1727+50 by Sandage (1967b). Table II gives a list of known OVV galaxies together with their properties.

Table II
Properties of the OVV Galaxies

Name	3C 371	3C 390 · 3	3C 120	VR 42.22.01	NGC 4151
B Magnitude	13^m—15^m	$14\frac{1}{2}^m$—16^m	14^m	13^m—16^m	12^m
Amplitude	2^m	$1^m_.5$	$0^m_.7$	3^m	$\sim 1^m$
Comments	brighter in 1890—1900 than present day		BW Tau	BL Lac	

We wish to thank Drs. T. D. Kinman and C. Barbieri and Mr. R. Angione for permission to quote their results prior to publication. We would like to thank our colleagues who have assisted us with our observations. Miss Rosemary Brett gave us valuable help in measuring our plates and preparing the diagrams. We are particularly grateful to Mr. C. A. Murray, Dr. D. Lynden-Bell and the Astronomer Royal for their support and encouragement.

REFERENCES

Angione, R., 1967, private communication.
Barbieri, C. and Erculandi, Laura A., 1968, private communication.
Cannon, R. D. and Penston, M. V., 1967, Nature 214. 256.
Cannon, R. D., Penston, M. V. and Penston, Margaret J., 1968, Nature **217**, 340.
Fitch, W. S., Pacholczyk, A. G. and Weymann, R. J., 1967, Astrophys. J. **150**, 167.
Hanley, C. M. and Shapley, H., 1940, Harvard Bull. No. 913.
Hoffmeister, C., 1930, Mitt. Sternw. Sonneberg, No. 17.
Kinman, T. D., et al. 1968, Astr. J. (in press).
Kukarkin, B. V. and Parenago, P. P., 1958, General Catalogue of Variable Stars (2nd edition), Moscow.
Oke, J. B., 1967, Astrophys. J. **150**, 15.
Penston, M. V., 1967, Inf. Bull. Var. Stars No. 255.

Penston, M. V. and Cannon, R. D., 1968, R. Obs. Bull. (in press).
Sandage, A., 1967 a, Astrophys. J. **150**, L9.
Sandage, A., 1967 b, Astrophys. J. **150**, L177.
Schmitt, J., 1968, Nature **218**, 663.
Schorn, R. A., Epstein, E. E., Oliver, J. P., Soter, S. L. and Wilson, W. J., 1968,
    Astrophys. J. **161**, 126.
Smith, H. J. and Hoffleit, D., 1963, Nature **198**, 650.

DISCUSSION

Feast: Do you regard the sources of large and low amplitudes as distinct
    species?

Penston: 1) There are several possible reasons why OVV Quasars should be
        different. They may be
        1. distinguished by such a property as total mass,
        2. an evolutionary stage for AU Quasars,
        3. an intermittent phase in the activity of any individual Quasar,
        4. an effect of the direction from which we view the Quasars (i. e.
            an aspect effect).
    2) The OVV Quasars are indeed distinguished by their radio proper-
        ties.

Rosino: 1. Which criteria do you use for the reality of small variations?
    2. How much are affected the observations by the atmospheric extin-
        tion?

Penston: If I may answer your second point first. We have used a refractor
    which acts as an extremely efficient filter to the ultra-violet. Thus effects
    of atmospheric extinction do not worry us much. To your first point we
    devised a statistical test to detect variability and also our results are
    in good agreement with those of Kinman in several cases.

69.67101 Akadémiai Nyomda, Budapest — Felelős vezető: Bernát György